Springer Series in Computational Mathematics

28

Springer
Berlin
Heidelberg
New York
Barcelona
Hong Kong
London
Milan
Paris
Singapore
Tokyo

Zhen Mei

Numerical Bifurcation Analysis for Reaction-Diffusion Equations

Springer

Zhen Mei
Department of Mathematics
University of Marburg
35032 Marburg, Germany
e-mail: meizhen@mathematik.uni-marburg.de

Cataloging-in-Publication Data applied for

Die Deutsche Bibliothek - CIP-Einheitsaufnahme
Mei, Zhen:
Numerical bifurcation analysis for reaction diffusion equations / Zhen Mei. - Berlin ; Heidelberg ; New York ; Barcelona ; Hong Kong ; London ; Milan ; Paris ; Singapore ; Tokyo : Springer, 2000
ISBN 3-540-67296-6

Mathematics Subject Classification (2000):
65P30, 37M20, 35K57, 65H17, 35B32, 34C37, 92-08, 92E20

ISSN 0179-3632
ISBN 3-540-67296-6 Springer-Verlag Berlin Heidelberg New York

Springer-Verlag is a company in the BertelsmannSpringer publishing group.

Printed in Germany

Typesetting: By the author using a Springer TEX macro package
Printed on acid-free paper   SPIN 10764452   46/3143LK - 5 4 3 2 1 0

Preface

Reaction-diffusion equations are typical mathematical models in biology, chemistry and physics. These equations often depend on various parameters, e.g. temperature, catalyst and diffusion rate, etc. Moreover, they form normally a nonlinear dissipative system, coupled by reaction among different substances. The number and stability of solutions of a reaction-diffusion system may change abruptly with variation of the control parameters. Correspondingly we see formation of patterns in the system, for example, an onset of convection and waves in the chemical reactions. This kind of phenomena is called bifurcation. Nonlinearity in the system makes bifurcation take place constantly in reaction-diffusion processes. Bifurcation in turn induces uncertainty in outcome of reactions. Thus analyzing bifurcations is essential for understanding mechanism of pattern formation and nonlinear dynamics of a reaction-diffusion process. However, an analytical bifurcation analysis is possible only for exceptional cases. This book is devoted to numerical analysis of bifurcation problems in reaction-diffusion equations. The aim is to pursue a systematic investigation of generic bifurcations and mode interactions of a class of reaction-diffusion equations. This is realized with a combination of three mathematical approaches: numerical methods for continuation of solution curves and for detection and computation of bifurcation points; effective low dimensional modeling of bifurcation scenario and long time dynamics of reaction-diffusion equations; analysis of bifurcation scenario, mode-interactions and impact of boundary conditions. The basic idea is to start from a nonsingular solution and continue it with respect to a control parameter. When a bifurcation is detected between two consecutive steps in the numerical continuation of solution curves, we undertake an accurate approximation of the bifurcation point and proceed to a local analysis of the bifurcation scenario and stability of bifurcating solution branches. Thereafter, we switch to a chosen solution branch: stationary, periodic or homoclinic, and continue it further. A systematic study in the interesting ranges of parameters reveals the nonlinear behavior of the problem.

This book is divided mainly into four parts. Chapters 1–4 serve as a general introduction to reaction-diffusion equations, symmetries and numerical solutions of parameter dependent problems. Chapters 6–8 describe low dimensional modeling of bifurcation problems, namely, techniques and their implementations for reducing bifurcation problems to equivalent low dimensional systems. In Chapters 9–13 we discuss generic bifurcations and mode interactions of a class of reaction-diffusion equations. Impact of boundary conditions on bifurcations is investigated in Chapters 14–16. Symmetries and elementary group-theoretic concepts are briefly discussed in Chapter 5. They are exploited throughout the book for analysis and numerics of bifurcations in reaction-diffusion equations.

Let us outline the content of this book chapter by chapter.

Chapter 1 gives a brief introduction of reaction-diffusion equations, pattern formations and boundary conditions.

Chapter 2 summarizes the numerical methods for continuation of solution branches. We consider at first parameterizations of solution curves with respect to control parameters, arclength and pseudo-arclength, etc. Thereafter, predictor-corrector methods are discussed for continuation of solution curves. Iterative methods for solving large sparse linear systems are incorporated in the continuation procedures.

In Chapter 3 we explore test functions for detecting bifurcations along continuation of solution curves. Direct methods are discussed for calculation of simple and Hopf bifurcation points.

Chapter 4 is concerned with branch switching at simple bifurcation points. One approach is to calculate at first a tangent of a bifurcating branch and then continue the solution with appropriate predictors. Convergence of this method is ensured in a conical region along the solution curve with the bifurcation point at vertex. Another method is to regularize the problem at a bifurcation point with an enlarged system and continue its solutions with predictor-corrector methods.

In Chapter 5 we use the elementary group-theoretic methods to formulate symmetries of nonlinear problems. Some straightforward properties of symmetries are discussed for utilization in the subsequent chapters. We consider an extension of the well-known equivariant branching lemma and illustrate it with an example.

With Chapter 6 we start the discussions on effective low dimensional modeling of bifurcation and dynamics of reaction-diffusion equations. We introduce the Liapunov-Schmidt method for bifurcation problems. This transforms the problem equivalently to a underlying low dimensional system of algebraic equations, which is responsible for the bifurcation scenario and can be analyzed often directly with the singularity theory. We formulate Liapunov-Schmidt reduction as an algorithm such that calculations of Taylor expansion of the reduced equations can be easily implemented. A scaling technique for branch switching at multiple bifurcation points is presented and illustrated with examples.

Reduction technique in the center manifold theory is the focus of Chapter 7. We describe the main results in the center manifold theory and combinations with normal form theory in calculations of the reduced equations. We apply the center manifold reduction to reaction-diffusion equations and discuss its relations to Liapunov-Schmidt method.

Chapter 8 discusses a numerical method for studying (quasi-) periodic solutions near homoclinic orbits of the parameter dependent autonomous differential equations, typically, spatial discretizations of reaction-diffusion equations. Similarly to the Liapunov-Schmidt reduction, this method derives from the dynamical system a bifurcation function for locating periodic orbits in the neighborhood of a homoclinic orbit. Calculations of the bifurcation function involve approximation of the homoclinic orbit and numerical solution of the adjoint variational equation. We discuss a class of methods for approximating the latter equation such that an conservation property is preserved. We consider implications of continuous symmetries and discuss how to handle bifurcations of relative periodic orbits from relative homoclinic orbits.

We begin in Chapter 9 the study of generic bifurcations and mode interactions of reaction-diffusion equations. We consider at first systems of one-dimensional reaction-diffusion equations, in particular, the Brusselator equations. Double bifurcations occur if both the diffusion rate and the size of domain vary as bifurcation parameters. To ensure a correct reflection of the bifurcation scenario in discretizations and to reduce imperfection of bifurcations, we consider a preservation of multiplicities of bifurcation points in the discrete problems. A continuation-Arnoldi algorithm is exploited to trace the solution branches and to detect secondary bifurcations.

In Chapter 10 we consider properties of reaction-diffusion equations on square domains, such as symmetries, eigenpairs of the Laplacian, bifurcation points, occurrence of mode interactions, etc. We derive kernels of the linearized operator and the adjoint operator. By exploring the bifurcation equations in Liapunov-Schmidt reduction we obtain the complete bifurcation scenarios at simple and double bifurcation points.

Chapter 11 develops normal forms for generic Hopf bifurcations of reaction-diffusion equations on a square. These normal forms cover both reducible and irreducible representations of the symmetry group. We discuss conditions on coefficients of the normal form to ensure generic branching behavior at Hopf bifurcation. Furthermore, we introduce an algorithm to find the number of branches and their symmetries. As an example, we use the normal form analysis to interpret the bifurcation diagram of the Brusselator equations with Robin boundary conditions.

Chapter 12 is concerned with steady/steady state mode interactions of reaction-diffusion equations. We distinguish two types of mode interactions, namely those induced by symmetries by nonlinearities of the problem, respectively. These correspond to linear degeneracies and nonlinear degeneracies in reaction terms, respectively. We use symmetries and group theoretic concepts to derive normal forms of the reduced bifurcation equations and to analyze bifurcation scenarios. A scaling technique is applied to problems with Z_2-symmetric reaction terms.

In Chapter 13 we examine Hopf/steady state mode interactions of reaction-diffusion equations. We are particularly interested in realization of the center manifold reduction near a Hopf/steady state mode interaction. We determine points where a Hopf/steady state mode interaction occurs, and derive symmetries of the reduced equations in the critical eigenspace. Thereafter, we consider normal form of the reduced equations on the center manifold and analyze bifurcation scenario for a special Hopf/steady state mode interaction. Calculations of coefficients in the normal form are illustrated.

The last three chapters deal with influence of boundary conditions on bifurcation scenarios. Chapter 14 discusses properties of boundary conditions and differential operators. We start with a few frequently used boundary conditions and summary the classical Sturm-Liouville theory. Then we vary the boundary conditions and investigate properties of eigenpairs of the Laplacian and solutions of a linear scalar elliptic differential equation.

Chapter 15 studies impact of boundary conditions on steady state bifurcations of a scalar reaction-diffusion equation. We analyze variation of bifurcation scenario along a homotopy from Neumann to Dirichlet boundary conditions. Changes of symmetries, numbers and directions of bifurcating solution branches reveal the role of boundary conditions in bifurcation problems. Furthermore, we use forced symmetry-breaking in boundary conditions to explore how symmetries of a problem are inherited by its solution manifold.

Finally, a steady/steady state mode interaction induced by variation of boundary conditions is discussed in Chapter 16. We treat a homotopy parameter in boundary conditions as the second bifurcation parameter and derive normal form for the reduced bifurcation equations. We discuss generic bifurcation scenario and stability of the bifurcating solution branches. By continuing solution branches along the homotopy of boundary conditions towards Dirichlet problem we see how these branches merge to the Dirichlet solutions, e.g. via primary or secondary bifurcations.

I want to thank Klaus Böhmer for inspiring me to make habilitation at the Philipps-University of Marburg. His constant support makes this book possible. I owe thanks to Bernhard Schmitt for many helpful discussions. I learnt a lot from Peter Ashwin. Several joint work with him are included in the text. Special thanks go to Andreas Schwarzer for discussions and an example on scaling techniques. This book is completed during my stay at the Fields Institute. I am grateful to Bill Langford and Don Dawson for their support.

I would like to thank Deutsche Forschungsgemeinschaft for financial support. I take this opportunity to thank both the Department of Mathematics, Philipps-University of Marburg, and the Fields Institute for Research in Mathematical Sciences for providing me excellent working environment and for their hospitality. I would also like to thank Martin Peters and Thanh-Ha Le Thi at Springer-Verlag for their effective assistance during the publication of the book.

Contents

1. Reaction-Diffusion Equations

1.1 Introduction

Reaction-diffusion equations are widely used for modeling chemical reactions, biological systems, population dynamics and nuclear reactor physics. They are of the form

$$\frac{\partial u}{\partial t} = D\Delta u + f(u, \lambda). \tag{1.1}$$

Here $u = (u_1, \ldots, u_k)$ represents various substances in a chemical reaction or species of a biological system; $\lambda \in \mathbf{R}^p$ is a vector of control parameters; Δ is the Laplace operator in the spatial variables and $D \in \mathbf{R}^{k\times k}$ is a symmetric and semi-positive definite matrix. They describe diffusion of different substances. The matrix D is often diagonal and corresponds to the diffusion rates of the substances. The function $f: \mathbf{R}^k \times \mathbf{R}^p \to \mathbf{R}^k$ is a vector of smooth functions and represents the reaction among the substances.

We consider (1.1) in bounded domains $\Omega \in \mathbf{R}^n (n \leq 3)$ with $C^{m,1}$-boundaries ($m \geq 0$), such as a vessel for chemical reactions, a yard for biological systems or an embryo integument for formation of animal coat patterns. We are interested in effects of geometry, scale of domains and boundary conditions on the type of spatial-temporal patterns in reaction-diffusion systems. For example, markings of an animal coat have been studied as a combination of pattern formation in different local parts (cf. Murrey [235]). We represent the tail and legs as tapering cylinders and consider the surface as the reaction-diffusion domain. In this case the surface of an animal body corresponds to a rectangular domain. Boundary conditions can be imposed for these problems differently along sides of domains. Furthermore, boundary conditions may differ among species of the system.

To ensure classical solutions we assume a polynomial growth of the reaction term $f(u, \lambda)$ for $||u||$ at infinity and domains $\Omega \in \mathbf{R}^n$ with $n \geq 2$, i.e., for any fixed parameter λ there are constants $c_1,\ c_2 > 0$ such that

$$||f(u, \lambda)|| \leq c_1 + c_2||u||^s, \tag{1.2}$$

where $1 < s < (n+2)/(n-2)$. We are particularly interested in the scalar equation

$$\frac{\partial u}{\partial t} = \Delta u + f(u, \lambda), \tag{1.3}$$

and the system of two equations

$$\frac{\partial}{\partial t}\begin{pmatrix} u_1 \\ u_2 \end{pmatrix} = \begin{pmatrix} \Delta & 0 \\ 0 & d\Delta \end{pmatrix}\begin{pmatrix} u_1 \\ u_2 \end{pmatrix} + \begin{pmatrix} f_1(u_1, u_2, \lambda) \\ f_2(u_1, u_2, \lambda) \end{pmatrix}. \tag{1.4}$$

Heat conduction in a homogeneous medium is described by (1.1) with $k = n = 1$ and

$$f(u, \lambda) = \lambda e^{-c/u},$$

see e.g. Henry [157, pp. 41]. By choosing $k = 2$ and

$$\begin{aligned} f_1(u_1, u_2, \lambda) &= -(\lambda + 1)u_1 + \alpha + u_1^2 u_2, \\ f_2(u_1, u_2, \lambda) &= \lambda u_1 - u_1^2 u_2, \end{aligned} \tag{1.5}$$

we obtain the well-known Brusselator equations proposed by Lefever and Prigogine [201], a model system for describing autocatalytic process in chemical reactions with diffusions, see also Alexander/Auchmuty [1], Nicolis/Prigogine [237], Nicolis/Gaspard [236]. The Brusselator equations are derived by suppressing less relevant behavior of a chemical system. They can be considered as a reaction involving two intermediates and a number of initial and final products with variable concentrations. Another example is the Fitzhugh-Nagumo equations

$$\begin{aligned} f_1(u_1, u_2, \lambda) &= \mu u_2 - \lambda u_1(u_1 + a)(a - u_1), \\ f_2(u_1, u_2, \lambda) &= \nu(u_1 - u_2) \end{aligned}$$

for pulse propagations along the nerve axon, in which the electric potential diffuses as chemicals (cf. Segel [275], Temam [290, pp.97]).

Reaction-diffusion equations have been also applied extensively to model population dynamics with migration (cf. Conway/Smoller [69], Hadeler/Freedman [147] and Murray [235]) and animal coat patterns. For example, the system

$$\begin{aligned} \frac{\partial u}{\partial t} &= \Delta u + r[a - u - h(u, v)], \\ \frac{\partial v}{\partial t} &= d\Delta v + r[\lambda(b - v) - h(u, v)], \\ h(u, v) &:= \frac{\rho u v}{1 + u + K u^2}, \end{aligned} \tag{1.6}$$

is a model for population dynamics. Here a, b, λ, ρ and K are positive parameters. The diffusion rate $d \geq 1$ makes the diffusion as a source for spatial variation and instability. The scale factor r is a measure for the domain size.

1.2 Bifurcations and Pattern Formations

In 1952 Alan Turing [293] predicted theoretically that different diffusion rates of species in a chemical reaction will produce spatial gradients in the concentration of participating species and lead an initially well-mixed, homogeneous state to some spatial-temporal patterns in the chemical system. This spontaneous development of spatial form, i.e., through local enhancements or depletion in the concentration of some species, is called *spontaneous pattern formation*, sometimes it is also said the self-organization and the emergence of dissipative structure in a non-uniform medium. Turing's patterns correspond to the local variations of colors caused by diffusions in Belousov-Zhabotinskii reactions.

Reaction-diffusion problems are considered as the underlying mechanism for many pattern formation processes. Interaction of diffusion and reaction, measured with the Laplace operator Δ, the diffusion rate D and the function f, can yield rich and unexpected phenomena. Often a process develops its own dynamic which is not predicated directly by the given physical, chemical or biological setting. It is the nonlinear stability and a series of bifurcations in the problem that induce sophisticated scenario in dynamics of the system. A bifurcation corresponds to abrupt change in behavior of solutions of a dynamics system. More precisely, a *bifurcation point* (u_0, λ_0) is a solution of (1.1) such that in all neighborhoods of (u_0, λ_0) the problem (1.1) has at least two different solution branches. Bifurcation theory studies how solutions of (1.1) and their stability change as the parameter λ varies.

The pioneering work of Turing [293] on dynamics of biological systems and Belousov [34], Zhabotinskii [308] on pattern formation ignited a rapid development of nonlinear analysis of reaction-diffusion equations. In particular, qualitative behavior of solutions has been studied extensively in various context, for examples, periodic solutions, traveling waves, spiral waves, long time behavior and bifurcations in the stationary problem

$$G(u, \lambda) := D\Delta u + f(u, \lambda) = 0. \tag{1.7}$$

Stationary bifurcations lead to time-independent patterns. A pattern formation in reaction-diffusion problems corresponds to a continuous interaction and selection of critical modes and change of stabilities of the system. Spatial patterns of modes are resulted from properties of eigenfunctions of the Laplacian, i.e., diffusion of the participating substances. The development of patterns in a chemical reaction or an embryo depends strongly on eigenpairs of the Laplacian for given domains and chosen boundary conditions.

A wide spectrum of the mathematical methods is applicable to the analysis of bifurcations and pattern formations in reaction-diffusion equations. They range from comparison of sub- and super-solutions, degree theory, Conley index, theory of critical points and singular perturbations to various types of maximum principals (cf. Amann [16], Barrow [33], Fiedler/Mallet-Paret [103], Furter/López-Gómez [115], Jones/Küpper [177], Li/Ghoreishi [204], López-Gómez et al [206], Mimura [229], Ruan/Pao [258], Smoller [281]). These methods are effective for analysis of positive stationary solutions, traveling waves and solution curves which branch at simple bifurcation points. Bifurcation theory has been used successfully in discussion of pattern formations in chemical reactions and in biology (cf. Alexander/Auchmuty [1], Duncan/Eilbeck [96], Elezgaray/Arneodo [98], Fiedler [100], Fields/Golubitsky

[104], Golubitsky/Stewart/Schaeffer [131], Goldwasser/Maini/Murray [127], Kernevez [183], Kirchgässner [189, 190], Malomed/Gedalin [208], Mimura [229], Murray [235], Price/Wambacq/Oosterlinck [245], Seydel [276]). A systematic study of highly complicated bifurcation scenario has been developed by applying group-theoretic methods to problems with symmetries (cf. Allgower/Böhmer/Golubitsky [6], Allgower/Böhmer/Mei [10, 11, 12], Allgower/Georg/Miranda [15], Ashwin/Böhmer/Mei [22], Böhmer/Mei [45], Brunovsky/Fiedler [50], Eilbeck/Furter [97], Golubitsky/Stewart/Schaeffer [131]). We want to make use of bifurcation theory and numerical methods for bifurcation problems to study nonlinear behavior of reaction-diffusion equations.

1.3 Boundary Conditions

Boundary and initial conditions are standard tools in specifying solutions of differential equations on their solution manifolds. In engineering and mathematics we often simplify boundary conditions to make a differential equation to be well-posed and tractable. Classical boundary conditions for reaction-diffusion equations in a given domain Ω are as follows.

- Dirichlet boundary conditions:

$$u|_{\partial\Omega} = g;$$

- Neumann boundary conditions:

$$\frac{\partial u}{\partial n}\Big|_{\partial\Omega} = g,$$

  where n is the outward normal direction of $\partial\Omega$;
- Robin boundary conditions:

$$\left(au + b\frac{\partial u}{\partial n}\right)\Big|_{\partial\Omega} = g,$$

  where $a,\ b \in \mathbf{R}$ are constants. Here additional consistency condition is required on g to ensure smoothness of solutions.

These conditions can be also imposed on different segments of boundary of the domain Ω. In fact, posing and realizing appropriate boundary conditions sometimes can be quite delicate. For examples, the types of boundary conditions of chemical reactions in open and large systems change constantly (cf. Gray/Scott [273]); the simply supported boundary condition for buckling problems is hard to realize experimentally (cf. Szilard [289]). It is still a challenge to model boundary conditions at walls for turbulent flows (cf. Rodi [252]).

Recently, it is observed that different boundary conditions of nonlinear problems can lead to different structures and bifurcations of solutions, see e.g. Ashwin/Böhmer/Mei [21, 22], Ashwin/Mei [25], Dillon/Maini/Othmer [86]. Experiments have confirmed this prediction. Putting the reaction vessel into electrical field, Münster/Watzl [234] have simulated various boundary conditions and observed spatial non-uniform steady state solutions. In particular, mixed boundary conditions seem to be favored by stable non-uniform stationary solutions. Patterns produced in such ways in polymers with large molecules have specific physical properties, such as conductivity and enforcement in the structure, etc. However, influence of boundary conditions on the pattern formation, e.g. target patterns of chemical reactions and stripes in animal coats, is sophisticated and involves many factors and processes.

Neumann and Dirichlet solutions of reaction-diffusion problems are sometimes coupled implicitly in proper domains Ω, for example, in a infinite stripe a Dirichlet solution can be transferred to a Neumann solution along a homotopy path. This is particularly true if one imposes periodic boundary conditions (cf. Castro [53]). Gardner [118, 117] has showed that as traveling waves the nonsingular solutions along a homotopy between Neumann and Dirichlet problems remain regular and the homotopy paths between different solutions don't intersect, see also Fielder [102] and Hale/Rocha [149, 150] for discussions on influence of boundary conditions on attractors of reaction-diffusion equations. Some Neumann and Dirichlet problems can be embedded into equations with periodic boundary conditions, which have then translation symmetries (cf. Crawford et al [77], Gomes/Stewart [133], Mei [217]). Thereafter, solutions of some Dirichlet problems can be simply shifted to those of Neumann problems via coordinate transformations (cf. Castro [53]). On the other hand, boundary conditions do have influence on bifurcation scenario of some problems, for examples, in Mei [219, 223] we have shown that the steady state problem

$$\Delta u + \lambda f(u) = 0, \qquad f(u) = u + O(\|u\|^2)$$

on a square domain has different bifurcation diagrams for Neumann and Dirichlet problems, even the numbers of bifurcating solution branches are different. Similar conclusions are derived for Hopf bifurcations, see e.g. Ashwin/Böhmer/Mei [21, 22]. Recently, Dillon/Maini/Othmer [86] considered a system of reaction-diffusion equations on a one-dimensional domain with mixed boundary conditions

$$\left(a_i u_i + b_i \frac{\partial u_i}{\partial n}\right)\Big|_{\partial\Omega} = g_i, \quad i = 1, \ldots, k,$$

i.e., the components may have different types of boundary conditions. By varying boundary conditions for different components they obtained interesting relations between solutions of Neumann and Dirichlet problems.

We consider a homotopy of boundary conditions

$$g_0(\mu)u + g_1(\mu)\frac{\partial u}{\partial n} = 0 \quad \text{on } \partial\Omega, \tag{1.8}$$

where $g_0,\ g_1 :\ \mathbf{R} \to \mathbf{R}$ are differentiable functions and

$$g_0(0) = g_1(1) = 0, \quad g_0(1)g_1(0) \neq 0.$$

Variation of bifurcation scenario along a homotopy from Neumann to Dirichlet boundary conditions, e.g. changes of numbers, symmetries and stabilities of solution branches as well as bifurcation directions, reveals impact of boundary conditions on bifurcation scenario and local dynamics of reaction-diffusion equations. Similarly, a study of symmetry-breaking and mode interactions resulted from boundary conditions sheds light on structure of the solution manifolds.

2. Continuation of Nonsingular Solutions

In this chapter we summarize some techniques for continuation of nonsingular solutions of the finite dimensional parameter-dependent problems

$$G(u, \lambda) = 0, \tag{2.1}$$

where $G: \mathbf{R}^n \times \mathbf{R}^p \to \mathbf{R}^n$ is a smooth mapping. The unknown u describes state of the system and λ represents parameters. Typically, this equation can be considered as spatial discretized reaction-diffusions equations, stationary problem of well-stirred reactions, population model in biological systems. Variation of a parameter in the system corresponds to continuation of a solution of the problem. In other words, whenever a parameter changes, the system will adjust itself correspondingly to the new state. There exists a large number of problems for which number and stability of solution curves changes abruptly and the structure of solutions varies dramatically when a parameter reaches some critical values, for examples, buckling of rods and plates, onset of convection and turbulence, pattern formation in reaction-diffusion processes, etc. This kind of phenomena is called *bifurcation*, i.e., a qualitative change in the behavior of solutions of a dynamics system. In such a situation the linear stability theory fails and does not given much information on the qualitative behavior of the nonlinear system. Bifurcation analysis and nonlinear stability have to be considered, and mostly in combinations with numerical simulations and approximations.

Continuation method is a fundamental tool for analyzing bifurcation scenario of nonlinear problems. Particularly, predictor-corrector method is widely used for numerical continuation of solution curves of one-parameter problems. This method has been discussed extensively in Allgower/Georg [14], Keller [179], Rheinboldt [249] and Seydel [276]. For multiple parameter problems, the moving frame algorithms in Rheinboldt [249, 247, 248], Hong [162] and Dai/Rheinboldt [78] can be used for approximating multi-dimensional solution manifolds, see also Allgower/Schmidt [3] for piecewise-linear approximation of implicitly defined manifolds.

We follow the discussions in Allgower/Georg [14] and examine at first local parameterizations of nonsingular solution branches. Thereafter, we consider several predictor-corrector methods for continuation of solution branches. In particular, we exploit iterative methods for solving large linear systems in the continuation methods. In the last section of this chapter we outline the moving frame scheme in Rheinboldt [249] for approximation of solution manifolds.

2.1 Parameterization of Solution Curves

In this section we characterize solution set of the finite dimensional problem (2.1) in the neighborhood of regular solutions.

Definition 2.1.1. *Let $G: \mathbf{R}^{n+p} \to \mathbf{R}^n$ be a smooth mapping. A point x in $\mathbf{R}^{n+p}$ is called a regular point of G, if the derivative $DG(x) \in \mathbf{R}^{(n+p)\times n}$ has full rank, i.e.,*

$$\mathrm{Rank}(DG(x)) = n.$$

A point y in $\mathbf{R}^n$ is said to be a regular value of G, if the inverse set

$$G^{-1}(y) = \{x \in \mathbf{R}^{n+p}; \qquad G(x) = y\}$$

consists only of regular points of G. A point $x \in \mathbf{R}^{n+p}$ (resp. $y \in \mathbf{R}^n$) which is not a regular point (resp. value) of G, is said to be singular.

We assume in the following that $0 \in \mathbf{R}^n$ is a regular value of G, i.e., the Jacobian $DG(x)$ has full rank at all solutions of $G(x) = 0$. We denote

$$\mathcal{M} := \{x \in \mathbf{R}^{n+1}; \qquad G(x) = 0, \;\; \mathrm{Rank}(DG(x)) = n\} \tag{2.2}$$

as the set of nonsingular solutions of (2.1). In the sequel we discuss how to parameterize the set $\mathcal{M}$.

2.1.1 Natural parameterization

In practice we often distinguish a parameter in nonlinear systems and examine variation of solutions with respect to the parameter. Namely we choose $k = 1$ and consider one-parameter problems with variables in $\mathbf{R}^{n+1}$ as

$$x = (u, \lambda) \in \mathbf{R}^{n+1},$$

where u represents the state variable and λ is the control parameter in the system. By the implicit function theorem the solution set $\mathcal{M}$ in (2.2) can be parameterized locally with respect to the parameter λ.

Lemma 2.1.1. *Let $(u_0, \lambda_0) \in \mathbf{R}^{n+1}$ satisfy $G(u_0, \lambda_0) = 0$ and $\mathrm{Rank}(D_uG(u_0, \lambda_0)) = n$. Then there exists locally a maximal constant $\epsilon > 0$, and a unique function $u: [\lambda_0 - \epsilon, \; \lambda_0 + \epsilon] \to \mathbf{R}^n$, such that*

$$\{(u(\lambda), \lambda); \qquad |\lambda - \lambda_0| \le \epsilon, \; u(\lambda_0) = u_0\} \subset \mathcal{M}. \tag{2.3}$$

Remark 2.1.1. The parameterization (2.3) is valid as long as the statement $\mathrm{Rank}(D_uG(u(\lambda),\lambda)) = n$ holds. Since ϵ is chosen to be maximal, the rank of $D_uG(u(\lambda+\epsilon),\lambda+\epsilon)$ or $D_uG(u(\lambda-\epsilon),\lambda-\epsilon)$ is less than n. We can also extend the parameterization differently at the left and right sides of λ_0 until singularities appear at both ends. If zero is a regular value of G and rank deficiency of D_uG occurs, changing the parameter λ with another variable in u allows a global continuation of the solution curve.

2.1.2 Parameterization with arclength

Let $x_0 \in \mathbf{R}^{n+1}$ be a nonsingular solution of $G(x) = 0$, i.e.,

$$G(x_0) = 0 \quad \text{and} \quad \mathrm{Rank}(DG(x_0)) = n. \tag{2.4}$$

It follows from the implicit function theorem that there exist an open interval $J \ni 0$ and a unique smooth curve $c:\ t \in J \subset \mathbf{R} \to c(t) \in \mathbf{R}^{n+1}$, such that

$$\begin{array}{lll} \text{i)} & G(c(t)) = 0, & \\ \text{ii)} & \mathrm{Rank}(DG(c(t))) = n & \text{for all } t \in J, \\ \text{iii)} & \dot{c}(t) \neq 0, \quad c(0) = x_0. & \end{array} \tag{2.5}$$

In particular, we have $\{c(t);\ \ t \in J\} \subset \mathcal{M}$ and

$$c(t) = x_0 + \dot{c}(0)t + \cdots .$$

Since $DG(c(t))\dot{c}(t) = 0$, we obtain from (2.5ii)

$$\mathrm{Ker}(DG(c(t))) = \mathrm{span}[\dot{c}(t)].$$

Based on the tangent $\dot{c}(t)$, we parameterize the solution curve with its arclength. Let s represent the arclength. Then

$$ds = \left[\sum_{i=1}^{n+1} \dot{c}_i(t)^2\right]^{1/2} dt.$$

If we replace the variable t by the arclength s, the following equality holds

$$\|\dot{c}(s)\| = \left[\sum_{i=1}^{n+1} \dot{c}_i(s)^2\right]^{1/2} = 1.$$

Lemma 2.1.2. *Let x_0 be a nonsingular solution of $G(x) = 0$. Then the solution curve can be parameterized with the arclength s in a neighborhood of x_0, i.e., there exist an open interval $J \subset \mathbf{R}$ with $0 \in J$ and a unique function $c:\ s \in J \to c(s) \in \mathbf{R}^{n+1}$, such that $c(0) = x_0$ and*

$$G(c(s)) = 0, \qquad \dot{c}^T(s)\dot{c}(s) = 1 \quad \text{for all } s \in J.$$

Furthermore, the tangent $\dot{c}(s)$ *is uniquely determined by the system*

$$\begin{array}{rcl} DG(c(s))\dot{c}(s) & = & 0, \\ \dot{c}^T(s)\dot{c}(s) & = & 1 \end{array} \tag{2.6}$$

with the property

$$\det \begin{pmatrix} DG(c(s)) \\ \dot{c}(s)^T \end{pmatrix} \geq 0, \quad (resp. \leq 0). \tag{2.7}$$

We call the matrix

$$\begin{pmatrix} DG(x) \\ t(DG(x))^T \end{pmatrix}$$

the extended Jacobian of G at x.

Definition 2.1.2. *Let* $A \in \mathbf{R}^{n\times(n+1)}$ *be a rectangular matrix with full rank, i.e.,* $\text{Rank}(A) = n$. *The unique vector* $t(A) \in \mathbf{R}^{n+1}$ *which satisfies the conditions*

$$At = 0; \qquad t^T t = 1 \quad and \quad \det \begin{pmatrix} A \\ t^T \end{pmatrix} > 0,$$

is called the tangent vector induced by A. *The vector* $\dot{c}(s)$, *determined by (2.6) and (2.7), is called the oriented tangent vector of the solution curve* $c(s)$.

If $c(s)$ is a solution of the differential equation $\dot{c} = t(DG(c))$, then we have

$$\frac{d}{ds}G(c(s)) = DG(c(s))\dot{c}(s) = DG(c(s))t(DG(c(s))) = 0.$$

Thus $G(c(s)) = \text{constant}$. If there is a point s_0 such that $G(c(s_0)) = 0$, then $c(s)$ is a solution curve. Therefore, the solution curve $c(s)$ in the neighborhood of x_0 can be described by the initial value problem

$$\dot{c} = t(DG(c)), \qquad c(0) = x_0. \tag{2.8}$$

This is the so called *Davidenko equation.* It is normally used to investigate the global properties of parameterization of a solution curve with respect to the arclength.

Theorem 2.1.1. *Let* (a, b) *be the maximal interval for the existence of the solution* $c(s)$ *of the initial value problem (2.8). If the left endpoint* a *is finite, then* $c(s)$ *converges to a singular point of* G *when* $s \to a^+$. *Similar conclusion holds for the right end point* b *if* $b < +\infty$.

Proof: see Allgower/Georg [14] and Mei [225].

Theorem 2.1.1 implies that a solution curve can be parameterized with respect to the arclength s until a rank deficiency of $DG(c(s))$ occurs.

Theorem 2.1.2. *If zero is a regular value of* G, *the solution curve* $c(s)$ *is defined for all* $s \in \mathbf{R}$ *and has one of the following properties.*

a) *The curve* $c(s)$ *is diffeomorphic to a circle, i.e., there exists a (smallest) constant* $T > 0$*, such that* $c(s) = c(s+T)$*;*
b) *The curve* $c(s)$ *is diffeomorphic to* $\mathbf{R}$*, i.e.,* $c(s)$ *is injective and has no accumulation point when* $s \to \pm\infty$*.*

Relation to the natural parameterization: We recall that in a natural parameterization the coordinates (u, λ) are chosen in $\mathbf{R}^{n+1}$ to distinguish the parameter λ. With respect to the arclength s a solution curve is of the form $(u(s), \lambda(s))$. The natural parameterization stops whenever

$$\text{Rank}(D_u G(u(s), \lambda(s)) < n$$

occurs, while the parameterization with arclength continues at this point. Furthermore, differentiating

$$G(u(s), \lambda(s)) = 0$$

with respect to s we obtain

$$\begin{aligned} D_u G(u(s), \lambda(s))\dot{u}(s) + D_\lambda G(u(s), \lambda(s))\dot{\lambda}(s) &= 0, \\ \|\dot{u}(s)\|^2 + |\dot{\lambda}(s)|^2 &= 1. \end{aligned}$$

2.1.3 Parameterization with pseudo-arclength

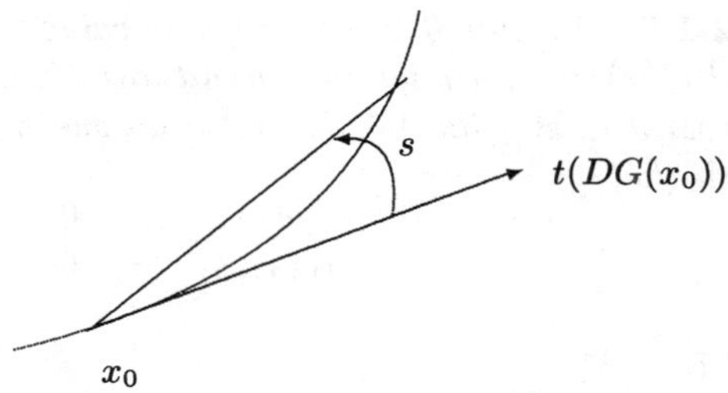

Fig. 2.1. $N(x, s) := t(DG(x_0))^T(x - x_0) - s$.

Let zero be a regular value of G and $x_0 \in \mathbf{R}^{n+1}$ be a solution of $G(x) = 0$. By Lemma 2.1.2 there exists a unique smooth function $c(s)$ such that

$$G(c(s)) = 0, \qquad c(0) = x_0$$

and $\dot{c}(s)^T \dot{c}(s) \equiv 1$. If we choose a smooth function

$$N: \ (x, s) \in \mathbf{R}^{n+1} \times \mathbf{R} \to N(x, s) \in \mathbf{R}$$

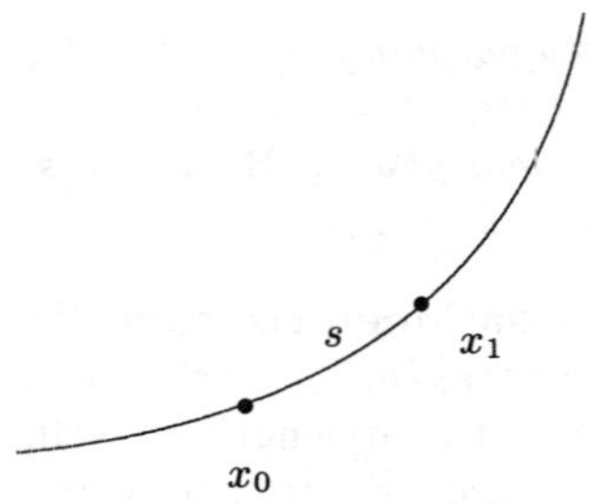

Fig. 2.2. $N(x,s) := \|x - x_0\| - s$.

such that the equality

$$N(x_0, 0) = 0 \tag{2.9}$$

holds and the matrix

$$\begin{pmatrix} DG(x_0) \\ D_x N(x_0, 0) \end{pmatrix} \quad \text{is nonsingular,} \tag{2.10}$$

i.e., $D_x N(x_0, 0) t(DG(x_0)) \neq 0$, then the equation

$$N(x, s) = 0 \tag{2.11}$$

defines a parameterization of solution curve of $G(x) = 0$ near x_0 as follows.

Lemma 2.1.3. *Assume 0 to be a regular value of G and $x_0 \in \mathbf{R}^{n+1}$ to be a solution of $G(x) = 0$. Under the conditions (2.9) and (2.10) there exist an open interval $J \subset \mathbf{R}$ with $0 \in J$, and a unique mapping $c: s \in J \to c(s) \in \mathbf{R}^{n+1}$, such that*

$$\begin{aligned} G(c(s)) &= 0, \\ N(c(s), s) &= 0. \end{aligned} \tag{2.12}$$

Remark 2.1.2.
a) The parameter s is called a pseudo-arclength.
b) Lemma 2.1.3 gives a local parameterization of solution curve. In particular, the function $N(x, s)$ is constructed usually with the current solution x_0 and will be updated at the next point on the solution curve. Figures 2.1-2.4 show some frequently used pseudo-arclength.

Relation to the arclength: Differentiating the equation (2.12) with respect to s yields

$$DG(c(s))\dot{c}(s) = 0,$$
$$D_x N(c(s), s)\dot{c}(s) + D_s N(c(s), s) = 0.$$

Hence, $c(s)$ is a solution of the initial value problem

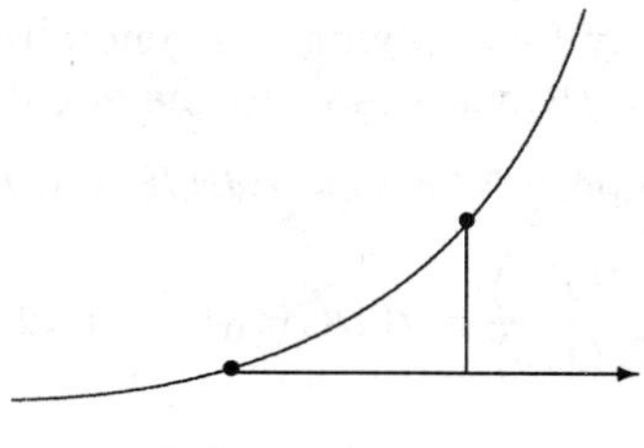

Fig. 2.3. $N(x,s) := e_i^T(x - x_0) - s;\ \ i \in \{1, \dots, n\},\ \ e_i$–unit vector.

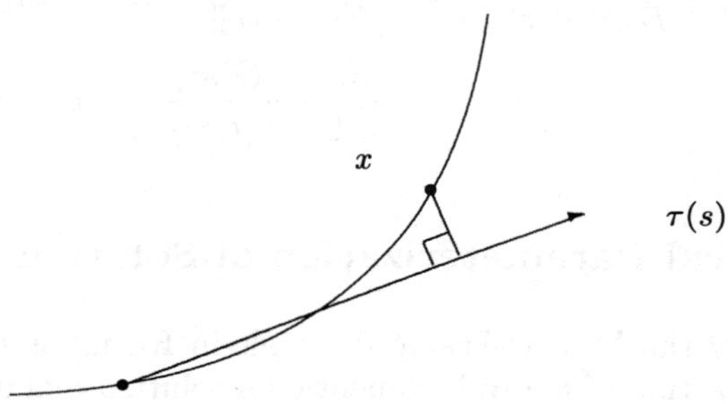

Fig. 2.4. $N(x,s) := \tau(s)^T(x - x(s));\ \tau(s)$–an approximation of the tangent $t(DG(x(s)))$.

$$\dot{c} = \begin{pmatrix} DG(c) \\ -\dfrac{D_xN(c,s)}{D_sN(c,s)} \end{pmatrix}^{-1} \begin{pmatrix} 0 \\ 1 \end{pmatrix}, \qquad c(0) = x_0. \tag{2.13}$$

As we have seen in (2.8), parameterization with the arclength corresponds to solution of the equation

$$\dot{c} = t(DG(c)) = \begin{pmatrix} DG(c) \\ -t(DG(c))^T \end{pmatrix}^{-1} \begin{pmatrix} 0 \\ 1 \end{pmatrix}, \qquad c(0) = x_0. \tag{2.14}$$

Denote the solutions of (2.13) and (2.14) with $d(s)$ and $c(s)$, respectively. We have the following estimate for the difference between $d(s)$ and $c(s)$.

Theorem 2.1.3. *If the pseudo-arclength is chosen such that*

$$\lim_{s \to 0} \|[D_sN(d(s),s)]^{-1} D_xN(d(s),s) + t(DG(x_0))^T\| \le \epsilon,$$

then

$$\|c(s) - d(s)\| \le O(\epsilon s) + O(s^2),$$
$$\|\dot{c}(s) - \dot{d}(s)\| \le O(\epsilon) + O(s).$$

Proof: It follows directly by applying the generalized implicit function theorem (cf. Brezzi/Rappaz/Raviart [48]) to the equation (2.14). ■

Example 2.1.1. If we choose $N(x, s) = t(DG(x_0))^T(x - x_0) - s$, then we have

$$-\frac{D_x N(x,s)}{D_s N(x,s)} = t(DG(x_0))^T \quad \text{and} \quad \epsilon = 0.$$

The difference between $d(s)$ and $c(s)$ is of the order $O(s^2)$. Similar conclusion holds for $N(x, s) = \|x - x_0\| - s$, because of

$$\begin{aligned} -\frac{D_x N(d(x),s)}{D_s N(d(x),s)} &= \frac{d(s) - x_0}{\|d(s) - x_0\|} = t(DG(x_0))^T + O(s) \\ &= \frac{\dot{d}(0)s + O(s^2)}{\|\dot{d}(0\| + O(s^2)} = t(DG(x_0))^T + O(s). \end{aligned}$$

2.2 Local Parameterization of Solution Manifolds

We apply the Moving-Frame-Principle in Rheinboldt [249, 247, 248] to parameterization of multiple dimensional solution manifolds.

Let 0 be a regular value of $G: \mathbf{R}^{n+p} \to \mathbf{R}^n$ and the solution set

$$\mathcal{M} := \{x = (u, \lambda) \in \mathbf{R}^{n+p}; \qquad G(x) = 0\}$$

be not empty. Then $\mathcal{M}$ is a p-dimensional manifold as smooth as G (cf. [249, pp.43]).

Definition 2.2.1. *For $x_0 \in \mathcal{M}$ the tangent space of $\mathcal{M}$ at x_0 is defined as*

$$\begin{aligned} T_{x_0}\mathcal{M} &:= \text{Ker}(DG(x_0)) \\ &= \left\{ \sum_{i=1}^{p} \alpha_i t_i(x_0); \quad \alpha_i \in \mathbf{R},\ \text{Ker}(DG(x_0)) = \text{span}[t_1, \ldots, t_p] \right\} \\ &= T(x_0)\mathbf{R}^p, \qquad T(x_0) = (t_1(x_0), \ldots, t_p(x_0)) \in \mathbf{R}^{(n+p)\times p}. \end{aligned}$$

The normal space of $\mathcal{M}$ at x_0 is the orthogonal complement of $T_{x_0}\mathcal{M}$ in $\mathbf{R}^{n+p}$,

$$N_{x_0}\mathcal{M} := (\text{Ker}(DG(x_0)))^{\perp} = \text{Im}(DG(x_0)^T).$$

Since 0 is a regular value of G, for all $x \in \mathcal{M}$ the tangent space $T_x\mathcal{M}$ is p-dimensional. Correspondingly, the normal space $N_x\mathcal{M}$ is n-dimensional.

Choosing p orthogonal vectors $\{b_i\}_{i=1}^p \subset \mathbf{R}^{n+p}$, such that $(b_i,\ b_j) = \delta_{ij}$, $i, j = 1, \ldots, p$ hold and the extended matrix

$$\begin{pmatrix} DG(x_0) \\ B^T \end{pmatrix} \quad \text{with} \quad B := (b_1, \ldots, b_p) \tag{2.15}$$

is nonsingular, we solve the systems

$$\begin{pmatrix} DG(x_0) \\ B^T \end{pmatrix} \tau = e_{n+i}, \quad i = 1, \dots, p$$

for τ_i, $i = 1, \dots, p$, and obtain a basis $\{\tau_1, \dots, \tau_p\}$ for $T_{x_0}\mathcal{M}$. This can be seen from the equations $DG(x_0)\tau_i = 0$ $i = 1, \dots, p$, which imply

$$\tau_i \in T_{x_0}\mathcal{M}, \quad i = 1, \dots, p,$$

and from the equality

$$B^T(\tau_1, \dots, \tau_p) = I_p,$$

which ensures the linear independence of $\tau_1, \dots, \tau_p$. The conditioning number of the matrix (2.15) depends on the choice of B, see e.g., Sebastian [274].

For a p-dimensional subspace $T \subset \mathbf{R}^{n+p}$, if

$$T \cap N_{x_0}\mathcal{M} = \{0\},$$

then we can decompose the space $\mathbf{R}^{n+p}$ as

$$\mathbf{R}^{n+p} = T \oplus N_{x_0}\mathcal{M} = T \oplus T^{\perp}.$$

Let $\{h_1, \dots, h_p\}$ be an orthogonal basis of the tangent space T, e.g. the Gram-Schmidt orthogonalization of $\tau_1, \dots, \tau_p$. Define the matrix

$$H := (h_1, \dots, h_p) \in \mathbf{R}^{(n+p)\times p}.$$

We write all elements $x \in \mathbf{R}^{n+p}$ uniquely as

$$x = x_0 + Ht + w, \qquad \text{with } \; t \in \mathbf{R}^p, \;\; w \in T^{\perp}.$$

Thereafter we consider an enlarged system defined by

$$F(w,t) := \begin{pmatrix} G(x_0 + Ht + w) \\ H^T w \end{pmatrix} = 0 \tag{2.16}$$

for the unknown $w \in \mathbf{R}^n$ with t as the parameter. The mapping $F: \mathbf{R}^{n+p} \times \mathbf{R}^p \to \mathbf{R}^{n+p}$ is obviously as smooth as G. Moreover,

$$F(0,0) = 0$$

and the matrix

$$D_w F(0,0) = \begin{pmatrix} DG(x_0) \\ H^T \end{pmatrix}$$

is invertible. By the implicit function theorem there exist neighborhoods $V_1 \subset \mathbf{R}^p$, $V_2 \subset \mathbf{R}^{n+p}$ of the origin and a unique mapping $w: \; t \in V_1 \to w(t) \in V_2 \subset T^{\perp}$, such that

$$F(t, w(t)) = 0 \quad \text{for all} \quad t \in V_1,$$
$$w(0) = 0.$$

Therefore, the p-dimensional manifold $\mathcal{M}$ has a local parameterization at x_0 as

$$\mathcal{M} = \left\{ c(t) = x_0 + Ht + w(t) \in \mathbf{R}^{n+p};\ t \in V_1,\ w(0) = 0,\ w(t) \in V_2 \subset T^{\perp} \right\}. \tag{2.17}$$

Remark 2.2.1.
a) The parameterization (2.17) depends on the choice of H.
b) The system (2.16) is nonsingular for w and can be used for continuation of the solution manifold.

2.3 Predictor-Corrector Methods

We consider the one-parameter problem defined by a smooth mapping $G:\ \mathbf{R}^{n+1} \to \mathbf{R}^n$

$$G(x) = 0. \tag{2.18}$$

Given an initial point x_0 on a solution curve, we can exploit formally the predictor-corrector method for solving the initial value problem

$$\dot{x} = t(DG(x)), \quad x(0) = x_0. \tag{2.19}$$

Equivalently, this continues the solution curve of (2.18) from x_0, see Fig. 2.5.

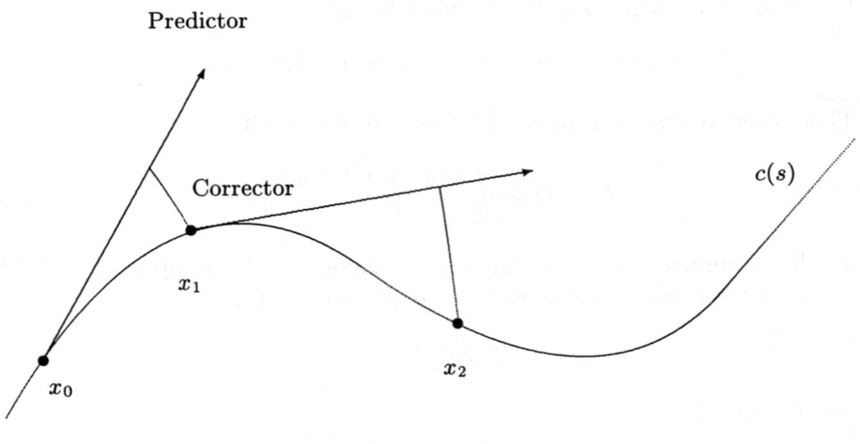

Fig. 2.5. The principle of a predictor-corrector method.

General Principles of a Predictor-Corrector Method:

- Predictors: First approximation along the solution curve, normally in the tangent direction. Different predictors can be derived via numerical methods for the initial value problem (2.19). Fig. 2.6 illustrates three possible predictors;

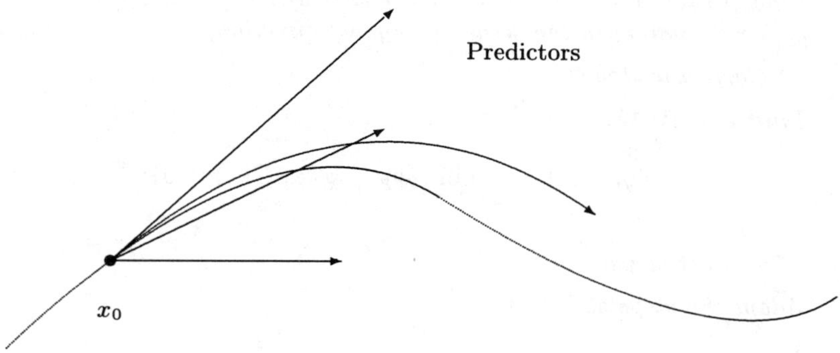

Fig. 2.6. Possible predictors.

- Correctors: Iterative approximation of the solution of $G(x) = 0$, e.g. using Newton-like methods to bring the predictor back to the solution curve. Fig. 2.7 shows three possible correctors.

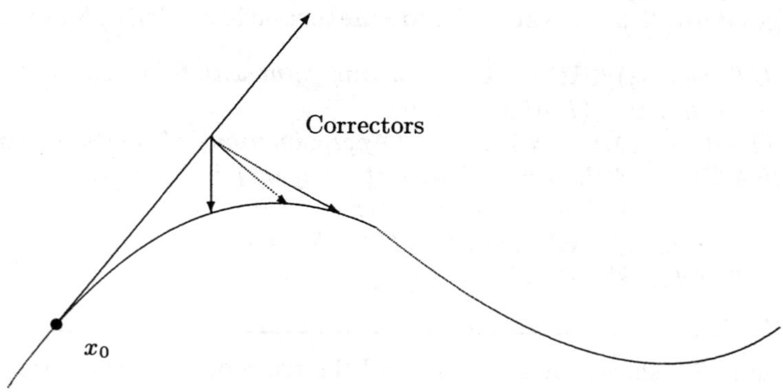

Fig. 2.7. Choices of correctors.

We collect these general principles into an algorithm.

Algorithm 2.3.1. Predictor-Corrector Methods

INPUT: $x_0 \in \mathbf{R}^{n+1}$ *(Initial approximation with* $G(x_0) \approx 0$*),*
$h > 0$ *(Initial stepsize);*
OUTPUT: x_i, $i = 1, 2, \ldots$ *(Approximation of the solution curve).*
ITERATION: FOR $(i = 1$*; Stop criterion; i++) {*
/ Predictor step */*
Find predictor y_i*, such that* $G(y_i) \approx 0$ *and* $\|y_i - x_{i-1}\| \approx h$*, and* $y_i - x_{i-1}$ *points to the secant (tangent) direction;*
/ Corrector step */*
Find $x_i \in \mathbf{R}^{n+1}$*, such that*

$$\|y_i - x_i\| = \min_{x \in \mathbf{R}^{n+1}} \{\|y_i - x\|; \quad G(x) = 0\};$$

/ Adjust stepsize* $h > 0$ **/*
Adapt the stepsize $h > 0$.
}

An efficient predictor-corrector method is made up of appropriate combinations of four aspects: *predictors, local parameterization of the solution curve, correctors and stepsize control.* Adapting the general principles of a predictor-corrector method to the various parameterizations of a solution curve in Section 2.1, we obtain the following frequently used types of continuation methods.

Algorithm 2.3.2. Natural parameterization *(Embedding method)*

INPUT: $(u_0, \lambda_0) \in \mathbf{R}^n \times \mathbf{R}$ *(Starting point with* $G(u_0, \lambda_0) \approx 0$*);*
$h > 0$ *(Initial stepsize);*
OUTPUT: (u_i, λ_i), $i = 1, 2, \ldots$ *(Approximations of solution curve);*
ITERATION: FOR $(i = 1$*; Stop criterion; i++) {*
i) $\lambda_i = \lambda_{i-1} + h$; $u_i = u_{i-1}$;
ii) Find $u_i \in \mathbf{R}^n$*, such that* $G(u_i, \lambda_i) \approx 0$*;*
iii) Adapt the stepsize $h > 0$.
}

Figure 2.8 shows the predictor and the trace of a few correction steps.

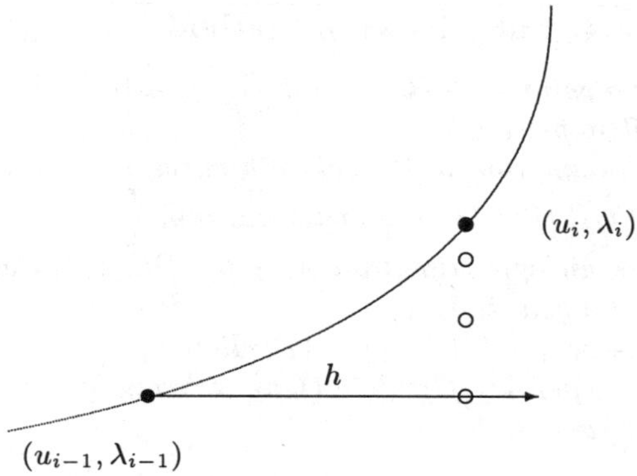

Fig. 2.8. Natural parameterization.

Algorithm 2.3.3. Pseudo-arclength parameterization

INPUT: $x_0 \in \mathbf{R}^{n+1}$ *(Starting point with* $G(x_0) \approx 0$*);*
$h > 0$ *(Initial stepsize);*
OUTPUT: $x_i,\ i = 1, 2, \ldots$ *(Approximation of the solution curve);*
ITERATION: FOR $(i = 1;$ *Stop criterion;* $i{+}{+})$ $\{$

i) Find $N : \mathbf{R}^{n+1} \times \mathbf{R} \to \mathbf{R}$, *a local pseudo-arclength, such that*

$$N(x_{i-1}, 0) = 0, \ D_x N(x_{i-1}, 0) t(DG(x_{i-1})) > 0;$$

ii) Find a predictor $y_i \in \mathbf{R}^{n+1}$ *such that* $G(y_i) \approx 0,\ N(y_i, h) \approx 0;$
iii) Start from y_i, *calculate* x_i *as numerical solution of*

$$\begin{pmatrix} G(x) \\ N(x,h) \end{pmatrix} = 0, \quad \text{e.g.theNewtonmethod.}$$

$$x_i \longleftarrow x_i - \begin{pmatrix} DG(x_i) \\ D_x N(x_i, h) \end{pmatrix}^{-1} \begin{pmatrix} G(x_i) \\ N(x_i, h) \end{pmatrix};$$

iv) Adapt the stepsize $h > 0$.
$\}$

2.3.1 Euler-Newton method

A simple and frequently used predictor-corrector method is the Euler-Newton continuation.

Algorithm 2.3.4. Euler-Newton Method

INPUT: *Starting point* $x_0 \in \mathbf{R}^{n+1}$ *with* $G(x_0) \approx 0$ *and*
Initial stepsize $h > 0$;
OUTPUT: *Approximation of the solution curve* x_i, $i = 0, 1, 2, \ldots$.
ITERATION: *FOR* ($i = 1$; *Stop criterion*; i++) $\{$
 i) *Choose an approximation* A_{i-1} *for* $DG(x_{i-1})$ *and calculate its induced tangent* $t(A_{i-1})$;
 ii) $x_i \longleftarrow x_{i-1} + ht(A_{i-1})$ (*Predictor*);
 iii) $x_i \longleftarrow x_i - A^+_{i-1} G(x_i)$ (*One correction*);
 iv) *Adapt stepsize* $h > 0$.
$\}$.

Here A^+ represents the generalized inverse of the matrix A. Let $c_h(s)$ be the polygonal curve, starting at x_0 and passing through the points x_i, $i = 1, 2, 3, \ldots$ which are calculated by the Euler-Newton method with a fixed stepsize $h > 0$. Denote the arclength parameterization of the solution curve as $c(s)$. We assume that $c(0) = x_0$.

Theorem 2.3.1. *Let* $G : \mathbf{R}^{n+1} \to \mathbf{R}^n$ *be a smooth mapping and 0 be a regular value of* G. *If* $x_0 \in \mathbf{R}^{n+1}$ *is a solution of* $G(x) = 0$ *and the matrices* $A_i \in \mathbf{R}^{n \times (n+1)}$, $i = 1, 2, \ldots$ *are chosen such that the conditions*

$$\|A_i - DG(x_i)\| = O(h)$$

are satisfied uniformly for all i, *then there exists a maximal* s_0 *such that the following estimate holds uniformly for all* $s \in [0, s_0]$

$$\|G(c_h(s))\| \leq O(h^2), \quad \|c_h(s) - c(s)\| \leq O(h^2).$$

Proof: see Allgower/Georg [14] and Mei [225]. ■

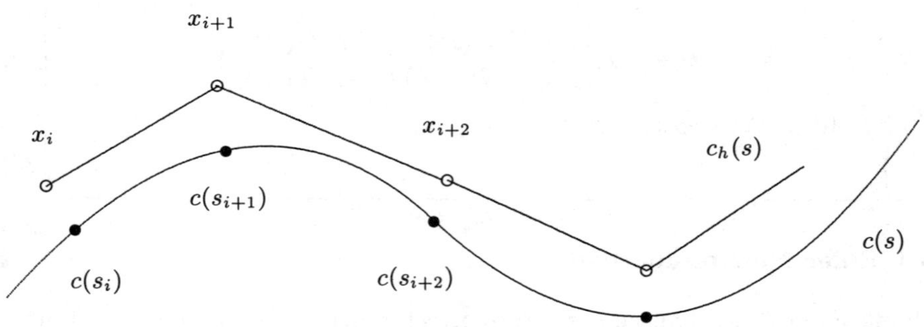

Fig. 2.9. Euler-Newton method.

Linear systems used in continuation methods are generally of the form

$$\begin{pmatrix} A & p \\ q^T & r \end{pmatrix} \begin{pmatrix} u \\ \lambda \end{pmatrix} = \begin{pmatrix} f \\ g \end{pmatrix}, \tag{2.20}$$

where $A \in \mathbf{R}^{n\times n}$ and $p,\ q,\ f \in \mathbf{R}^n$, $r,\ g \in \mathbf{R}$. If the matrix A is nonsingular, these bordered linear systems can be solved with the following techniques of block eliminations, see e.g. Chan [57, 58], Govaerts [134, 135].

Scheme 1: Solve the systems $Av = p$ and $Aw = f$. Then $\lambda = (g - q^T w)/(r - q^T v)$ and $u = w - \lambda v$.

Alternatively we can also solve (2.20) by

Scheme 2: Solve the system $A^T w = q^T$ and compute $\lambda = (g - w^T f)/(r - w^T p)$. We obtain u by solving u from $Au = f - \lambda p$.

Both schemes solve the system (2.20), especially for matrices with reasonable conditioning numbers. If A is nearly singular, then the former gives a better approximation of u, while the latter yields a more accurate approximation for λ. A combination of these two approaches in Algorithm 2.3.5 is considered by Govaerts in [134, 135] and shown to be stable, even for nearly singular matrix A.

Algorithm 2.3.5. *(Govaerts [135])* **Mixed Elimination Method**

1. *Solve* $A^T w = q^T$*;*
2. *Compute* $\lambda = (g - w^T f)/(r - w^T p)$*;*
3. *Solve* $Av = p$*;*
4. *Calculate* $\delta = r - q^T v$, $f_1 = f - \lambda p$ *and* $g_1 = g - \lambda r$*;*
5. *Solve* $Av_1 = f_1$*;*
6. *Compute* $\lambda_1 = (g_1 - q^T v_1)/\delta$, *then* $u = v_1 - v\lambda_1$, $\lambda = \lambda + \lambda_1$.

Recall that in a predictor-corrector continuation method the tangent vector is calculated by solving linear systems of the form

$$\begin{pmatrix} D_u G & p \\ q^T & r \end{pmatrix} t = \begin{pmatrix} 0 \\ 1 \end{pmatrix}, \tag{2.21}$$

while the Newton corrector involves the linear systems

$$\begin{pmatrix} D_u G & p \\ q^T & r \end{pmatrix} w = \begin{pmatrix} -G(u,\lambda) \\ 0 \end{pmatrix}. \tag{2.22}$$

By Algorithm 2.3.5 we need to solve the system $D_u G\, u = f$ two or three time for a solution of (2.21) and (2.22), respectively. For parameter-dependent differential equations, discretization of the differential operator yields usually large systems. Correspondingly, the linear system in the correction step is large and sparse. We want to incorporate some well established iterative methods for solving large linear systems into continuation methods.

2.3.2 A continuation-Lanczos algorithm

Some continuation-Lanczos algorithms have been applied to trace solutions of the bifurcation problems

$$G(u, \lambda) = 0, \tag{2.23}$$

see e.g., Chien/Weng/Shen [61], Chien/Lin/Mei [60], Schwarzer [272] and Sebastian [274]. The main idea is to exploit the symmetric Lanczos method for solving the associated linear systems in continuation of solution branches, for detecting bifurcation points and for approximating tangents of solution branches. Furthermore, the band Lanczos method with reorthogonalization can be used to handle multiple bifurcations.

a) Lanczos scheme

Suppose the matrix $A \in \mathbf{R}^{n\times n}$ is large, sparse and symmetric. Algorithm 2.3.6 summarizes the major characteristics of the Lanczos method for solving the linear system $Ax = b$.

Algorithm 2.3.6. Lanczos algorithm for solving the symmetric system $Ax = b$.

Input

$x_0,\ r_0 = b - Ax_0;\quad \beta_0 = \|r_0\|_2;\quad q_0 = 0;\quad j = 0;\quad s_0 = r_0.$

while $(\|s_j\|_\infty \geq \epsilon$ *(Error tolerance)*$)$ $\{$

$q_{j+1} = r_j/\beta_j;\quad j = j+1;\quad \alpha_j = q_j^T A q_j;$

$r_j = (A - \alpha_j I)q_j - \beta_j q_{j-1};\quad \beta_j = \|r_j\|_2;$

if $(j = 1)$ **then**

$\{\ d_1 = \alpha_1;\quad c_1 = q_1;\quad \rho_1 = \beta_0/\alpha_1;\quad x_1 = \rho_1 q_1;\ \}$

else

$\{\mu_{j-1} = \beta_{j-1}/d_{j-1};\quad d_j = \alpha_j - \beta_{j-1}\mu_{j-1};\quad c_j = q_j - \mu_{j-1}c_{j-1};$

$\rho_j = -\mu_{j-1}d_{j-1}\rho_{j-1}/d_j;\quad x_j = x_{j-1} + \rho_j c_j;\quad s_j = b - Ax_j;$

$\}$

$\}$

Output $x = x_j.$

Starting with a normalized vector q_1, the Lanczos algorithm generates a sequence of orthonormal vectors $q_1, \ldots, q_j$, called *Lanczos vectors*, for the Krylov subspace $\boldsymbol{K}(A, q_1, j)$ such that

$$\mathrm{span}[q_1, \ldots, q_j] = \mathrm{span}[q_1, Aq_1, \ldots, A^{j-1}q_1] =: \boldsymbol{K}(A, q_1, j).$$

Let $Q_j = (q_1, q_2, \ldots, q_j) \in \mathbf{R}^{n\times j}$. After j steps of the Lanczos algorithm we obtain

$$AQ_j = Q_j T_j + r_j e_j^T,$$

where $q_{j+1} = r_j/\beta_j$ with $\beta_j = \pm\|r_j\|_2$, and e_j is the jth standard unit vector. An approximate solution of $Ax = b$ is given as $x_j = Q_j y_j \in \boldsymbol{K}(A, q_1, j)$. Here y_j is the solution of the following tridiagonal linear system

$$T_j y_j = Q_j^T b, \tag{2.24}$$

where

$$T_j := Q_j^T A Q_j = \begin{bmatrix} \alpha_1 & \beta_1 & & & 0 \\ \beta_1 & \alpha_2 & \ddots & & \\ & \ddots & \ddots & \ddots & \\ & & \ddots & \ddots & \beta_{j-1} \\ 0 & & & \beta_{j-1} & \alpha_j \end{bmatrix}$$

and $\alpha_j = q_j^T A q_j$. For the eigenvalue problem $Ax = \lambda x$ we expect that eigenvalues of the reduced matrix T_j approximate the extremal eigenvalues of A.

If the coefficient matrix A is indefinite, as those encountered in the bifurcation problems, then β_j may fail to converge to zero (cf. Golub/Van Loan [128]). As an alternative Paige and Saunders [240] use a sequence of Givens rotations to transform T_j to a lower triangular matrix L_j. This improves the convergence. Furthermore, note that for the linear system $Ax = b$ and the eigenvalue problem $Ax = \lambda x$ the the residual vectors $s_j = Ax_j - b$ and r_j may have different convergence behavior. In fact, they are related as

$$\|s_j\|_2 = \|r_j\|_2 |e_j^T y_j|.$$

Choosing the norm of s_j as the stopping criterion often yields convergent approximate solution of $Ax = b$, see e.g. Chien/Lin/Mei [60]. On the other hand, both A and $A - \sigma I$ yield the same Lanczos vectors for all $\sigma \in \mathbf{R}$, as pointed out by Parlett in [241]. For an indefinite matrix A if a shift makes $A - \sigma I$ definite for some $\sigma \in \mathbf{R}$, then we apply the Lanczos algorithm to the shifted matrix. This happens typically for matrices in the discretization of a second order semilinear elliptic eigenvalue problem.

b) Unsymmetric Lanczos scheme

If the matrix A is nonsymmetric, but nearly symmetric and nonsingular, the unsymmetric Lanczos method can be used to approximate the solution of the system $Ax = b$ (cf. Saad [260]). In unsymmetric Lanczos algorithm two biorthogonal systems $V_j = [v_1, v_2, \cdots, v_j]$ and $W_j = [w_1, w_2, \cdots, w_j]$ are generated such that

$$W_j^T A V_j = T_j = \begin{bmatrix} \alpha_1 & \delta_1 & & & 0 \\ \beta_1 & \alpha_2 & \ddots & & \\ & \ddots & \ddots & \ddots & \\ & & \ddots & \ddots & \delta_{j-1} \\ 0 & & & \beta_{j-1} & \alpha_j \end{bmatrix}.$$

The entries α_j, β_j and δ_j are calculated as in the Algorithm 2.3.7. Thereafter, the reduced matrix T_j is used to generate approximations of solutions of the linear system $Ax = b$ and the eigenvalue problem $Ax = \lambda x$. The eigenvalues $\theta_1, \dots, \theta_j$ of T_j approximate particularly well the extremal eigenvalues of the matrix A see e.g. Trefethen/Bau [292], Barret et al [32], Golub/Van Loan[128]. They are called *Ritz values* of the matrix A. The eigenvectors $y_1, \dots, y_j$ of H_j yields approximations $x_i := V_j y_i,\ i = 1, \dots, j$ of the corresponding eigenvectors of A. We call $V_j y_i,\ i = 1, \dots, j$ *Ritz vectors* of A.

Let x_0 be the initial guess, and $r_0 = b - Ax_0$ be the initial residual. We seek an approximate solution of the form

$$x_j = x_0 + z_j, \tag{2.25}$$

where $z_j = V_j y_j \in \text{span}[v_1, v_2, \cdots, v_j]$ for some $y_j \in \mathbf{R}^j$, and satisfies

$$\langle r_0 - Az_j,\ w_i \rangle = 0, \quad i = 1, 2, \dots, j, \tag{2.26}$$

i.e.,

$$W_j^T (r_0 - Az_j) = 0. \tag{2.27}$$

Here and in the sequel $\langle \cdot,\ \cdot \rangle$ denotes the inner product in $\mathbf{R}^n$. Replacing z_j by $V_j y_j$ in (2.27) yields an equation for y_j

$$T_j y_j = W_j^T r_0. \tag{2.28}$$

By the biorthogonality of V_j and W_j we have $W_j^T r_0 = \beta W_j^T v_1 = \beta e_1$, where $\beta = \|r_0\|_2$ and $e_1 = [1, 0, \cdots, 0]^T \in \mathbf{R}^j$ is the first standard unit vector. It follows from (2.28) that the approximate solution in (2.25) is given as

$$x_j = x_0 + V_j y_j = x_0 + \beta V_j T_j^{-1} e_1. \tag{2.29}$$

Algorithm 2.3.7. Lanczos algorithm for solving the unsymmetric linear system $Ax = b$.

i) Input x_0 *and*
$i = 0;\quad r_0 = b - Ax_0;\quad v_0 = w_0 = 0;\quad \beta_0 = 1;\quad p_0 = r_0.$

ii) While $(i < j)$ $\{$
 $\delta_i = r_i^T p_i;\quad v_{i+1} = r_i/\beta_i;\quad w_{i+1} = p_i/\delta_i;$
 $i = i + 1;$
 $\alpha_i = w_i^T A v_i;$
 $r_i = (A - \alpha_i I)v_i - \delta_{i-1} v_{i-1};\quad p_i = (A - \alpha_i I)w_i - \beta_{i-1} w_{i-1};$
 $\beta_i = \|r_i\|_2;$
$\}$

iii) Form the approximate solution

$$x_j = x_0 + \beta V_j T_j^{-1} e_1 \quad \textit{with} \quad y_j = T_j^{-1} \beta e_1 \quad \textit{and} \quad \beta = \|r_0\|_2.$$

As a stopping criterion for Algorithm 2.3.7 we use

$$||b - Ax_j|| = ||r_{j+1}|| \cdot |e_j^T y_j| < \varepsilon \tag{2.30}$$

for some $\varepsilon > 0$, see e.g. Saad [260]. Obviously, if the case $\delta_i = r_i^T p_i = 0$ or $\beta_i = 0$ occurs, but $r_i \neq 0$ and $p_i \neq 0$, the unsymmetric Lanczos method will break down. As a remedy we refer to some variations of the quasi-minimal residual algorithms (QMR) discussed in Freund *et al* [108, 109].

We integrate the Lanczos method for solving linear systems into continuation methods for parameter-dependent problems. To detect secondary bifurcations we compute the minimum eigenvalue of T_j and monitor its variation along the solution curve.

Algorithm 2.3.8. A continuation-Lanczos algorithm

Input :

$x_0 \in \mathbf{R}^{n+1}$, *such that* $G(x_0) \approx 0$ {*an approximate solution*}
$h > 0$ {*initial stepsize*}
$t > 0$ {*initial tangent vector*}
$\varepsilon^* > 0$ { *approximate eigenvalue*}

Step 1) $x := x_0 + ht$ {*predictor step*}

Step 2) (i) Use Algorithms 2.3.5 and 2.3.6 or 2.3.7 to solve (2.22) until convergence, and update x. {*Newton corrector*}
(ii) Adapt the stepsize h. {*stepsize control*}

Step 3) Monitor bifurcations and compute the tangent vector.
(i) Determine the Ritz value θ *of* T_j *by solving* $T_j v = \theta v$.
(ii) If $|\theta| < \varepsilon^*$ *or* $\mathrm{Re}(\theta)\cdot \mathrm{Re}(\tilde{\theta}) \leq 0$ *with* $\tilde{\theta}$ *as the Ritz value of* T_j *at the previous continuation step, then*
(a) Compute the Ritz vectors $w := Q_j v$;
(b) Set $w' = \begin{pmatrix} w \\ 1 \end{pmatrix} \in \mathbf{R}^{n+1}$;
(c) Normalize the tangent vector $t := w'/||w'||_2$ *and goto Step 1), else*
Solve (2.21) for the tangent t *by Algorithms 2.3.5 and 2.3.6; goto Step 1) until traversing is stopped.*

2.3.3 A continuation-Arnoldi algorithm

Compared with the unsymmetric Lanczos scheme, Arnoldi method is more robust. We exploit the Arnoldi method for the continuation of a solution curve and the detection of bifurcations. Namely, we consider a continuation-Arnoldi algorithm, based on the discussions in Huitfeldt/Ruhe [165], see also Schwarzer [272] and Sebastian [274].

Suppose that $A = D_uG \in \mathbf{R}^{n\times n}$ is nonsymmetric and nonsingular, and the eigenvalues of A are ordered as

$$\mathrm{Re}(\lambda_1) \geq \mathrm{Re}(\lambda_2) \geq \cdots \geq \mathrm{Re}(\lambda_n).$$

Saad [259, 260, 263] shows that the eigenvalues which are in the outmost part of the spectrum $\sigma(A)$, i.e., the extremal eigenvalues, are likely to be well approximated by the Arnoldi method (cf. Arnoldi [17]). This property will be used in solving the linear system $Ax = b$ and the eigenvalue problem $Ax = \lambda x$.

Let $x_0 \in \mathbf{R}^n$ be an initial guess, and $r_0 := b - Ax_0$ be the initial residual. The Arnoldi iteration builds an orthonormal basis $v_1, \ldots, v_m$ for the Krylov subspace $K(A, r_0, m) = \mathrm{span}[r_0, Ar_0, \cdots, A^{m-1}r_0]$ by a modified Gram-Schmidt process.

Algorithm 2.3.9. The Arnoldi Method

1. *Choose a starting vector* r_0 ($r_0 = b - Ax_0$ *for solving linear system*); *set* $\beta := \|r_0\|$ *and* $v_1 = r_0/\beta$;
2. *For* $k = 1, \ldots, m$ *Do*

   $$w := Av_k - \sum_{i=1}^{k} h_{ik}v_i \quad \text{with} \quad h_{i,k} := \langle Av_k,\ v_i\rangle$$

   $h_{k+1,k} := \|w\|, \quad v_{k+1} := w/h_{k+1,k}$.
3. *Solve the linear system*

   $$H_m y = \beta e_1,$$

   for y_m *and calculate the approximate solution* $x_m = x_0 + V_m y_m$ *of* $Ax = b$. *Here* $H_m = (h_{ij})$ *is an upper Hessenberg matrix,* $V_m = (v_1, \ldots, v_m)$, *and* $e_1 \in \mathbf{R}^m$ *is the first standard unit vector.*
4. *Calculate eigenpairs* $(\sigma_i,\ y_i)$, $i = 1, \ldots, m$ *of the reduced matrix* H_m. *We obtain approximations* $(\sigma_i,\ V_m y_i)$, $i = 1, \ldots, m$ *of the first* m *eigenpairs of* A.

In discretization of reaction-diffusion equations the spectrum of the matrix D_uG normally lies in one side of the imaginary axis. Instability and bifurcation occur if variations of control parameters cause some eigenvalues to cross the imaginary axis. Thus we are interested in the situation that $A = D_uG$ has a small number of eigenvalues near the imaginary axis, while the rest of the spectrum of A is well away from it.

To monitor simple bifurcations during continuation of solution curves, we assume that at the kth continuation step an approximate solution x_k is close to the simple bifurcation point x^*. Since 0 is the minimum of absolute value of eigenvalues and at (right) end of the spectrum of $D_uG(x^*)$, its the eigenpair $(0,\ \phi)$, i.e., $D_uG(x^*)\phi = 0$ can be approximated by the Ritz pair the upper Hessenberg matrix generated by Arnoldi process with the matrix $D_uG(x^*)$.

Consequently, if x_k is sufficiently close to x^*, the Ritz pair $(\theta,\ v)$ of H_m, the matrix generated by Arnoldi process with $D_uG(x_k)$, yields an approximation $(\theta,\ V_m v)$ for $(0,\ \phi)$. If $|\theta| < \varepsilon^*$ for some $\varepsilon^* > 0$ small enough, a simple bifurcation point is signaled on the solution curve. Furthermore, the Ritz vector can be used as the approximating tangent vector at the bifurcation point for branch-switching.

Algorithm 2.3.10. A Continuation-Arnoldi algorithm

Input :

$x_0 \in \mathbf{R}^{n+1}$, *such that* $G(x_0) = 0$ {*an initial solution*}
$\delta > 0$ {*initial stepsize*}
$t \in \mathbf{R}^{n+1}$ {*initial tangent vector*}
$\varepsilon^* > 0$ {*tolerance* }

Step 1) $x := x_0 + \delta t$ {*predictor step*}

Step 2) (i) Use Algorithms 2.3.5 and 2.3.9 to solve the linear systems until convergence, and then update x; {*Newton corrector*}
(ii) adapt stepsize. {*stepsize control*}

Step 3) Test for bifurcation point and compute tangent vector.
(i) Solve $H_m v = \theta v$;
(ii) If $|\theta| < \varepsilon^*$, *or* $\mathrm{Re}(\theta)\cdot \mathrm{Re}(\tilde{\theta}) \le 0$ *with* $\tilde{\theta}$ *as the Ritz value of* H_m *at the previous continuation step, then signal occurrence of bifurcation and do local bifurcation analysis*
else
adapt and go to Step 1), until traversing is stopped.

2.4 Computation of Multi-Dimensional Solution Manifolds

In this section we consider some numerical methods for computation of regular multi-dimensional solution manifolds developed by Rheinboldt [249, 247, 248].

Let $G:\ \mathbf{R}^{n+p} \to \mathbf{R}^n$ $(p \ge 1)$ be a C^r-continuous differentiable mapping. We consider solutions of the problem

$$G(x) = 0, \quad \text{with} \quad x = (u, \lambda) \in \mathbf{R}^{n+p}.$$

It is well-known that if the regular solution set

$$\mathcal{M} := \left\{ x \in \mathbf{R}^{n+p} \;\middle|\; G(x) = 0,\ x \text{ regular} \right\}$$

is non-empty, then $\mathcal{M}$ is a p-dimensional manifold in $\mathbf{R}^{n+p}$ as smooth as G. We assume that $\mathcal{M}$ is non-empty. The tangent space $T_x\mathcal{M}$ at any point $x \in \mathcal{M}$ is the kernel of $DG(x)$ and the normal space $N_x\mathcal{M}$ at $x \in \mathcal{M}$ is the orthogonal complement of the tangent space in $\mathbf{R}^{n+p}$, i.e.,

$$T_x\mathcal{M} := \mathrm{Ker}(DG(x)), \quad N_x\mathcal{M} := (\mathrm{Ker}(DG(x))^{\perp} = \mathrm{Im}(DG(x)^T).$$

Let x_0 be a point in $\mathcal{M}$ and $\{h_1, \ldots, h_p\}$ be an orthonormal basis of the tangent space $T_{x_0}\mathcal{M}$. Since the system

$$F(w,t) := \begin{pmatrix} G(x_0 + Ht + w) \\ H^T w \end{pmatrix} = 0 \tag{2.31}$$

with $H := (h_1, \ldots, h_p)$ is nonsingular at $(0,0)$, applying the implicit function theorem to F $(0,0)$ yields a local parameterization of $\mathcal{M}$ at x_0

$$\mathcal{M} \cap V_2 = \{c(t) = x_0 + Ht + w(t) \in \mathbf{R}^{n+p}; \quad t \in V_1 \}, \tag{2.32}$$

where the subsets $V_1 \subset \mathbf{R}^p$ and $V_2 \subset \mathbf{R}^{n+p}$ contain the origin and x_0, respectively. To continue x_0 in the manifold $\mathcal{M}$, we determine points near x_0 by computing $w(t)$ from the system (2.31).

First of all, we choose t small to ensure validity of the local parameterization (2.32). For $c(t) = x_0 + Ht + w(t)$ we derive formally from the equations

$$\begin{aligned} G(c(t)) &= 0, \\ H^T w(t) &= 0 \end{aligned}$$

that the tangent $c'(t) = H + w'(t)$ should satisfy

$$\begin{pmatrix} DG(c(t)) \\ H^T \end{pmatrix} c'(t) = \begin{pmatrix} 0 \\ -I_{p\times p} \end{pmatrix}. \tag{2.33}$$

With the given point $c(0) = x_0$ we solve $c'(0)$ from (2.33) and obtain immediately the tangent of $w(t)$ at $t = 0$,

$$w'(0) = c'(0) - H.$$

Let $\mathcal{M}_0$ be an open subset of the manifold $\mathcal{M}$. A moving frame of class C^s on $\mathcal{M}_0$ consists of C^s-continuous functions $u_1, \ldots, u_p$, which make up a basis of $T_x\mathcal{M}$ and have the property $u_i(x) \in T_x\mathcal{M}$ for all $x \in \mathcal{M}_0$, $i = 1, \ldots, p$. In Algorithm 2.4.1 we update the tangent space T by choosing its basis as solutions $\tau_1, \ldots, \tau_p$ of the systems (2.37).

Theorem 2.4.1. *If the matrix H is chosen such that*

$$\begin{pmatrix} DG(x) \\ H^T \end{pmatrix}$$

is nonsingular for all $x \in \mathcal{M}_0$, *then the solutions* $\tau_i(x),\ i = 1, \ldots, p$ *of*

$$\begin{pmatrix} DG(x) \\ H^T \end{pmatrix} \tau_i = e_{n+i}, \; i = 1, \ldots, p \tag{2.34}$$

is C^{r-1}*-continuous and make up a moving frame for* $\mathcal{M}_0$.

Proof: The existence and linear independence as well as the C^{r-1}-continuity follow directly from an application of the implicit function theorem to the mapping (2.34) at x_0 and from the C^r-continuity of G. ■

Algorithm 2.4.1. Predictor-Corrector Method for Solution Manifold

INPUT: $x_0 \in \mathcal{M}$; (*Initial point on* $\mathcal{M}$)
t; (*Initial grid near* x_0)
$H = (h_1, \ldots, h_p)$; (*Initial tangent vectors*)
$\epsilon > 0$; (*Tolerance for correctors*)
OUTPUT: $c^i,\ i = 1, 2, \ldots$ (*Approximation of points in* $\mathcal{M}$ *near* x_0)

ITERATION: FOR $(i = 1$*; Stop criterion;* $i{+}{+})$ $\{$

Step 1) Calculate the predictor

$$w^0 := w(0) + w'(0)t = (c'(0) - H)t, \quad t > 0 \tag{2.35}$$

as a linear approximation of $w(t)$;

Step 2) Calculate corrections for w^0 *successively for* $k = 0, 1, \ldots$

$$\begin{pmatrix} DG(x_0 + Ht + w^k) \\ H^T \end{pmatrix} \Delta w = -\begin{pmatrix} G(x_0 + Ht + w^k) \\ H^T w^k \end{pmatrix},$$

$$w^{k+1} = w^k + \Delta w; \tag{2.36}$$

Stop if $\|G(c^k)\| \le \epsilon$ *for the approximation* $c^k = x_0 + Ht + w^k$ *of* $c(t)$;

Step 3) Set $x_0 := c^k$, $c^i := c^k$ *and solve the systems*

$$\begin{pmatrix} DG(x_0) \\ H^T \end{pmatrix} \tau_i = e_{n+i}, \; i = 1, \ldots, p. \tag{2.37}$$

Orthonormalize $\tau_1, \ldots, \tau_p$ *to generate new tangent vectors* $h_1, \ldots, h_p$ *and to form new matrix* H.

$\}$

3. Detecting and Computing Bifurcation Points

Exploring nonlinear phenomena has become a major challenge in physics, chemistry, biology, engineering, medicine and social science. We consider nonlinear problems of the form

$$\frac{\partial u}{\partial t} = G(u, \lambda), \tag{3.1}$$

where $G : X \times \mathbf{R}^p \to Y$ is a "smooth" mapping and $\lambda \in \mathbf{R}^p$ represents various control parameters, e.g. Reynolds number, catalyst, temperature, density, initial or final products, etc. Bifurcation theory studies how solutions of (3.1) and their stability change as the parameter λ varies. A point (u_0, λ_0) in $X \times \mathbf{R}^p$ is called a *bifurcation point* if it satisfies (3.1) and in all neighborhoods of (u_0, λ_0) the problem (3.1) has at least two different (stationary or time-dependent) solution branches. A bifurcation problem is *generic* if for all $\tilde{G}$ in a small neighborhood of G, the problem

$$\frac{\partial u}{\partial t} = \tilde{G}(u, \lambda)$$

has the same number of solution branches with similar stability behavior as (3.1) in the neighborhood of (u_0, λ_0).

In this chapter we choose X, Y to be the finite dimensional space $\mathbf{R}^n$ and use elementary knowledge of calculus and linear algebra to give a brief introduction of detection and computation of some frequently encountered bifurcation points. The discussion can be generalized easily for differential equations in appropriate function spaces, e.g. Sobolev spaces. We note that the finite dimensional problem can be also considered as a discretization of differential equations.

3.1 Generic Bifurcation Points

Let

$$\mathcal{M} := \{(u, \lambda) \in X \times \mathbf{R}^p \quad | \quad G(u, \lambda) = 0\} \tag{3.2}$$

be the manifold of stationary solutions of (3.1) in $X \times \mathbf{R}^p$. A singularity of the system (3.1) on $\mathcal{M}$ origins either from a loss of linear stability or from a nonlinear degeneracy. A generic singular point (u_0, λ_0) on the manifold $\mathcal{M}$ can be classified with codimension of the singularity and number of parameters in the problem, see e.g. Golubitsky *et al* [129, 131] and Kuznetsov [197] for more detailed discussion on this topic. We consider in this section a few generic bifurcations.

3.1.1 One-parameter problems

For one-parameter problems the solution manifold $\mathcal{M}$ consists of a curve $(u(s), \lambda(s))$. We assume

$$(u(0), \lambda(0)) = (u_0, \lambda_0)$$

to be a singular point of (3.1). Denote evaluations of the derivatives $D_uG,\ D_\lambda G, \ldots$ at (u_0, λ_0) as $D_uG_0,\ D_\lambda G_0, \ldots$. In generic case the singular point $(u_0, \lambda_0) \in \mathcal{M}$ is one of the following two types.

1) Turning points: This is a singularity in which zero is a simple eigenvalue of D_uG_0 and $D_\lambda G_0 \notin \mathrm{Im}(D_uG_0)$. Denote

$$\mathrm{Ker}(D_uG_0) = \mathrm{span}[\phi], \qquad \mathrm{Ker}(D_uG_0^*) = \mathrm{span}[\phi^*]. \tag{3.3}$$

Here and in the sequel the star * represents the adjoint operator, or the (complex conjugate) transpose matrix, respectively. From the statement

$$\frac{d}{ds}G(u(s), \lambda(s))\Big|_{s=0} = D_uG_0\dot{u}(0) + D_\lambda G_0\dot{\lambda}(0) = 0,$$

we obtain

$$\dot{u}(0) = \mathrm{const} \cdot \phi \quad \text{and} \quad \dot{\lambda}(0) = 0. \tag{3.4}$$

The *nondegeneracy condition* for a turning point is

$$\langle \phi^*,\ D_{uu}G_0\phi\phi \rangle \neq 0. \tag{3.5}$$

2) Hopf bifurcation points: This singularity results from the fact that the linearized operator D_uG_0 has exactly one pair of eigenvalues $\pm i\omega_0,\ \omega_0 > 0$ on the imaginary axis. Moreover, in this case $(u(s), \lambda(s))$ is the unique stationary solution curve across (u_0, λ_0). Let

$$\sigma(s) = \alpha(s) + i\omega(s)$$

be the eigenvalue of $D_uG(u(s), \lambda(s))$ such that

$$\alpha(0) = 0, \qquad \omega(0) = \omega_0.$$

If $\alpha(s)$ satisfies

$$\alpha'(0) \neq 0, \tag{3.6}$$

i.e., the eigenvalue $\sigma(s)$ crosses the imaginary axis at $s = 0$ with a nonvanishing speed, we say that (u_0, λ_0) is a *simple Hopf bifurcation point.* This condition is equivalent to

$$\left\langle \chi_0, \left(\frac{d}{ds} D_u G(u(s), \lambda(s)) \right) \bigg|_{s=0} \xi_0 \right\rangle \neq 0, \tag{3.7}$$

where χ_0, ξ_0 are null vectors of $(D_u G_0)^2 + \omega_0^2 I$ and $(D_u G_0^*)^2 + \omega_0^2 I$, respectively, chosen by the conditions

$$\begin{aligned} \mathrm{Ker}((D_u G_0)^2 + \omega_0^2 I) &= \mathrm{span}[\xi_0,\ D_u G_0 \xi_0], \\ \mathrm{Ker}((D_u G_0^*)^2 + \omega_0^2 I) &= \mathrm{span}[\chi_0,\ D_u G_0^* \chi_0] \end{aligned}$$

and

$$\langle \chi_0,\ \xi_0 \rangle = 0.$$

The condition (3.7) can be also reformulated as

$$\mathrm{Re} \left(\bar{d}^T \left(\frac{d}{ds} D_u G(u(s), \lambda(s)) \right) \bigg|_{s=0} c \right) \neq 0, \tag{3.8}$$

where $c,\ d \in \mathbf{C}^n$ are the eigenvectors of $D_u G_0$ and $D_u G_0^*$ corresponding to the eigenvalues $-i\omega_0$ and $i\omega_0$ respectively, i.e.,

$$D_u G_0 c = -i\omega_0 c, \qquad D_u G_0^* d = i\omega_0 d.$$

Moreover, the eigenvector $c,\ d$ are normalized with respect to the inner product $\langle u,\ v \rangle = \bar{u}^T v$, such that

$$\langle c,\ c \rangle = \langle d,\ c \rangle = 2, \qquad \langle \bar{d},\ c \rangle = 0,$$

see e.g. Golubitsky/Schaeffer [129, pp. 346], Mei [221]. The normal form for Hopf bifurcation problems, i.e., to which any Hopf bifurcation problem can be reduced equivalently, is

$$\dot{z} = (\alpha(s) + i\omega(s))z + l_1(s) z^2 \bar{z} + h.o.t., \quad z \in \mathbf{C}.$$

Here and in the sequel *h.o.t.* represents high order terms. The constant $l_1(s)$ is called the *first Liapunov coefficient*, see e.g. Khibnik [185], Kuznetsov [197, pp. 92]. The nondegeneracy condition for a simple Hopf bifurcation is

$$l_1(0) \neq 0.$$

The following formulation of this condition for the system (3.1)

$$\mathrm{Re} \left\langle d,\ D_{uu} G_0 b_0 c + D_{uu} G_0 b_2 \bar{c} + \frac{1}{4} D_{uuu} G_0 c^2 \bar{c} \right\rangle \neq 0,$$

is based on the reduced equations in the center manifold, see e.g. Chapters 6, 7 and Golubitsky/Schaeffer [129, pp. 346]. Here b_0, b_2 are defined implicitly by the equations

$$D_u G_0 b_0 = -\frac{1}{2} D_{uu} G_0 c\bar{c},$$
$$D_u G_0 b_2 = -\frac{1}{4} D_{uu} G_0 cc.$$

The following degenerate bifurcation is also encountered frequently.
Simple bifurcation points: Zero is a simple eigenvalue of $D_u G_0$ and $D_\lambda G_0 \in \text{Im}(D_u G_0)$. Let $v_0 \in X$ be a solution of

$$D_u G_0 v_0 + D_\lambda G_0 = 0. \tag{3.9}$$

Under the transversal condition

$$\langle \phi^*,\ D_u D G_0(v_0, 1)\phi_1 \rangle^2 - \langle \phi^*,\ D_u D_u G_0 \phi_1 \phi_1 \rangle \cdot \langle \phi^*,\ D^2 G_0(v_0, 1)(v_0, 1) \rangle > 0 \tag{3.10}$$

the problem (3.1) has exactly two different stationary solution branches intersecting at (u_0, λ_0) (cf. Crandall/Rabinowitz [71], Brezzi/Rappaz/Raviart [49], Mei [210, 213], Moore [232] and Weber/Werner [300]). Here ϕ, ϕ^* are defined in (3.3).

3.1.2 Two-parameter problems

Let $p = 2$, i.e., the problem (3.1) depends two parameters $\lambda = (\lambda_1,\ \lambda_2)$. If (3.1) is singular at (u_0, λ_0), there exists a curve of singular points across (u_0, λ_0). These are generic turning points or Hopf points for a fixed parameter. However, high singularities can occur if both parameters vary.
1) Cusp bifurcation points:

This singularity occurs when zero is a simple eigenvalue of $D_u G_0$ and the statement $D_{\lambda_i} G_0 \notin \text{Im}(D_u G_0)$ holds for $i = 1$ or 2. In addition, the condition (3.5) is violated, i.e.,

$$\langle \phi^*,\ D_{uu} G_0 \phi\phi \rangle = 0 \Longleftrightarrow D_{uu} G_0 \phi\phi \in \text{Im}(D_u G_0),$$

while the following nondegeneracy condition is satisfied

$$\langle \phi^*,\ D_{uuu} G_0 \phi\phi\phi + 3 D_{uu} G_0 \phi v_1 \rangle \neq 0. \tag{3.11}$$

Here $v_1 \in X$ is defined implicitly by the equations

$$D_u G_0 v_1 = -D_{uu} G_0 \phi\phi, \qquad \langle \phi,\ v \rangle = 0.$$

2) Generalized Hopf bifurcation points: This is the situation where the linearized operator $D_u G$ has exactly one pair eigenvalues across the imaginary axis at (u_0, λ_0) with nonvanishing speed, as the case of simple Hopf bifurcations. However, there is a degeneracy in nonlinear terms of the problem

such that the first Liapunov coefficient vanishes, see e.g. Golubitsky/Schaeffer [129, pp. 372] and Kuznetsov [197, pp. 258].
3) Takens-Bogdanov bifurcation points: D_uG_0 has one and only one eigenvalue $\sigma = 0$ on the imaginary axis. The geometric multiplicity of this eigenvalue is one, while its algebraic multiplicity is two. In particular, for finite dimensional problems the Jordan form of D_uG_0 has exactly one block of the form

$$\begin{pmatrix} 0 & 1 \\ 0 & 0 \end{pmatrix}.$$

Bifurcation and dynamics in the neighborhood of a TB-point have been investigated thoroughly. Especially it is known that a branch of Hopf points and homoclinic orbits can emanate from a TB-point (cf. Guckenheimer/Holmes [144], Kuznetsov [197], Werner/Janovsky [306]).
4) Hopf/steady state mode interactions: (Gavrilov-Guckenheimer bifurcation points) The system (3.1) loses linear stability at (u_0, λ_0), in which intersection of the spectrum of D_uG_0 and the imaginary axis consists of zero and a conjugate pair $\pm i\omega_0$, $\omega_0 > 0$.
5) Hopf/Hopf mode interactions: D_uG_0 has two different pairs $\pm i\omega_i$, $\omega_i > 0$, $i = 1, 2$ of eigenvalues on the imaginary axis.

Dynamics in neighborhoods of Hopf/steady state and Hopf/Hopf mode interactions can be sophisticated, particularly global bifurcations such as homoclinic and/or heteroclinic orbits may be induced by such interactions. For details, see Guckenheimer/Holmes [144], Golubitsky/Stewart/Schaeffer [131] and Kuznetsov [197], Steindl [284, 285].

Let $\sigma(D_uG_0)$ be the spectrum of D_uG at the bifurcation point (u_0, λ_0). We summary in Table 3.1 the generic bifurcations for one- and two-parameter problems.

3.2 Test Functions

In numerical continuation of a solution curve we are interested in monitoring occurrence of singularities. Suppose that (u_0, λ_0) is a singular point on the solution curve $(u(s), \lambda(s))$ and

$$(u(s_0), \lambda(s_0)) = (u_0, \lambda_0). \tag{3.12}$$

Definition 3.2.1. *Let U be a neighborhood of $(u(s), \lambda(s))$ and $\tau : U \to \mathbf{R}$ be a continuous function. If $\tau(u(s), \lambda(s))$ changes signs strictly at $s = s_0$, then $\tau(u, \lambda)$ is called a test function for the singular point (u_0, λ_0).*

codim	one-parameter problems	two-parameter problems
1	**Turning point:** $\sigma(D_uG_0) \cap i\mathbf{R} = \{0\}$, $D_\lambda G_0 \notin \mathrm{Im}(D_uG_0)$, $\langle \phi^*,\ D_{uu}G_0\phi\phi \rangle \neq 0$.	curve of turning points
1	**Hopf bifurcation:** $\sigma(D_uG_0) \cap i\mathbf{R} = \{\pm i\omega_0 \neq 0\}$, $\alpha'(0) \neq 0$; the first Liapunov coefficient satisfies $l_1(0) \neq 0$.	curve of Hopf points
2		**cusp bifurcation:** $\sigma(D_uG_0) \cap i\mathbf{R} = \{0\}$, $D_{\lambda_i}G_0 \notin \mathrm{Im}(D_uG_0)$ for $i = 1$ or 2; $\langle \phi^*,\ D_{uu}G_0\phi\phi \rangle = 0$; $\langle \phi^*,\ D_{uuu}G_0\phi\phi\phi + 3D_{uu}G_0\phi v_1 \rangle \neq 0$.
2		**Generalized Hopf bifurcation:** $\sigma(D_uG_0) \cap i\mathbf{R} = \{\pm i\omega_0,\ \omega_0 > 0\}$ and $\alpha'(0) \neq 0$; the first Liapunov coefficient $l_1(0)$ vanishes, while the second not.
2		**Takens-Bogdanov bifurcation:** $\sigma(D_uG_0) \cap i\mathbf{R} = \{0\}$; the geometric multiplicity of zero is one, while its algebraic multiplicity is two.
2		**Hopf/steady state mode interaction:** $\sigma(D_uG_0) \cap i\mathbf{R} = \{0,\ \pm i\omega_0\}$, $\omega_0 > 0$.
2		**Hopf/Hopf mode interaction:** $\sigma(D_uG_0) \cap i\mathbf{R} = \{\pm i\omega_1, \pm i\omega_2\}$, $\omega_k > 0$.

Table 3.1. Generic local bifurcations of one- and two-parameter problems.

A test function is normally chosen to be continuous differentiable such that the bifurcation point (u_0, λ_0) can be monitored directly with the properties

$$\begin{aligned} \tau(u(s_0), \lambda(s_0)) &= 0, \\ \frac{d}{ds}\tau(u(s), \lambda(s))\Big|_{s=s_0} &\neq 0. \end{aligned} \tag{3.13}$$

3.2.1 Test functions for turning points

We consider here three frequently used test functions.

a) Determinant

Since the matrix D_uG_0 is singular at a turning point, it is natural to use determinant of D_uG

$$\tau(u,\lambda) = \det(D_u G(u,\lambda)) \tag{3.14}$$

as a test function. This is efficient as long as the determinant of $D_u G(u,\lambda)$ is easily available, e.g. as a by-product of direct numerical solution of the linearized problem in Newton iterations of the corrector step. Normally it is applicable only to problems of small to medium sizes.

b) Approximation of critical eigenvalues

In predictor-corrector methods for numerical continuation of a solution curve, if we use iterative methods in the corrector, e.g. (un-) symmetric Lanczos and Arnoldi methods, extreme eigenpairs of linearized operator $D_u G$ are well approximated in the Krylov-subspace. A singularity is signaled on the solution branch between two successive continuation steps when an eigenvalue of the condensed matrix H_k of $D_u G$ in the Krylov-subspace, i.e., the Ritz value of $D_u G$, crosses the imaginary axis, see sections 2.3.2 and 2.3.3. This is an efficient method for detecting both steady state and Hopf bifurcations of problems in which the instability of the current solution is resulted from eigenvalues in extreme parts of spectrum of the linearized operator, for examples, nonlinear elliptic and parabolic differential equations, in particular, reaction-diffusion problems. If the critical eigenvalue is inside the spectrum, but within a cluster which is well separated from others, then appropriate transformations of the spectrum, e.g. by Chebyshev, Faber polynomials, Cayley transformations, can be used for accurate approximation of the critical eigenvalues, we refer to Garratt/Moore/Spence [119, 230] and Saad [261] for more detail. Other sophisticate techniques for preconditioning has been developed by Sebastian [274] and Schwarzer [272].

c) Test function via bordered matrices

Along a solution curve $(u(s), \lambda(s))$ a turning point at $s = s_0$ is characterized by Griewank and Reddien [143] with the root of a function $\tau : D \subset \mathbf{R}^n \times \mathbf{R} \to \mathbf{R}$, defined implicitly via the following bordered system

$$\begin{pmatrix} D_u G(u,\lambda) & r \\ l^T & 0 \end{pmatrix} \begin{pmatrix} v \\ \tau \end{pmatrix} = \begin{pmatrix} 0 \\ 1 \end{pmatrix}. \tag{3.15}$$

Here $r,\ l \in \mathbf{R}^n$ are normalized vectors, $\|r\| = \|l\| = 1$. They are chosen generically such that the bordered matrix in (3.15) is nonsingular in the neighborhood of the turning point. Properties of the bordered matrix in (3.15) can be described by the following lemma.

Lemma 3.2.1. (Keller [181]) *Let X be a Banach space and*

$$\mathcal{A} := \begin{pmatrix} A & B \\ C & D \end{pmatrix} \in L(X \times \mathbf{R}^p,\ X \times \mathbf{R}^p) \tag{3.16}$$

be a linear bounded operator from $X \times \mathbf{R}^p$ into itself. Here $A \in L(X,X)$, $B \in L(\mathbf{R}^p, X)$, $C \in L(X, \mathbf{R}^p)$ and $D \in \mathbf{R}^{p\times p}$ are all bounded operators.

i) If A is nonsingular then $\mathcal{A}$ is nonsingular if and only if the matrix

$$D - CA^{-1}B$$

is regular.

ii) If A is singular and $\dim\Big(\mathrm{Ker}(A)\Big) = \mathrm{codim}\Big(\mathrm{Im}(A)\Big) = p$, *then $\mathcal{A}$ is nonsingular if and only if*

$$\begin{aligned} &\dim\Big(\mathrm{Im}(B)\Big) = p, \qquad \mathrm{Im}(B) \cap \mathrm{Im}(A) = \{0\}, \\ &\dim\Big(\mathrm{Im}(C)\Big) = p, \qquad \mathrm{Ker}(A) \cap \mathrm{Ker}(C) = \{0\}. \end{aligned} \tag{3.17}$$

iii) If $\dim\ \mathrm{Ker}(A) > p$, *then $\mathcal{A}$ is singular.*

Applying Lemma 3.2.1 to

$$\begin{pmatrix} D_uG(u,\lambda) & r \\ l^T & 0 \end{pmatrix}, \tag{3.18}$$

we see that if $D_uG(u,\lambda)$ is nonsingular, then the bordered matrix (3.18) is regular if and only if

$$\langle l,\ D_uG(u,\lambda)^{-1}r\rangle \neq 0. \tag{3.19a}$$

If $D_uG(u,\lambda)$ is singular, then the bordered matrix (3.18) is nonsingular if and only if the following two conditions are satisfied

$$\begin{aligned} \langle \phi^*,\ r\rangle \neq 0 &\Longleftrightarrow r \notin \mathrm{Im}(D_uG), \\ \langle \phi,\ l\rangle \neq 0 &\Longleftrightarrow l \notin \mathrm{Im}(D_uG^*). \end{aligned} \tag{3.19b}$$

Here ϕ and ϕ^* are given as in (3.3).

Theorem 3.2.1. *The bordered matrix (3.18) is nonsingular if and only if the matrix $(I - rr^T)D_uG + rl^T$ is nonsingular. Furthermore, solution of the system (3.15) can be given as*

$$\begin{aligned} v(u,\lambda) &= \left[(I - rr^T)D_uG(u,\lambda) + rl^T\right]^{-1} r, \\ \tau(u,\lambda) &= -r^T D_uG(u,\lambda)v(u,\lambda). \end{aligned} \tag{3.20}$$

Thereafter, (u_0,λ_0) is a turning point if and only if

$$\tau(u_0,\lambda_0) = 0. \tag{3.21}$$

Moreover,

$$\frac{d}{ds}\tau(u(s),\lambda(s))\Big|_{s=s_0} = \mathrm{const}\cdot r^T D_{uu}G_0\phi\phi. \tag{3.22}$$

Remark 3.2.1. A similar result is obtained by Seydel in [277]. From (3.22) and (3.19b) we see that the nondegeneracy condition (3.5) ensures a simple root of $\tau(u(s),\lambda(s))$ at $s = s_0$.

Proof: Let $(I-rr^T)D_uG+rl^T$ be nonsingular. We consider the homogeneous system

$$\begin{aligned} D_uG\, x + r\alpha &= 0, \\ l^T x &= 0. \end{aligned} \tag{3.23}$$

for the unknown (x, α). From the first equation of (3.23) and $\|r\| = 1$ we derive

$$\alpha = -r^T D_uG\, x$$

Substituting this into the first equation of (3.23) and adding the zero vector $r(l^T x)$ to it yields

$$D_uG\, x - rr^T D_uG\, x + rl^T x = [(I - rr^T)D_uG + rl^T]x = 0.$$

Since the coefficient matrix $(I - rr^T)D_uG + rl^T$ is nonsingular, it has only the trivial solution $x = 0$. Hence, $(x,\ \alpha) = (0,\ 0)$ is the unique solution of (3.23) and the bordered matrix (3.18) is nonsingular.

Conversely, if the bordered matrix in (3.18) is nonsingular, we consider the homogeneous equation

$$[(I - rr^T)D_uG + rl^T]x = D_uG\, x - r(r^T D_uG\, x - l^T x) = 0. \tag{3.24}$$

The inner product of r with the equation (3.24) implies

$$l^T x = 0.$$

Together with (3.24) we obtain

$$\begin{pmatrix} D_uG(u,\lambda) & r \\ l^T & 0 \end{pmatrix} \begin{pmatrix} x \\ -r^T D_uG\, x \end{pmatrix} = 0.$$

It follows from regularity of the coefficient matrix that $x = 0$. Hence, the equation (3.24) has only the trivial solution, which in turn implies that the matrix $(I - rr^T)D_uG + rl^T$ is nonsingular.

It is easy to verify that (3.20) is the unique solution of (3.15). If (u_0, λ_0) is a turning point, then $\mathrm{Rank}(D_uG_0) = n - 1$. The first equation

$$D_uG_0 v + r\tau = 0$$

in (3.15) and the statement (3.19a) show

$$D_uG_0 v = 0 \quad \text{and} \quad \tau = 0.$$

Hence, $v = \text{const} \cdot \phi$. Furthermore, we consider the definition (3.15) of the test function τ along the solution curve $(u(s), \lambda(s))$

$$\begin{aligned} D_uG(u(s), \lambda(s))v + r\tau &= 0, \\ l^T v &= 1. \end{aligned}$$

Differentiating it with respect to s at $s = s_0$ yields

$$\begin{aligned} D_u DG_0(\dot{u}(s_0), \dot{\lambda}(s_0))v + D_u G_0 \dot{v} + r\dot{\tau} &= 0, \\ l^T \dot{v} &= 0. \end{aligned}$$

Since $\dot{u}(s_0) = \text{const} \cdot \phi$, $\dot{\lambda}(s_0) = 0$, see (3.4), we have

$$\frac{d}{ds}\tau(u(s), \lambda(s))\Big|_{s=s_0} = \text{const} \cdot r^T D_{uu} G_0 \phi\phi.$$

Hence, at a quadratic turning point, i.e., $r^T D_{uu} G_0 \phi\phi \neq 0$, the test function $\tau(u, \lambda)$ changes signs strictly along the solution curve. ■

Evidently, during path following of a solution curve, one can calculate the test function either by directly solving the bordered system (3.15) or by first solving

$$[(I - rr^T)D_u G + rl^T]v = r \tag{3.25}$$

and then calculating

$$\tau = -r^T D_u G v. \tag{3.26}$$

Via the implicit function theorem it is easy to see that the test function defined in (3.15) is as smooth as G in (u, λ) and depends also smoothly on the data r, l.

Remark 3.2.2. Choosing $r = e_j$, $l = e_k$, j, $k \in \{1, 2, \ldots, n\}$ as two unit vectors in $\mathbf{R}^n$, we reduce the equation (3.25) to

$$[(I - e_j e_j^T)D_u G + e_j e_k^T]v = e_j. \tag{3.27}$$

It corresponds to replacing the j-th equation of the system

$$D_u G v = 0 \tag{3.28}$$

with $v_k = 1$. This system and the corresponding test function are identical to those in Seydel [276, pp.154]. In effect, removing some equations from (3.28) and considering the reduced system for a subset of unknowns is equivalent to a special bordering of the system and leads to the same type of test functions, see also Seydel [277]. Nevertheless, if the matrix $D_u G$ is of special structure, e.g. band or sparse, the systems (3.27) and (3.15) with proper l, $r \in \mathbf{R}^T$ which preserves the structure of $D_u G$ are preferred, though an iterative solution of (3.25) may requires only matrix-vector product and the rank-1 update of the coefficient matrix does not cause too much additional computational cost.

3.2.2 Test functions for simple bifurcation point

As we have seen in Section 3.1.1, a simple bifurcation point (u_0, λ_0) of the one-parameter problem $G(u, \lambda) = 0$ is characterized by the properties

$$\operatorname{Rank}(D_u G_0) = n - 1, \qquad D_\lambda G_0 \in \operatorname{Im}(D_u G_0)$$

and a nondegeneracy condition of nonlinear terms. Thus all the test functions for turning points can be used to detect simple bifurcations. Nevertheless, we need to verify additionally whether $D_\lambda G_0$ is in the range of $D_u G_0$. In this section we consider test functions which can be used directly for detecting simple bifurcation points during the numerical continuation of a solution curve.

If we use the enlarged system

$$F(u, \lambda, s) = \begin{pmatrix} G(u, \lambda) \\ N(u, \lambda, s) \end{pmatrix} = 0 \tag{3.29}$$

and (pseudo-) arclength to parameterize the solution curve, a turning point becomes nonsingular and can be calculated as any other nonsingular point of $G(u, \lambda) = 0$ in the continuation of solution curves. However, at a simple bifurcation point of $G(u, \lambda) = 0$ the enlarged system (3.29) is still singular. By the (pseudo-) arclength parameterization, the derivative $D_{(u,\lambda)} F(u, \lambda, s)$ reveals simple bifurcation points on the solution curve.

Lemma 3.2.2. *Let $(u(s), \lambda(s))$ be a solution curve parameterized via the enlarged system (3.29). We choose the local pseudo-arclength such that $D_{(u,\lambda)} N(u(s), \lambda(s), s)^T$ is not orthogonal to the tangent $(\dot{u}(s), \dot{\lambda}(s))$. If $(u(s_0), \lambda(s_0))$ is a simple bifurcation point, then the $(n+1) \times (n+1)$ matrix has the property*

$$\begin{aligned} &\operatorname{Rank}\Big(D_{(u,\lambda)} F(u(s_0), \lambda(s_0), s_0)\Big) \\ &= \operatorname{Rank}\begin{pmatrix} D_{(u,\lambda)} G(u(s_0), \lambda(s_0)) \\ D_{(u,\lambda)} N(u(s_0), \lambda(s_0), s_0) \end{pmatrix} \\ &= n. \end{aligned} \tag{3.30}$$

Conversely, if (3.30) is satisfied, then $\operatorname{Rank}(DG(u(s_0), \lambda(s_0))) = n - 1$ *and $(u(s_0), \lambda(s_0))$ is either a simple bifurcation point, or a double turning point of $G(u, \lambda) = 0$ which corresponds to* $\dim(\operatorname{Ker}(D_u G_0)) = 2$ *and* $D_\lambda G_0 \notin \operatorname{Im}(D_u G_0)$.

Proof: If $(u(s_0), \lambda(s_0))$ is a simple bifurcation point, it follows from (3.3) and (3.9) that the kernel

$$\operatorname{Ker}(DG_0) = \operatorname{span}[(\phi, 0), \ (v_0, 1)]$$

is two dimensional. Note that $D_{(u,\lambda)} N(u(s_0), \lambda(s_0), s_0)^T$ is not orthogonal to $(\dot{u}(s_0), \dot{\lambda}(s_0))$, in other words, it intersects $\operatorname{Ker}(DG_0)$ transversally. Hence, $\operatorname{Ker}(D_{(u,\lambda)} F(u(s_0), \lambda(s_0), s_0))$ is one dimensional, the statement (3.30) holds.

Conversely, under the assumption on $D_{(u,\lambda)} N$ along the solution curve if (3.30) is satisfied at the point $s = s_0$, we have immediately

$$\text{Rank}\Big(D_{(u,\lambda)}G(u(s_0),\lambda(s_0))\Big) = n-1.$$

The conclusion follows directly. ∎

The equality

$$D_sF(u,\lambda,s) = \begin{pmatrix} 0 \\ D_sN(u,\lambda,s) \end{pmatrix} \neq 0$$

holds for a pseudo-arclength parameterization. Together with Lemma 3.2.2 we obtain generically $D_sF(u_0,\lambda_0,s_0) \notin \text{Im}(D_{(u,\lambda)}F(u_0,\lambda_0,s_0))$, i.e., (u_0,λ_0,s_0) is a turning point of (3.29). Conversely, a turning point of the enlarged system (3.29) is necessarily a bifurcation point of the problem $G(u,\lambda)=0$. Hence, we exploit the test functions for turning points of (3.29) to detect simple bifurcation points of (3.1).

a) Determinant and approximation of eigenvalues

The statement (3.30) implies that zero is a simple eigenvalue of $D_{(u,\lambda)}F_0$. Consequently, determinant of $D_{(u,\lambda)}F(u(s),\lambda(s),s)$ changes sign strictly at $s=s_0$ and can be used to monitor bifurcations of $G(u,\lambda)=0$. Similarly, approximation of the critical eigenvalue of $D_{(u,\lambda)}F$ reveals simple bifurcations of the original problem.

b) Test function via bordered system

Choosing two normalized vectors $l,\ r \in \mathbf{R}^{n+1}$ such that the matrix

$$\begin{pmatrix} D_{(u,\lambda)}F(u,\lambda,s) & r \\ l^T & 0 \end{pmatrix}$$

is nonsingular, we obtain a test function $\tau(u,\lambda,s)$ via the system

$$\begin{pmatrix} D_{(u,\lambda)}F & r \\ l^T & 0 \end{pmatrix}\begin{pmatrix} v \\ \tau \end{pmatrix} = \begin{pmatrix} 0 \\ 1 \end{pmatrix}.$$

In particular, if the pseudo-arclength $N(u,\lambda,s)$ is chosen as

$$N(u,\lambda,s) = \langle \tilde{l},\ (u,\lambda)\rangle - s = 0$$

for some $\tilde{l} \in \mathbf{R}^{n+1}$, the last component of the solution v of the system

$$\begin{pmatrix} D_{(u,\lambda)}F & r \\ l^T & 0 \end{pmatrix}\begin{pmatrix} v \\ \tau \end{pmatrix} = \begin{pmatrix} e_n \\ 0 \end{pmatrix}.$$

can be used also as a test function.

c) Orientation of tangent of the solution curve (cf. Allgower/Georg [14])

By arclength parameterization of a solution curve we do continuation with the system

$$F(u,\lambda,s) = \begin{pmatrix} G(u(s),\lambda(s)) \\ t(s)^T \end{pmatrix} = 0,$$

where $t(s)$ is the tangent of $(u(s),\lambda(s))$ and defined by

$$\begin{aligned} DG(u(s),\lambda(s))\ t(s) &= 0, \\ t(s)^T t(s) &= 1 \end{aligned}$$

with the orientation

$$\det \begin{pmatrix} DG(u(s),\lambda(s)) \\ t(s)^T \end{pmatrix} \geq 0, \tag{3.31}$$

see Lemma 2.1.2. At a simple bifurcation point (u_0,λ_0) for $s = s_0$, the statement (3.30) as well as the definition of the tangent show that the determinant vanishes

$$\det \begin{pmatrix} DG_0 \\ t(s_0)^T \end{pmatrix} = 0$$

and changes sign strictly. Correspondingly the condition (3.31) requires a change of orientation of the tangent in the continuation process. Therefore, if the inequality

$$t(s_1)^T t(s_2) < 0$$

is satisfied for two consecutive continuation steps, a simple bifurcation point is located between $(u(s_1),\lambda(s_1))$ and $(u(s_2),\lambda(s_2))$. Note that we have not distinguished the state variable u and the parameter λ in the arclength parameterization. To exclude double turning points we need to choose an appropriate parameter among the components of u and λ.

3.2.3 Test functions for Hopf bifurcations

A Hopf bifurcation of the problem

$$\frac{du}{dt} = G(u,\lambda) \tag{3.32}$$

occurs on a stationary solution curve $(u(s),\lambda(s))$ when a pair of complex conjugate eigenvalues of $D_uG(u(s),\lambda(s))$ crosses the imaginary axis with nonzero speed. In this subsection we briefly review some frequently used test functions for Hopf bifurcations.

a) Characteristic polynomials

For small systems, Khibnik [185] introduces the test function

$$\tau(u,\lambda) = \text{Res}(P(\sigma), P(-\sigma)) \tag{3.33}$$

for detecting Hopf bifurcations, where $P(\sigma)$ is the characteristic polynomial of $D_uG(u,\lambda)$ and $\text{Res}(p(\sigma),q(-\sigma))$ is the *resultant* of two polynomials p and q, see also Guckenheimer/Myers/Sturmfels [145]. Along the solution curve

$(u(s), \lambda(s))$ the equality $\tau(u(s), \lambda(s)) = 0$ holds if and only if $P(\sigma)$ and $P(-\sigma)$ have common root, i.e., $D_uG(u(s), \lambda(s))$ has either a pair imaginary eigenvalues $\pm i\omega_0$, $\omega > 0$ or a pair of real eigenvalues $\pm\sigma$, $\sigma > 0$. The former corresponds to Hopf bifurcation, the later to the imaginary Hopf bifurcation (cf. Werner/Janovsky [306]) or neutral saddle (Khibnik [185]). If we decompose the polynomial $P(\sigma)$ into even and odd parts and rewrite it as $P(\sigma) = P_e(\sigma) + \sigma P_o(\sigma)$, the test function (3.33) can be replaced by

$$\tau(u, \lambda) = \text{Res}(P_o(\sigma), P_e(\sigma)).$$

For detailed discussions and numerical implementations, see Guckenheimer *et al* [145], Khibnik [186] and Werner [305].

Equivalently, Kuznetsov [197] considers the following form of (3.33)

$$\tau(u, \lambda) = \prod_{i>j} \Big(\sigma_i(u, \lambda) - \sigma_j(u, \lambda)\Big)$$

where $\sigma_i(u, \lambda)$, $i = 1, \ldots, n$ are eigenvalues of $D_uG(u, \lambda)$. This covers also the imaginary Hopf bifurcation and in particular, TB-point at which the curves of real and imaginary Hopf points merge. Both methods require knowledge of all eigenvalues of D_uG. Similarly, AUTO [94] detects Hopf bifurcations by computing all eigenvalues of D_uG with unsymmetric QR method along the solution curve.

b) Approximation of (critical) eigenvalues

Similarly to the detection of steady state bifurcations we can trace a few potential critical eigenvalues along the path following and observe their variations. By applying iterative methods in the correctors of continuation methods, we approximate a small number of eigenvalues in Krylov subspaces with appropriate preconditioners and transformations, e.g. Arnoldi method with Chebyshev preconditioning (cf. Garratt/Moore/Spence [119, 230], Schwarzer [272] and Sebastian [274]).

c) Test function via bialternative product

Definition 3.2.2. *A bialternative product $C := A \odot B$ of two matrices $A = (a_{ij})$ and $B = (b_{ij}) \in \mathbf{R}^{n\times n}$ is a $m \times m$ matrix with $m = n(n-1)/2$. The elements of C are ordered with row index (p, q) and column index (r, s) as follows*

$$C_{(p,q)(r,s)} = \frac{1}{2}\left\{\begin{vmatrix} a_{pr} & a_{ps} \\ b_{qr} & b_{qs} \end{vmatrix} + \begin{vmatrix} b_{pr} & b_{ps} \\ a_{qr} & a_{qs} \end{vmatrix}\right\}$$
$$p = 2, 3, \ldots, n \qquad q = 1, \ldots p-1,$$
$$r = 2, 3, \ldots, n \qquad s = 1, \ldots r-1,$$

Theorem 3.2.2. *Let $\sigma_1, \ldots, \sigma_n$ be eigenvalues of a matrix $A \in \mathbf{R}^{n\times n}$. We have*

i) eigenvalues of $A \odot A$ are $\sigma_i \cdot \sigma_j \quad i = 2, 3, \ldots, n,\ j = 1, 2, \ldots, i-1$.
ii) eigenvalues of $2A \odot I$ are $\sigma_i + \sigma_j \quad i = 2, 3, \ldots, n,\ j = 1, 2, \ldots, i-1$.

If A has a pair of pure imaginary eigenvalues, zero is an eigenvalue of $2A \odot I$. Thus a test function for Hopf bifurcation can be defined as

$$\tau(u, \lambda) = \det\left(2D_uG(u, \lambda) \odot I\right), \tag{3.34}$$

see e.g. Govaerts *et al* [136, 137] and Kuznetsov [197]. Therefore, similar as test functions for turning points, we monitor the eigenvalue zero of the bialternate product $2D_uG(u, \lambda) \odot I$, for example, via the function $\tau(u, \lambda)$ defined by the bordered system

$$\begin{pmatrix} 2D_uG(u, \lambda) \odot I & r \\ l^T & 0 \end{pmatrix} \begin{pmatrix} \nu \\ \tau \end{pmatrix} = \begin{pmatrix} 0 \\ 1 \end{pmatrix}.$$

Here $r,\ l \in \mathbf{R}^m$ are chosen such that the coefficient matrix is nonsingular.

Bifurcation Points (u_0, λ_0)	Test Functions $\tau(u, \lambda)$
turning points: $\sigma(D_uG_0) \cap i\mathbf{R} = \{0\}$, $D_\lambda G_0 \notin \mathrm{Im}(D_uG_0)$, $\langle \phi^*,\ D_{uu}G_0\phi\phi \rangle \neq 0$.	a) $\det(D_uG(u, \lambda))$; b) approximation of eigenvalues; c) $\begin{pmatrix} D_uG(u, \lambda) & r \\ l^T & 0 \end{pmatrix} \begin{pmatrix} v \\ \tau \end{pmatrix} = \begin{pmatrix} 0 \\ 1 \end{pmatrix}$.
Hopf points: $\sigma(D_uG_0) \cap i\mathbf{R} = \{\pm i\omega_0, \omega_0 > 0\}$, $\alpha'(0) \neq 0$;	a) resultant b) approximation of eigenvalues c) bordered systems

Table 3.2. Test function for turning and Hopf points.

d) Test function via bordered system

At a Hopf bifurcation point (u_0, λ_0), the matrix $A^2 + \nu I$ with $A := D_uG_0,\ \nu = \omega_0^2$ has a rank deficiency two. Therefore, along a stationary solution curve $(u(s), \lambda(s))$ if a rank deficiency two in the matrix $A(s)^2 + \nu I,\ A(s) := D_uG(u(s), \lambda(s))$ is detected, we obtain a Hopf bifurcation for $\nu > 0$, a TB-point for $\nu = 0$ and an imaginary Hopf bifurcation for $\nu < 0$ (cf. Werner/Janovsky [306]). Occurrence of such a rank deficiency can be observed by monitoring whether two eigenvalues of $A(s)^2 + \nu I$ cross the imaginary axis at zero simultaneously. Alternatively, Chu/Govaerts/Spence [65] introduce a bordered system

$$\begin{pmatrix} A(s)^2 + \nu I & r_1 & r_2 \\ l_1^T & 0 & 0 \\ l_2^T & 0 & 0 \end{pmatrix} \begin{pmatrix} v(s, \nu) \\ g_1(s, \nu) \\ g_2(s, \nu) \end{pmatrix} = \begin{pmatrix} 0 \\ 0 \\ 1 \end{pmatrix}, \tag{3.35}$$

where $r_i,\ l_i \in \mathbf{R}^n$ are chosen to make the bordered matrix nonsingular. At a (imaginary) Hopf bifurcation point or TB-point $s = s_0$ the equalities

$$g_1(s_0, \nu_0) = 0 \quad \text{and} \quad g_2(s_0, \nu_0) = 0 \tag{3.36}$$

hold for appropriate *Hopf number* ν_0. The statements (3.36) are necessary and sufficient for real Hopf bifurcation. But only necessary for imaginary Hopf and TB-points. Werner [305] makes them also sufficient by choosing $r_1 = A(s)r_2,\ l_1 = A(s)l_2$ for the regularization of the bordered matrix (3.35). Furthermore, solving

$$g_2(s, \nu) = 0,$$

formally for ν as a function of s, Werner [305] takes

$$\tau(s) := g_1(s, \nu(s)). \tag{3.37}$$

as a test function for Hopf bifurcations, since this function changes signs strictly when $(u(s), \lambda(s))$ passes a Hopf point or TB-point for which the critical eigenvalues cross the imaginary axis with a nonzero speed. Alternatively, Govaerts and Spence detect Hopf bifurcation points in [138] by counting sectors in the complex plane.

We summarize the discussion on test functions for local bifurcations in Table 3.2.

3.2.4 Minimally extended systems for computing bifurcation points

Test functions are also used in various minimally extended systems for calculations of bifurcation points, see e.g. Allgower/Böhmer/Hoy/Janovsky [7], Griewank/Reddien [143], Khibnik [185], Kuznetsov [197] and Werner [304]. A minimally extended system is of the form

$$F(u, \lambda) := \begin{pmatrix} G(u, \lambda) \\ \tau(u, \lambda) \end{pmatrix} = 0 \tag{3.38}$$

in the space $R^n \times \mathbf{R}^p$. The operator F maps $R^n \times \mathbf{R}^p$ into itself and $\tau(u, \lambda)$ is a test function for the interesting bifurcation points. For examples, for turning points, we may use the test function

$$\tau(u, \lambda) = \det D_u G(u, \lambda)$$

or

$$\tau(u, \lambda) = -r^T D_u G(u, \lambda) v$$

with the accompanying system

$$[(I - rr^T) D_u G + r l^T] v = r.$$

Here r, $l \in X$ are chosen such that the coefficient matrix $(I - rr^T)D_uG + rl^T$ is nonsingular, see Section 3.2.1.

Similarly, for Hopf points the function $\tau(u, \lambda)$ in (3.38) can be chosen as

$$\tau(u, \lambda) = \mathrm{Res}(P(\sigma), P(-\sigma))$$

see (3.33), or

$$\tau(u, \lambda) = \det(2D_uG(u, \lambda) \odot I)$$

as in (3.34) or the one defined in (3.37) via bordered systems. Alternatively we can also treat the Hopf number as an additional unknown and use the extended system

$$F(u, \lambda, \nu) = \begin{pmatrix} G(u, \lambda) \\ g_1(u, \lambda, \nu) \\ g_2(u, \lambda, \nu) \end{pmatrix} = 0 \tag{3.39}$$

to calculate Hopf points, where $g_i(u, \lambda, \nu)$, $i = 1, 2$ are defined implicitly by the bordered system (3.35). Since the extended systems (3.38) and (3.39) are nonsingular for turning and Hopf points, respectively (cf. Khibnik [185], Kuznetsov [197] and Werner [305]), Newton-like methods can be applied to calculate these points. However, the function $\tau(u, \lambda)$ as well as $g_i(u, \lambda, \nu)$ are defined implicitly. Evaluations of these functions and their derivatives in the iterations need to be done in separate steps. For details and special techniques, see Griewank/Reddien [143] and Werner [305]. In the following sections we consider direct methods in Mei [214] and Roose/Hlavaček [257] for computing simple bifurcation points and Hopf bifurcation points.

3.3 Computing Simple Bifurcation Points

Accurate approximation of bifurcation points is required in several situations, e.g. for a systematic analysis of bifurcation scenario and classification of singularities and identification of critical parameters in dynamical systems.

If the primary interest is branch switching and continuation of a solution curve, indirect methods often provide sufficiently accurate locations of bifurcation points. These methods interpolate the bifurcation point with available information from the test functions and from the computed points on the solution curve, see e.g. Allgower/Georg [14], Keller [180, 181] Rheinboldt [249] and Seydel [276] for more details and references.

Direct methods are usually employed for accurate approximation of bifurcation points and kernels of some relevant linearized operators. A direct method consists of regularization of the original problem and approximation of the bifurcation point and the null vectors of the linearized operators. Several extended systems have been proposed for these purposes, see e.g. Allgower/Böhmer [4], Beyn [36], Böhmer/Mei [43], Mei [215, 214, 211], Moore [232], Moore/Spence [231], Jepson/Spence [174, 283], Weber/Werner [300],

Werner [301] and Werner/Spence [307], etc. The common feature of all these methods is that they augment the original system with additional equations, such that the extended system becomes nonsingular at the bifurcation point. The main difference among the various approaches is how to make effectively use of structure of the extended system, aiming for reduction of the computational cost.

We consider in this section a special extended system in Mei [214] for the simple bifurcation points. The block structure of this system allows an approximation of the solution with a simplified Newton-like method. A rank-1 correction of the singular Jacobian is discussed for large sparse problems. At the end of this section we consider a simplified buckling problem to show convergence of the numerical approximations. Since this method is applicable to operator equations in Hilbert space, we describe it in an abstract form.

3.3.1 Simple bifurcation points

Let X be a Hilbert space and $G: X \times \mathbf{R} \to X$ be a C^3-continuous mapping. We consider the one-parameter dependent equation

$$G(u, \lambda) = 0 \tag{3.40}$$

and assume that (u_0, λ_0) is a bifurcation point. Let D_uG_0 be a Fredholm operator of index 0 and let zero be a simple eigenvalue of D_uG_0 with

$$a) \quad \dim(\mathrm{Ker}(D_uG_0)) = 1, \qquad b) \quad D_\lambda G_0 \in \mathrm{Im}(D_uG_0). \tag{3.41}$$

Hence there are elements $\phi,\ \phi^* \in X$, such that

$$\mathrm{Ker}(D_uG_0) = \mathrm{span}[\phi], \qquad \langle \phi,\ \phi \rangle = 1, \tag{3.42a}$$
$$\mathrm{Ker}(D_uG_0^*) = \mathrm{span}[\phi^*], \qquad \langle \phi^*,\ \phi^* \rangle = 1 \tag{3.42b}$$

and

$$\mathrm{Im}(D_uG_0) = \{u \in X \ \mid \ \langle \phi^*,\ u \rangle = 0\}, \tag{3.43a}$$
$$\mathrm{Im}(D_uG_0^*) = \{u \in X \ \mid \ \langle \phi,\ u \rangle = 0\}, \tag{3.43b}$$

where $\langle \cdot,\ \cdot \rangle$ represents the inner product in X and the adjoint operator $D_uG_0^*$ is also defined with respect to this inner product. Moreover, we have the orthogonal decompositions (cf. Hirzebruch/Scharlau [159])

$$X = \mathrm{Ker}(D_uG_0) \oplus \mathrm{Im}(D_uG_0^*) = \mathrm{Ker}(D_uG_0^*) \oplus \mathrm{Im}(D_uG_0). \tag{3.44}$$

The statement (3.41b) and (3.44) show that there is a unique element v_0 in $\mathrm{Im}(D_uG_0^*)$ such that

$$D_uG_0v_0 + D_\lambda G_0 = 0. \tag{3.45}$$

Define

$$\begin{aligned} a &:= \langle \phi^*,\ D_{uu}G_0\phi\phi\rangle, \\ b &:= \langle \phi^*,\ D_u DG_0(v_0,1)\phi\rangle, \\ c &:= \langle \phi^*,\ D^2G_0(v_0,1)^2\rangle. \end{aligned} \tag{3.46}$$

$$\tag{3.47}$$

If the inequality

$$b^2 - ac > 0 \tag{3.48}$$

holds, then (u_0, λ_0) is called a *simple bifurcation point* of (3.40), see e.g. Brezzi/Rappaz/Raviart [49], Mei [212, 213], Moore [232] and Weber/Werner [300].

Lemma 3.3.1. *The condition (3.48) is independent of the bases of* $\mathrm{Ker}(D_uG_0)$, $\mathrm{Ker}(D_uG_0^*)$ *and the constant* $\mu \in \mathbf{R}$ *in the general solution* $v_0 + \mu\phi$ *of (3.45).*

Proof: Since the kernels $\mathrm{Ker}(D_uG_0)$ and $\mathrm{Ker}(D_uG_0^*)$ are one dimensional, any new basis elements are of the form $\alpha\phi$, $\beta\phi^*$ for some constants α, $\beta \in \mathbf{R}$, $\alpha\cdot\beta \neq 0$, respectively. For these new bases and the general solution $v_0+\mu\phi$ of (3.45), the condition (3.48) becomes

$$\begin{aligned} &\langle \beta\phi^*,\ D_uG_0(v_0+\mu\phi,1)\alpha\phi\rangle^2 \\ &\quad - \langle \beta\phi^*,\ D_{uu}G_0(\alpha\phi)^2\rangle \cdot \langle \beta\phi^*,\ D^2G_0(v_0+\mu\phi,1)^2\rangle \\ &= \alpha^2\beta^2[(b+\mu a)^2 - a(c+2\mu b+\mu^2 a)] \\ &= \alpha^2\beta^2(b^2 - ac). \end{aligned}$$

This proves our conclusion. ■

Consequently, the solution v_0 of (3.45) can be chosen such that

$$\langle \phi,\ v_0\rangle = 0. \tag{3.49}$$

3.3.2 Extended systems

To approximate the bifurcation point (u_0, λ_0), we set up an extended system. Its nonsingular solution contains (u_0, λ_0) and the null vectors ϕ, ϕ^*, v_0. These elements are used in the analysis of the bifurcation scenario at (u_0, λ_0), see Chapters 2, 6 and 7.

We define a function space

$$Y := X \times \mathbf{R} \times X^3, \qquad y := (u, \lambda, u_1, u_2, u_3) \in Y \tag{3.50}$$

with the (product) norm

$$\|y\|_Y := \left(\|u\|^2 + |\lambda|^2 + \|u_1\|^2 + \|u_2\|^2 + \|u_3\|^2\right)^{1/2}. \tag{3.51}$$

Here $\|\cdot\|$ is a norm in X. We consider the following system in Y

$$F(y) := \begin{pmatrix} G(u,\lambda) + \langle u_2, D_uGu_1\rangle u_2 \\ \langle u_2, D_uG \cdot u_3 + D_\lambda G\rangle \\ D_uG \cdot u_1 + [\langle u_1, u_1\rangle - 1]u_2/2 \\ D_uG^* \cdot u_2 + [\langle u_2, u_2\rangle - 1]u_1/2 \\ D_uG \cdot u_3 + D_\lambda G + \langle u_1, u_3\rangle u_2 \end{pmatrix} = 0. \tag{3.52}$$

Since G is C^3-continuous, the mapping F is well defined and C^2-continuous. Let

$$y_0 := (u_0,\ \lambda_0,\ \phi,\ \phi^*,\ v_0). \tag{3.53}$$

Theorem 3.3.1. *If (u_0, λ_0) is a simple bifurcation point of (3.40), then y_0 is a nonsingular solution of (3.52).*

Proof: It follows from the definition of F and the statements (3.42), (3.45) and (3.49) that

$$F(y_0) = 0. \tag{3.54}$$

A simple calculation shows a lower triangular structure of $DF(y_0)$, more precisely,

$$\begin{aligned} &DF(y_0) \\ &= \begin{pmatrix} A(y_0) & 0 & 0 & 0 \\ D(D_uG_0\phi)\cdot & D_uG_0 + \phi^*\langle\phi,\cdot\rangle & 0 & 0 \\ (DD_uG_0)^*\phi^*\cdot & 0 & D_uG_0^* + \phi\langle\phi^*,\cdot\rangle & 0 \\ D^2G_0(v_0,1)\cdot & \phi^*\langle v_0,\cdot\rangle & 0 & D_uG_0 + \phi^*\langle\phi,\cdot\rangle \end{pmatrix}, \end{aligned} \tag{3.55}$$

where

$$A(y) := \begin{pmatrix} D_uG + \langle u_2, D_{uu}Gu_1\cdot\rangle u_2 & D_\lambda G + \langle u_2, D_{u\lambda}Gu_1\rangle u_2 \\ \langle u_2, D_uDG(u_3,1)\rangle & \langle u_2, D_\lambda DG(u_3,1)\rangle \end{pmatrix}. \tag{3.56}$$

We claim that the operators $A(y_0)$ and $D_uG_0 + \phi^*\langle\phi,\cdot\rangle$ are nonsingular in the spaces $X \times \mathbf{R}$ and X, respectively, see Lemma 3.3.2 below. Then the operator $DF(y_0)$ is nonsingular from Y onto Y. ■

Lemma 3.3.2. *Let (u_0, λ_0) be a simple bifurcation point of (3.40) and the conditions (3.41), (3.48) be satisfied. Then the operators $D_uG_0 + \phi^*\langle\phi,\cdot\rangle$ and $A(y_0)$ are nonsingular in X and $X \times \mathbf{R}$, respectively.*

Proof: Let $w \in X$ be given arbitrarily. We consider the linear equation

$$(D_uG_0 + \phi^*\langle\phi,\ \cdot\rangle)v = w. \tag{3.57}$$

By (3.42) and (3.44) every element v in X is of the form $v = x\phi + \tilde{v}$ with $x \in \mathbf{R}$, $\tilde{v} \in \operatorname{Im}(D_uG_0^*)$. Substituting this into (3.57) and taking the inner product with ϕ^*, we obtain

$$x = \langle\phi^*, w\rangle.$$

Thereafter, the equation (3.57) reduces to

$$D_u G_0 \tilde{v} = w - \langle \phi^*, w \rangle \phi^*.$$

Thanks to (3.44) and (3.43a), this equation is uniquely solvable for $\tilde{v}$ in $\mathrm{Im}(D_u G_0^*)$. Hence, the equation (3.57) has a unique solution for any given right hand side. In other words, the linear mapping $D_u G_0 + \phi^* \langle \phi, \cdot \rangle$ is nonsingular from X onto X.

The nonsingularity of $A(y_0)$ can be proven in a similarly way. Given an arbitrary $(w, \alpha) \in X \times \mathbf{R}$, we consider the equation

$$A(y_0) \begin{pmatrix} u \\ \lambda \end{pmatrix} = \begin{pmatrix} w \\ \alpha \end{pmatrix} \tag{3.58}$$

for the unknown (u, λ) in $X \times \mathbf{R}$, i.e.,

$$D_u G_0 u + D_\lambda G_0 \lambda + \langle \phi^*, D_u D G_0(u, \lambda)\phi \rangle \phi^* = w, \tag{3.59a}$$
$$\langle \phi^*, D^2 G_0(v_0, 1)(u, \lambda) \rangle = \alpha. \tag{3.59b}$$

Note that $D_\lambda G_0 = -D_u G_0 v_0 \in \mathrm{Im}(D_u G_0)$ and the element $w \in X$ has a unique decomposition

$$w = \langle \phi^*, w \rangle \phi^* + \hat{w}, \qquad \hat{w} = \hat{Q} w \in \mathrm{Im}(D_u G_0).$$

Here and in the sequel $Q: \; X \to \mathrm{Im}(D_u G_0^*)$ and $\hat{Q}: \; X \to \mathrm{Im}(D_u G_0)$ are orthogonal projections defined by

$$Q u := u - \langle \phi, \; u \rangle \phi, \qquad \hat{Q} u := u - \langle \phi^*, \; u \rangle \phi^*, \tag{3.60}$$

respectively, see (3.42) and (3.44). The equation (3.59a) is equivalent to

$$D_u G_0 (u - \lambda v_0) = \hat{Q} w, \tag{3.61a}$$
$$\langle \phi^*, \; D_u D G_0(u, \lambda)\phi \rangle = \langle \phi^*, w \rangle. \tag{3.61b}$$

From (3.61a) we obtain

$$u = \lambda v_0 + \mu \phi + \tilde{u}, \tag{3.62}$$

where $\lambda, \; \mu \in \mathbf{R}$ are to be determined, and $\tilde{u} \in \mathrm{Im}(D_u G_0^*)$ is the unique solution of

$$D_u G_0 \tilde{u} = \hat{Q} w$$

in the subspace $\mathrm{Im}(D_u G_0^*)$. Due to (3.49) and $\tilde{u} \in \mathrm{Im}(D_u G_0^*)$, we have $\tilde{u} = Qu$, moreover, $D_u G_0 \tilde{u} = (D_u G_0 + \phi^* \langle \phi, \; \cdot \rangle)\tilde{u}$. Hence

$$\tilde{u} = Qu = B_0 w, \qquad B_0 := Q(D_u G_0 + \phi^* \langle \phi, \cdot \rangle)^{-1} \hat{Q}.$$

Substituting (3.62) into (3.61b) and (3.59b) yields a system for the unknowns μ, λ:

$$a\mu + b\lambda = \langle \phi^*, w \rangle - \langle \phi^*, D_{uu} G_0 \tilde{u} \phi \rangle, \tag{3.63a}$$
$$b\mu + c\lambda = \alpha - \langle \phi^*, D_u D G_0(v_0, 1)\tilde{u} \rangle. \tag{3.63b}$$

The transcritical condition (3.48) ensures that this system is uniquely solvable for μ, λ. Thus the equation (3.58) has a unique solution. Equivalently, the operator $A(y_0)$ is nonsingular. ■

In fact, via the notation

$$d := b^2 - ac, \tag{3.64}$$

the solution of (3.63) can be written as

$$\begin{aligned}\mu &= d^{-1}[-c\alpha + b\langle\phi^*, (I - D_{uu}G_0\phi B_0)w\rangle + c\langle\phi^*, D_uDG_0(v_0,1)B_0w\rangle],\\ \lambda &= d^{-1}[b\alpha - b\langle\phi^*, D_uDG_0(v_0,1)B_0w\rangle - a\langle\phi^*, (I - D_{uu}G_0\phi B_0)w\rangle].\end{aligned}$$

Together with (3.62) we obtain the solution of (3.58)

$$\begin{aligned}u &= d^{-1}\Big\{ - v_0\Big\langle\phi^*,\ \Big[bD_uDG_0(v_0,1)B_0 + a(I - D_{uu}G_0\phi B_0)\Big]w\Big\rangle\\ &\quad + \ \phi\Big\langle\phi^*,\ \Big[b(I - D_{uu}G_0\phi B_0) + cD_uDG_0(v_0,1)B_0\Big]w\Big\rangle\Big\}\\ &\quad + \ B_0w + d^{-1}(bv_0 - c\phi)\alpha,\\ \lambda &= d^{-1}\Big\{ - \Big\langle\phi^*,\ \Big[bD_uDG_0(v_0,1)B_0 + a(I - D_{uu}G_0\phi B_0)\Big]w\Big\rangle\Big\} + b\alpha.\end{aligned}$$

From this we derive the inverse of $A(y_0)$ as

$$A(y_0)^{-1} = \begin{pmatrix} v_0l_1 + \phi l_2 + B_0 & (bv_0 - c\phi)/d \\ l_1 & b/d \end{pmatrix}, \tag{3.65}$$

where $l_1,\ l_2$ are linear functional from X into $\mathbf{R}$,

$$\begin{aligned}l_1(u) &:= -d^{-1}\langle\phi^*, [bD_uDG_0(v_0,1)B_0 + a(I - D_{uu}G_0\phi B_0)]u\rangle,\\ l_2(u) &:= d^{-1}\langle\phi^*, [b(I - D_{uu}G_0\phi B_0) + cD_uDG_0(v_0,1)B_0]u\rangle.\end{aligned}$$

The operator $DF(y_0)$ in (3.55) is a lower triangular block matrix, its inverse has the same structure and can be expressed explicitly by the inverses of $A(y_0)$ and $D_uG_0 + \phi^*\langle\phi,\ \cdot\rangle$.

Remark 3.3.1. If the operator D_uG_0 is self-adjoint, then $\phi = \phi^*$ and the extended system (3.52) reduces to

$$F(y) := \begin{pmatrix} G(u,\lambda) + (u_1, D_uG\cdot u_1)u_1 \\ \langle u_1, D_uG\cdot u_2 + D_\lambda G\rangle \\ D_uG\cdot u_1 + [\langle u_1,u_1\rangle - 1]u_1/2 \\ D_uG\cdot u_2 + D_\lambda G + \langle u_1,u_2\rangle u_1 \end{pmatrix} = 0, \tag{3.66}$$

where $y := (u, \lambda, u_1, u_2) \in Y := X \times \mathbf{R} \times X^2$.

There are other extended systems with similar properties, e.g.

$$\tilde{F}(y) := \begin{pmatrix} G(u,\lambda) + \langle u_3, DG \cdot (u_2, 1) \rangle u_1 \\ \langle u_3, D_u G \cdot u_1 \rangle \\ D_u G \cdot u_1 + [\langle u_1, u_1 \rangle - 1] u_1 / 2 \\ D_u G \cdot u_2 + D_\lambda G + \langle u_1, u_2 \rangle u_1 \\ D_u G^* \cdot u_3 + [\langle u_1, u_3 \rangle - 1] u_1 \end{pmatrix} = 0.$$

For turning points, Seydel's extended system

$$\begin{pmatrix} G(u,\lambda) \\ D_u G(u,\lambda)\phi \\ l(\phi) - \xi \end{pmatrix} = 0 \qquad (3.67)$$

provides efficient approximations and easy combination with a class of branch switching methods, where l is a linear functional that imposes a value $\xi \neq 0$ on ϕ, see e.g. Seydel [276]. This system can also be used to approximate simple bifurcation points if the problem (3.40) has symmetries and the extended system (3.67) is restricted in fixed point subspaces of proper subgroups, see e.g. Werner [301] and Werner/Spence [307].

A small extended system for computations of simple bifurcation points is introduced by Moore in [232]

$$\begin{pmatrix} G(u,\lambda) + \mu u_1 \\ D_u G(u,\lambda)^* u_1 \\ \langle u_1, D_\lambda G(u,\lambda) \rangle \\ [\langle u_1, u_1 \rangle - 1]/2 \end{pmatrix} = 0,$$

see also Mei [213].

3.3.3 Newton-like methods

Newton method is powerful for solving nonlinear equations. In neighborhoods of exact solution this method generates quadratically convergent approximations. To maintain the quadratic convergence and to enlarge the convergent domain, one may use various Newton-like methods. Here we consider this kind of methods for approximating nonsingular solutions of the extended systems. An important factor for numerical approximation is to minimize the computational work and to maintain the fast convergence of Newton method. To this end, we make use of the lower triangular structure of $DF(y_0)$ and construct an Newton-like method.

For $\delta > 0$, define a neighborhood of y_0 in Y as

$$S(y_0, \delta) := \{y \in Y; \quad \|y - y_0\| \leq \delta\}.$$

Let $M_1 : Y \to GL(X \times \mathbf{R})$ and $M_2 : Y \to GL(X)$ be two C^2-continuous mappings such that

$$\text{a)} \quad M_1(y_0) = A(y_0)^{-1}, \qquad \text{b)} \quad M_2(y_0) = (D_u G_0 + \phi^* \langle \phi, \cdot \rangle)^{-1}. \qquad (3.68)$$

We consider a linear mapping $M:\ Y \to GL(Y)$

$$M(y) := \begin{pmatrix} M_1 & 0 & 0 & 0 \\ -M_2 DD_u G u_1 M_1 & M_2 & 0 & 0 \\ -M_2^*(D_u DGM_1 \cdot)^* u_2 & 0 & M_2^* & 0 \\ -M_2[D^2G(u_2,1)M_1 - u_2\langle u_3, M_2 DD_u G u_1 M_1 \cdot\rangle] & -M_2 u_2 \langle u_2, M_2 \cdot\rangle & 0 & M_2 \end{pmatrix}.$$

Lemma 3.3.3. *If the mappings M_1, M_2 satisfy the condition (3.68), then*

$$M(y_0) = DF(y_0)^{-1}. \tag{3.69}$$

Consider the following Newton-like method for approximation of the non-singular solution y_0 of the extended system (3.52)

$$\begin{cases} \text{Choose } \ y^0 \in S(y_0, \delta), \\ \text{FOR } \ k = 0, 1, \ldots, \ \text{ DO } \ y^{k+1} = y^k - M(y^k) \cdot F(y^k). \end{cases} \tag{3.70}$$

Theorem 3.3.2. *Let (u_0, λ_0) be a bifurcation point of (3.40) and let $D_u G_0$ be a Fredholm operator of index 0. If the conditions (3.41), (3.48) and (3.68) are satisfied, then the iteration (3.70) converges locally quadratically to y_0.*

Proof: Define a C^1-continuous mapping $H: Y \to Y$

$$H(y) := y - M(y) \cdot F(y).$$

It follows from (3.54) that

$$H(y_0) = y_0.$$

Furthermore, Lemma 3.3.3 shows

$$DH(y_0) = 0.$$

Therefore, the fixed point iteration (3.70) for H is locally quadratically convergent (cf. Ortega/Rheinboldt [239]). ∎

Implementation of the iteration (3.70) is realized as follows.

Algorithm 3.3.1. A Newton-like method for simple bifurcation points

Choose a starting point $y^0 \in S(y_0, \delta)$.

FOR($k = 0$; Stop criterion; k++) $\{$

1) *Compute*

$$(\Delta u^k, \Delta\lambda^k) = -M_1(y^k)\Big(G^k + \langle u_2^k, D_u G^k u_1^k\rangle u_2^k,\ \langle u_2^k, DG^k(u_3^k, 1)\rangle\Big);$$

2) *Using $(\Delta u^k, \Delta\lambda^k)$, calculate*

$$\Delta u_1^k = -M_2(y^k)\Big[D_u G^k u_1^k + \Big(\langle u_1^k, u_1^k\rangle - 1\Big)u_2^k/2 + D_u DG^k(\Delta u^k, \Delta\lambda^k)u_1^k\Big];$$

3) *Using* $(\Delta u^k, \Delta\lambda^k)$, Δu_1^k, *update* $u_1^{k+1} = u_1^k + \Delta u_1^k$ *and compute*

$$\begin{aligned}\Delta u_2^k = &-(M_2(y^k))^*\Big[(D_uG^k)^* u_2^k \\ &+ \Big(\langle u_2^k, u_2^k\rangle - 1\Big)u_1^{k+1}/2 + (D_uDG^k(\Delta u^k, \Delta\lambda^k))^* u_2^k\Big];\end{aligned}$$

and

$$\Delta u_3^k = -M_2(y^k)\Big[D_uG^k(u_3^k, 1) + \langle u_1^{k+1}, u_3^k\rangle u_2^k + D_uDG^k(\Delta u^k, \Delta\lambda^k)u_3^k\Big];$$

4) *Update* $y^{k+1} = y^k + \Delta y^k$ *with* $\Delta y^k := (\Delta u^k,\ \Delta\lambda^k,\ \Delta u_1^k,\ \Delta u_2^k,\ \Delta u_3^k)$.

}

Depending on the properties of the original problem (3.40), we choose the mappings $M_1(y)$ and $M_2(y)$ to make the terms $M_1(y)(v,\beta)^T$ and $M_2(y)v$ easily available. For example, If we choose

$$M_1(y) := A(y)^{-1}, \quad M_2(y) := (D_uG(u,\lambda) + u_2\langle u_1,\ \cdot\rangle)^{-1},$$

Algorithm 3.3.1 reduces to

Algorithm 3.3.2. *Given a starting point* $y^0 \in S(y_0, \delta)$, *we do* $k = 0, 1, \ldots$

$$\begin{aligned}(u^{k+1},\ \lambda^{k+1}) &= (u^k, \lambda^k) - M_1(y^k)\Big(G^k + \langle u_2^k, D_uG^k u_1^k\rangle u_1^k,\ \ \langle u_2^k, DG^k(u_3^k, 1)\rangle\Big);\\ u_1^{k+1} &= M_2(y^k)\Big[(1 + \langle u_1^k, u_1^k\rangle)u_2^k/2 - D_uDG^k(\Delta u^k, \Delta\lambda^k)u_1^k\Big];\\ u_2^{k+1} &= -M_2(y^k)^*\Big[(1 + \langle \Delta u_2^k, u_2^k\rangle)u_1^k + (D_uDG^k(\Delta u^k, \Delta\lambda^k))^* u_2^k\Big];\\ u_3^{k+1} &= -M_2(y^k)\Big[D_\lambda G^k + (\Delta u_1^k, u_3^k))u_2^k + D_uDG^k(\Delta u^k, \Delta\lambda^k)u_3^k\Big].\end{aligned}$$

Remark 3.3.2. The terms involving the factors Δu, $\Delta\lambda$, Δu_1 in Algorithm 3.3.2 tend to zero during the iteration process. Neglecting these terms in the computations of u_i^{k+1}, $i = 1, 2, 3$ does not affect the quadratical convergence of u^k, λ^k, see e.g. Hoy [163, 164] and Mei [212, 211, 215].

To exploit the explicit form of $A(y_0)^{-1}$ in (3.65), we introduce a mapping $B:\ Y \to L(X, X)$

$$B(y) := (I - u_1\langle u_1,\ \cdot\rangle)(D_uG(u,\lambda) + u_2\langle u_1,\ \cdot\rangle)^{-1}(I - u_1^*\langle u_1^*, \cdot\rangle)$$

and six functionals defined as

$$\begin{aligned}a(y) &:= \langle u_2,\ D_{uu}Gu_1u_1\rangle, & b(y) &:= \langle u_2,\ D_uDG(u_3, 1)u_1\rangle,\\ c(y) &:= \langle u_2,\ D^2G(u_3, 1)^2\rangle, & d(y) &:= b(y)^2 - a(y)\cdot c(y)\end{aligned}$$

$$L_1(y, v) := -d(y)^{-1}\Big\langle u_2,\ \Big[b(y)D_uDG(u,\lambda)(u_3, 1)B(y)$$

$$+ a(y)(I - D_{uu}G(u,\lambda)u_1B(y))\Big]v\Big\rangle,$$

$$L_2(y,v) := d(y)^{-1}\Big\langle u_2,\ \Big[b(y)(I - D_{uu}G(u,\lambda)u_1B(y)) + c(y)D_uDG(u,\lambda)(u_3,1)B(y)\Big]v\Big\rangle.$$

At $y = y_0$ these mappings have the properties

$$B(y_0) = B_0,\quad a(y_0) = a,\quad b(y_0) = b,\quad c(y_0) = c,\quad d(y_0) = d,$$
$$L_1(y_0,v) = l_1(v),\quad L_2(y_0,v) = l_2(v)\quad \text{for all } v \in X .$$

Thus we choose the operator $M_1(y) \in L(X \times \mathbf{R}, X \times \mathbf{R})$ as

$$M_1(y) := \begin{pmatrix} u_3L_1(y,\cdot) + u_1L_2(y,\cdot) + B(y) & (b(y)u_3 - c(y)u_1)/d(y) \\ L_1(y,\cdot) & b(y)/d(y) \end{pmatrix}. \tag{3.71}$$

This operator satisfies the condition (3.68a). Therefore, the term

$$\begin{pmatrix} v \\ b \end{pmatrix} := M_1(y)\begin{pmatrix} w \\ c \end{pmatrix}$$

in Algorithm 3.3.2 can be calculated with the following three steps. Firstly, we solve the system

$$(D_uG + u_2\langle u_1,\cdot\rangle)\tilde{v} = w \tag{3.72}$$

for $\tilde{v} \in X$. Then we calculate the functionals $a(y)$, $b(y)$, $c(y)$, $\langle u_2, D_uDG(u_3,1)\tilde{v}\rangle$ and $\langle u_2, D_{uu}Gu_1\tilde{v}\rangle$, etc. These values can be also approximated by difference quotient as in Griewank [140, 141]. Thereafter, following the definition (3.71) of $M_1(y)$ we obtain the element $(v,b)^T$ with simple multiplications and additions. In this way the main computational work of Algorithm 3.3.2 reduces to four times of solution of the system (3.72).

3.3.4 Rank-1 corrections for sparse problems

Many problems, in particular, discretizations of differential equations, have special structure such as banded, sparse Jacobian matrix D_uG. To exploit these properties in Newton-like methods for the extended systems we consider rank-1 corrections of D_uG. For simplicity we restrict the discussion to finite dimensional problems, i.e., $X = \mathbf{R}^n$, n may be large.

Rank-1 corrections have been used for approximating simple bifurcation points in Moore [232], Moore/Spence [231], Rheinboldt [249], Rheinboldt/Burkadt [250] and Seydel [276], where one suitable column (row) of D_uG is dropped or replaced with a proper vector in $\mathbf{R}^n$. We consider here a priori rank-1 correction of D_uG such that the fast solvers for sparse, banded, ... matrices in the available software packages, e.g. LINPACK and NAG, can be utilized directly.

Denote

$$\phi = (\mu_1, \mu_2, \ldots, \mu_n)^T, \qquad \phi^* = (\nu_1, \nu_2, \ldots, \nu_n)^T \in \mathbf{R}^n$$

and $e_1, \ldots, e_n$ the standard Cartesian unit vectors in $\mathbf{R}^n$.

Lemma 3.3.4. *If zero is a simple eigenvalue of D_uG_0 and the condition (3.41) is satisfied, then the index set*

$$I\!D := \{i \in \mathbf{N} \quad | \quad \mu_i \nu_i \neq 0\}$$

is nonempty. Furthermore, if $j \in I\!D$ and $\omega \neq 0$, the matrix $D_uG_0 + \omega e_j \cdot e_j^T$ is nonsingular.

Proof: The fact $I\!D \neq \emptyset$ follows directly from the assumption (3.41) and the property of simple eigenvalues. For the nonsingularity of $D_uG_0 + \omega e_j \cdot e_j^T$, we consider the equation

$$(D_uG_0 + \omega e_j \cdot e_j^T)u = 0 \tag{3.73}$$

for u in $\mathbf{R}^n$. Taking an inner product with ϕ^*, one gets $\omega\nu_j(e_j^T \cdot u) = 0$. Since $j \in I\!D$ and $\omega \neq 0$, we have $\omega\nu_j \neq 0$. Hence

$$e_j^T u = 0. \tag{3.74}$$

Substituting it back into (3.73) yields $u = c\phi$, $c \in \mathbf{R}$, which in turn shows $c\mu_j = 0$ in (3.74). The statement $j \in I\!D$ implies that $c = 0$, i.e., $u = 0$ is the unique solution of (3.73). This proves our conclusion. ■

The rank-1 corrections $D_uG_0 + \omega e_j \cdot e_j^T$ change merely one element on the diagonal of D_uG and has exactly the same structure as D_uG, e.g. symmetry, band, sparsity, etc. Moreover, since G is the C^3-continuous, there is a constant $\delta_0 > 0$, such that $D_uG(u, \lambda) + \omega e_j \cdot e_j^T$ is nonsingular in the neighborhood $S(y_0, \delta_0)$.

Remark 3.3.3. If D_uG_0 is semi-positive definite, then $D_uG_0 + \omega e_j \cdot e_j^T$ is positive definite for $\omega > 0$.

Lemma 3.3.5. *Assume that zero is a simple eigenvalue of D_uG_0 and the condition (3.41) is satisfied. Then for all $0 \neq \omega \in \mathbf{R}$ and $j \in I\!D$, we have*

$$e_j^T \cdot (D_uG_0 + \omega e_j \cdot e_j^T)^{-1} \cdot (I - \phi^* \cdot \phi^{*T}) = 0, \tag{3.75a}$$

$$e_j^T \cdot (D_uG_0^T + \omega e_j \cdot e_j^T)^{-1} \cdot (I - \phi \cdot \phi^T) = 0. \tag{3.75b}$$

Proof: Since $D_uG_0 + \omega e_j \cdot e_j^T$ is invertible for all $\omega \in \mathbf{R}$ ($\omega \neq 0$) and $j \in I\!D$, for arbitrary $v \in \mathbf{R}^n$ we define

$$u := (D_uG_0 + \omega e_j \cdot e_j^T)^{-1} \cdot (I - \phi^* \cdot \phi^{*T})v.$$

Then,

$$(D_uG_0 + \omega e_j \cdot e_j^T)u = (I - \phi^* \cdot \phi^{*T})v.$$

Taking an inner product at its both sides with ϕ^*,

$$\omega\nu_j(e_j^T u) = 0.$$

Owing to $\omega \neq 0$ and $\nu_j \neq 0$, we have $e_j^T u = 0$. In other words,

$$e_j^T(D_uG_0 + \omega e_j \cdot e_j^T)^{-1} \cdot (I - \phi \cdot \phi^{*T})v = 0, \qquad \text{for all } v \in \mathbf{R}^n.$$

This proves (3.75a). The statement (3.75b) follows similarly. ■

Let

$$M_2(y) := (I - u_1 \cdot u_1^T)(D_uG + \omega e_j \cdot e_j^T)^{-1}(I - u_2 \cdot u_2^T) + u_1 \cdot u_2^T. \quad (3.76)$$

Theorem 3.3.3. *Under the conditions (3.41) the matrix $M_2(y)$ in (3.76) satisfies (3.68b), i.e.,*

$$M_2(y_0) = (D_uG_0 + \phi^* \cdot \phi^T)^{-1}.$$

Proof: We verify the conclusion directly as

$$\begin{aligned}
&(D_uG_0 + \phi^* \cdot \phi^T) \cdot M_2(y_0)\\
&= (D_uG_0 + \phi^* \cdot \phi^T)[(I - \phi \cdot \phi^T)(D_uG_0 + \omega e_j \cdot e_j^T)^{-1}(I - \phi^* \cdot \phi^{*T}) + \phi \cdot \phi^{*T}]\\
&= D_uG_0(D_uG_0 + \omega e_j \cdot e_j^T)^{-1}(I - \phi^* \cdot \phi^{*T}) + \phi^* \cdot \phi^{*T} \qquad \text{(Due to (3.42}a\text{))}\\
&= (I - \phi^* \cdot \phi^{*T}) - \omega e_j \cdot e_j^T(D_uG_0 + \omega e_j \cdot e_j^T)^{-1}(I - \phi^* \cdot \phi^{*T}) + \phi^* \cdot \phi^{*T}\\
&= I. \qquad \text{(Due to (3.75}a\text{))}
\end{aligned}$$

■

Consequently, calculating the product $M_2(x) \cdot w$ becomes solving linear systems of the form

$$(D_uG + \omega e_j \cdot e_j^T)v = \tilde{w}.$$

If the matrix D_uG has sparse or banded structure, this system can be solved efficiently with special methods. The parameter $\omega \in \mathbf{R}$ will be chosen to minimize the condition number of $D_uG_0 + \omega e_j \cdot e_j^T$.

Remark 3.3.4. The index set $\mathbb{D}$ is usually a priori unknown. It is approximated and updated during the iteration by

$$\mathbb{D}_k := \{i \in \mathbf{N} \mid u_{1i}^k \cdot u_{2i}^k \neq 0\}, \quad k = 0, 1, \ldots.$$

At the same time, the constant ω will also be adapted to improve the condition number of $D_uG^k + \omega e_j \cdot e_j{}^T$. Initial values for Algorithms 3.3.1 and 3.3.2 are taken along continuation of the solution curve.

3.3.5 A numerical example

We consider a simplified buckling problem:

$$\Delta u + \lambda \sin(u) = 0 \qquad \text{in } \Omega := [0,1] \times [0,1], \tag{3.77a}$$

$$u = 0 \qquad \text{on } \partial\Omega. \tag{3.77b}$$

It is known that $(u_0, \lambda_0) := (0, 2\pi^2)$ is a simple bifurcation point of (3.77) and

$$\phi = \phi^* = 2 \sin \pi x \sin \pi y, \qquad v_0 = 0.$$

Discretizing (3.77) with five-point star difference scheme yields

$$G(u, \lambda) := Au + \lambda f(u) = 0,$$

where $h = 1/(N+1)$ and

$$u := (u_{11}, \ldots, u_{N1}, \ldots, u_{NN})^T,$$
$$f(u) := h^2(\sin(u_{11}), \ldots, \sin(u_{NN}))^T,$$

and

$$A := \begin{pmatrix} D & I & & & \\ I & D & I & & \\ & \ddots & \ddots & \ddots & \\ & & \ddots & \ddots & I \\ & & & I & D \end{pmatrix}, \quad D := \begin{pmatrix} -4 & 1 & & & \\ 1 & -4 & 1 & & \\ & \ddots & \ddots & \ddots & \\ & & \ddots & \ddots & 1 \\ & & & 1 & -4 \end{pmatrix}.$$

Since the matrix $D_uG(u, \lambda)$ is symmetric, we take the extended system (3.66)

$$H(x) := \begin{pmatrix} Au + \lambda f(u) + u_1 \cdot u_1^T(A + \lambda f'(u))u_1 \\ u_1^T(Au_2 + \lambda Df(u)u_2 + f(u)) \\ Au_1 + \lambda f'(u)u_1 + (u_1^T u_1 - 1)u_1/2 \\ Au_2 + \lambda f'(u)u_2 + f(u) + u_1^T u_2 \end{pmatrix} = 0.$$

The functionals $u_1^T f''(u)vw, \ldots$ in Algorithm 3.3.1 are computed directly and the index set $\mathbb{D}$ is chosen as $\mathbb{D} = \{1, 2, \ldots, N^2\}$. Taking discrete values of the functions

$$u^0 = x(1-y), \qquad u_1^0 = 2 \sin \pi x \sin \pi y + 0.5x,$$
$$u_2^0 = 0.5x, \qquad \lambda^0 = 18,$$

as starting point and $N = 20$, $j = 105$, we approximate $y_0 := (0, 2\pi^2, \phi, 0)$ with Algorithm 3.3.1. Table 3.3 shows convergent behavior of the approximate solutions.

k	λ^k	$\|u^k - 0\|$	$\|u_1^k - \phi\|$	$\|u_2^k - 0\|$
1	14.557284	0.062868	0.070813	0.027669
2	17.270507	0.024838	0.029792	0.003089
3	18.691284	0.008693	0.005503	0.000125
4	19.494918	0.000917	0.000320	0.000009
5	19.605178	0.000006	0.000004	0.000001

Table 3.3. Approximations of a simple bifurcation point, $N = 20$.

3.4 Computing Hopf Bifurcation Points

A Hopf bifurcation point is characterized as a point on a branch of stationary solutions where the Jacobian D_uG has a pair of simple imaginary eigenvalues and a branch of time-periodic solutions emanates. A Hopf bifurcation point can be detected along a continuation of steady state solutions by examining variation of a few (critical) eigenvalues of D_uG, or number of eigenvalues with positive real part or solving some bordered systems for test functions, see Section 3.2.3.

Indirect methods based on interpolations of steady state solutions and eigenvalues of the Jacobian D_uG are often sufficient for locating Hopf points, if the primary interest is path following and branch switching of solution curves. For details, see e.g. Hassard/Kazarinoff/Wan [152], Heinenman/Poore [156], Seydel [276] and the references therein. We are concerned with accurate approximation of a Hopf bifurcation point after it has been detected between two consecutive continuation steps. Similarly to the computation of a simple bifurcation point, we consider extended systems and discuss the method of Roose/Hlavaček [255, 257] and Griewank/Reddien [142] for computation of simple Hopf bifurcation points. For computations of periodic solutions near Hopf points we refer to Lust *et al* [207], Shroff/Keller [270].

3.4.1 Hopf points

Assume that (u_0, λ_0) is a simple Hopf bifurcation point, i.e., a solution to $G(u, \lambda) = 0$ and

H1) The intersection of the spectrum of $D_uG(u_0, \lambda_0)$ with the imaginary axis is a pair of simple eigenvalues $\pm\omega_0 i$, $\omega_0 > 0$;

H2) The pair of simple eigenvalues $\sigma(\lambda) \pm i\omega(\lambda)$ of the Jacobian $D_uG(u(\lambda), \lambda)$ on the steady state solution branch $(u(\lambda), \lambda)$ crosses the imaginary axis at the point $\lambda = \lambda_0$, $u(\lambda_0) = u_0$ with a nonvanishing speed, i.e.,

$$\sigma(\lambda_0) = 0, \quad \omega(\lambda_0) = \omega_0 \quad \text{and} \quad \frac{d\sigma}{d\lambda}(\lambda_0) \neq 0.$$

Remark 3.4.1. The existence of the steady state solution $(u(\lambda), \lambda)$ in **H2**) follows from the fact that under the assumption **H1**) the operator $\frac{\partial G}{\partial u}(u_0, \lambda_0)$

is nonsingular and the implicit function theorem is applicable to $G(u, \lambda) = 0$ at (u_0, λ_0).

Let $c = c_r + ic_i$ be the eigenvector of $D_u G(u_0, \lambda_0)$ associated with the eigenvalue $i\omega_0 \neq 0$, i.e.,

$$\text{a)} \quad D_u G_0 c_r = -\omega_0 c_i, \qquad \text{b)} \quad D_u G_0 c_i = \omega_0 c_r. \tag{3.78}$$

Since $\omega_0 \neq 0$, the vectors c_r, c_i are linearly independent. In fact, if $c_r = x c_i$ for some $x \neq 0$, then multiplying (3.78a) with x and adding it to (3.78b) we obtain

$$0 = D_u G_0 (x c_r + c_i) = D_u G_0 (x^2 + 1) c_i = (x^2 + 1)\omega_0 c_r \quad \Rightarrow \quad c_r = c_i = 0,$$

which leads to a contradiction. On the other hand, we derive from (3.78) that

$$\text{a)} \quad (D_u G_0)^2 c_r = -\omega_0^2 c_r, \qquad \text{b)} \quad (D_u G_0)^2 c_i = -\omega_0^2 c_i.$$

Thus c_r, c_i are eigenvectors of $(D_u G_0)^2$ associated to the eigenvalue $-\omega_0^2$.

Lemma 3.4.1. *Let* $\pm i\omega_0$ *be simple eigenvalues of* $D_u G_0$ *and* c, $\bar{c}$ *be the associated eigenvectors. Then*

$$\operatorname{Ker}\left((D_u G_0)^2 + \omega_0^2 I\right) = \operatorname{span}[c_r,\ c_i]. \tag{3.79}$$

Proof: Obviously, $c_r,\ c_i \in \operatorname{Ker}\left((D_u G_0)^2 + \omega_0^2 I\right)$,

$$\operatorname{span}[c_r,\ c_i] \subset \operatorname{Ker}\left((D_u G_0)^2 + \omega_0^2 I\right).$$

On the other hand, for any $a \in \operatorname{Ker}\left((D_u G_0)^2 + \omega_0^2 I\right)$, we have

$$\begin{aligned} 0 &= \left((D_u G_0)^2 + \omega_0^2 I\right) a \\ &= (D_u G_0 - i\omega_0 I)(D_u G_0 + i\omega_0 I) a \\ &= (D_u G_0 + i\omega_0 I)(D_u G_0 - i\omega_0 I) a. \end{aligned}$$

Since $\pm i\omega_0 I$ are simple eigenvalues of $D_u G_0$, there is $x \in \mathbf{C}$, such that

$$(D_u G_0 + i\omega_0 I) a = x c.$$

The operator $D_u G_0$ and the element a are all real. Hence

$$(D_u G_0 - i\omega_0 I) a = \overline{x c}$$

and

$$a = \frac{x c - \overline{x c}}{2\omega_0 i} = \frac{\operatorname{Im}(x c)}{\omega_0} \in \operatorname{span}[c_r,\ c_i],$$

in other words,

$$\operatorname{Ker}\left((D_u G_0)^2 + \omega_0^2 I\right) \subset \operatorname{span}[c_r,\ c_i],$$

and the statement (3.79) holds. ■

Similarly, for the adjoint operator $(D_u G_0 - i\omega_0 I)^*$, there exists $d = d_r + id_i$ in X, such that

$$\text{Ker}\left(((D_u G_0)^2 + \omega_0^2 I)^*\right) = \text{span}[d_r,\ d_i].$$

Moreover, the eigenvectors c and d can be chosen such that

$$\langle d,\ c\rangle = 2, \quad \langle d_r,\ c_r\rangle = \langle d_i,\ c_i\rangle = 1, \quad \langle d_r,\ c_i\rangle = \langle d_i,\ c_r\rangle = 0. \tag{3.80}$$

3.4.2 Extended systems

Let

$$Y := X \times X \times \mathbf{R} \times \mathbf{R}. \tag{3.81}$$

We define a mapping $F:\ Y \to Y$ by

$$F(y) := \begin{pmatrix} G(u,\lambda) \\ \left((D_u G(u,\lambda))^2 + \omega^2 I\right)v \\ \langle v,\ v\rangle - 1 \\ \langle q,\ v\rangle \end{pmatrix}, \quad y := (u, v, \lambda, \omega) \in Y. \tag{3.82}$$

Here $\langle\cdot,\ \cdot\rangle$ denotes the inner product in X, and $q \in X$ is chosen such that

$$|\langle q, c_r\rangle| + |\langle q, c_i\rangle| > 0. \tag{3.83}$$

Theorem 3.4.1. *Let (u_0, λ_0) be a Hopf bifurcation point and let the conditions* **H1**), **H2**) *and (3.83) be satisfied. Then the extended system*

$$F(y) = 0 \tag{3.84}$$

has a nonsingular solution of the form $y_0 = (u_0, \phi, \lambda_0, \omega_0)$, $\phi \in \text{span}[c_r,\ c_i]$.

Remark 3.4.2. Obviously, the Hopf bifurcation point (u_0, λ_0) and the reference frequency ω_0 of the bifurcating periodic solution are included in the nonsingular solution $y_0 = (u_0, \phi, \lambda_0, \omega_0)$ of (3.84).

Proof: Without loss of generality, we assume $\langle q, c_i\rangle \neq 0$. An element v in the space $\text{Ker}\left((D_u G_0)^2 + \omega_0^2 I\right)$ is of the form $v = \alpha c_r + \beta c_i$ for some $\alpha, \beta \in \mathbf{R}$. We fix α, β with the equalities $\langle v, v\rangle - 1 = 0$, $\langle q, v\rangle = 0$, i.e.,

$$\alpha^2\langle c_r, c_r\rangle + 2\alpha\beta\langle c_r, c_i\rangle + \beta^2\langle c_i, c_i\rangle = 1 \tag{3.85a}$$

$$\alpha\langle q, c_r\rangle + \beta\langle q, c_i\rangle = 0. \tag{3.85b}$$

Since $\langle q, c_i\rangle \neq 0$, we solve $\beta = \langle q,\ c_r\rangle / \langle q,\ c_i\rangle\alpha$ from (3.85b) and substitute it into (3.85a)

$$\alpha^2 = \left\|\langle q,\ c_i\rangle c_r - \langle q,\ c_r\rangle c_i\right\|^{-2}.$$

Here the inequality $\langle q, c_i \rangle c_r - \langle q, c_r \rangle c_i \neq 0$ follows from the linear independence of c_r and c_i, and the statement $\langle q, c_i \rangle \neq 0$. Let

$$(\alpha_0,\ \beta_0) := \pm \Big(\frac{\langle q, c_i \rangle}{\|\langle q, c_i \rangle c_r - \langle q, c_r \rangle c_i\|},\ \frac{\langle q, c_r \rangle}{\|\langle q, c_i \rangle c_r - \langle q, c_r \rangle c_i\|} \Big) \tag{3.86}$$

and

$$\phi := \alpha_0 c_r + \beta_0 c_i. \tag{3.87}$$

Then $y_0 := (u_0, \phi, \lambda_0, \omega_0)$ satisfies the equation (3.84). To show it is also a nonsingular solution, we prove nonsingularity of the operator $\frac{dF}{dy}(y_0)$. Alternatively, we show that $\frac{dF}{dy}(y_0)$ maps Y one-to-one onto Y. Let $z = (w_1, w_2, a_1, a_2) \in Y$ be fixed arbitrarily. We consider the equation

$$\frac{dF}{dy}(y_0)x = z \tag{3.88}$$

for the unknown $x = (u, v, \alpha, \beta)$. Rewrite (3.88) as

$$D_u G_0 u + D_\lambda G_0 \alpha = w_1, \tag{3.89a}$$

$$\begin{aligned} &D_u G_0 (D_{uu} G_0 \phi u) + D_{uu} G_0 (D_u G_0 \phi) u + [(D_u G_0)^2 + \omega_0^2 I] v \\ &+ \alpha [D_u G_0 D_{u\lambda} G_0 \phi + D_{u\lambda} G_0 D_u G_0 \phi] + 2\beta \omega_0 \phi = w_2 \end{aligned} \tag{3.89b}$$

$$\langle \phi, v \rangle = a_1 \tag{3.89c}$$

$$\langle q, v \rangle = a_2. \tag{3.89d}$$

Since $D_u G_0$ is nonsingular, there is a unique $v_0 \in X$, such that

$$D_u G_0 v_0 = D_\lambda G_0. \tag{3.90}$$

The general solution of (3.89a) is of the form

$$u = \alpha v_0 + \tilde{u} \tag{3.91}$$

where $\tilde{u}$ is uniquely determined by

$$D_u G_0 \tilde{u} = w_1.$$

Recall the definitions of α_0, β_0 and ϕ in (3.86), (3.87), respectively. We choose two elements $p_1,\ p_2 \in \mathrm{Ker}\Big((D_u G_0^2 + \omega_0^2 I)^* \Big)$ as

$$p_1 := \frac{1}{\alpha_0^2 + \beta_0^2} (\alpha_0 d_r + \beta_0 d_i), \quad p_2 := \frac{1}{\alpha_0^2 + \beta_0^2} (\beta_0 d_r - \alpha_0 d_i) \tag{3.92}$$

The statements in (3.80) imply

$$\begin{aligned} \langle p_1,\ \phi \rangle &= \frac{1}{\alpha_0^2 + \beta_0^2} \langle \alpha_0 d_r + \beta_0 d_i,\ \alpha_0 c_r + \beta_0 c_i \rangle = 1, \\ \langle p_2,\ \phi \rangle &= \frac{1}{\alpha_0^2 + \beta_0^2} \langle \alpha_0 d_r - \beta_0 d_i,\ \alpha_0 c_r + \beta_0 c_i \rangle = 0. \end{aligned}$$

Substituting (3.91) into (3.89b) turns (3.89) into a system for $(v, \alpha, \beta) \in X \times \mathbf{R} \times \mathbf{R}$

$$\mathbf{A}\begin{pmatrix} v \\ \alpha \\ \beta \end{pmatrix} = \begin{pmatrix} w_2 - D_uG_0D_uDG_0(\tilde{u},1)\phi \\ a_1 \\ a_2 \end{pmatrix} \tag{3.93}$$

with

$$\mathbf{A} := \begin{pmatrix} (D_uG_0)^2 + \omega_0^2 I & D_uG_0D_uDG_0(v_0,1)\phi + D_uDG_0(v_0,1)D_uG_0\phi & 2\omega_0\phi \\ 2\langle\phi,\cdot\rangle & 0 & 0 \\ \langle q,\cdot\rangle & 0 & 0 \end{pmatrix}.$$

We show that this system is uniquely solvable for (v, α, β), which in turn implies the nonsingularity of the coefficient matrix. Taking an inner product of the first equation in (3.93) with p_2, we obtain an equation for α in $\mathbf{R}$

$$\begin{aligned} &\alpha\langle p_2,\ D_uG_0D_uDG_0(v_0,1)\phi + D_uDG_0(v_0,1)D_uG_0\phi\rangle \\ &\qquad = \langle p_2,\ w_2\rangle - \langle p_2,\ D_uG_0D_uDG_0(\tilde{u},1)\phi\rangle. \end{aligned} \tag{3.94}$$

We claim

$$\langle p_2,\ D_uG_0D_uDG_0(v_0,1)\phi + D_uDG_0(v_0,1)D_uG_0\phi\rangle = 2\omega_0\dot{\sigma}(\lambda_0) \neq 0, \tag{3.95}$$

see Lemma 3.4.2 below. Thus α is uniquely determined by (3.94). This in turn implies that the unknowns u, α in (3.89a), (3.89b) are solved uniquely by (3.91) and (3.94), respectively. Similarly, the inner product of the first equation in (3.93) with p_1 yields uniquely

$$\begin{aligned} \beta = \frac{1}{2w_0}\Big[&\langle p_1,\ w_2\rangle - \langle p_1,\ D_uG_0(D_{uu}G_0\phi u) + D_{uu}G_0(D_uG_0\phi)u\rangle \\ &- \alpha\langle p_1,\ D_uG_0D_{u\lambda}G_0\phi + D_{u\lambda}G_0D_uG_0\phi\rangle\Big]. \end{aligned}$$

Finally, let us consider (3.89b) as an equation for v. Its general solution is of the form

$$v = \mu c_r + \nu c_i + \tilde{v} \tag{3.96}$$

where $\mu, \nu \in \mathbf{R}$ are arbitrarily, $\tilde{v} \in \operatorname{Im}\Big((D_uG_0)^2 + \omega_0^2 I)^*\Big)$ is the unique solution of the equation

$$\begin{aligned} [(D_uG_0)^2 + \omega_0^2 I]\tilde{v} = w_2 &- D_uG_0(D_{uu}G_0\phi u) - D_{uu}G_0(D_uG_0\phi)u - \\ &- \alpha[D_uG_0D_{u\lambda}G_0\phi + D_{u\lambda}G_0D_uG_0\phi] - 2\beta\omega_0\phi \\ &\in \operatorname{Im}\left((D_uG_0)^2 + \omega_0^2 I\right). \end{aligned}$$

Substituting the general solution (3.96) into (3.89c), (3.89d) yields a linear system for the unknowns μ, ν:

$$\begin{pmatrix} \langle \phi, c_r \rangle & \langle \phi, c_i \rangle \\ \langle q, c_r \rangle & \langle q, c_i \rangle \end{pmatrix} \begin{pmatrix} \mu \\ \nu \end{pmatrix} = \begin{pmatrix} a_1/2 - \langle \phi, \tilde{v} \rangle \\ a_2 - \langle q, \tilde{v} \rangle \end{pmatrix}. \tag{3.97}$$

Since $y_0 = (u_0, \phi, \lambda_0, \omega_0)$ satisfies the extended system (3.84), we have $(\alpha_0, \beta_0) \neq 0$ and

$$\langle q, \phi \rangle = \alpha_0 \langle q, c_r \rangle + \beta_0 \langle q, c_i \rangle = 0. \tag{3.98}$$

Moreover, $q \in X$ is chosen such that

$$|\langle q, c_r \rangle| + |\langle q, c_i \rangle| > 0.$$

Hence, if $\beta_0 \neq 0$, then $\langle q, c_r \rangle \neq 0$. On the other hand, $\beta_0 = 0$ implies that $\langle q, c_r \rangle = 0$, which in turn yields $\langle q, c_i \rangle \neq 0$. Similarly, if $\alpha_0 \neq 0$, we have $\langle q, c_i \rangle \neq 0$. Thereafter,

$$\begin{aligned}
& \det \begin{pmatrix} \langle \phi, c_r \rangle & \langle \phi, c_i \rangle \\ \langle q, c_r \rangle & \langle q, c_i \rangle \end{pmatrix} \\
= \; & \Big[\alpha_0 \langle c_r, c_r \rangle + \beta_0 \langle c_r, c_i \rangle\Big] \langle q, c_i \rangle - \Big[\alpha_0 \langle c_r, c_i \rangle + \beta_0 \langle c_i, c_i \rangle\Big] \langle q, c_r \rangle \\
\overset{(3.98)}{=} \; & \alpha_0 \langle c_r, c_r \rangle \langle q, c_r \rangle - 2\alpha_0 \langle c_r, c_i \rangle \langle q, c_r \rangle - \beta_0 \langle c_i, c_i \rangle \langle q, c_r \rangle \\
= \; & \begin{cases} -\beta_0{}^{-1} [\alpha_0^2 \langle c_r, c_r \rangle + 2\alpha_0 \beta_0 \langle c_r, c_i \rangle + \beta_0^2 \langle c_i, c_i \rangle] \langle q, c_r \rangle & \text{for } \beta_0 \neq 0 \\ \alpha_0 \langle c_r, c_r \rangle \langle q, c_i \rangle & \text{for } \beta_0 = 0 \end{cases} \\
= \; & \begin{cases} -\beta_0{}^{-1} \langle \alpha_0 c_r + \beta_0 c_i, \alpha_0 c_r + \beta_0 c_i \rangle \langle q, c_r \rangle & \text{for } \beta_0 \neq 0 \\ \alpha_0 \langle c_r, c_r \rangle \langle q, c_i \rangle & \text{for } \beta_0 = 0 \end{cases} \\
\neq \; & 0.
\end{aligned}$$

The last inequality follows from the facts that $|\alpha_0| + |\beta_0)| > 0$ and the vectors c_r, c_i are linearly independent. Hence, the system (3.97) is uniquely solvable for μ, ν. We conclude that for any given $z \in Y$, the system (3.88) has a unique solution x, i.e., the linear operator $\frac{dF}{dy}(y_0)$ is nonsingular. ∎

To complete the proof of Theorem 3.4.1, let us now verify the statement (3.95).

Lemma 3.4.2. *Let (u_0, λ_0) be a Hopf bifurcation point and let the conditions* **H1**), **H2**) *be satisfied. Then for v_0 in (3.90), ϕ in (3.87) and p_2 in (3.92), we have*

$$\Big\langle p_2, \; D_u G_0 D_u DG_0(v_0, 1)\phi + D_u DG_0(v_0, 1) D_u G_0 \phi \Big\rangle = 2\omega_0 \dot{\sigma}(\lambda_0) \neq 0. \tag{3.95}$$

Proof: Let $\Sigma(\lambda) = \sigma(\lambda) + i\omega(\lambda)$, $c(\lambda) = c_r(\lambda) + ic_i(\lambda)$ be the eigenpair of $\frac{\partial G}{\partial u}(u(\lambda), \lambda)$ with $\Sigma(\lambda_0) = i\omega_0$, $c(\lambda_0) = c_r + ic_i$ $(= c)$. Differentiating the equations

$$G(u(\lambda), \lambda) = 0,$$
$$\frac{\partial G}{\partial u}(u(\lambda), \lambda) c(\lambda) = \Sigma(\lambda) c(\lambda)$$

with respect to λ at $\lambda = \lambda_0$

$$D_u G_0 \dot{u}(\lambda_0) + D_\lambda G_0 = 0,$$
$$D_{u\lambda} G_0 c + D_{uu} G_0 \dot{u}(\lambda_0) c + D_u G_0 \dot{c}(\lambda_0) = \dot{\Sigma}(\lambda_0) c + i\omega_0 \dot{c}(\lambda_0).$$

Since $\dot{u}(\lambda_0) = v_0$, and $d \in \operatorname{Ker}(D_u G_0^* - i\omega_0)$, we have

$$\dot{\Sigma}(\lambda_0) = \frac{1}{2}\langle d,\ D_{u\lambda} G_0 c + D_{uu} G_0 v_0 c\rangle.$$

Keeping in mind of the complex conjugacy in the inner product, we calculate its real part

$$\begin{aligned}\dot{\sigma}(\lambda_0) &= \frac{1}{2}\langle d_r,\ D_{u\lambda} G_0 c_r + D_{uu} G_0 v_0 c_r\rangle + \frac{1}{2}\langle d_i,\ D_{u\lambda} G_0 c_i + D_{uu} G_0 v_0 c_i\rangle \\ &= \frac{1}{2}\langle d_r,\ D_u DG_0(v_0,1) c_r\rangle + \frac{1}{2}\langle d_i,\ D_u DG_0(v_0,1) c_i\rangle. \end{aligned} \tag{3.99}$$

Furthermore, we note the definition of ϕ in (3.87) and p_2 in (3.92) and rewrite the left hand side of (3.95) as

$$\begin{aligned}&\langle p_2,\ D_u G_0 D_u DG_0(v_0,1)\phi + D_u DG_0(v_0,1) D_u G_0 \phi\rangle \\ &= \langle D_u G_0^* p_2,\ D_u DG_0(v_0,1)\phi\rangle + \langle p_2,\ D_u DG_0(v_0,1) D_u G_0 \phi\rangle \\ &= \frac{1}{\alpha_0^2+\beta_0^2}\Big[\langle D_u G_0^*(\beta_0 d_r - \alpha_0 d_i),\ D_u DG_0(v_0,1)(\alpha_0 c_r + \beta_0 c_i)\rangle \\ &\quad + \langle \beta_0 d_r - \alpha_0 d_i,\ D_u DG_0(v_0,1) D_u G_0(\alpha_0 c_r + \beta_0 c_i)\rangle\Big] \\ &= \frac{\omega_0}{\alpha_0^2+\beta_0^2}\Big[\langle (\beta_0 d_i + \alpha_0 d_r),\ D_u DG_0(v_0,1)(\alpha_0 c_r + \beta_0 c_i)\rangle \\ &\quad + \langle \beta_0 d_r - \alpha_0 d_i,\ D_u DG_0(v_0,1) D_u G_0(-\alpha_0 c_i + \beta_0 c_r)\rangle\Big] \\ &= \omega_0[\langle d_r,\ D_u DG_0(v_0,1) c_r\rangle + \langle d_i,\ D_u DG_0(v_0,1) c_i\rangle] \\ &= 2\omega_0 \dot{\sigma}(\lambda_0) \quad \text{(due to (3.99))}.\end{aligned}$$

This completes the proof. ■

We refer to Jepson [169] and Griewank/Reddien [142] for other extended systems, such as

$$F(u,\phi,v_2,\lambda,\omega) := \begin{pmatrix} G(u,\lambda) \\ D_u G(u,\lambda)\phi + \omega v_2 \\ D_u G(u,\lambda) v_2 - \omega\phi \\ N(\phi, v_2) \end{pmatrix} = 0. \tag{3.100}$$

Here the normalization $N(\phi, v_2)$ is chosen, e.g., in Griewank/Reddien [142], as

$$N(\phi, v_2) := \begin{pmatrix} \langle q,\phi\rangle \\ \langle q, v_2\rangle - 1 \end{pmatrix}, \tag{3.101}$$

where $q \in X$ is fixed and satisfies $|\langle q, c_r\rangle| + |\langle q, c_i\rangle| > 0$. Another normalization is given in Roose/Hlavaček [257] as

$$N(\phi, v_2) := \begin{pmatrix} \langle q, \phi\rangle \\ \langle v_2, v_2\rangle - 1 \end{pmatrix}. \tag{3.102}$$

Extended systems have been also used to compute Hopf points of multiple parameter problems, see e.g. Pönisch [244], Roose/De Dier [256] and Li/Mei [203] for more details.

3.4.3 Newton method for extended systems

The nonsingular solution $y_0 = (u_0, \phi, \lambda_0, \omega_0)$ of the Roose/Hlavaček's extended system (3.84) can be approximated with Newton method. In the continuation of steady state solution, if the test function indicates that there is possibly a Hopf bifurcation point between to continuation steps, we switch to the extended systems for accurate approximation of the bifurcation point. We start from one continuation step (u^0, λ^0) and calculate those eigenpairs of $D_uG(u^0, \lambda^0)$ with eigenvalues closest to the imaginary axis. Then we set up an initial approximation $y^0 = (u^0, v^0, \lambda^0, \omega^0)$ of $y_0 = (u_0, \phi, \lambda_0, \omega_0)$ and apply the following Newton iteration for refining the approximations of y_0

$$\frac{dF}{dy}(y^k)\Delta y = -F(y^k), \tag{3.103a}$$

$$y^{k+1} := y^k + \Delta y, \quad k = 0, 1, 2 \ldots, \tag{3.103b}$$

where $y^k = (u^k, v^k, \lambda^k, \omega^k)$ and $\Delta y = (\Delta u, \Delta v, \Delta\lambda, \Delta\omega)$. The main computational work is the numerical solution of the linear system (3.103a). We utilize the special structure of $\frac{dF}{dy}$ to break (3.103a) into small problems. More precisely, we write (3.103a) as

$$A\Delta u + d_1\Delta\lambda = -r_1, \tag{3.104a}$$

$$B\Delta u + [A^2 + (\omega^k)^2 I]\Delta v + (Ad_2 + d_3)\Delta\lambda + 2\omega^k v^k \Delta\omega = -r_2, \tag{3.104b}$$

$$2\langle v^k, \Delta v\rangle = -r_3, \tag{3.104c}$$

$$\langle q, \Delta v\rangle = -r_4, \tag{3.104d}$$

where

$$\begin{aligned}
&A := D_uG(u^k, \lambda^k), \quad B := AD_{uu}G(u^k, \lambda^k)v^k + D_{uu}G(u^k, \lambda^k)D_uG(u^k, \lambda^k)v^k \\
&d_1 := D_\lambda G(u^k, \lambda^k), \quad d_2 := D_{u\lambda}G(u^k, \lambda^k)v^k, \\
&d_3 := D_{u\lambda}G(u^k, \lambda^k)D_uG(u^k, \lambda^k)v^k \\
&r_1 := G(u^k, \lambda^k), \qquad r_2 := [D_uG^2(u^k, \lambda^k) + (\omega^k)^2 I]v^k, \\
&r_3 := \langle v^k, v^k\rangle - 1, \qquad r_4 := \langle q, v^k\rangle.
\end{aligned}$$

At the Hopf bifurcation point (u_0, λ_0), the Jacobian $D_u G_0$ is nonsingular. Thus for (u^k, λ^k) near (u_0, λ_0), the increment Δu can be calculated in terms of $\Delta\lambda$ via

$$\Delta u = h_1 + \Delta\lambda h_2, \tag{3.105}$$

where h_1, h_2 are solutions of the nonsingular systems

$$Ah_1 = -r_1, \qquad Ah_2 = -d_1.$$

After calculating Δu, we rewrite (3.104b)-(3.104d) as

$$\begin{pmatrix} A^2 + (\omega^k)^2 I & Bh_2 + Ad_2 + d_3 & 2\omega^k v^k \\ 2\langle v^k, \cdot\rangle & 0 & 0 \\ \langle q, \cdot\rangle & 0 & 0 \end{pmatrix} \begin{pmatrix} \Delta v \\ \Delta\lambda \\ \Delta\omega \end{pmatrix} = \begin{pmatrix} -r_2 - Bh_1 \\ -r_3 \\ -r_4 \end{pmatrix}. \tag{3.106}$$

This system is nonsingular for $(u^k, v^k, \lambda^k, \omega^k)$ near $(u_0, \phi, \lambda_0, \omega_0)$, see (3.93) in the proof of Theorem 3.4.1. If $X = \mathbf{R}^n$ and the derivatives of G are easily available, a single step in Newton iteration consists mainly of the following steps:

- evaluation of the residual r_1 and the vectors d_1, d_2 and d_3;
- evaluation of the matrices A, B;
- computation of the matrix product A^2 (not necessary if iterative methods are used);
- computation of the matrix-vector products, $Av^k, Bh_1, Bh_2, Ad_2, [A^2 + (\omega^k)^2 I]v^k$;
- LU-decompositions of A and $A^2 + \omega^2 I$;
- 3 backward substitutions.

Altogether $5n^3/3 + O(n^2)$ operations are required at each step. If the matrix A is transformed into an upper Hessenberg form at the beginning, the computational work can be reduced to $n^3 + O(n^2)$, see e.g. Griewank/Reddien [142]. We note the second derivatives of G involved in (3.106) are all in the form of directional derivatives and can be approximated by appropriate difference quotient. Furthermore, splitting of the iterations for u^k and v^k and other techniques can also be used to approximate the nonsingular solution $(u_0, \phi, \lambda_0, \omega_0)$, see e.g. Li/Mei [203].

4. Branch Switching at Simple Bifurcation Points

Analytical analysis of bifurcations reveals qualitative behavior of a nonlinear system. Quantitatively, we need numerical approximations of bifurcating solution curves to gain insight how one physical state transits to another as control parameter changes and how sensitive such a transition is with respect to the parameter. Often very interesting scenario occurs as the solution moves from one branch to another. Branch switching and path following across bifurcation points are essential for numerical analysis of bifurcation problems. Since different solution branches intersect at the bifurcation point (u_0, λ_0), the linearized problem becomes singular at this point and path following with the standard predictor-corrector methods often fails. In this chapter we incorporate some branch switching techniques into the numerical continuation methods. We shall concentrate on an approach by Decker and Keller [82] which switches the branches by prescribing tangent of the bifurcating solution curves and on an approach by Mei [210, 213, 218] which regularizes the problem at bifurcation points. Other branch switching techniques, for examples, unfolding the singularity via perturbations, constructing predictors and correctors with selective properties, are discussed in Allgower/Georg [14], Allgower/Chien [13], Decker/Keller [82], Deuflhard/Fiedler/Kunkel [85], Keller [179], Rheinboldt [249] and Seydel [276].

Though the enlarged systems in [82, 85, 179, 276] are still singular at bifurcation points, convergence of correctors has been established in a conical domain such that solutions can be continued from the bifurcation point effectively with these methods. Alternatively, a regularized system is constructed in Jepson/Keller [171] and Mei [210, 213, 218] for accurate analysis of local behavior of solutions near bifurcation points.

Section 4.1 describes general structure of solution branches at a bifurcation point. In Section 4.2 we study behavior of the linearized operator along bifurcating solution branches. We discuss branch switching with a classical Euler-Newton continuation method and with a regularization of the original problem in Sections 4.3 and 4.4, respectively. These branch switching techniques are extended in Chapter 6 for handling multiple bifurcations.

4.1 Structure of Bifurcating Solution Branches

Let X, Y be Banach spaces of functions with a continuous embedding $X \hookrightarrow Y$. We assume that both X and Y can be embedded into a Hilbert space Z (e.g. $X = H_0^1(\Omega)$, $Y = Z = L^2(\Omega)$) and $G: X \times \mathbf{R} \to Y$ is a C^k-continuous ($k \geq 2$) nonlinear mapping. Consider the one-parameter problem

$$G(u, \lambda) = 0. \tag{4.1}$$

Let (u_0, λ_0) be a bifurcation point of (4.1), characterized by the properties

$$\dim\Big(\mathrm{Ker}(D_uG(u_0, \lambda_0))\Big) = n \geq 1, \quad D_\lambda G(u_0, \lambda_0) \in \mathrm{Im}(D_uG(u_0, \lambda_0)). \tag{4.2}$$

Here we denote the dimension of $\mathrm{Ker}(D_uG(u_0, \lambda_0))$ with n to include multiple bifurcations ($n > 1$) discussed in the later chapters. We are interested in how to continue a solution from a bifurcation point towards different branches.

Let $G_0, D_uG_0, \ldots$ denote evaluations of $G, D_uG, \ldots$ at (u_0, λ_0). We assume that D_uG_0 is a Fredholm operator of index 0 and zero is a semi-simple eigenvalue of it, i.e., the eigenvalue zero has the same geometric and algebraic multiplicities. Furthermore, we assume that bases of the subspaces

$$\mathrm{Ker}(D_uG_0) = \mathrm{span}[\phi_1, \ldots, \phi_n], \qquad \mathrm{Ker}(D_uG_0^*) = \mathrm{span}[\phi_1^*, \ldots, \phi_n^*] \tag{4.3}$$

have been calculated either analytically or numerically, see e.g. Allgower/Böhmer [4], Böhmer/Mei [43], Kunkel [193] and Mei [211]. Multiple bifurcations ($n > 1$) occur generically in problems with symmetries and with multi-parameters. If the multiple bifurcation is induced by symmetries, the basis vectors ϕ_i, $i = 1, \ldots, n$ are generically on the same group orbit. Same conclusion holds for ϕ_i^*, $i = 1, \ldots, n$. They can be normalized such that

$$\langle \phi_i,\ \phi_j \rangle = \langle \phi_i^*,\ \phi_j \rangle = \delta_{ij}, \quad i, j = 1, \ldots, n.$$

Here $\langle \cdot,\ \cdot \rangle$ is the inner product in the space Z. Alternatively we can also use the dual product in $Y \times Y^*$ due to $X \hookrightarrow Y$.

Due to the Fredholm properties of D_uG_0 and the semi-simplicity of the eigenvalue zero (cf. Hirzebruch/Scharlau [159]), we have

$$X = \mathrm{Ker}(D_uG_0) \oplus \mathrm{Im}(D_uG_0^*), \qquad Y = \mathrm{Ker}(D_uG_0^*) \oplus \mathrm{Im}(D_uG_0). \tag{4.4}$$

If $D_\lambda G(u_0, \lambda_0) \neq 0$, there is a unique element $w_1 \neq 0$ and $w_1 \in \mathrm{Im}(D_uG_0^*)$, such that

$$D_uG_0w_1 + D_\lambda G_0 = 0. \tag{4.5}$$

Let M be the complementary of $\mathrm{span}[w_1]$ in the subspace $\mathrm{Im}(D_uG_0^*)$, i.e.,

$$\mathrm{Im}(D_uG_0^*) = \mathrm{span}[w_1] \oplus M.$$

We write elements $(u, \lambda) \in X \times \mathbf{R}$ as

$$\begin{aligned} u &= u_0 + v + \mu w_1 + w, \qquad & v \in \mathrm{Ker}(D_u G_0),\ \mu \in \mathbf{R},\ w \in M, \\ \lambda &= \lambda_0 + \beta, & \beta \in \mathbf{R}. \end{aligned}$$

If $\{(u(s), \lambda(s)) \in X \times \mathbf{R};\ s \in \mathbf{R}\}$ is a solution curve passing through (u_0, λ_0), by Puiseux's theorem (cf. Fischer [106, pp.104]) for one-parameter problem we can scale it such that

$$\text{a)}\quad (u(0),\ \lambda(0)) = (u_0,\ \lambda_0) \qquad \text{and} \qquad \text{b)}\quad (\dot{u}(0),\ \dot{\lambda}(0)) \neq (0,0) \tag{4.6}$$

in a neighborhood $\mathcal{I}$ of $0 \in \mathbf{R}$, see also Jepson/Spence [172]. Thus $(u(s), \lambda(s))$ is of the form

$$\begin{aligned} u(s) &= u_0 + \sum_{i=1}^{n} s\alpha_i(s)\phi_i + s\mu(s)w_1 + sw(s) \\ \lambda(s) &= \lambda_0 + s\beta(s), \qquad s \in \mathcal{I} \subset \mathbf{R}. \end{aligned}$$

where $\alpha_i(s),\ i = 1, \ldots, n,\ \mu(s),\ \beta(s) \in \mathbf{R}$ are real functions and $w(s) \in M$ for $s \in \mathcal{I}$.

Differentiating the equation

$$G(u(s), \lambda(s)) = 0 \tag{4.7}$$

with respect to s at $s = 0$ yields

$$\begin{aligned} & D_u G_0 \Big(\sum_{i=1}^{n} \alpha_i(0)\phi_i + \mu(0)w_1 + w(0) \Big) + D_\lambda G_0 \beta(0) \\ &= D_u G_0 \Big[\sum_{i=1}^{n} \alpha_i(0)\phi_i + (\mu(0) - \beta(0))w_1 + w(0) \Big] \\ &= 0. \end{aligned}$$

Therefore,

$$\sum_{i=1}^{n} \alpha_i(0)\phi_i + (\mu(0) - \beta(0))w_1 + w(0) \in \mathrm{Ker}(D_u G_0).$$

Since

$$(\mu(0) - \beta(0))w_1 + w(0) \in \mathrm{Im}(D_u G_0^*),$$

we have

$$(\mu(0) - \beta(0))w_1 + w(0) \in \mathrm{Ker}(D_u G_0) \cap \mathrm{Im}(D_u G_0^*) = \{0\}.$$

Recall the statements $\mathrm{span}[w_1] \cap M = \{0\}$ and $w_1 \neq 0$. We obtain

$$\mu(0) = \beta(0), \qquad w(0) = 0.$$

Thus $\mu(s)$ can be decomposed as

$$\mu(s) = \beta(s) + s\nu(s), \ s \in \mathcal{I}.$$

As a conclusion the solution branch $(u(s), \lambda(s))$ is necessarily of the form

$$\begin{aligned} u(s) &= u_0 + \sum_{i=1}^{n} s\alpha_i(s)\phi_i + s\beta(s)w_1 + s^2 w(s), \quad w(s) \in \mathrm{Im}(D_u G_0^*), \\ \lambda(s) &= \lambda_0 + s\beta(s), \quad s \in \mathcal{I} \subset \mathbf{R}. \end{aligned} \tag{4.8}$$

Here $w(\cdot)$ maps $\mathcal{I} \subset \mathbf{R}$ into the subspace $\mathrm{Im}(D_u G_0^*)$. We derive from the statement (4.6b)

$$\sum_{i=1}^{n} |\alpha_i(0)| + |\beta(0)| > 0. \tag{4.9}$$

Example 4.1.1. Let us consider solution branches at a simple bifurcation point (u_0, λ_0). Since the kernel $\mathrm{Ker}(D_u G_0)$ is one-dimensional for simple bifurcations, by (4.8) a bifurcating solution branch is of the form

$$\begin{aligned} u(s) &= u_0 + s\alpha(s)\phi_1 + s\beta(s)w_1 + s^2 w(s), \quad w(s) \in \mathrm{Im}(D_u G_0^*), \\ \lambda(s) &= \lambda_0 + s\beta(s), \quad s \in \mathcal{I} \subset \mathbf{R}. \end{aligned} \tag{4.10}$$

Furthermore,

$$|\alpha(0)| + |\beta(0)| > 0. \tag{4.11}$$

Evidently, if $\beta(0) \neq 0$ (or $\alpha(0) \neq 0$), the solution branch (4.10) can be parameterized directly by λ (or the coefficient α of ϕ_1 in (4.10)). To distinguish these two cases, we utilize the structure (4.10) and differentiate $G(u(s), \lambda(s)) = 0$ twice with respect to s at $s = 0$

$$\begin{aligned} &D_{uu}G_0\phi_1\phi_1\alpha(0)^2 + 2D_u\ DG_0(w_1, 1)\phi_1\alpha(0)\beta(0) \\ &\quad + \ D^2G_0(w_1, 1)^2\beta(0)^2 + D_u G_0\ddot{u}(0) + D_\lambda G_0\ddot{\lambda}(0) = 0. \end{aligned}$$

Taking the inner product with ϕ_1^* at the both sides of this equation yields the following equation for the unknown $\beta(0)$

$$a\alpha(0)^2 + 2b\alpha(0)\beta(0) + c\beta(0)^2 = 0. \tag{4.12}$$

Here the constants $a, \ b, \ c$ are defined as

$$\begin{aligned} a &:= \langle \phi_1^*, \ D_{uu}G_0\phi_1\phi_1 \rangle, \\ b &:= \langle \phi_1^*, \ D_u DG(w_1, 1)\phi_1 \rangle, \\ c &:= \langle \phi_1^*, \ D^2G(w_1, 1)^2 \rangle \end{aligned}$$

see also (3.10) in Chapter 3 and (6.19) in Chapter 6. Under the nondegeneracy condition

$$b^2 - ac > 0$$

for simple bifurcations, the equation (4.12) has two solutions $\beta_i(0),\ i = 1, 2$. Hence, $\beta(0)$ in a solution branch (4.10) coincides necessarily with one of $\beta_i(0)$, $i = 1, 2$.

If $a \neq 0$, the statements (4.9) and (4.12) imply $\beta_i(0) \neq 0$ for both $i = 1, 2$. Thereafter, the corresponding solution branch in (4.10) can be parameterized as

$$\begin{aligned} u(s) &= u_0 + s\alpha(s)\phi_1 + sw_1 + s^2 w(s) \\ \lambda(s) &= \lambda_0 + s. \end{aligned} \tag{4.13}$$

If $a = 0$, then (4.12) becomes

$$(2b\alpha(0) + c\beta(0))\beta(0) = 0.$$

Thanks to (4.9), we obtain one solution

$$\beta(0) = \begin{cases} -2b\alpha(0)/c \neq 0 & \text{for } \ c \neq 0; \\ \beta(0) \neq 0 \ \text{ and } \alpha(0) = 0 & \text{for } c = 0, \end{cases}$$

leading to the same parameterization as (4.13). The other solution $\beta(0) = 0,\ \alpha(0) \neq 0$ leads to the following parameterization of $(u(s),\ \lambda(s))$

$$\begin{aligned} u(s) &= u_0 + s\phi_1 + s^2\beta(s)w_1 + s^2 w(s) \\ \lambda(s) &= \lambda_0 + s^2\beta(s), \quad \beta(0) \neq 0. \end{aligned} \tag{4.14}$$

Concluding above, we have proved that

Theorem 4.1.1. *Let X, Y with $X \hookrightarrow Y$ be Banach spaces . We assume that X, Y can be embedded into a Hilbert space Z and $G:\ X \times \mathbf{R} \to Y$ is a C^k-continuous ($k \geq 2$) mapping. If (u_0, λ_0) is simple bifurcation point of $G(u, \lambda) = 0$ and D_uG_0 is a Fredholm operator of index 0, then a solution branch of $G(u, \lambda) = 0$ passing through (u_0, λ_0) with a nonvanishing speed is necessarily of the form (4.13) or (4.14).*

4.2 Behavior of the Linearized Operator

Let $(u(s), \lambda(s))$ be a solution curve and $(u(0), \lambda(0))$ be a simple bifurcation point. The operator $D_uG(u(s), \lambda(s))$ tends to be singular as s approaches zero. Assume that the eigenvalue zero of D_uG_0 is well separated from others in the spectrum of D_uG_0, i.e., the singularity of $D_uG(u, \lambda)$ in the neighborhood of (u_0, λ_0) is caused merely by the eigenvalue zero of D_uG_0. We study variations of this "critical" eigenvalue along the solution curve $(u(s), \lambda(s))$.

Lemma 4.2.1. *Let the conditions in Theorem 4.1.1 hold. If $(u(s), \lambda(s))$ is a C^k-continuous ($k \geq 2$) solution curve and $(u(0), \lambda(0)) = (u_0, \lambda_0)$ is a simple bifurcation point, then there exists a C^{k-1}-continuous eigenpair $(\sigma(s),\ \phi(s))$, such that*

$$D_uG(u(s), \lambda(s))\phi(s) = \sigma(s)\phi(s) \tag{4.15}$$

and $\sigma(0) = 0,\ \phi(0) = \phi_1$.

Proof: Consider the enlarged system

$$H(v,\sigma,s) := \begin{pmatrix} D_uG(u(s),\lambda(s))v - \sigma v \\ \langle \phi_1^*,\ v\rangle - 1 \end{pmatrix} = 0$$

for the unknowns v, σ. Here we treat s as a parameter. Obviously $H(\phi_1,0,0) = 0$. Since the operator

$$D_{(v,\sigma)}H(\phi_1,0,0) = \begin{pmatrix} D_uG_0 & -\phi_1 \\ \langle \phi_1^*,\cdot\rangle & 0 \end{pmatrix}$$

is nonsingular, applying the implicit function theorem to H at $(v,\sigma,s) = (\phi_1,0,0)$ we obtain the conclusions. ■

Since a solution curve $(u(s),\lambda(s))$ with $(u(0),\lambda(0)) = (u_0,\lambda_0)$ is of the form (4.8), differentiating (4.15) with respect to s at $s = 0$ yields

$$D_{uu}G_0\dot{u}(0)\phi(0) + D_{u\lambda}G_0\dot{\lambda}(0)\phi(0) + D_uG_0\dot{\phi}(0) = \dot{\sigma}(0)\phi(0).$$

By $(\dot{u}(0),\ \dot{\lambda}(0)) = (\alpha(0)\phi_1 + \beta(0)w_1,\ \beta(0))$ and (4.11), (4.12), we have

$$\begin{aligned}\dot{\sigma}(0) &= \langle \phi_1^*, D_{uu}G_0\phi_1\phi_1\rangle\alpha(0) + \langle \phi_1^*, D_{uu}G_0\phi_1 w_1 + D_{u\lambda}G_0\phi_1\rangle\beta(0) \\ &= a\alpha(0) + b\beta(0).\end{aligned}$$

Lemma 4.2.2. *Let the conditions in Theorem 4.1.1 be satisfied. If $(u(s),\lambda(s))$ is a C^k-continuous ($k \geq 2$) solution curve of $G(u,\lambda) = 0$, passing through the simple bifurcation point (u_0,λ_0) at $s = 0$ and $|\alpha(0)| + |\beta(0)| > 0$, then we have*

$$\dot{\sigma}(0) = \begin{cases} 0 & \text{for } \beta(0) = \dot{\lambda}(0) = 0; \\ \neq 0 & \text{otherwise.} \end{cases} \tag{4.16}$$

Proof: Recall the nondegeneracy condition $b^2 - ac > 0$ for simple bifurcation points. Since $\alpha(0)$, $\beta(0)$ satisfy the equation (4.12), if $\beta(0) = 0$, we derive from (4.12) $a\alpha(0)^2 = 0$. Hence, $\dot{\sigma}(0) = a\alpha(0) + b\beta(0) = 0$. Otherwise for $\beta(0) \neq 0$ and $a = 0$, the nondegeneracy condition implies $b \neq 0$ and $\dot{\sigma}(0) = b\beta(0) \neq 0$. Finally, let us consider the case $\beta(0) \neq 0$ and $a \neq 0$. If

$$\dot{\sigma}(0) = a\alpha(0) + b\beta(0) = 0,$$

we derive $\alpha(0) = -b\beta(0)/a$. Substituting it into (4.12) yields

$$-\frac{1}{a}\beta(0)^2(b^2 - ac) = 0.$$

Since $b^2 - ac > 0$, we have $\beta(0) = 0$, contradicting to the assumption. Therefore, $\dot{\sigma}(0)$ does not vanish. ■

The case $a = 0$ corresponds to a pitchfork bifurcation and is generic for problems with symmetries. If the eigenvalue zero is well separated from other points in the spectrum of D_uG_0, the following estimate holds for s sufficiently small

$$\|D_uG(u(s),\lambda(s))^{-1}\| = O\left(|\sigma(s)^{-1}|\right).$$

By the Taylor expansion

$$\sigma(s) = \dot{\sigma}(0)s + \frac{1}{2}\ddot{\sigma}(0)s^2 + O(s^3),$$

we see

$$\|D_uG(u(s),\lambda(s))^{-1}\| = O(|s|^{-r}), \tag{4.17}$$

where

$$r = \begin{cases} 1 & \text{for } \beta(0) \neq 0; \\ 2 & \text{otherwise.} \end{cases}$$

4.3 Euler-Newton Continuation across Simple Bifurcation Points

Assume that the simple bifurcation point (u_0, λ_0) is known and the null elements ϕ_1, ϕ_1^*, w_1 have been calculated. We consider the Euler-Newton method for a continuation of solution branches bifurcating at (u_0, λ_0).

To continue a solution curve we recall that the Euler-Newton method starts at a known solution $(u(s_0), \lambda(s_0))$ with a predictor

$$(u^0, \lambda^0) = (u(s_0), \lambda(s_0)) + (s - s_0)\tau,$$

where τ is tangent of the solution curve at $s = s_0$ and determined by the system

$$\begin{pmatrix} DG(u(s_0), \lambda(s_0))\tau \\ \langle \tau,\ \tau \rangle - 1 \end{pmatrix} = 0.$$

Thereafter, the iteration

$$A(u^k, \lambda^k)\left[\begin{pmatrix} u^{k+1} \\ \lambda^{k+1} \end{pmatrix} - \begin{pmatrix} u^k \\ \lambda^k \end{pmatrix}\right] = -F(u^k, \lambda^k, s) \tag{4.18}$$

generates a sequence $(u^k,\ \lambda^k)$, $k = 0, 1, 2, \ldots$, approaching a new point on the solution curve. Here we parameterize the solution curve via the system

$$F(u, \lambda, s) := \begin{pmatrix} G(u, \lambda) \\ N(u, \lambda, s) \end{pmatrix} = 0 \tag{4.19}$$

with a local pseudo-arclength s in

$$N : (u, \lambda, s) \in X \times \mathbf{R} \times \mathbf{R} \to \langle \tau,\ (u - u(s_0),\ \lambda - \lambda(s_0)) \rangle + s - s_0 \in \mathbf{R}, \tag{4.20}$$

see Chapter 2. The operator $A(u, \lambda)$ is an approximation of the derivative $D_{(u,\lambda)}F(u, \lambda, s)$. We obtain the Newton method for $A(u, \lambda) = D_{(u,\lambda)}F(u, \lambda, s)$, and the chord method for $A(u, \lambda) = D_{(u,\lambda)}F(u^0, \lambda^0, s_0)$.

Remark 4.3.1. The assumption that both X, Y can be embedded into a Hilbert space allows the use of tangent τ in the definition of the mapping N. Extension of the discussion in this section to general Banach spaces is straightforward by choosing appropriate elements τ^* in the dual space $X^* \times \mathbf{R}$, such that $\|\tau^*\| = 1$ and $\langle \tau^*, \tau \rangle = 1$, see e.g. Decker/Keller [82].

It is known that the problem (4.19) is nonsingular at regular solutions and turning points of $G(u, \lambda) = 0$. This ensures convergence of the correction (4.18) in neighborhoods of the solution $(u(s), \lambda(s))$ and allows an effective Euler-Newton continuation of the solution curve. However, at the bifurcation point (u_0, λ_0) the enlarged problem (4.19) is still singular. Thus this enlarged system does not remove singularities of the original problem at bifurcation points. This can be seen from the derivative

$$D_{(u,\lambda)}F(u_0, \lambda_0, 0) = \begin{pmatrix} DG_0 \\ \langle \tau, \cdot \rangle \end{pmatrix}$$

and the fact that the kernel of DG_0 is two dimensional. Elementary calculations show that (u_0, λ_0) is a simple bifurcation point of the system (4.19). Nevertheless, by exploiting the constraint $N(u, \lambda, s) = 0$ and the pseudo-arclength we show that the system (4.19) is appropriate for branch switching and continuation of solutions at simple bifurcation points.

Consider the so-called *algebraic bifurcation equations*

$$\begin{aligned} a\alpha^2 + 2b\alpha\beta + c\beta^2 &= 0, \\ \alpha^2 + \beta^2 - 1 &= 0, \end{aligned} \tag{4.21}$$

see e.g. Decker/Keller [82]. These equations can be also derived from the reduced bifurcation equations in the Liapunov-Schmidt reduction, see (6.7) and (6.18) in Chapter 6. Let (α_0, β_0) be an isolated solution of the equations (4.21). We choose

$$N(u, \lambda, s) := \langle \alpha_0 \phi_1 + \frac{\beta_0}{\sqrt{2}} w_1,\ u - u_0 \rangle + \frac{\beta_0}{\sqrt{2}}(\lambda - \lambda_0) + s. \tag{4.22}$$

Theorem 4.3.1. (Decker/Keller[82]) *Let $(u(s), \lambda(s))$ be a C^k-continuous ($k \geq 2$) solution curve of $G(u, \lambda) = 0$ and $(u(0), \lambda(0)) = (u_0, \lambda_0)$. If (α_0, β_0) is an isolated solution of (4.21) and the mapping F is defined by (4.19) and (4.22), there exist constants $\delta > 0$, $C_0 > 0$, such that the following estimate holds for $s \in (-\delta, \delta)$*

$$\|D_{(u,\lambda)}F(u(s), \lambda(s), s)^{-1}\| \leq C_0 |s|^{-r}, \qquad r = \begin{cases} 1 & \textit{for } \beta(0) \neq 0; \\ 1/2 & \textit{otherwise.} \end{cases}$$

We see that the singularity is removed, but decreased, comparing with (4.17). Let

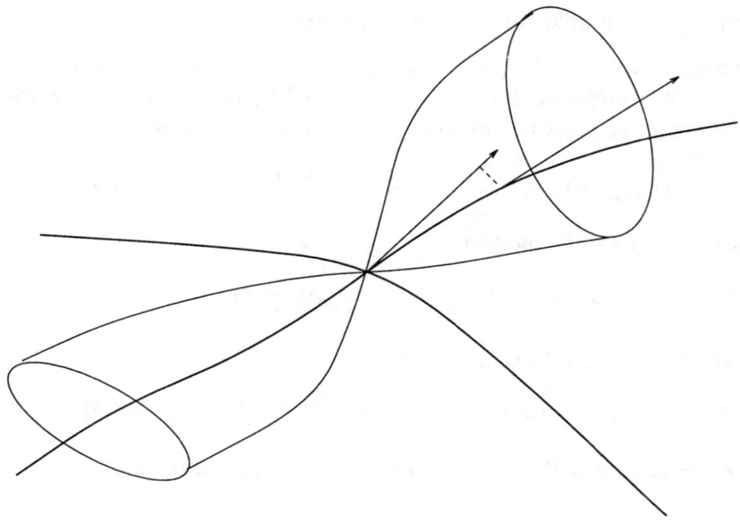

Fig. 4.1. Conical convergence domain of the Newton (chord) method.

$$\kappa(t) := \max_{|s|<t} \|(\ddot{u}(s),\ \ddot{\lambda}(s))\|$$

and $L(s) > 0$ be the Lipschitz constant such that

$$\|D_{(u,\lambda)}F(u,\lambda,s) - D_{(u,\lambda)}F(u(s),\lambda(s),s)\| \leq L(s)\|(u,\lambda) - (u(s),\lambda(s))\|$$

holds for all (u,λ) in the cone

$$\|(u,\lambda) - (u(s),\lambda(s))\| \leq \frac{1}{2}s^2\kappa(s).$$

By slightly modifying the discussions in Decker/Keller [82] we obtain convergence of the chord method in a conical domain.

Theorem 4.3.2. (Decker/Keller [82]) *Let (α_0, β_0) be an isolated solution of (4.21) and the mapping $N(u,\lambda,s)$ be defined by (4.22). If the constant $\delta > 0$ is chosen such that for all $|s| \leq \delta$,*

$$\kappa(s)L(s)C_0|s|^r < \frac{2}{3}, \qquad r := \begin{cases} 1 & \text{for } \beta_0 \neq 0; \\ 3/2 & \text{for } \beta_0 = 0, \end{cases}$$

then for each $s \in [-\delta, \delta]$ and the initial point

$$(u^0(s), \lambda^0(s)) := (u_0, \lambda_0) + s\left(\alpha_0\phi_1 + \frac{\beta_0}{\sqrt{2}}w_1, \frac{\beta_0}{\sqrt{2}}\right) \tag{4.23}$$

the chord method (4.18) with $A(u,\lambda) = D_{(u,\lambda)}F(u^0(s_0), \lambda^0(s_0))$ converges linearly to $(u(s), \lambda(s))$.

Similar results hold for Newton method.

Theorem 4.3.3. (Decker/Keller [82]) *Let (α_0, β_0) be an isolated solution of (4.21) and the mapping $N(u, \lambda, s)$ be defined by (4.22). For the initial point (4.23) there exist positive constants δ, $C_1 > 0$, such that*

$$\|D_{(u,\lambda)}F(u^0(s), \lambda^0(s), s)^{-1}F(u^0(s), \lambda^0(s), s)\| \le C_1 s^2.$$

Furthermore, if the inequality

$$\|D_{(u,\lambda)}F(u_1, \lambda_1, s) - D_{(u,\lambda)}F(u_2, \lambda_2, s)\| \le L(s)\|(u_1, \lambda_1) - (u_2, \lambda_2)\|$$

holds for all (u_1, λ_1), (u_2, λ_2) in the cone

$$\|(u, \lambda) - (u^0(s), \lambda^0(s))\| \le 2C_1 s^2 \quad \textit{for} \quad s \in [-\delta, \delta],$$

and the constant $\delta > 0$ is chosen sufficiently small such that

$$\kappa(s)L(s)C_0C_1|s|^r < \frac{2}{3}, \qquad r := \begin{cases} 1/2 & \textit{for } \beta_0 \ne 0 \\ 1 & \textit{for } \beta_0 = 0, \end{cases}$$

then the Newton iteration (4.18) with $A(u, \lambda) = D_{(u,\lambda)}F(u, \lambda, s)$ converges quadratically to $(u(s), \lambda(s))$ for all $s \in [-\delta,\ \delta]$.

In conclusion branch switching at a simple bifurcation point is done in three steps. Firstly we determine tangents of the bifurcating solution branches by choosing the corresponding isolated solutions of the algebraic bifurcation equations (4.21); then we construct a pseudo-arclength parameterization and a predictor with a chosen tangent and an appropriate stepsize; thereafter we continue the solution curve with the Euler-Newton (chord) method. We summarize these three aspects with the following algorithm.

Algorithm 4.3.1. Euler-Newton (chord) method at simple bifurcation points

Input: The bifurcation point (u_0, λ_0) and the null vectors ϕ_1, ϕ_1^*, w_1;
Step 1) Define the algebraic bifurcation equations

$$\begin{aligned} a\alpha^2 + 2b\alpha\beta + c\beta^2 &= 0, \\ \alpha^2 + \beta^2 - 1 &= 0, \end{aligned}$$

and determine the tangent of a bifurcating solution branch via its isolated solution $(\alpha_0,\ \beta_0)$ and

$$\left(\alpha_0\phi_1 + \frac{\beta_0}{\sqrt{2}}w_1,\ \frac{\beta_0}{\sqrt{2}}\right)$$

Step 2) Define a pseudo-arclength parameterization of the bifurcating solution branch by the system

$$F(u, \lambda, s) := \begin{pmatrix} G(u, \lambda) \\ N(u, \lambda, s) \end{pmatrix} = 0$$

with

$$N(u, \lambda, s) := \langle \alpha_0 \phi_1 + \frac{\beta_0}{\sqrt{2}} w_1, \; u - u_0 \rangle + \frac{\beta_0}{\sqrt{2}} (\lambda - \lambda_0) + s;$$

Determine the predictor by

$$(u^0(s), \lambda^0(s)) := (u_0, \lambda_0) + s \left(\alpha_0 \phi_1 + \frac{\beta_0}{\sqrt{2}} w_1, \; \frac{\beta_0}{\sqrt{2}} \right)$$

with an appropriate stepsize s;

Step 3) Continue the solution branch of $F(u, \lambda, s) = 0$ with Euler-Newton (chord) method.

The conical convergence region with vertex at the bifurcation point for the Newton (chord) corrector allows continuation step to jump over bifurcations, see e.g. Fig. 4.2. Hence, test functions, as we have discussed in Chapter 3, are necessary for detecting bifurcations along the continuation. A decrease of convergence rate of correctors is also an indication of singularity of the linearized operator.

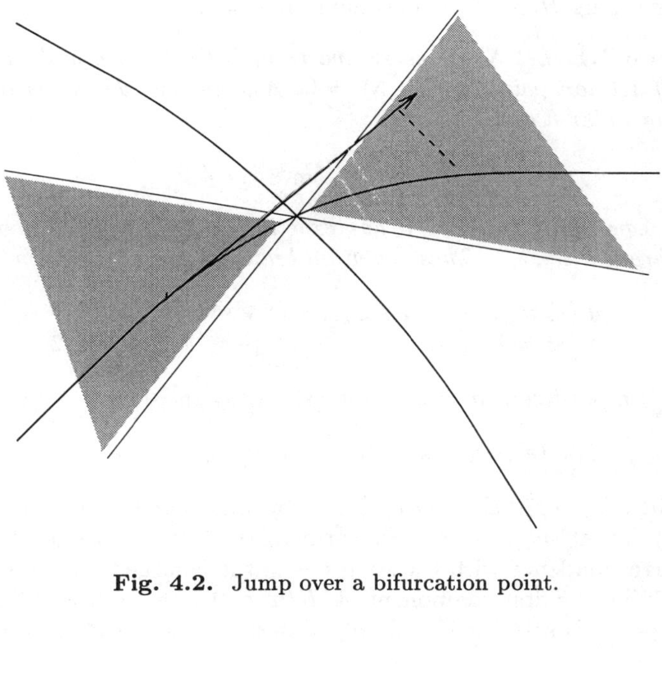

Fig. 4.2. Jump over a bifurcation point.

4.4 Branch Switching via Regularized Systems

We consider in this section a regularization of the problem $G(u, \lambda) = 0$ at the bifurcation point (u_0, λ_0) with slightly enlarged systems. Depending on whether the equality $a = \langle \phi_1^*, D_{uu}G_0\phi_1\phi_1\rangle = 0$ holds, we choose the corresponding enlarged systems to trace $\alpha(s)$, $w(s)$ and $\beta(s), w(s)$ in the solution branches (4.13) and (4.14) separately.

First of all, for $a \neq 0$ solution branches passing through (u_0, λ_0) are of the form (4.13). To determine these solution branches, we define an enlarged system by

$$H(w, \alpha, s) := \begin{pmatrix} G(u_0 + s\alpha\phi_1 + sw_1 + s^2 w, \lambda_0 + s)/s^2 \\ \langle \phi_1, w\rangle \end{pmatrix} = 0. \tag{4.24}$$

Obviously, H maps $X \times \mathbf{R}^2$ into $Y \times \mathbf{R}$. At $s = 0$ the mapping H is defined by its limit. If $G(u, \lambda)$ is C^k-continuous ($k \geq 3$), its Taylor expansion at (u_0, λ_0) is

$$G(u_0+s\alpha\phi_1+sw_1+s^2w, \lambda_0+s) = s^2\left[D_uG_0w+\frac{1}{2}D^2G_0(\alpha\phi_1+w_1,\ 1)^2\right]+o(s^2).$$

Hence

$$H(w, \alpha, 0) := \begin{pmatrix} D_uG_0w + \frac{1}{2}D^2G_0(\alpha\phi_1 + w_1,\ 1)^2 \\ \langle \phi_1, w\rangle \end{pmatrix} = 0 \tag{4.25}$$

and the mapping H is C^{k-2}-continuous at $s = 0$.

Theorem 4.4.1. *Let X, Y be subspaces of Hilbert space and (u_0, λ_0) be a simple bifurcation point of $G(u, \lambda) = 0$. Assume that D_uG_0 is a Fredholm operator of index 0 and*

$$a = \langle \phi_1^*, D_{uu}G_0\phi_1\phi_1\rangle \neq 0.$$

Then the equation $G(u, \lambda) = 0$ has exactly two different solution branches passing through (u_0, λ_0). These solution branches are of the form

$$\begin{aligned} u_i(s) &= u_0 + s(\alpha_i(s)\phi_1 + w_1) + s^2 w_i(s), \\ \lambda_i(s) &= \lambda_0 + s, \qquad\qquad s \in [-s_0, s_0],\quad i = 1, 2 \end{aligned}$$

where s_0 is a positive constant and $w_i(s)$, $\alpha_i(s)$ satisfy

$$H(w_i(s), \alpha_i(s), s) = 0, \qquad s \in [-s_0, s_0],\quad i = 1, 2.$$

Proof: Since $a = \langle \phi_1^*, D_{uu}G_0\phi_1\phi_1\rangle \neq 0$, by Theorem 4.1.1 solution branches of $G(u, \lambda) = 0$ at (u_0, λ_0) are necessarily of the form (4.13) and have a one-to-one correspondence with the solutions of the enlarged system (4.24).

Examining the first component of $H(w, \alpha, 0)$, we see from (4.4) that the equation (4.25) is solvable if and only if there exists $\alpha \in \mathbf{R}$, such that

$$D^2G_0\ (\alpha\phi_1 + w_1, 1)^2 \in \mathrm{Im}(D_uG_0).$$

This gives an equation for α

$$\langle \phi_1^*,\ D^2G_0(\alpha\phi_1 + w_1, 1)^2\rangle = a\alpha^2 + 2b\alpha + c = 0, \tag{4.26}$$

see also (4.12). Under the nondegeneracy condition $b^2 - ac > 0$ this quadratic equation has exactly two different solutions $\alpha_1^0,\ \alpha_2^0$. Correspondingly, there are two elements $w_1^0,\ w_2^0 \in \mathrm{Im}(D_uG_0^*)$, such that

$$D_uG_0w_i^0 + \frac{1}{2}D^2G_0(\alpha_i^0\phi_1 + w_1, 1)^2 = 0, \quad i = 1, 2.$$

In other words,

$$H(w_i^0, \alpha_i^0, 0) = 0.$$

On the other hand, direct calculations show

$$D_{(w,\alpha)}\ H(w_i^0, \alpha_i^0, 0) = \begin{pmatrix} D_uG_0 & D_u\ DG_0(\alpha_i^0\phi_1 + w_1, 1)\phi_1 \\ \langle \phi_1,\ \cdot\rangle & 0 \end{pmatrix}.$$

We claim that this linear operator is nonsingular from $X \times \mathbf{R}$ onto $Y \times \mathbf{R}$. To show this, we note first that $D_{(w,\alpha)}\ H(w_i^0, \alpha_i^0, 0)$ is bounded. Then, for any element $(f, g) \in Y \times \mathbf{R}$, we consider the equation

$$D_{(w,\alpha)}\ H(w_i^0, \alpha_i^0, 0)\begin{pmatrix} w \\ \alpha \end{pmatrix} = \begin{pmatrix} f \\ g \end{pmatrix},$$

namely,

$$D_uG_0w + D_uDG_0(\alpha_i^0\phi_1 + w_1, 1)\phi_1\alpha = f, \tag{4.27a}$$
$$\langle \phi_1,\ w\rangle = g. \tag{4.27b}$$

Since

$$\langle \phi_1^*,\ D_uDG_0(\alpha_i^0\phi_1 + w_1,\ 1)\phi_1\rangle = \alpha_i^0a + b = \pm\sqrt{b^2 - ac} \neq 0,$$

the unknown α is solved uniquely from (4.27a) as

$$\alpha = \frac{\langle \phi_1^*,\ f\rangle}{\langle \phi_1^*,\ D_u\ DG_0(\alpha_i^0\phi_1 + w_1, 1)\phi_1\rangle}.$$

Thereafter, we see

$$f - D_uDG_0(\alpha_i^0\phi_1 + w_1, 1)\phi_1\alpha \in \mathrm{Im}(D_uG_0).$$

Now the equation (4.27a) is solvable for w and the general solution is of the form

$$w = \gamma\phi_1 + \tilde{w},$$

where $\tilde{w} \in \operatorname{Im}(D_u G_0^*)$ is uniquely determined and $\gamma \in \mathbf{R}$ is arbitrary. Substituting it into (4.27b) yields

$$\gamma = \alpha / \langle \phi_1, \ \phi_1 \rangle.$$

Thus the system (4.27) has unique solution for any given right hand side. In other words, the operator $D_{(w,\alpha)}H(w_i^0, \alpha_i^0, 0)$ is nonsingular for $i = 1, 2$. Applying the implicit function theorem to the mapping H at $(w_i^0, \alpha_i^0, 0)$, we obtain a constant $s_0 > 0$ and a unique mapping $(w_i(s), \ \alpha_i(s))$, $s \in [-s_0, \ s_0]$, such that

$$H(w_i(s), \alpha_i(s), s) = 0,$$
$$(w_i(0), \alpha_i(0)) = (w_i^0, \alpha_i^0), \quad i = 1, 2.$$

Since the enlarged system (4.24) has no other solution branches across the axis $s = 0$, in correspondence with $(w_i(s), \alpha_i(s), t)$, $i = 1, 2$, the problem (4.1) has exactly two solution branches at (u_0, λ_0) of the form (4.13). ■

If $\langle \phi_1^*, \ D_{uu} G_0 \phi_1 \phi_1 \rangle = 0$, Theorem 4.1.1 shows that solution branches at (u_0, λ_0) are of the form either (4.13) or (4.14). The solution branch (4.13) can be studied as above except that the quadratic equation for α in the proof of Theorem 4.4.1 becomes linear and leads to one solution branch.

To trace the branch (4.14), we consider the system

$$H(w, \beta, s) := \begin{pmatrix} G(u_0 + s\phi_1 + s\beta w_1 + s^2 w, \lambda_0 + s\beta)/s^2 \\ \langle \phi_1, \ w \rangle \end{pmatrix} = 0 \tag{4.28}$$

for the unknowns w, β. Here $H(w, \beta, 0)$ is defined by its limit. It follows from a similar discussion as above that this system has exactly one solution branch $(w(s), \beta(s), t)$. Correspondingly the original problem $G(u, \lambda) = 0$ has one solution curve of the parameterization (4.14).

Remark 4.4.1. Instead of (4.14) solution curves in the case $\beta(0) = 0$, $\alpha(0) \neq 0$ with $a = \langle \phi_1^*, \ D_{uu} G_0 \phi_1 \phi_1 \rangle = 0$ can be also described by

$$\begin{aligned} u(s) &= u_0 + s\alpha(s)\phi_1 + s^2\beta(s)w_1 + s^2 w_2 + s^3 w(s) \\ \lambda(s) &= \lambda_0 + s^2 \beta(s) \end{aligned} \tag{4.29}$$

where $w_2 \in \operatorname{Im}(D_u G_0^*)$ is the solution of the equation

$$D_u G_0 w + \frac{1}{2} D_{uu} G_0 \phi_1 \phi_1 = 0.$$

If we scale $\alpha(s) \equiv 1$ or $\beta(s) \equiv 1$ in according to the third derivative of $G(u(s), \lambda(s)) = 0$ with respect to s at $s = 0$, and modify the enlarged system (4.24) by replacing the factor s^{-2} with s^{-3} and the arguments u, λ in G with

$$u = u_0 + s\alpha\phi_1 + s^2\beta w_1 + s^2 w_2 + s^3 w(s),$$
$$\lambda = \lambda_0 + s^2 \beta,$$

the existence of the solution branches (4.29) can be verified similarly as in Theorem 4.4.1.

As we have seen before, different solutions of (4.25) lead to different solutions of (4.24), which in turn gives different bifurcating solution branches of the original problem. We start with a solution of (4.24) at $s = 0$ and trace the corresponding solution curve of this nonsingular problem with continuation methods. The following algorithm concludes the discussion on branch switching at simple bifurcation points via regularized systems.

Algorithm 4.4.1. (Mei [213]) **Branch switching at a simple bifurcation point** ($a \neq 0$)

Step 1) Calculate the bifurcation point (u_0, λ_0) *and the null vectors* ϕ_1, ϕ_1^*, w_1*;*

Step 2) Find solutions of (4.26), i.e., $a\alpha^2 + 2b\alpha + c = 0$*;*

Step 3) For each solution α^0 *of (4.26), solve* w^0 *from the nonsingular system*

$$\begin{pmatrix} D_u G_0 & \phi_1 \\ \langle \phi_1^*, \cdot \rangle & 0 \end{pmatrix} \begin{pmatrix} w \\ c_1 \end{pmatrix} = \begin{pmatrix} -\frac{1}{2} D^2 G_0 (\alpha^0 \phi_1 + w_1, 1)^2 \\ 0 \end{pmatrix}.$$

Step 4) Continue the solution $(w^0, \alpha^0, 0)$ *of (4.24) with Euler-Newton method.*

FOR $k = 1, 2, \ldots,$ *DO*

1. *Choose* h *as the current stepsize. We take* $s^k = s^{k-1} + h$ *and the following predictor for* $(w(s^k), \alpha(s^k))$

$$(\tilde{w}^0,\ \tilde{\alpha}^0) := (w^{k-1}, \alpha^{k-1}) + h\tau.$$

Here τ *is the tangent induced by* $D_{(w,\alpha)} F(w^0, \alpha^0, 0)$*;*

2. *Use Newton method to correct the approximation* $(\tilde{w}^0,\ \tilde{\alpha}^0)$*, i.e., FOR* $l = 0, 1, \ldots,$ *DO*

$$D_{(w,\alpha)} H(\tilde{w}^l, \tilde{\alpha}^l, s^k) \left((\tilde{w}^{l+1}, \tilde{\alpha}^{l+1}) - (\tilde{w}^l, \tilde{\alpha}^l) \right) = -H(\tilde{w}^l, \tilde{\alpha}^l, s^k),.$$

If $l = N_k$ *— a fixed number of corrections at* s^k *or the tolerance of error estimates is arrived, then*

$$(w^k, \alpha^k) := (\tilde{w}^l, \tilde{\alpha}^l);$$

Step 5) Set up approximate solution of $G(u, \lambda) = 0$ *via*

$$\begin{aligned} u^k &= u_0 + s^k \alpha^k \phi_1 + s^k w_1 + (s^k)^2 w^k, \\ \lambda^k &= \lambda_0 + s^k. \end{aligned}$$

Remark 4.4.2. Once continuation of a solution branch is outside of neighborhoods of the bifurcation point, i.e., $|s|$ is sufficiently large, we switch back to the original problem.

4.5 Other Branch Switching Techniques

We have restricted the discussion to the case where the bifurcation point (u_0, λ_0) is known and the null vectors ϕ_1, ϕ_1^*, w_1 are available. These values are also essential for analysis of bifurcation scenarios. There are situations where the primary interest of continuation is to explore where the solution goes and to approach efficiently special points on the solution curve. To this end, one might be mainly interested in how to jump from the current solution curve to the others if a bifurcation is observed between two successive continuation steps. The main idea is to find a point on the new solution curve and to continue it further. The following three related approaches are frequently used for this purpose.

a) Construct a predictor for the new branch with knowledge of the current solution curve, so that Newton (chord) method corrects it towards a point on the new branch;

b) Construct appropriate correctors with selective properties, e.g., symmetries of the new branch, so that the corrections are directed to the new branch;

For these two approaches we refer to Seydel [276] for more details.

c) Choose an element p in Y, a small constant $\tau \in \mathbf{R}$, and trace solution curves of the perturbed problem

$$G(u, \lambda) = \tau p.$$

This perturbation unfolds bifurcations of the original problem such that continuation of solution curves goes around the bifurcation point. Solution branches of the original problem near a bifurcation point are approximated by choosing appropriate element $p \in Y$, e.g. via constraints or symmetry properties, see e.g. Allgower/Chien [13], Allgower/Georg [14] and Chien/Chen [59] for further details about this technique.

5. Bifurcation Problems with Symmetry

Symmetry and symmetry-breaking are features widely involved in nonlinear phenomena in nature and science. Reaction-diffusion equations have rich underlying symmetries, which origins, e.g., from the Euclidean symmetry of the Laplace operator. The continuous symmetry of a differential operator is often subjected to symmetries of domains, boundary conditions and reaction terms. We observe it normally in a discrete form. But its existence as underlying symmetries has profound impact on behavior of the system. Symmetries often induce multiple bifurcations and sophisticated scenarios in solution manifolds. For examples, symmetry-breaking, symmetry-increasing and twist of spatial and temporal symmetries are typical bifurcations in pattern formation of reaction-diffusion problems.

Recent developments in the bifurcation theory with symmetries have demonstrated that highly complicated bifurcation structure can be studied systematically with group theoretic concepts. Equivariant branching lemma and bifurcation subgroups are the fundamental tools for simplifying bifurcation analysis, see e.g. [11, 84, 129, 131, 295]. By restricting the nonlinear problem into proper fixed point subspace we can reduce a multiple bifurcation to simple one. Thereafter, we can analyze the bifurcation scenario in the subspace with the classical theory of Crandall/Rabinowitz [71, 72]. Furthermore, various numerical methods (cf. Allgower/Georg [14], Seydel [276], Mei [210, 213]) becomes applicable to these kinds of multiple bifurcations. Nevertheless, not all solutions can be determined in such a way, there are "submaximal" solutions which have less, even no symmetry at all (cf. Ashwin/Böhmer/Mei [21, 22], Fiedler [101], Lauterbach [200], Mei [218]). To obtain such solutions, Liapunov-Schmidt methods and center manifold reductions are effective approaches, which reduce the problem (1.1) at a bifurcation point to a lower dimensional system of algebraic, and ordinary differential equations, respectively.

Exploiting symmetries in numerical analysis is an efficient technique to reduce computational costs and to improve conditioning numbers in numerical solution of linear and nonlinear problems. In fact, group theoretic methods have been applied to develop efficient numerical schemes for both algebraic and differential equations, see e.g. [5, 10, 12, 46, 84, 95, 121, 125, 153, 154, 286, 287, 302] and [217, 220, 222, 224].

In this chapter we review some fundamental results of bifurcation theory with symmetry. We start with some basic group concepts and introduce the reduced problems in fixed point subspaces. Thereafter we discuss an extension of the equivariant branching lemma and its application to a semi-linear elliptic problem.

5.1 Basic Group Concepts

Let X, Y be Banach spaces and $GL(X,Y)$ be the space of nonsingular bounded linear mappings from X onto Y. We abbreviate $GL(X,X)$ to $GL(X)$ and $GL(\mathbf{R}^n,\mathbf{R}^n)$ to $GL(n)$. In the sequel we consider a *Lie group* as a subgroup Γ of $GL(n)$ which is closed in the sense as a subset of $\mathbf{R}^{n^2}$. A Lie group is said to be compact or connected if it is compact or connected as a subset of $\mathbf{R}^{n^2}$. Two subgroups Σ, $\tilde{\Sigma}$ of Γ are said to be *conjugate* if there is an element $\gamma \in \Gamma$ such that

$$\Sigma = \gamma\tilde{\Sigma}\gamma^{-1} = \{\gamma\sigma\gamma^{-1} \mid \sigma \in \tilde{\Sigma}\}.$$

The *conjugacy class* of a subgroup Σ consists of all subgroups of Γ that are conjugate to Σ.

Example 5.1.1. 1. The n-dimensional *orthogonal group* $O(n)$ consists of all $n \times n$ orthogonal matrices.
2. The *special orthogonal group* $SO(n)$ consists of all $A \in O(n)$ with $\det(A) = 1$. It is also called n-dimensional rotation group, e.g. $SO(2)$ consists of the rotations

$$R_\theta = \begin{pmatrix} \cos\theta & -\sin\theta \\ \sin\theta & \cos\theta \end{pmatrix}, \quad \theta \in \mathbf{R}.$$

3. The discrete *cyclic group* is defined as

$$C_n := \{R_{2\pi/n}, R_{4\pi/n}, \ldots, R_{2(n-1)\pi/n}, I\} \quad \text{for some } n \in \mathbf{N},$$

see e.g. Fig. 5.1.
4. The *dihedral group* D_n is the symmetry group of the regular n-gon and is generated by C_n and the flip κ

$$\kappa = \begin{pmatrix} -1 & 0 \\ 0 & 1 \end{pmatrix} \Longleftrightarrow S_1,$$

see e.g. Fig. 5.2 and Fig. 5.3, where the center of the n-gon and the origin coincide.

Let Γ be a Lie group and X be a vector space. The (linear) *action* of Γ on X is a continuous mapping $\rho:\ (\gamma, v) \in \Gamma \times X \to \gamma \cdot v \in X$ such that

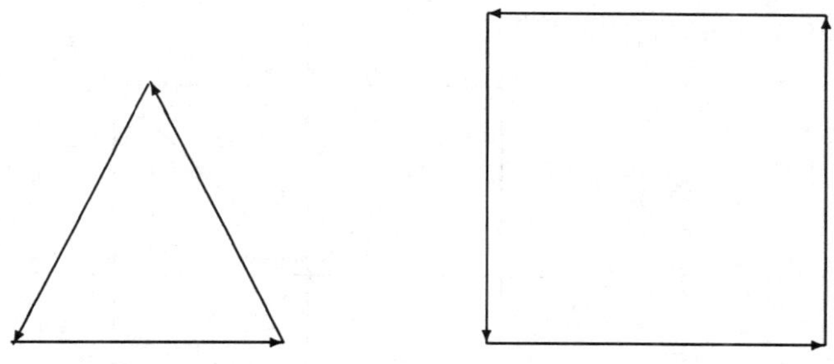

Fig. 5.1. The rotation groups C_3, C_4.

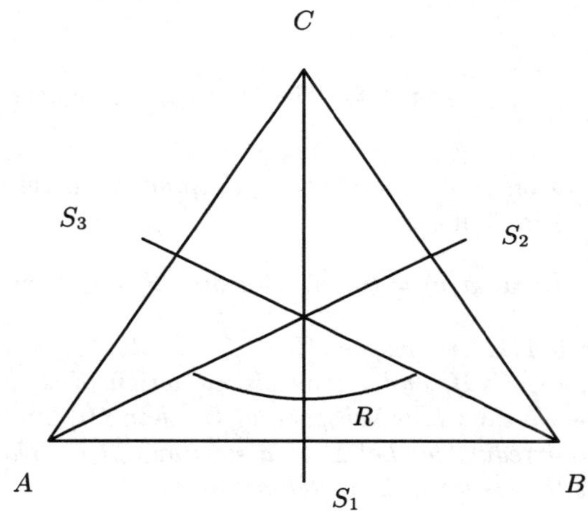

Fig. 5.2. The dihedral group D_3 with the reflections S_1, S_2, S_3 and the rotations $R_{2\pi/3}$, $R_{4\pi/3}$.

- For each fixed $\gamma \in \Gamma$, the mapping $\rho_\gamma : X \to X$ defined by

$$\rho_\gamma(v) := \gamma \cdot v$$

  is linear.
- If $\gamma_1, \gamma_2 \in \Gamma$, then $\rho_{\gamma_1} \cdot \rho_{\gamma_2} = \rho_{\gamma_1 \gamma_2}$.

The mapping $\rho : \gamma \in \Gamma \to \rho_\gamma \in GL(X)$ is called a *linear representation* of Γ in X. The representation ρ of Γ in X is said to be *isomorphic* to the representation $\tilde{\rho}$ of Γ in Y if X and Y are isomorphic and there is a linear isomorphism $A : X \to Y$ such that $\rho_\gamma = A^{-1}\tilde{\rho}_\gamma A$ for all $\gamma \in \Gamma$.

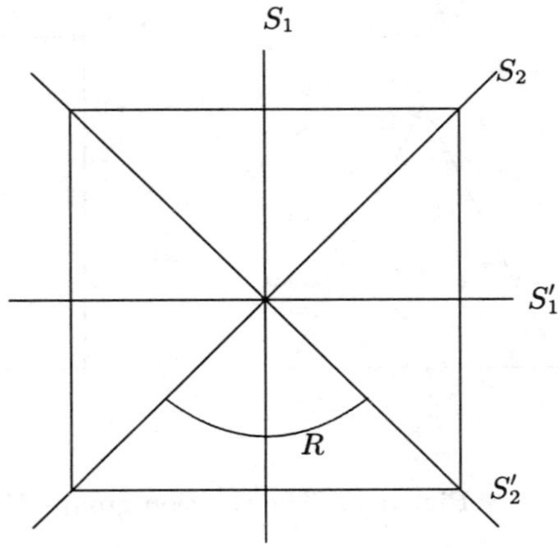

Fig. 5.3. The D_4-group of a square.

A representation ρ of Γ is called *orthogonal* with respect to a given inner product $\langle \cdot, \ \cdot \rangle$ in X if

$$\langle \rho_\gamma u, \ \rho_\gamma v \rangle = \langle u, \ v \rangle \quad \text{for all} \quad u, \ v \in X \text{ and } \gamma \in \Gamma.$$

Definition 5.1.1. *A subspace $W \subset X$ is called Γ-invariant if $\gamma w \in W$ for all $w \in W$, $\gamma \in \Gamma$. If, under the induced action of Γ in X, the subspace W has no other Γ-invariant subspace of W than $\{0\}$ and W itself, then W is said to be Γ-irreducible. Let Σ be a subgroup of Γ. The fixed point subspace X^Σ of X with respect to Σ is defined as*

$$X^\Sigma := \{u \in X \quad | \quad \sigma u = u \quad \text{for all} \quad \sigma \in \Sigma\}.$$

The isotropy group Σ_u of an element $u \in X$ is a subgroup of Γ,

$$\Sigma_u := \{\gamma \in \Gamma \quad | \quad \gamma u = u\}.$$

We see that the fixed point subspace X^Σ collects all the elements in X, which has at least the symmetry Σ. On the other hand, the isotropy group Σ_u describes all the symmetries of the element u.

The *group orbit* of an element $u \in X$ under the action of Γ is the set

$$\Gamma(u) := \{\gamma u \in X \quad | \quad \gamma \in \Gamma\}. \tag{5.1}$$

Lemma 5.1.1. *For any given element $u \in X$, points on the orbit $\Gamma(u)$ have conjugate isotropy subgroups. More precisely,*

$$\Sigma_{\gamma u} = \gamma \Sigma_u \gamma^{-1}.$$

Proof: Let $\gamma \in \Gamma$ and $\sigma \in \Sigma_u$. We have

$$\gamma\sigma\gamma^{-1}(\gamma u) = \gamma\sigma u = \gamma u.$$

Hence $\gamma\sigma\gamma^{-1} \in \Sigma_{\gamma u}$ for all $\sigma \in \Sigma_u$, i.e.,

$$\gamma\Sigma_u\gamma^{-1} \subset \Sigma_{\gamma u}.$$

Conversely, for any $\sigma \in \Sigma_{\gamma u}$ we have

$$\sigma\gamma u = \gamma u \Longrightarrow \gamma^{-1}\sigma\gamma u = u.$$

Therefore,

$$\tilde{\sigma} := \gamma^{-1}\sigma\gamma \in \Sigma_u.$$

This implies

$$\sigma = \gamma\tilde{\sigma}\gamma^{-1} \in \gamma\Sigma_u\gamma^{-1}$$

and

$$\Sigma_{\gamma u} \subset \gamma\Sigma_u\gamma^{-1},$$

which completes the proof. ∎

Definition 5.1.2. *Let $f : \Gamma \to \mathbf{R}$ be a continuous real-valued function. The Haar integral $\int_\Gamma f\, d\gamma \in \mathbf{R}$ of the function f on the group Γ is defined by the following three conditions*

Linearity: $\int_\Gamma (\lambda f + \mu g)\, d\gamma = \lambda \int_\Gamma f\, d\gamma + \mu \int_\Gamma g\, d\gamma$ *where $f,\ g :\ \Gamma \to \mathbf{R}$ are continuous and $\lambda,\ \mu \in \mathbf{R}$.*

Positivity: *If $f(\gamma) \geq 0$ for all $\gamma \in \Gamma$, then* $\int_\Gamma f\, d\gamma \geq 0$.

Translation-Invariance. $\int_{\gamma\in\Gamma} f(\delta\gamma)\, d\gamma = \int_{\gamma\in\Gamma} f(\gamma)\, d\gamma$ *for any fixed element $\delta \in \Gamma$.*

For a compact group the Haar integral is invariant under right translation, i.e.,

$$\int_{\gamma\in\Gamma} f(\gamma\delta)\, d\gamma = \int_{\gamma\in\Gamma} f(\gamma)\, d\gamma \quad \text{for all } \delta \in \Gamma.$$

Furthermore, the integral $\int_\Gamma 1 d\gamma$ is finite. We scale the Haar integral such that

$$\int_\Gamma 1 d\gamma = 1$$

which corresponds to the so-called *normalized Haar integral.* The existence and uniqueness of the Haar integral is shown in Hochschild [160, pp. 9]. A vector-valued mapping $f :\ \Gamma \to \mathbf{R}^n$ is integrated by performing the integration separately on each component.

Lemma 5.1.2. *Let Γ be a compact Lie group acting on a Hilbert space V and let ρ_γ be the linear mapping associated with $\gamma \in \Gamma$. There exists an inner product on V such that for all $\gamma \in \Gamma$, the representation ρ_γ is orthogonal.*

Proof: Let $\langle \cdot, \cdot \rangle$ be any inner product on V. We define

$$\langle v, w \rangle_\Gamma := \int_\Gamma \langle \rho_\gamma v, \rho_\gamma w \rangle \, d\gamma. \tag{5.2}$$

This is also an inner product by linearity of the integral on Γ. By translation invariance of the Haar integral we have

$$\langle \rho_\delta v, \rho_\delta w \rangle_\Gamma = \langle v, w \rangle_\Gamma \tag{5.3}$$

for all $v, w \in V, \delta \in \Gamma$. ■

Example 5.1.2.

(a) If Γ is a finite Lie Group of the order $\#(\Gamma)$, then the normalized Haar integral on Γ is

$$\int_\Gamma f \, d\gamma \equiv \frac{1}{\#(\Gamma)} \sum_{\gamma \in \Gamma} f(\gamma).$$

(b) Let $\Gamma = \mathbf{SO}(2)$ and $f: R_\theta \in \mathbf{SO}(2) \to f(R_\theta) \in \mathbf{R}$ be a continuous function. Define a continuous 2π-periodic function $\tilde{f}: \mathbf{R} \to \mathbf{R}$ via

$$\tilde{f}(\theta) := f(R_\theta) \quad \text{for all} \quad \theta \in \mathbf{R}.$$

The normalized Haar integral on $\mathbf{SO}(2)$ is

$$\int_\Gamma f \, d\gamma \equiv \frac{1}{2\pi} \int_0^{2\pi} \tilde{f}(\theta) \, d\theta.$$

5.2 Equivariant Bifurcation Problems

Consider the one-parameter problem

$$G(u, \lambda) = 0. \tag{5.4}$$

Suppose that Γ acts linearly on X, Y. A mapping $G: X \times \mathbf{R} \to Y$ is called *Γ-equivariant* if it satisfies

$$G(\gamma u, \lambda) = \gamma G(u, \lambda) \quad \text{for all} \quad (u, \lambda) \in X \times \mathbf{R}, \ \gamma \in \Gamma. \tag{5.5}$$

Obviously, if G is Γ-equivariant and (u_0, λ_0) is a solution of (5.4), the equality

$$G(\gamma u_0, \lambda_0) = \gamma G(u_0, \lambda_0) = 0$$

implies that every element on the *group orbit* of (u_0, λ_0)

$$\Gamma(u_0, \lambda_0) := \{(\gamma u_0, \lambda_0) \quad | \quad \gamma \in \Gamma\}$$

is also a solution.

One advantage of the group-theoretic approach for bifurcation problems is that we can embed the problem into the fixed point subspaces of isotropy subgroups. In many cases this will reduce the multiplicity of a bifurcation point. This feature is in certain sense beyond specific problems, since it is derived directly from symmetry and group theory.

Let G be Γ-equivariant and Σ be a subgroup of Γ. If $u \in X^{\Sigma}$, then we have

$$\sigma G(u, \lambda) = G(u, \lambda) = G(\sigma u, \lambda) \quad \text{for all } \sigma \in \Sigma.$$

Thus G maps $X^{\Sigma} \times \mathbf{R}$ into Y^{Σ}. Define

$$G^{\Sigma} := G\,|_{X^{\Sigma} \times \mathbf{R}} : \; X^{\Sigma} \times \mathbf{R} \mapsto Y^{\Sigma}.$$

We derive a reduced problem

$$\textit{Find } (u, \lambda) \in X^{\Sigma} \times \mathbf{R} \quad \textit{such that } G^{\Sigma}(u, \lambda) = 0. \tag{5.6}$$

Obviously, all solutions of (5.6) belong to the solution set of (5.4). Conversely, a solution of (5.4) with Σ-symmetry satisfies also (5.6). Instead of the whole space $X \times \mathbf{R}$ we use the reduced problem (5.6) in the subspace $X^{\Sigma} \times \mathbf{R}$ to determine Σ-symmetric solution branches of (5.4). In nondegenerate cases, the reduced problem has simple bifurcation behavior. Thus if we consider all isotropy subgroups of Γ and the reduced problem of (5.4) in the corresponding fixed point subspaces, we are often able to decompose the complicated scenario at a multiple bifurcation point to a cluster of simple bifurcations.

Let G_0, D_uG_0, $DG_0, \ldots$ represent the evaluations of G, D_uG, $DG, \ldots$ at (u_0, λ_0), respectively. If $v \in \operatorname{Im}(D_uG_0)$, i.e., there is $w \in X$ such that $v = D_uG_0w$, then the equality $D_uG_0\sigma w = \sigma v$ holds for all $\sigma \in \Sigma_{u_0}$. Namely, $\sigma v \in \operatorname{Im}(D_uG_0)$ for all $\sigma \in \Sigma_{u_0}$. Similarly, $v \in \operatorname{Ker}(D_uG_0)$ and $\sigma \in \Sigma_{u_0}$ implies $D_uG_0(\sigma v) = \sigma D_uG_0v = 0$. As a conclusion we see that the kernel $\operatorname{Ker}(D_uG_0)$ and the image $\operatorname{Im}(D_uG_0)$ are Σ_{u_0}-invariant.

Denote the kernel of the derivative DG_0 as

$$N_0 := \operatorname{Ker}(DG_0). \tag{5.7}$$

The characteristic of *a singular solution* (u_0, λ_0) is $\dim(N_0) > 1$.

Lemma 5.2.1. *For the Γ-equivariant operator G, let (u_0, λ_0) be a singular point and let $\Sigma^0 := \Sigma_{u_0}$. For all subgroups Σ of Σ^0 we have*

(i) $D_\lambda G_0^{\Sigma} = D_\lambda G_0 \in X^{\Sigma^0}$;
(ii) $N_0 = \operatorname{Ker}(DG_0)$ *is Σ^0-invariant;*

(iii) $\mathrm{Ker}(DG_0^\Sigma) = N_0^\Sigma = N_0 \cap (X^\Sigma \times \mathbf{R})$.

Proof: For all $(v,\beta) \in X^\Sigma \times \mathbf{R}$, the smoothness of G yields

$$\begin{aligned} D_uG_0v + D_\lambda G_0\beta &= \lim_{t\to 0}\frac{1}{t}[G(u_0+tv,\lambda_0+t\beta) - G(u_0,\lambda_0)] \\ &= \lim_{t\to 0}\frac{1}{t}[G^\Sigma(u_0+tv,\lambda_0+t\beta) - G^\Sigma(u_0,\lambda_0)] \\ &= D_uG_0^\Sigma v + D_\lambda G_0^\Sigma \beta. \end{aligned}$$

Hence, statement (iii) becomes obvious. From

$$DG_0\sigma(v,\beta) = \sigma DG_0(v,\beta) \quad \text{for all} \quad \sigma \in \Sigma^0,\ (v,\beta) \in X \times \mathbf{R}$$

with $(v,\beta) = (0,1)$, we obtain (i) and (ii). ■

A sufficient condition for existence of a solution curve C_{Σ^0} with the full symmetry Σ^0 of u_0 is as follows.

Lemma 5.2.2. *Let (u_0,λ_0) be a singular point for (5.4) and let Σ^0 be the isotropy group of u_0. If*

$$\dim[N_0 \cap (X^{\Sigma^0} \times \mathbf{R})] = 1, \tag{5.8}$$

then (5.4) has exactly one Σ^0-symmetric curve C_{Σ^0} passing through (u_0,λ_0).

Proof: Combining the conclusion (ii) in Lemma 5.2.1 with the assumption (5.8) we see $\dim(N_0^{\Sigma^0}) = 1$. Let $N_0^{\Sigma^0} = (\phi,\beta)$. If $\beta = 0$, then (u_0,λ_0) is a turning point of (5.4) and there is exactly one solution curve passing through it. Otherwise, for $\beta \neq 0$ the operator $D_uG_0^{\Sigma^0}$ is nonsingular. The implicit function theorem yields directly the assertion. ■

5.3 Equivariant Branching Lemma

The well-known equivariant branching lemma of Vanderbauwhede [295], Cicogna [67] and Golubitsky/Stewart/Schaeffer [131] allows us to deduce existence of many symmetric solution branches simply by examining dimensions of intersection of $\mathrm{Ker}(DG_0)$ with the corresponding fixed point subspaces. We study here this lemma in a slightly modified version as in Allgower/Böhmer/Mei [11].

We assume that the spaces X, Y can be embedded into a Hilbert space whose inner product is Γ-invariant, see Lemma 5.1.2. Let the representations of the compact group Γ in X, Y be orthogonal under the induced inner product and let D_uG_0 be Fredholm of index 0. For the bifurcation point (u_0,λ_0) we denote Σ^0 the isotropy group of u_0.

As a modification of the definition of a bifurcation subgroup in Dellnitz/Werner [84], we consider, see Allgower/Böhmer/Mei [11],

Definition 5.3.1. *Let $\Sigma \preceq \Sigma^0$ be a subgroup of Γ and let V be a subspace of $X \times \mathbf{R}$ such that*

$$\begin{array}{ll} \text{(i)} & V \textit{ is } \Sigma^0 - \textit{invariant;} \\ \text{(ii)} & \textit{there is a } \bar{y} = (y, c) \in V, \textit{ such that } \Sigma_{\bar{y}} := \Sigma_y = \Sigma; \\ \text{(iii)} & \dim[V \cap (X^\Sigma \times \mathbf{R})] = 2. \end{array} \tag{5.9}$$

Then Σ is called a bifurcation subgroup of Σ^0 in Y.

The defining properties (5.9) for a bifurcation subgroup have been formulated so that the Σ-reduced problem relates directly to a simple bifurcation, for which the established theory is applicable, see Chapter 4 and [71, 49, 210, 213].

Let $\dim[\mathrm{Ker}(DG_0)] = n + 1 > 1$. We choose a basis as

$$N_0 = \mathrm{Ker}(DG_0) = \mathrm{span}[(\psi_1, \mu_1), \ldots, (\psi_{n+1}, \mu_{n+1})], \quad n \geq 1.$$

Except $\mu_j^0 = 0$ for all $j = 1, \ldots, n + 1$, which corresponds to $D_\lambda G_0 \notin \mathrm{Im}(D_u G_0)$, there are appropriate linear combinations such that

$$N_0 = \mathrm{Ker}(DG_0) = \mathrm{span}[(\phi_1, 0), \ldots, (\phi_n, 0), \ (\phi_{n+1}, 1)]. \tag{5.10}$$

Under assumptions that X, Y are subspaces of a Hilbert space and $D_u G_0$ is Fredholm of index 0, we have

$$X = \mathrm{Ker}(D_u G_0) \oplus \mathrm{Im}((D_u G_0)^*), \quad Y = \mathrm{Ker}((D_u G_0)^*) \oplus \mathrm{Im}(D_u G_0), \tag{5.11}$$

where $\oplus$ denotes the orthogonal sum. The adjoint operator $(D_u G_0)^*$ is taken under the induced inner product.

Lemma 5.3.1. *The subspaces $\mathrm{Im}((D_u G_0)^*)$ and $\mathrm{Ker}((D_u G_0)^*)$ are Σ^0-invariant.*

Proof: Since $\mathrm{Ker}(D_u G_0)$ is Σ^0-invariant and $\mathrm{Im}((D_u G_0)^*)$ is its orthogonal complement, with the orthogonal representation of the compact group Γ it follows from Proposition 2.1 in Section XII.2 of Golubitsky/Stewart/Schaeffer [131] that $\mathrm{Im}((D_u G_0)^*)$ is also Σ^0-invariant. Similar conclusion holds for $\mathrm{Ker}((D_u G_0)^*)$. ■

Since $D_u G_0$ is nonsingular from $\mathrm{Im}(D_u G_0^*)$ onto $\mathrm{Im}(D_u G_0)$ and commutates with Γ, the representations of Γ in $\mathrm{Im}(D_u G_0^*)$ and $\mathrm{Im}(D_u G_0)$ are isomorphic. This, together with the decomposition (5.11), implies that the induced representations of Γ in $\mathrm{Ker}(D_u G_0)$ and $\mathrm{Ker}(D_u G_0^*)$ are isomorphic (cf. Golubitsky/Schaeffer [129]).

By assumptions the induced inner product in X, Y is Γ-invariant. Thus the subspaces X^Σ, Y^Σ can be also embedded in a Hilbert space, respectively. For any subgroup Σ of Σ^0, the operator $D_u G_0$ maps X^Σ into Y^Σ. Let G^Σ be the restriction of G to $X^\Sigma \times \mathbf{R}$. The proof of (ii) in Lemma 5.2.1 shows

$$D_u G_0^\Sigma v = D_u G_0 v \quad \text{for all} \quad v \in X^\Sigma.$$

Hence, from the equality

$$\langle v,\ (D_u G_0^\Sigma)^* w\rangle = \langle D_u G_0^\Sigma v,\ w\rangle = \langle D_u G_0 v,\ w\rangle = \langle v,\ (D_u G_0)^* w\rangle$$

for all $v \in X^\Sigma,\ w \in Y^\Sigma$ we obtain

$$(D_u G_0^\Sigma)^* = (D_u G_0)^*|_{Y^\Sigma}. \tag{5.12}$$

Lemma 5.3.2. *Let Σ be a subgroup of Σ^0. Then*

$$X^\Sigma = \operatorname{Ker}(D_u G_0^\Sigma) \oplus \operatorname{Im}((D_u G_0^\Sigma)^*), \quad Y^\Sigma = \operatorname{Ker}((D_u G_0^\Sigma)^*) \oplus \operatorname{Im}(D_u G_0^\Sigma). \tag{5.13}$$

Proof: Similarly to the proof of (iii) in Lemma 5.2.1, we have

$$\operatorname{Ker}(D_u G_0^\Sigma) = \operatorname{Ker}(D_u G_0) \cap X^\Sigma = \big(\operatorname{Ker}(D_u G_0)\big)^\Sigma.$$

Therefore, it is finite dimensional. We show now that $\operatorname{Im}(D_u G_0^\Sigma)$ is closed. Let $\overline{\operatorname{Im}(D_u G_0^\Sigma)}$ denote the closure of $\operatorname{Im}(D_u G_0^\Sigma)$ in X^Σ. Then for any $v \in \overline{\operatorname{Im}(D_u G_0^\Sigma)}$, there is a sequence $\{u_n\}_{n=1}^\infty \subset X^\Sigma$, such that

$$v_n = D_u G_0^\Sigma u_n = D_u G_0 u_n \to v.$$

Since $\operatorname{Im}(D_u G_0^\Sigma) \subset \operatorname{Im}(D_u G_0)$ and $\operatorname{Im}(D_u G_0)$ is closed in X ($D_u G_0$ is a Fredholm operator), there is a $u \in X$ such that $v = D_u G_0 u$. Consider the element

$$\tilde{u} := \int_\Sigma \sigma u \, d\sigma \in X^\Sigma.$$

We have

$$\begin{aligned}
D_u G_0^\Sigma \tilde{u} &= D_u G_0 \tilde{u} \\
&= \int_\Sigma D_u G_0 \sigma u \, d\sigma \quad (\ D_u G_0 \text{ is a linear operator}) \\
&= \int_\Sigma \sigma D_u G_0 u \, d\sigma \quad (\ D_u G_0 \text{ is } \Sigma^0\text{-equivariant}) \\
&= \int_\Sigma \sigma v \, d\sigma \\
&= \int_\Sigma v \, d\sigma = v \quad (v \in X^\Sigma),
\end{aligned}$$

i.e., $v \in \operatorname{Im}(D_u G_0^\Sigma)$ and $\operatorname{Im}(D_u G_0^\Sigma)$ is closed. Furthermore, it follows from (5.12) that

$$\operatorname{Ker}((D_u G_0^\Sigma)^*) = \operatorname{Ker}((D_u G_0)^*) \cap Y^\Sigma = \big(\operatorname{Ker}(D_u G_0^*)\big)^\Sigma.$$

Hence $\dim \operatorname{Ker}((D_u G_0^\Sigma)^*) \leq \dim \operatorname{Ker}((D_u G_0)^*) < \infty$. The operator $D_u G_0^\Sigma$ is Fredholm from X^Σ into Y^Σ. Moreover, the isomorphism between the representations of Γ in $\operatorname{Ker}(D_u G_0)$ and $\operatorname{Ker}(D_u G_0^*)$ implies that

$$\dim\left(\operatorname{Ker}(D_u G_0^\Sigma)\right) = \dim\left(\operatorname{Ker}(D_u G_0^{\Sigma^*})\right).$$

Thus $D_u G_0^\Sigma$ is a Fredholm operator of index 0. The conclusion follows from classical results concerning Fredholm operators in Hilbert spaces, see e.g. Hirzebruch/Scharlau [159, Lemma 25.6]). ■

Lemma 5.3.3. *Let* $\operatorname{Ker}(DG_0) \cap (X^\Sigma \times \mathbf{R})$ *be two dimensional and*

$$\operatorname{Ker}(DG_0) \cap (X^\Sigma \times \mathbf{R}) = \operatorname{span}[(\psi_1, \mu_1),\ (\psi_2, \mu_1)]. \tag{5.14}$$

Then the statement $\mu_1 = \mu_2 = 0$ *is equivalent to* $D_\lambda G_0 \notin \operatorname{Im}(D_u G_0^\Sigma)$.

Proof: Let us consider the complementary relation $D_\lambda G_0 \in \operatorname{Im}(D_u G_0^\Sigma)$, i.e., there is $w_1 \in X^\Sigma$ such that

$$(D_u G_0^\Sigma) w_1 + D_\lambda G_0 = 0.$$

Thus $(w_1, 1) \in \operatorname{Ker}(DG_0) \cap (X^\Sigma \times \mathbf{R})$ and this proves our conclusion. ■

In fact, if $\mu_1 = \mu_2 = 0$, then (u_0, λ_0) is a turning point for all Σ-symmetric solution curves of (5.4). We shall exclude this special case, since by an appropriate parameter transformation, for example, replacing the parameter λ with an appropriate coordinate of $u \in X$ with respect to ψ_i, $i = 1, 2$ or using (pseudo-) arclength parameterization, see Chapter 2 and Allgower/Georg [14, pp. 91], we can ensure

$$|\mu_1| + |\mu_2| > 0.$$

In the sequel we assume this inequality, equivalently, the relation

$$D_\lambda G_0 \in \operatorname{Im}(D_u G_0) \quad \text{and} \quad D_\lambda G_0 \neq 0. \tag{5.15}$$

By the decomposition of X in (5.11), there is a unique element $w_1 \in \operatorname{Im}((D_u G_0)^*)$ such that

$$D_u G_0 w_1 + D_\lambda G_0 = 0. \tag{5.16}$$

Owing to the Γ-equivariance of G, the subspaces $\operatorname{Ker}(D_u G_0)$, $\operatorname{Im}((D_u G_0)^*)$ are Σ^0-invariant, and

$$\sigma(D_u G_0 w_1 + D_\lambda G_0) = D_u G_0 \sigma w_1 + D_\lambda G_0 = 0 \qquad \text{for all } \sigma \in \Sigma^0.$$

The uniqueness of w_1 in $\operatorname{Im}((D_u G_0)^*)$ shows

$$\sigma w_1 = w_1 \qquad \text{for all } \sigma \in \Sigma^0.$$

Therefore, $w_1 \in X^{\Sigma^0} \cap \operatorname{Im}((D_u G_0)^*) = \operatorname{Im}((D_u G_0^{\Sigma^0})^*)$ and $D_\lambda G_0 \in \operatorname{Im}(D_u G_0^{\Sigma^0})$.

Lemma 5.3.4. *Let Σ be a bifurcation subgroup of Σ^0 in $Y = \text{Ker}(DG_0)$. If the statement (5.15) holds, there are $(v_1, 0)$, $(w_1, 1) \in Y$, such that*

$$\text{Ker}(DG_0^{\Sigma}) = Y \cap (X^{\Sigma} \times \mathbf{R}) = \text{span}[(v_1, 0),\ (w_1, 1)]. \tag{5.17}$$

Proof: It follows from the definition of bifurcation subgroup that there are elements $(\psi_i, \mu_i) \in X^{\Sigma} \times \mathbf{R}$, $i = 1, 2$, such that (5.14) holds and $|\mu_1| + |\mu_2| > 0$. If $\mu_1 = 0$, we choose $v_1 := \psi_1$ and $w_1 := (\psi_2 - (\psi_1, \psi_2)\psi_1)/\mu_2$. Otherwise, $v_1 := \mu_2\psi_1 - \mu_1\psi_2$ and $w_1 := (\psi_1 - (\psi_1, v_1)v_1)/\mu_1$ satisfy (5.17). ■

Combining (5.17) with (5.13), we see

$$\text{Ker}(D_u G_0^{\Sigma}) = \text{span}[v_1] \quad \text{and} \quad X^{\Sigma} = \text{span}[v_1] \oplus \text{Im}((D_u G_0^{\Sigma})^*). \tag{5.18}$$

For w_1, v_1 in (5.17) and arbitrary nonvanishing element $v_1^* \in \text{Ker}((D_u G_0^{\Sigma})^*)$ we define

$$\begin{aligned} a &:= \langle v_1^*,\ D_{uu}^2 G_0^{\Sigma} v_1 v_1 \rangle, \\ b &:= \langle v_1^*,\ D_u D G_0^{\Sigma}(w_1, 1) v_1 \rangle, \\ c &:= \langle v_1^*,\ D^2 G_0^{\Sigma}(w_1, 1)^2 \rangle, \end{aligned} \tag{5.19}$$

see also (3.46), (6.19). An extension of the *Equivariant Branching Lemma* is as follows.

Theorem 5.3.1. (Allgower/Böhmer/Mei [11]) *Suppose that a Γ-invariant inner product is imposed in the spaces X, Y and the representations of the compact group Γ in X, Y are orthogonal. Let G be Γ-equivariant and (u_0, λ_0) be a bifurcation point such that $D_\lambda G_0 \in \text{Im}(D_u G_0)$ and $D_u G_0$ is Fredholm of index 0. If Σ is a bifurcation subgroup of $\Sigma^0 := \Sigma_{u_0}$ in* $\text{Ker}(DG_0)$ *and*

$$b^2 - ac > 0, \tag{5.20}$$

the reduced problem

$$\textit{Find } (u, \lambda) \in X^{\Sigma} \times \mathbf{R}, \quad \textit{such that} \quad G^{\Sigma}(u, \lambda) = 0$$

has exactly two different solution curves intersecting at (u_0, λ_0).

Proof: For a bifurcation subgroup Σ in $N_0 := \text{Ker}(DG_0)$, the statement (5.18) holds. Together with (5.20) we see that (u_0, λ_0) is a simple bifurcation point of the reduced problem in $X^{\Sigma} \times \mathbf{R}$. According to the discussions in Section 4.4 and Crandall/Rabinowitz [71], see also Section 6.5 of Chapter 6, the reduced problem has exactly two different solution curves intersecting at (u_0, λ_0). ■

A systematic analysis of all isotropy subgroups of Γ reveals a classification of symmetric solution branches of $G(u, \lambda) = 0$ bifurcating at (u_0, λ_0). In particular, subgroups and their structure of the commonly used groups are well known and easily available in the literature. Once we known symmetries in a problem, we can derive possible bifurcation scenarios. In certain sense, this kind of bifurcation is symmetry driven and independent of the given problem. We call it *symmetry induced linear degeneracy.* Nonlinearities in the problem interact with symmetries normally in a more sophisticated way. They may break, increase or twist symmetries and lead the system to manifest in a peculiar way.

5.4 A Semi-linear Elliptic PDE on Unit Square

We consider the semi-linear elliptic partial differential equation

$$G(u,\lambda) := \Delta u + f(u,\lambda) = 0 \tag{5.21}$$

in the unit square $\Omega := [0,1]\times[0,1]$ with the Dirichlet boundary conditions $u = 0$ on $\partial\Omega$. Here $f:\ \mathbf{R}\times\mathbf{R}\mapsto\mathbf{R}$ is a smooth function and satisfies

$$f(0,\lambda) = 0 \quad \text{for all} \quad \lambda\in\mathbf{R}.$$

Thus the problem (5.21) has a trivial solution curve $\{(0,\lambda);\ \lambda\in\mathbf{R}\}$.

Let $C^{k,s}(\Omega)$ be the space of k-times differentiable functions u on the closure of Ω, such that u and its derivatives are Hölder continuous with the exponent $s\in(0,1)$. Define

$$X := \{u\in C^{2,s}(\Omega)\mid u=0 \ \text{ on } \partial\Omega\}, \qquad Y := C^{0,s}(\Omega), \qquad Z := L^2(\Omega)$$

with the Hölder norms $\|\cdot\|_{2,s}$ and $\|\cdot\|_{0,s}$ respectively. The discussions below will be based on the L^2-product in X, Y. The classical theory of elliptic partial differential equations shows that the derivative

$$D_uG(0,\lambda) = \Delta + \frac{\partial f}{\partial u}(0,\lambda):\ X\mapsto Y$$

is self-adjoint and Fredholm of index zero (cf. Golubitsky/Schaeffer [129, pp. 335], Healey/Kielhöfer [155]). Therefore, we have

$$\operatorname{Im}(D_uG(0,\lambda)) = \operatorname{Ker}(D_uG(0,\lambda))^{\perp}$$

and

$$X = \operatorname{Ker}(D_uG(0,\lambda)) \oplus \operatorname{Ker}(D_uG(0,\lambda))^{\perp},$$

where $\oplus$ represents the orthogonal sum under the inner product in $L^2(\Omega)$. Since $D_\lambda G(0,\lambda)=0$, we have $\operatorname{Im}(DG(0,\lambda)) = \operatorname{Im}(D_uG(0,\lambda))$.

If the equality $n = \dim[\operatorname{Ker}(D_uG(0,\lambda))]$ (=codim[$\operatorname{Im}(D_uG(0,\lambda))$]) holds at a point $(0,\lambda)=(0,\lambda_0)$, then $(0,\lambda_0)$ is called *a corank-n bifurcation point* of (5.21). A corank-n bifurcation point $(0,\lambda_0)$ of (5.21) is characterized by the parameter λ_0 such that $\frac{\partial f}{\partial u}(0,\lambda_0)$ is an n-fold eigenvalue of $-\Delta$. Since eigenvalues and eigenfunctions of $-\Delta$ with the Dirichlet boundary conditions on the unit square are well known (cf. Isaacson/Keller[168, pp. 458]), at a corank-n bifurcation point there are exactly n different (ordered) pairs of $(p_i,\ q_i) \in \mathbf{N} \times \mathbf{N}$, $i = 1, \ldots, n$, such that

$$\frac{\partial f}{\partial u}(0, \lambda_0) = (p_i^2 + q_i^2)\pi^2, \quad i = 1, \ldots, n.$$

Correspondingly,

$$\begin{aligned} \mathrm{Ker}(D_u G_0) &= \mathrm{span}[\phi_1, \ldots, \phi_n], \qquad \langle \phi_i, \phi_j \rangle = \delta_{ij}, \quad i,\ j = 1, \ldots, n, \\ \mathrm{Im}(D_u G_0) &= \{u \in X \quad | \quad \langle \phi_i, u \rangle = 0, \quad i = 1, \ldots, n\}, \end{aligned}$$

where

$$\phi_i := 2 \sin p_i \pi x \sin q_i \pi y, \qquad i = 1, \ldots, m.$$

Let

$$D_4 := \{S_1,\ S_2;\ \ S_1',\ S_2',\ I,\ R^1,\ R^2,\ R^3\} \subset O(2)$$

be the symmetry group of the unit square Ω, consisting of the reflections S_i, S_i', $i = 1, 2$ and the rotations R^i, $i = 1, 2, 3$, see Fig. 5.3.

We separate the generators of the group from other elements by ";". For D_4 and $Z_2 := \{1, -1\}$, let

$$Z_2 \times D_4 = \{\pm S_1,\ \pm S_2;\ \ \pm I,\ \pm S_1',\ \pm S_2',\ \pm R^1,\ \pm R^2,\ \pm R^3\}.$$

We define group actions of $Z_2 \times D_4$ in $Y\ (\supset X)$ by

$$\gamma u = \pm u(\delta^{-1}(x, y)) \qquad \text{for all } \gamma = \pm\delta \in \Gamma,\ \delta \in D_4 \text{ and } u \in Y. \tag{5.22}$$

Moreover, let

$$\Gamma := \begin{cases} Z_2 \times D_4 & \text{if the function } f(u, \lambda) \text{ is odd in } u; \\ D_4 & \text{otherwise.} \end{cases}$$

The spaces X, Y are obviously Γ-invariant and the L^2-product is also Γ-invariant. The *Γ-equivariance* of the mapping G

$$G(\gamma u, \lambda) = \gamma G(u, \lambda) \qquad \text{for all} \quad \gamma \in \Gamma,\ u \in X,\ \lambda \in \mathbf{R}.$$

can be verified directly on the generator $\pm S_1, \pm S_2$ of Γ.

As an example, we consider

$$f(u, \lambda) = \lambda(u + u^2). \tag{5.23}$$

Owing to the quadratic term in $f(u, \lambda)$, we have $\Gamma = D_4$. Furthermore, a corank-n bifurcations occurs at $(0, \lambda_0)$ if and only if λ_0 is a n-fold eigenvalue of the Laplacian $-\Delta$. Let us consider the corank-2 bifurcation point $(0, \lambda_0) = (0, 26\pi^2)$ and

$$N_0 := \mathrm{Ker}(DG_0) = \mathrm{span}[(\phi_1, 0),\ (\phi_2, 0),\ (0, 1)]$$

with $\phi_1 = 2\sin\pi x \sin 5\pi y$, $\phi_2 = 2\sin 5\pi x \sin\pi y$. To determine D_4-symmetric solution curves of (5.21) we study its restriction to $X^{D_4} \times \mathbf{R}$

$$\text{Find } (u, \lambda) \in X^{D_4} \times \mathbf{R} \quad \text{such that} \quad G^{D_4}(u, \lambda) = 0. \tag{5.24}$$

Now, it is easy to verify by Definition 5.3.1 that D_4 itself is a bifurcation subgroup. Furthermore, $(0, 26\pi^2)$ is a simple bifurcation point of (5.24) with a two dimensional kernel

$$N_0^{D_4} := \mathrm{Ker}(DG_0^{D_4}) = \mathrm{span}[(\phi_1 + \phi_2, 0),\ (0, 1)].$$

The constants a, b, c in (5.19) can be calculated directly

$$\begin{aligned} a &= \langle \phi_1 + \phi_2,\ D^2_{uu} G_0^{D_4} (\phi_1 + \phi_2)^2 \rangle = \frac{692224}{3465}, \\ b &= \langle \phi_1 + \phi_2,\ D^2_{u\lambda} G_0^{D_4} (\phi_1 + \phi_2) \rangle = 2, \\ c &= 0. \end{aligned}$$

Hence, $b^2 - a \cdot c = 4$, i.e., the transcritical condition (5.20) holds. By Theorem 5.3.1 the reduced problem (5.24) has two different solution branches passing through $(0, 26\pi^2)$. One of them is the trivial solution curve $\{(0, \lambda);\ \lambda \in \mathbf{R}\}$ and the other one is of the form

$$\begin{aligned} u(t) &= t\alpha(t)(\phi_1 + \phi_2) + t^2 w(t), \\ \lambda(t) &= \lambda_0 + t\beta(t), \qquad w(t) \in \mathrm{Im}(D_u G_0^{D_4}), \end{aligned}$$

where $\alpha(0) = 2b(a^2 + 4b^2)^{-1/2}$ and $\beta(0) = a(a^2 + 4b^2)^{-1/2}$.

Let $\Sigma_M := \{S_1, S_1';\ I, R^2\}$. We consider the reduced problem

$$\text{Find } (u, \lambda) \in X^{\Sigma_M} \times \mathbf{R} \quad \text{such that} \quad G^{\Sigma_M}(u, \lambda) = 0. \tag{5.25}$$

For this subgroup $(0, 26\pi^2)$ is still a corank-2 bifurcation point of (5.25) and

$$N_0^{\Sigma_M} := \mathrm{Ker}(DG_0^{\Sigma_M}) = \mathrm{span}[(\phi_1, 0),\ (\phi_2, 0),\ (0, 1)].$$

If we choose

$$X^0 := \{u \in X \ \mid \ u(x, \frac{1}{5} + y) = u(x, \frac{1}{5} - y)\},$$

then Σ_M is a bifurcation subgroup of D_4 in the subspace $N_0^{\Sigma_M} \cap (X^0 \times \mathbf{R}) = \mathrm{span}[(\phi_1, 0), (0, 1)]$. Moreover,

$$a = \langle \phi_1,\ D^2_{uu} G_0^{\Sigma_M} \phi_1 \phi_1 \rangle = \frac{6656}{45}, \qquad b = \langle \phi_1,\ D^2_{u\lambda} G_0^{\Sigma_M} \phi_1 \rangle = 1, \qquad c = 0.$$

Theorem 5.3.1 becomes applicable to (5.25) in the space $X^{\Sigma_M} \cap X^0 \times \mathbf{R}$. Thus there are two solution branches of (5.21) and (5.23), passing through $(0, 26\pi^2)$. Besides the trivial branch there is a nontrivial branch in the form

$$\begin{aligned} u(t) &= t\alpha(t)\phi_1 + t^2 w(t), \\ \lambda(t) &= \lambda_0 + t\beta(t), \qquad w(t) \in \operatorname{Im}(D_u G_0^{\Sigma_M}) \cap X^0. \end{aligned}$$

Similarly, if we choose

$$\tilde{X}^0 := \{u \in X \mid u(\frac{1}{5} + x, y) = u(\frac{1}{5} - x, y)\},$$

Σ_M becomes a bifurcation subgroup of D_4 in $N_0^{\Sigma_M} \cap (\tilde{X}^0 \times \mathbf{R}) = \operatorname{span}[(\phi_2, 0), (0, 1)]$. Furthermore,

$$a = \langle \phi_2, D^2_{uu} G_0^{\Sigma_M} \phi_2 \phi_2 \rangle = \frac{6656}{45}, \qquad b = \langle \phi_2, D^2_{u\lambda} G_0^{\Sigma_M} \phi_2 \rangle = 1, \qquad c = 0$$

and the transcritical condition (5.20) is satisfied. Theorem 5.3.1 shows existence of two solution branches in $X^{\Sigma_M} \cap \tilde{X}^0 \times \mathbf{R}$. One of them is the trivial branch and and the other one is nontrivial and of the form

$$\begin{aligned} u(t) &= t\alpha(t)\phi_2 + t^2 w(t), \\ \lambda(t) &= \lambda_0 + t\beta(t), \qquad w(t) \in \operatorname{Im}(D_u G_0^{\Sigma_M}) \cap \tilde{X}^0. \end{aligned}$$

We note that the problem (5.21) has the so-called hidden symmetries induced by the Euclidean symmetry of the Laplacian. These symmetries become visible if we embed the problem (5.21) into equations with periodic boundary conditions. Exploiting these symmetries allow in some cases a direct application of equivariant branching lemma, see e.g. Healey/Kielhöfer [155] and Mei [217] for more details.

6. Liapunov-Schmidt Method

To study steady state and Hopf bifurcations, we introduce the well-known Liapunov-Schmidt method and derive an underlying low dimensional system of algebraic equations. This system is responsible for the bifurcation scenario and normally is easy to analyze. Another advantage of this approach is that the established singularity theory can be utilized directly to determine normal forms of these algebraic equations and their bifurcation scenario, see e.g. the monographes Golubitsky *et al* [129, 131] and Vanderbauwhede [295]. For numerical purposes Jepson *et al* generalize the Liapunov-Schmidt method in several aspects in a series of papers [173, 174, 175, 176]. We adapter in this chapter the discussion of Liapunov-Schmidt method in Ashwin/Böhmer/Mei [22] and a scaling technique in Mei/Schwarzer [226]. Center manifold reduction is another approach for analyzing local bifurcations and will be discussed in Chapter 7.

In Section 6.1 we introduce the Liapunov-Schmidt reduction for bifurcation problems with symmetries. We show in Section 6.2 that the reduced bifurcation equations inherit symmetries of the original problem. Derivatives and Taylor expansion of the reduced bifurcation equations are calculated in Section 6.3. In Section 6.4 we review the concepts of equivalence, determinacy and stability. A simple example is discussed in Section 6.5. For finitely determined problems, we use an algorithm in Section 6.6 to calculate Taylor expansion of the reduced bifurcation equations. We discuss in Section 6.7 branch switching at multiple bifurcation points and illustrate the techniques in Section 6.8 with a semi-linear elliptic differential equation.

6.1 Liapunov-Schmidt Reduction

To combine the Liapunov-Schmidt reduction for steady state and Hopf bifurcations, we consider a general operator equation

$$\Phi(u, \lambda) = 0, \tag{6.1}$$

where $\Phi: \mathcal{X} \times \mathbf{R}^p \to \mathcal{Y}$ is a nonlinear mapping as smooth as required in the context below, and $\mathcal{X}$, $\mathcal{Y}$ are Banach spaces.

For steady state bifurcation problems, we choose the spaces $\mathcal{X} = X$, $\mathcal{Y} = Y$ and the operator $\Phi(u, \lambda) = G(u, \lambda)$ as in the previous chapters. For Hopf bifurcations, we scale the time $t \to (1+\tau)\omega_0 t$ with $\omega_0 > 0$, determined at the Hopf bifurcation point, see Section 3.4. Thereafter, we consider the problem

$$\Phi(u,\tilde{\lambda}) := (1+\tau)\omega_0 \frac{\partial u}{\partial t} + G(u,\lambda) = 0$$

in the periodic function spaces

$$\mathcal{X} = \left\{u(t) \in X \quad | \quad u(t) = u(t+2\pi),\ \frac{\partial u}{\partial t} \in X\right\}, \tag{6.2a}$$

$$\mathcal{Y} = \{u(t) \in Y \quad | \quad u(t) = u(t+2\pi)\}\,. \tag{6.2b}$$

Here and in the sequel we consider τ as an additional parameter and include it into $\tilde{\lambda} := (\lambda, \tau)$. We are interested in periodic solutions bifurcating at a Hopf point.

Let Γ be a compact Lie group acting on $\mathcal{X}$, $\mathcal{Y}$ and let Φ be Γ-equivariant

$$\Phi(\gamma u, \lambda) = \gamma\Phi(u,\lambda) \quad \text{for all} \quad (u,\lambda) \in \mathcal{X} \times \mathbf{R}^p,\ \gamma \in \Gamma.$$

Furthermore, we assume that $\mathcal{X}$, $\mathcal{Y}$ can be embedded into a Hilbert space with a Γ-invariant inner product $\langle \cdot,\ \cdot \rangle$.

Let (u_0, λ_0) be a bifurcation point of (6.1). We want to study the behavior of solutions of (6.1) in the neighborhood of (u_0, λ_0). Without loss of generality we assume $(u_0, \lambda_0) = (0,0)$, otherwise a shift of Φ by $\tilde{\Phi}(u,\lambda) = \Phi(u_0+u, \lambda_0+\lambda)$ yields this effect. Denote

$$L := \frac{\partial \Phi}{\partial u}(0,0). \tag{6.3}$$

We assume in the sequel that L is a Fredholm operator of index zero. In this case the spaces $\mathcal{X}$, $\mathcal{Y}$ are decomposed as

$$\mathcal{X} = \mathrm{Ker}(L) \oplus M, \qquad \mathcal{Y} = \mathrm{Im}(L) \oplus N$$

with $M = (\mathrm{Ker}(L))^{\perp}$, $N = (\mathrm{Im}(L))^{\perp}$ and $\dim(N) = \dim(\mathrm{Ker}(L)$. The orthogonality is defined under the Γ-invariant inner product $\langle \cdot,\ \cdot \rangle$. An element u in $\mathcal{X}$ is of the form $u = v + w$ with $v \in \mathrm{Ker}(L)$, $w \in M$. Since the spaces $\mathrm{Ker}(L)$ and $\mathrm{Im}(L)$ are Γ-invariant, so are M, N.

Lemma 6.1.1. *Let Q be a projection from $\mathcal{X}$ onto M and $\hat{Q}$ be a projection from $\mathcal{Y}$ onto* $\mathrm{Im}(L)$. *Then Q, $\hat{Q}$ commute with Γ.*

Proof: For all $u \in \mathcal{X}$ there are $v \in \mathrm{Ker}(L)$ and $w \in M$ such that $u = v + w$. By the linearity of Q, the action of Γ and the invariance of $\mathrm{Ker}(L)$ and M we have

$$\gamma Q u = \gamma Q(v+w) = \gamma Q w = \gamma w = Q\gamma w = Q\gamma u.$$

Therefore,

$$\gamma Q = Q\gamma \qquad \text{for all} \quad \gamma \in \Gamma.$$

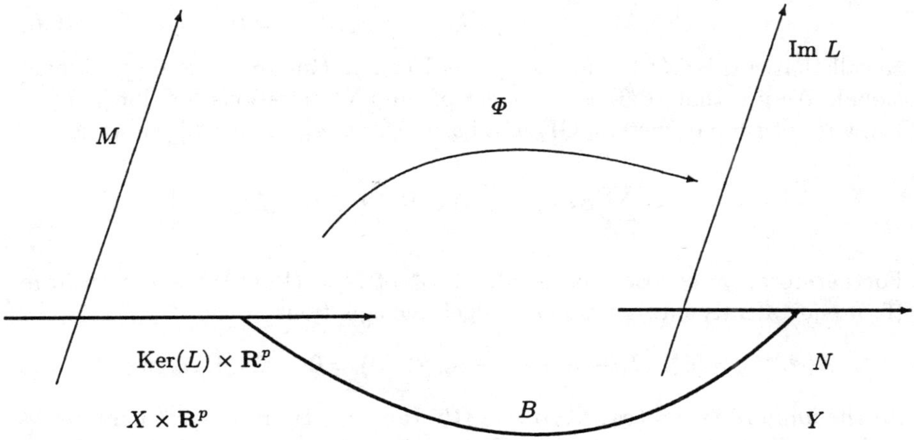

Fig. 6.1. Liapunov-Schmidt reduction

Similarly, $\hat{Q}$ and $I - \hat{Q}$ are commutative with Γ. This completes the proof. ■

We write now the equation (6.1) equivalently into a system of two equations

$$\hat{Q}\Phi(v+w,\lambda) = 0, \tag{6.4a}$$

$$(I-\hat{Q})\Phi(v+w,\lambda) = 0 \tag{6.4b}$$

for the unknowns v and w. Define a mapping $F:\ \mathrm{Ker}(L)\times M\times \mathbf{R}^p \to \mathrm{Im}(L)$ by

$$F(v,w,\lambda) := \hat{Q}\Phi(v+w,\lambda).$$

We have

$$F(0,0,0) = 0.$$

and

$$D_wF(0,0,0) = \hat{Q}\ D_u\Phi(0,0) = \hat{Q}L = L.$$

Since M, $\mathrm{Im}(L)$ are both closed subspaces and L maps M one-to-one onto $\mathrm{Im}(L)$, the operator $D_wF(0,0,0)$ is nonsingular. The implicit function theorem is applicable to F at $(0,0,0)$, i.e., there is a unique function $w(v,\lambda):\mathrm{Ker}(L)\times\mathbf{R}^p\to M$, such that

$$F(v,w(v,\lambda),\lambda) = 0, \qquad w(0,0) = 0. \tag{6.5}$$

Substituting $w(v,\lambda)$ into (6.4b) yields the so-called *reduced bifurcation equations*

$$B(v,\lambda) := (I-\hat{Q})\ \Phi(v+w(v,\lambda),\lambda) = 0. \tag{6.6}$$

Recall that $v \in \mathrm{Ker}(L)$, and the spaces $\mathrm{Ker}(L)$, $(\mathrm{Im}(L))^{\perp}$ are finite dimensional. We see that (6.6) is a system of $\dim(N)$ equations for $\dim(N)$ unknowns with p parameters. GIven a basis $\phi_1,\ldots,\phi_n$ of $\mathrm{Ker}(L)$ we write

$$v = \sum_{i=1}^{n} \alpha_i \phi_i, \qquad \alpha_i \in \mathbf{R},\ \ v \in \mathrm{Ker}(L).$$

Furthermore, we choose a basis $\phi_1^*,\ldots,\phi_n^*$ of $N = (\mathrm{Im}(L))^{\perp}$ and transform (6.6) equivalently into a system of algebraic equations

$$f_i(\alpha,\lambda) := \langle \phi_i^*,\ B(\alpha_1\phi_1+\cdots+\alpha_n\phi_n,\lambda)\rangle = 0, \quad i=1,\ldots,n \tag{6.7}$$

for the unknown $\alpha := (\alpha_1,\ldots,\alpha_n) \in \mathbf{R}^n$. Here $\lambda \in \mathbf{R}^p$ represents parameters and $\langle\cdot,\ \cdot\rangle$ is the inner product in Z or dual product in $Y \times Y^*$.

Theorem 6.1.1. *If $L := D_u\Phi(0,0)$ is a Fredholm operator of index zero, then the problem (6.1) in the neighborhood of $(0,0)$ is equivalent to the finite algebraic system (6.7), i.e., there is a one-to-one correspondence between the solutions (6.1) and (6.7).*

6.2 Equivariance of the Reduced Bifurcation Equations

We consider inheritance of the symmetry Γ by the reduced bifurcation equations.

Lemma 6.2.1. *The mapping $w:\ \mathrm{Ker}(L)\times\mathbf{R}^p \to M$ is Γ-equivariant.*

Proof: For $\gamma \in \Gamma$ let

$$w_\gamma(v,\lambda) := \gamma^{-1}w(\gamma v,\lambda).$$

This is well-defined since both $\mathrm{Ker}(L)$ and M are Γ-invariant. On the other hand, for all (v,λ) in the neighborhood of $(0,0) \in \mathrm{Ker}(L)\times\mathbf{R}^p$ we have

$$\begin{aligned}
F(v,w_\gamma(v,\lambda),\lambda) &= \hat{Q}\Phi(v+\gamma^{-1}w(\gamma v,\lambda),\lambda)\\
&= \hat{Q}\Phi(\gamma^{-1}(\gamma v+w(\gamma v,\lambda)),\lambda)\\
&= \gamma^{-1}\hat{Q}\Phi(\gamma v+w(\gamma v,\lambda),\lambda)\\
&= \gamma^{-1}F(\gamma v,w(\gamma v,\lambda),\lambda)\\
&= 0
\end{aligned}$$

Uniqueness of the solution $w(v,\lambda)$ implies

$$w_\gamma(v,\lambda) = w(v,\lambda) \iff \gamma w(v,\lambda) = w(\gamma v,\lambda),$$

i.e., w is Γ- equivariant. ∎

Theorem 6.2.1. *The mapping* $B: \ \mathrm{Ker}(L) \times \mathbf{R}^p \to N \left(= (\mathrm{Im}(L))^\perp\right)$, *defined by the reduced bifurcation equation (6.6), is* Γ*-equivariant. Namely, we have*

$$\gamma B(v, \lambda) = B(\gamma v, \lambda) \qquad \textit{for all} \ \ \gamma \in \Gamma, \ (v, \lambda) \in \mathrm{Ker}(L) \times \mathbf{R}^p.$$

Proof: Recall that the mappings Φ, $w(v, \lambda)$, $I - \hat{Q}$ are Γ-equivariant, and actions of Γ in X and Y are linear. We derive

$$\begin{aligned} B(\gamma v, \lambda) &= (I - \hat{Q})\Phi(\gamma v + w(\gamma v, \lambda), \lambda) \\ &= (I - \hat{Q})\Phi(\gamma(v + w(v, \lambda)), \lambda) \\ &= \gamma(I - \hat{Q})\Phi(v + w(v, \lambda), \lambda) \\ &= \gamma B(v, \lambda) \end{aligned}$$

for all $\gamma \in \Gamma$ and $(v, \lambda) \in \mathrm{Ker}(L) \times \mathbf{R}^p$ in the neighborhood of the origin. ■

Owning to the Γ-invariance of $\mathrm{Ker}(L)$ and the linearity of the action of Γ, for each $\gamma \in \Gamma$ there is a matrix $A(\gamma) = \left(a_{ij}(\gamma)\right) \in \mathbf{R}^{n \times n}$, such that

$$\gamma \phi_i = \sum_{j=1}^{n} a_{ji}(\gamma) \phi_j. \tag{6.8}$$

Thus the induced action of Γ on α in $\mathbf{R}^n$ via $v = \alpha_1 \phi_1 + \cdots + \alpha_n \phi_n$ is given by

$$\gamma \alpha = A(\gamma) \alpha.$$

On the other hand, if we choose a basis $\phi_1^*, \ldots, \phi_n^*$ for $(\mathrm{Im}(L))^\perp$ to be *consistent* with the basis in $\mathrm{Ker}(L)$ in the sense

$$\gamma \phi_i^* = \sum_{j=1}^{n} a_{ji}(\gamma) \phi_j^*,$$

then the mapping $f: \ (\alpha, \lambda) \in \mathbf{R}^n \times \mathbf{R}^p \to (f_1(\alpha, \lambda), \ldots, f_n(\alpha, \lambda))^T \in \mathbf{R}^n$ defined by the reduced bifurcation equations (6.7) is Γ-equivariant with respect to the induced actions of Γ in $\mathbf{R}^n$

$$f(A(\gamma)\alpha, \lambda) = A(\gamma) f(\alpha, \lambda) \qquad \text{for all} \ \ \gamma \in \Gamma \ \ \text{and} \ \ (\alpha, \lambda) \in \mathbf{R}^n \times \mathbf{R}^p.$$

6.3 Derivatives and Taylor Expansion

Let the mapping Φ be sufficiently smooth. We consider the Taylor expansions of $w(\alpha, \lambda) := w(\alpha_1 \phi_1 + \cdots + \alpha_n \phi_n, \lambda)$ and $f(\alpha, \lambda)$ in the neighborhood of $(0, 0)$

$$\begin{aligned} w(\alpha, \lambda) &= w(0,0) + w'(0,0)(\alpha, \lambda) + \frac{1}{2} w''(0,0)(\alpha, \lambda)^2 + \cdots, \\ f(\alpha, \lambda) &= f(0,0) + f'(0,0)(\alpha, \lambda) + \frac{1}{2} f''(0,0)(\alpha, \lambda)^2 + \cdots. \end{aligned}$$

In particular, via the implicit function theorem for the existence of $w(\alpha, \lambda)$ and the definition of $f(\alpha, \lambda)$ we have

$$w(0,0) = 0, \quad \frac{\partial w}{\partial \alpha_i}(0,0) = 0, \tag{6.10}$$

$$f(0,0) = 0, \quad \frac{\partial f_i}{\partial \alpha_j}(0,0) = 0. \tag{6.11}$$

Denote by $\frac{\partial \Phi_0}{\partial \lambda}, \frac{\partial^2 \Phi_0}{\partial u \partial \lambda}, \ldots$ the evaluations of $\frac{\partial \Phi}{\partial \lambda}, \frac{\partial^2 \Phi}{\partial u \partial \lambda}, \ldots$ at $(0,0)$. By successively differentiating the equation (6.4a) we obtain derivatives of $w(\alpha, \lambda)$ as projected solutions in the subspace M of linear equations:

$$\begin{aligned}
L \, \frac{\partial w}{\partial \lambda_i}(0,0) &= -\hat{Q} \frac{\partial \Phi_0}{\partial \lambda_i}, \quad i = 1, \ldots, p, \\
L \, \frac{\partial^2 w}{\partial \alpha_i \partial \alpha_j}(0,0) &= -\hat{Q} \frac{\partial^2 \Phi_0}{\partial u^2} \phi_i \phi_j, \\
L \, \frac{\partial^2 w}{\partial \alpha_i \partial \lambda_j}(0,0) &= -\hat{Q} \Big(\frac{\partial^2 \Phi_0}{\partial u^2} \phi_i \frac{\partial w}{\partial \lambda_j}(0,0) + \frac{\partial^2 \Phi_0}{\partial u \partial \lambda_j} \phi_j \Big), \\
L \, \frac{\partial^2 w}{\partial \lambda_i \partial \lambda_j}(0,0) &= -\hat{Q} \Big(\frac{\partial^2 \Phi_0}{\partial u^2} \frac{\partial w}{\partial \lambda_i}(0,0) \frac{\partial w}{\partial \lambda_j}(0,0) + \frac{\partial^2 \Phi_0}{\partial \lambda_i \partial \lambda_j} \\
&\quad + \frac{\partial^2}{\partial u \partial \lambda_i} \frac{\partial w}{\partial \lambda_j}(0,0) + \frac{\partial^2}{\partial u \partial \lambda_j} \frac{\partial w}{\partial \lambda_i}(0,0) \Big).
\end{aligned} \tag{6.12}$$

All these equations are of the form

$$Lw = \hat{Q} g$$

for the unknown $w \in M$. This problem is well-defined and nonsingular in the subspaces M and $\mathrm{Im}(L)$. Alternatively, for numerical approximation of the solution we consider it in the whole spaces as

$$\begin{aligned}
Lw + \sum_{i=1}^{n} c_i \phi_i^* &= \hat{Q} g, \\
\langle \phi_i, \, w \rangle = 0, &\quad i = 1, \ldots, n.
\end{aligned} \tag{6.13}$$

Here $c_1, \ldots, c_2 \in \mathbf{R}$ are unknowns introduced artificially and will vanish in the solution. This enlarged system is an extension of the bordered systems for finite dimensional problems, see Chapter 3. Sassmannshausen [266] relaxes the projection $\hat{Q}$ at the right hand side and considers the system

$$\begin{aligned}
Lw + \sum_{i=1}^{n} c_i \phi_i^* &= g, \\
\langle \phi_i, \, w \rangle = 0, &\quad i = 1, \ldots, n.
\end{aligned} \tag{6.14}$$

The component w in solutions of the systems (6.13) and (6.14) are identical, while the components c_i, $i = 1, \ldots, n$ in the solution of (6.14) can be solved separately via the system

$$\begin{pmatrix} \langle \phi_1^*, \phi_1^* \rangle & \cdots & \langle \phi_1^*, \phi_n^* \rangle \\ \vdots & & \vdots \\ \langle \phi_n^*, \phi_1^* \rangle & \cdots & \langle \phi_n^*, \phi_n^* \rangle \end{pmatrix} \begin{pmatrix} c_1 \\ \vdots \\ c_n \end{pmatrix} = \begin{pmatrix} \langle \phi_1^*, g \rangle \\ \vdots \\ \langle \phi_n^*, g \rangle \end{pmatrix}.$$

Apparently, $c_1, \ldots, c_n$ need not to vanish.

Derivatives of the mapping $f(\alpha, \lambda)$ are calculated by

$$\begin{aligned} \frac{\partial f_i}{\partial \lambda_j}(0,0) &= \Big\langle \phi_i^*, \ \frac{\partial \Phi_0}{\partial \lambda_j} \Big\rangle, \\ \frac{\partial^2 f_i}{\partial \alpha_j \partial \alpha_k}(0,0) &= \Big\langle \phi_i^*, \ \frac{\partial^2 \Phi_0}{\partial u^2} \phi_j \phi_k \Big\rangle, \\ \frac{\partial^2 f_i}{\partial \alpha_j \partial \lambda_k}(0,0) &= \Big\langle \phi_i^*, \ \frac{\partial^2 \Phi_0}{\partial u \partial \lambda_k} \phi_j + \frac{\partial^2 \Phi_0}{\partial u^2} \phi_j \frac{\partial w}{\partial \lambda_k}(0,0) \Big\rangle, \\ \frac{\partial^2 f_i}{\partial \lambda_j \partial \lambda_k}(0,0) &= \Big\langle \phi_i^*, \ \frac{\partial^2 \Phi_0}{\partial u^2} \frac{\partial w}{\partial \lambda_j}(0,0) \frac{\partial w}{\partial \lambda_k}(0,0) + \frac{\partial^2 \Phi_0}{\partial \lambda_j \partial \lambda_k} \\ &\quad + \frac{\partial^2}{\partial u \partial \lambda_j} \frac{\partial w}{\partial \lambda_k}(0,0) + \frac{\partial^2}{\partial u \partial \lambda_k} \frac{\partial w}{\partial \lambda_j}(0,0) \Big\rangle. \end{aligned} \tag{6.15}$$

Since the function $w(\alpha, \lambda)$ is usually not analytically and explicitly available, Taylor expansion of the bifurcation equations (6.6) are calculated term by term numerically and successively. At which order it is safe to truncate the Taylor expansion, i.e., the truncation does not destroy the bifurcation scenario of original problem, is a question of *determinacy.* This measures the effect of a perturbation with terms of the order $k+1$ or higher on a polynomial bifurcation equations with terms up to the k-th order. If such a perturbation does not result in qualitative changes in the behavior, the given system of bifurcation is said to be *k-determined.* For a k-determined problem, behavior of the reduced bifurcation equations is qualitatively reflected in their Taylor expansion truncated at the order $k + 1$. For one-parameter problems, the truncated bifurcation equations can be calculated with an iterative method, see Section 6.6 below.

6.4 Equivalence, Determinacy and Stability

We summary briefly some classical concepts of determinacy and stability for one-parameter problems.

Let $E_{n,1}^m$ denote the module of germs at $(0,0)$ of C^∞-continuous vector– or matrix–valued functions from $\mathbf{R}^n \times \mathbf{R}$ into $\mathbf{R}^m$, ($m = n$ or $n \times n$ respectively) over the ring $E_{n,1}^1$ of germs of scalar–valued functions. To compare two systems of bifurcation equations, we start with a set of germs

$$\mathcal{F} := \{f \in E^n_{n,1} \mid f(0,0) = 0,\ Df(0,0) = 0\} \tag{6.16}$$

and a set of Γ-equivariant germs

$$\mathcal{F}_\Gamma := \{f \in \mathcal{F} \mid \gamma f(u,\lambda) = f(\gamma u, \lambda) \quad \text{for all} \quad \gamma \in \Gamma\}.$$

Here we have implicitly assumed that the symmetry group Γ is represented orthogonally in $\mathbf{R}^n$.

Definition 6.4.1. (Golubitsky/Stewart/Schaefer [131]) *Consider germs $f,\ g$ in $\mathcal{F}_\Gamma$. If there exist germs $T \in E^n_{n,1}$ and $S \in E^{n\times n}_{n,1}$ with nonsingular operators $D_uT(0,0)$, $S(0,0)$, and a scalar function $\Lambda : \mathbf{R} \to \mathbf{R}$ with $\Lambda(0) = 0,\ \Lambda'(0) > 0$, such that*

$$\begin{aligned} &f(u,\lambda) = S(u,\lambda)g(T(u,\lambda), \Lambda(\lambda)) \quad \textit{and} \\ &T(\gamma u, \lambda) = \gamma T(u,\lambda), \qquad S(\gamma u, \lambda)\gamma = \gamma S(u,\lambda) \quad \textit{for all} \quad \gamma \in \Gamma, \end{aligned}$$

then f and g are said to be equivalent (we write $f \sim g$).

An important consequence of this definition is that if $f \sim g$, then for fixed λ the equations

$$f(u,\lambda) = 0$$

and

$$g(T(u,\lambda), \Lambda(\lambda)) = 0$$

have the same number of solutions. We write j_kg for the truncated Taylor expansion of g with respect to all its arguments up to and including the homogeneous terms of order k (the so called k-jet).

Definition 6.4.2. *A germ $g \in \mathcal{F}_\Gamma$ is said to be k-determined if $k \in N$ is the minimum integer such that $j_kg \sim g$. Furthermore, if g is k-determined for some $k < \infty$, then g is said to be finitely determined.*

Definition 6.4.3. *For any $g \in \mathcal{F}_\Gamma$ we define the pseudo-norm $\|\cdot\|^0_k$ (a norm on the subspace of $k-$jets) by*

$$\|g\|^0_k = \sum_{|i|+j\leq k} \left| \frac{\partial^{|i|+j} g}{\partial u^i \partial \lambda^j}(0,0) \right|,$$

using a multi-index i. If there exists $\varepsilon > 0$ such that all perturbations $f \in \mathcal{F}_\Gamma$ of g with

$$\|f - g\|^0_k < \varepsilon$$

satisfy $f \sim g$, then we say g is stable with respect to $\|\cdot\|^0_k$.

As consequence the group Γ must be absolutely irreducibly represented on $\mathbf{R}^n$ (see [131, pp. 40]). We shall work with the set of germs of bifurcation problems that are k–determined and stable with respect to $\|\cdot\|^0_k$ (or just simply *stable*).

To justify this restriction, firstly note that the set of germs of bifurcations which are *not* $k-$determined is of codimension infinity in $\mathcal{F}_\Gamma$. Namely, infinite number of parameters are required to set up a *universal unfolding* of these germs (cf. [129]). Secondly, we examine only generic structurally stable bifurcations that persist under sufficiently small $\Gamma-$equivariant perturbations, e.g. pitchfork bifurcations under $\mathbf{Z}_2$-symmetry or corank-1 or 2 bifurcations of nonlinear elliptic problems under D_4-equivariance. For both examples, singularity theory can be used to show that any stable problems must be 3−determined. The finite determinacy and stability of a bifurcation problem together imply that it is stable with respect to the smooth norm in the following sense: let $\bar{f}$, $\bar{g}$ be two functions whose corresponding germs f, g are in $\mathcal{F}_\Gamma$ with $\|f-g\|_k^0 < \varepsilon$. Then there is a neighborhood U of the origin such that in the Sobolev-norm

$$\|\bar{g}\|_k := \sum_{|i|+j\leq k} \max_{(u,\lambda)\in U} \left| \frac{\partial^{|i|+j}\bar{g}}{\partial u^i \partial \lambda^j}(u,\lambda) \right| \tag{6.17}$$

we have $\|\bar{f}-\bar{g}\|_k < 2\varepsilon$.

For many problems we can employ the established approaches in singularity theory to determine a priori normal forms of the bifurcation equations (6.7). Then we calculate the necessary coefficients in the normal form. In this way singularity theory not only simplifies analysis of the bifurcation behavior, but also reduces computational effort by avoiding calculations of many unnecessary terms in Taylor expansion of the reduced bifurcation equations. We refer to Golubitsky/Schaefer [129] and Golubitsky/Stewart/Schaefer [131] for further discussions on normal forms.

6.5 Simple Bifurcation Points

Let $p=1$ and $n=1$, i.e., the problem (6.1) depends on one parameter and (u_0,λ_0) is a simple bifurcation point. Denote

$$\mathrm{Ker}(D_u\Phi_0) = \mathrm{span}[\phi_1], \qquad \mathrm{Ker}(D_u\Phi_0^*) = \mathrm{span}[\phi_1^*]$$

and $w_1 \in M$ as solution of the equation

$$D_u\Phi_0 w_1 + D_\lambda\Phi_0 = 0.$$

The reduced bifurcation equation is, see also Section 4.1,

$$f(\alpha,\lambda) = a\alpha^2 + 2b\alpha\lambda + c\lambda^2 + \cdots = 0. \tag{6.18}$$

where

$$\begin{aligned} a &:= \langle \phi_1^*,\ D_{uu}\Phi_0\phi_1\phi_1 \rangle, \\ b &:= \langle \phi_1^*,\ D_u D\Phi_0(w_1,1)\phi_1 \rangle, \\ c &:= \langle \phi_1^*,\ D^2\Phi_0(w_1,1)^2 \rangle. \end{aligned} \tag{6.19}$$

Under the transversal condition

$$b^2 - ac > 0, \tag{6.20}$$

see also (3.10), the equation (6.18) is 2-determined. Perturbations of higher order $O(\|(\alpha,\lambda)\|^3)$ do not change the local structure of $f(\alpha,\lambda)$ and the solution of (6.18). The normal form of (6.18) is

$$\varepsilon(\alpha^2 - \lambda^2), \quad \varepsilon = \pm 1,$$

i.e.,

$$f(\alpha,\lambda) \sim \varepsilon(\alpha^2 - \lambda^2).$$

Thus the reduced bifurcation equation (6.18) has two solution branches $\alpha_i(\lambda)$, $i = 1, 2$, see e.g. Golubitsky/Schaefer [129, pp. 196]. Correspondingly, there are exactly two different solution branches of (6.1) passing through (u_0, λ_0), as described in Chapter 4.

6.6 Truncated Liapunov-Schmidt Method

As we have seen before, at steady state and Hopf bifurcations the Liapunov-Schmidt method reduces the operator equation (6.1) to a low dimensional system of bifurcation equations

$$B(v,\lambda) = 0.$$

Here (v,λ) is an element in $\mathrm{Ker}(D_u\Phi_0) \times \mathbf{R}^p$. The mapping $B(v,\lambda)$ is singular and usually not available via analytical methods. What we can do is to calculate its Taylor expansion at $(v,\lambda) = (0,0)$. The truncated Taylor expansion of $B(v,\lambda)$ describe qualitatively local bifurcation behavior of finitely determined problem. We consider an iterative method in [21, 22, 24] for computing Taylor expansion of $B(v,\lambda)$.

Define a subspace of polynomials of the degree $\leq k$

$$\Pi_k := \{p: \ \mathrm{Ker}(D_u\Phi_0) \times \mathbf{R}^p \to \mathcal{X} \mid p\text{– a polynomial of } \alpha, \lambda \text{ of degree } \leq k\}.$$

Let $\phi_1, \ldots, \phi_n$ be a basis of $\mathrm{Ker}(D_u\Phi_0)$. We identify an element $v \in \mathrm{Ker}(D_u\Phi_0)$ with its coordinates $\alpha = (\alpha_1, \ldots, \alpha_n) \in \mathbf{R}^n$ via the relation

$$v = \sum_{i=1}^{n} \alpha_i \phi_i.$$

For a nonlinear mapping $B: \ \mathrm{Ker}(D_u\Phi_0) \times \mathbf{R}^p \to \mathcal{X}$ with $B(0) = 0, \ DB(0) = 0$, we denote by $j_k B \in \Pi_k$ the truncated Taylor expansion of B at 0 up to the k-th order term.

We exclude turning point bifurcations (they are regular points of the mapping Φ) and assume, reparametrizing if necessary, that

$$D_\lambda \Phi(0,0) = 0.$$

This corresponds to the assumption of a trivial solution from which the branches bifurcate. We rewrite the mapping $\Phi(u,\lambda)$ into linear and nonlinear parts:

$$\Phi(u,\lambda) =: D_u\Phi(0,0)u + R(u,\lambda),$$

where $R(u,\lambda) = O(\|(u,\lambda)\|^2)$. With the decomposition $u = v + w$, $v \in \mathrm{Ker}(D_u\Phi_0)$ and $w \in M$, we recall, see the Liapunov-Schmidt reduction in Section 6.1, that w is determined uniquely as a function of v, λ and $w(v,\lambda) = O(\|(v,\lambda)\|^2)$. For v, λ we have the reduced bifurcation equation

$$B(v,\lambda) := (I - \hat{Q})\Phi(v + w(v,\lambda),\lambda) = (I - \hat{Q})R(v + w(v,\lambda),\lambda) = 0. \quad (6.21)$$

For $p = 1$ we calculate the functions $w(v,\lambda)$ and $B(v,\lambda)$ with the following iterative method.

Algorithm 6.6.1. Truncated Liapunov-Schmidt Method.

Step 1) Set $w_1(v,\lambda) = 0$;
Step 2) FOR $k = 2, 3, \ldots,$ *until determinacy; DO*
1. *Calculate the truncated bifurcation equations*

$$B_k(v,\lambda) := (I - \hat{Q})j_k R(v + w_{k-1}(v,\lambda),\lambda) = 0; \quad (6.22)$$

2. *Solve* $w_k(v,\lambda) \in M$ *from the equation*

$$D_u\Phi_0 w_k(v,\lambda) = \hat{Q} j_k R(v + w_{k-1}(v,\lambda),\lambda). \quad (6.23)$$

Here $B_k(v,\lambda)$ is a polynomial of k-th order from $\mathrm{Ker}(D_u\Phi_0) \times \mathbf{R}$ into $\mathrm{Ker}(D_u\Phi_0{}^*)$, so is $w_k(v,\lambda)$, but from $\mathrm{Ker}(D_u\Phi_0) \times \mathbf{R}$ into M.

Theorem 6.6.1. *For the iterations defined in (6.22), (6.23), we have*

$$w_k(v,\lambda) = j_k w(v,\lambda) \quad \textit{and} \quad B_k(v,\lambda) = j_k B(v,\lambda).$$

Proof: Consider

$$\begin{aligned} &w_{k+1} - w_k \quad (6.24)\\ &= Q(D_u\Phi_0)^{-1}\hat{Q}\left[j_{k+1}R(v + w_k(v,\lambda),\lambda) - j_k R(v + w_{k-1}(v,\lambda),\lambda)\right]. \end{aligned}$$

The equality $j_1(R(v,\lambda)) = 0$ implies that

$$j_k(R(v + \tilde{w}(v,\lambda),\lambda)) = j_k(R(v + j_{k-1}\tilde{w}(v,\lambda),\lambda))$$

for any $\tilde{w}(v,\lambda)$ with $j_1(\tilde{w}(v,\lambda)) = 0$. By induction we get that for all $k \geq 2$

$$\begin{aligned} & j_k[R(v + w_{k+1}(v,\lambda),\lambda) - R(v + w_k(v,\lambda),\lambda)] = 0 \\ \Rightarrow\ & j_k(w_{k+1} - w_k) = 0 \\ \Rightarrow\ & j_k w_{k+1} = w_k. \end{aligned}$$

From the equalities $j_k\hat{Q}R(v + w_k(v,\lambda),\lambda) = j_k\hat{Q}R(v + w(v,\lambda),\lambda) = 0$ we derive that the solution $w(v,\lambda)$ of $\hat{Q}R(v + w(v,\lambda),\lambda) = 0$ has Taylor series $j_k w(v,\lambda) = w_k(v,\lambda)$. This implies that $j_k B(v,\lambda) = B_k(v,\lambda)$. ∎

Example 6.6.1. For $v = \sum_{i=1}^{n} \alpha_i \phi_i \in \mathrm{Ker}(D_u\Phi_0)$ we write

$$\begin{aligned} w(v,\lambda) = w_{01}^{(1)}\lambda + \sum_{i,j=1}^{n} w_{ij0}^{(2)}\alpha_i\alpha_j + \sum_{i=1}^{n} w_{0i1}^{(2)}\alpha_i\lambda + w_{002}^{(2)}\lambda^2 \\ + O(\|(\alpha,\lambda)\|^3), \qquad w_{ijk}^{(2)} \in M. \end{aligned}$$

The coefficients $w_{ijk}^{(2)} = w_{jik}^{(2)} \in M$ are solutions of the following equations respectively

$$\begin{aligned} D_u\Phi_0 w_{ij0}^{(2)} &= -\frac{1}{2}\hat{Q}D_{uu}\Phi_0\phi_i\phi_j,\ i,j = 1,\dots,n, \\ D_u\Phi_0 w_{0i1}^{(2)} &= -\hat{Q}D_{u\lambda}\Phi_0\phi_i,\ i = 1,\dots,n, \\ D_u\Phi_0 w_{002}^{(2)} &= -\frac{1}{2}\hat{Q}D_{\lambda\lambda}\Phi_0. \end{aligned}$$

The k-jets ($k = 2,3$) of the reduced bifurcations equation are

$$j_2 B = \frac{1}{2}(I - \hat{Q})\Big(D^2\Phi_0(\sum_{i=1}^{n}\alpha_i\phi_i + w_{01}^{(1)}\lambda,\ \lambda)^2\Big) \tag{6.25a}$$

and

$$\begin{aligned} j_3 B = j_2 B + (I - \hat{Q})\Big[\frac{1}{6}D^3\Phi_0(\sum_{i=1}^{n}\alpha_i\phi_i + w_{01}^{(1)}\lambda,\ \lambda)^3 \qquad\qquad (6.25b) \\ + D_u D\Phi_0(\sum_{i=1}^{n}\alpha_i\phi_i + w_{01}^{(1)}\lambda,\ \lambda)\Big(\sum_{i,j=1}^{n} w_{ij0}^{(2)}\alpha_i\alpha_j + \sum_{i=1}^{n} w_{0i1}^{(2)}\alpha_i\lambda + w_{002}^{(2)}\lambda^2\Big)\Big]. \end{aligned}$$

6.7 Branch Switching at Multiple Bifurcation Points

Let (u_0,λ_0) be a multiple bifurcation point of the stationary problem

$$G(u,\lambda) = 0. \qquad [illegible])$$

Assume that (u_0,λ_0) and ϕ_i, ϕ_i^*, $i = 1,\dots,n$, w_1 are knov[illegible] are interested in branch switching at (u_0,λ_0).

6.7.1 Branch switching with prescribed tangents

We extend the simple and direct approach for branch switching with prescribed tangents to multiple bifurcation points. Following the discussions of Section 4.1, a solution curve $(u(s), \lambda(s))$ with $(u(0), \lambda(0)) = (u_0, \lambda_0)$ is of the form

$$\begin{aligned} u(s) &= u_0 + \sum_{i=1}^{n} s\alpha_i(s)\phi_i + s\beta(s)w_1 + s^2 w(s), \quad w(s) \in M, \\ \lambda(s) &= \lambda_0 + s\beta(s), \quad s \in \mathcal{I} \subset \mathbf{R}. \end{aligned} \tag{6.27}$$

Furthermore, $\sum_{i=1}^{n} |\alpha_i(0)| + |\beta(0)| > 0$. The tangent of this solution curve at $s = 0$ is

$$(\dot{u}(0),\ \dot{\lambda}(0)) = \Big(\sum_{i=1}^{n} \alpha_i(0)\phi_i + \beta(0)w_1,\ \beta(0)\Big).$$

We determine this tangent via an isolated solution of k-jet of the reduced bifurcation equations, e.g. the 2-jet

$$j_2(2f_i(\alpha,\beta)) = \sum_{j,k=1}^{n} a_{ijk}\alpha_j\alpha_k + 2\sum_{j=1}^{n} b_{ij}\alpha_i\beta + c_i\beta^2 = 0, \ i = 1, \ldots, n, \tag{6.28a}$$

with the normalization

$$\sum_{i=1}^{n} \alpha_i^2 + \beta^2 = 1, \tag{6.28b}$$

where

$$\begin{aligned} a_{ijk} &:= \langle \phi_i,\ D_{uu}G_0\phi_j\phi_k \rangle, \\ b_{i,j} &:= \langle \phi_i^*,\ D_uDG_0(w_1, 1)\phi_j \rangle, \quad i,j,k = 1, \ldots, n, \\ c_i &:= \langle \phi_i^*,\ D^2G_0(w_1, 1)^2 \rangle. \end{aligned}$$

Choosing an isolated solution $(\alpha_1^0, \ldots, \alpha_n^0,\ \beta^0)$ of (6.28), correspondingly, the k-jet of the reduced bifurcation equations, we define a pseudo-arclength parameterization of the bifurcating solution branch with the system

$$F(u, \lambda, s) := \begin{pmatrix} G(u, \lambda) \\ N(u, \lambda, s) \end{pmatrix} = 0$$

and

$$N(u, \lambda, s) := \Big\langle \sum_{i=1}^{n} \alpha_i^0\phi_i + \beta^0 w_1,\ u - u_0 \Big\rangle + \beta^0(\lambda - \lambda_0) + s.$$

Thereafter, we choose a predictor

$$(u^0,\ \lambda^0) := \left(u_0 + s\Big(\sum_{i=1}^{n} \alpha_i^0 \phi_i + w_1\Big),\ \lambda_0 + \beta^0 s\right)$$

and continue the solution branch with Euler-Newton (chord) method similarly as at simple bifurcation points. Convergence of the correctors is expected in a conical neighborhood of the solution curve. Vertex of the cone is the bifurcation point, see Mei [216].

6.7.2 Branch switching with scaling techniques

Generally 2-jet (3-jet) of the reduced bifurcation equations describes sufficiently the stationary (Hopf) bifurcation scenario. Singularity of higher orders occurs when the problem (6.1) involves symmetries. In that case higher order terms in the Taylor expansion of the reduced bifurcation equations need to be calculated. In the Liapunov-Schmidt reduction this is realized with two coupled iteration processes as in Algorithm 6.6.1. Determinacy of the truncated bifurcation equations is monitored along the reduction, see e.g. [22, 42]. As an alternative of this method, we exploit in this section a blowing-up technique which scales different solution branches and unify the two successive steps in the Liapunov-Schmidt method, see Mei/Schwarzer [226] and Schwarzer [272]. We consider a numerical algorithm for numerical verification of existence of the bifurcating solution branches and for branch switching at the bifurcation point. In this algorithm we check the determinacy of truncated Taylor expansion of the reduced bifurcation equations order by order and calculate simultaneously directions for branch switching along the path following of solution curves.

Recall that a solution curve $(u(s), \lambda(s))$ across the bifurcation point (u_0, λ_0) is necessarily of the form (6.27). We assume $\dot{u}(0) \neq 0$. If $\dot{u}(0) = 0$, then $\dot{\lambda}(0) \neq 0$ holds. Replacing the parameter λ with α_j for a generic $j \in \{1 \dots n\}$ will yield similar conclusions as below.

In (6.27) we expand $w(s)$ and $\beta(s)$ into Taylor series with respect to s at $s = 0$ and consider the following generic form of the solution curve

$$u = u_0 + \sum_{i=1}^{n} s\alpha_i \phi_i + \sum_{i=1}^{k} s^i w_i + s^{k+1} w(\alpha_1, \dots, \alpha_n, s),\ w_i \in M$$

$$\lambda = \lambda_0 + \sum_{i=1}^{k-1} s^i \beta_i + s^k \beta(s).$$

Here the integer k depends on the determinacy of the given problem and will be specified later. We will determine $\alpha = (\alpha_1, \dots, \alpha_n) \in \mathbf{R}^n$, $\beta \in \mathbf{R}$ and $w \in M$ in the solutions of (6.26) via an enlarged system.

To this end, we consider all $(u, \lambda) \in X \times \mathbf{R}$ of the form

$$u = u_0 + \sum_{i=1}^{n} s\alpha_i \phi_i + \sum_{i=1}^{k} s^i w_i + s^{k+1} w, \tag{6.29a}$$

$$\lambda = \lambda_0 + \sum_{i=1}^{k-1} s^i \beta_i + s^k \beta \tag{6.29b}$$

and take the Taylor expansion of $G(u, \lambda)$ at (u_0, λ_0)

$$\begin{aligned} G(u,\lambda) &= DG_0(u-u_0, \lambda-\lambda_0) + R(u-u_0, \lambda-\lambda_0) \\ &= \sum_{i=1}^{k} s^i \Big(D_u G_0 w_i + D_\lambda G_0 \beta_i + r_i(\alpha, w_1, \ldots, w_{i-1}, \beta_1, \ldots, \beta_{i-1}) \Big) \\ &+ O(s^{k+1}). \end{aligned}$$

Here $DG_0(u - u_0, \lambda - \lambda_0)$ is the linearization of $G(u, \lambda)$ at the bifurcation point (u_0, λ_0), while $R(u - u_0, \lambda - \lambda_0)$ collects all nonlinear terms

$$\begin{aligned} &R(u-u_0, \lambda-\lambda_0) \\ &= \sum_{m=2}^{\infty} \frac{1}{m!} D^m G_0 \left(\sum_{i=1}^{n} s\alpha_i \phi_i + \sum_{i=1}^{k} s^i w_i + s^{k+1} w, \ \sum_{i=1}^{k-1} s^i \beta_i + s^k \beta \right)^m . \end{aligned}$$

The term $r_i(\alpha, w_1, \ldots, w_{i-1}, \beta_1, \ldots, \beta_{i-1})$ is the coefficient of s^i in $R(u - u_0, \lambda - \lambda_0)$ and independent of w_i, β_i, because $R(u - u_0, \lambda - \lambda_0)$ starts quadratically in $u - u_0$ and $\lambda - \lambda_0$.

Remark 6.7.1. Let Γ be a compact Lie group acting linearly in X, Y and G is Γ-equivariant. Then the mapping $r_i(\alpha, w_1, \ldots, w_{i-1}, \beta_1, \ldots, \beta_{i-1})$ is Σ_{u_0}-equivariant, i.e., the statement

$$r_i(\sigma\alpha, \sigma w_1, \ldots, \sigma w_{i-1}, \beta_1, \ldots, \beta_{i-1}) = \sigma r_i(\alpha, w_1, \ldots, w_{i-1}, \beta_1, \ldots, \beta_{i-1})$$

holds for all $\sigma \in \Sigma_{u_0}$, $\alpha \in \mathbf{R}^n$ and $w_j \in X$, $j = 1, \ldots, i-1$.

Along a solution curve of (6.26) the coefficient $D_u G_0 w_i + D_\lambda G_0 \beta_i + r_i$ of s^i vanishes for all $i \in N$, leading to the following equations for the unknowns $w_i \in M$ and $\alpha_1, \ldots, \alpha_n$, $\beta_i \in \mathbf{R}$

$$D_u G_0 w_i + D_\lambda G_0 \beta_i = -r_i, \tag{6.30a}$$

$$\langle \phi_j, \ w_i \rangle = 0, \quad j = 1, \ldots, n, \tag{6.30b}$$

$$\langle \phi_j^*, \ r_i(\alpha, w_1, \ldots, w_{i-1}, \beta_1, \ldots, \beta_{i-1}) \rangle = 0. \tag{6.30c}$$

The right hand side r_i of (6.30a) is a polynomial in the variables $\alpha_1, \ldots, \alpha_n$ and $\beta_1, \ldots, \beta_{i-1}$. Similarly, the solution w_i of the linear equations (6.30a)-(6.30b) is also a polynomial in these variables. In particular, owing to $r_1 = 0$ we derive $w_1 = \tilde{w}_1 \beta_1$ from the system (6.30), where $\tilde{w}_1 \in M$ satisfies

$$D_uG_0\tilde{w}_1 + D_\lambda G_0 = 0.$$

If the statement

$$D_uDG_0(\tilde{w}_1, 1)\phi_1 \notin \mathrm{Im}(D_uG_0) \tag{6.31}$$

holds, we can solve β_{i-1} from the equation (6.30c) as a homogeneous polynomial of the degree $i-1$ in the variables $\alpha_1, \ldots, \alpha_n$. Substituting it into w_i shows that w_i is a homogeneous polynomial of the degree i in $\alpha_1, \ldots, \alpha_n$. In the sequel we restrict the discussion to the case (6.31).

The system (6.30c) with the unknowns $\alpha_1, \ldots, \alpha_n$ and the parameter β_{i-1} serves as a *test for existence of the bifurcating solution branches* of the problem $G(u, \lambda) = 0$. If there is an isolated solution α as a function of β_i, the corresponding solution of (6.30a)-(6.30b) leads to a solution branch of (6.26) passing through (u_0, λ_0). We use the following algorithm to find such a solution branch.

Algorithm 6.7.1. Branch switching at multiple bifurcation points.
Initial: *For $k = 1$, solve $w_1 \in M$ from the equation:*

$$D_uG_0w_1 = -D_\lambda G_0\beta_1. \tag{6.32}$$

Iteration: *FOR $k = 2, \ldots,$ DO*

Step 1) Check whether the system (6.30c), i.e.,

$$\langle \phi_j^*, r_k \rangle = 0, \quad j = 1, \ldots, n \tag{6.30c}$$

has a real and isolated solution $\alpha(\beta_{k-1})$ for β_{k-1} in an open interval. If not, we go to Step 2). Otherwise, we test existence of solution branches and switch the branches as follows;

1. *Define an enlarged system for (w, α, β) via the mapping* $F: \mathcal{X} := X \times \mathbf{R}^{n+1} \times \mathbf{R} \longmapsto \mathcal{Y} := Y \times \mathbf{R}^{n+1}$

$$F(w, \alpha, \beta, s) := \begin{pmatrix} G(u_0 + \sum_{i=1}^{n} s\alpha_i\phi_i + \sum_{i=1}^{k-1} s^iw_i + s^kw,\ \lambda_0 + \sum_{i=1}^{k-2} s^i\beta_i + s^{k-1}\beta)/s^k \\ \langle \phi_1, w \rangle \\ \vdots \\ \langle \phi_n, w \rangle \\ \beta^2 - 1 \end{pmatrix} = 0. \tag{6.33}$$

For $s = 0$, the mapping F is defined via its limit

$$F(w, \alpha, \beta, 0) := \begin{pmatrix} D_uG_0w + r_k \\ \langle \phi_1, w \rangle \\ \vdots \\ \langle \phi_n, w \rangle \\ \beta^2 - 1 \end{pmatrix} = 0. \tag{6.34}$$

Here the intermediate terms $w_i, i = 1, \ldots, k-1$ and β_j, $j = 1, \ldots, k-2$ are chosen successively as solutions of the equations (6.30).

2. *Find nonsingular solutions (α^0, β^0) of (6.30c) with the normalization $(\beta^0)^2 = 1$. Thereafter, we calculate the corresponding solution $w^0 \in M$ of (6.30a)-(6.30b) and set up nonsingular solutions $(w^0, \alpha^0, \beta^0, 0)$ of (6.33) at $s = 0$. Applying the implicit function theorem to the enlarged problem (6.33) at $(w^0, \alpha^0, \beta^0, 0)$ we obtain a unique solution branch $(w(s), \alpha(s), \beta(s), s)$, correspondingly, a solution branch of (6.26) across (u_0, λ_0).*
3. *Start at $(w^0, \alpha^0, \beta^0, 0)$ and continue the solution curve of (6.33) with the Euler-Newton method.*
4. *Go back to Step 1).*

Step 2) Solve β_{k-1} as a polynomial of $\alpha_1, \ldots, \alpha_n$; substitute it into the system

$$D_u G_0 w_k + D_\lambda G_0 \beta_k = -r_k$$
$$\langle \phi_j, w_k \rangle = 0, \quad j = 1, \ldots, n,$$

and solve it for w_k. If no real solutions are available, we stop the process, since (6.26) has no solution branches of the form (6.29) at (u_0, λ_0). Otherwise, let $k = k + 1$ and go back to Step 1).

For simple algebraic problems w_i may be calculated analytically, e.g., via computer algebra packages. For differential equations numerical solution is much more efficient and often the only possibility. On the other hand, note that as polynomials of α, the structure of r_i and w_i are identical. By examining the structure of r_i we obtain from (6.30) linear equations for the coefficients of the polynomial w_i, which are then solved numerically. Thus solving (6.30) effectively requires combinations of numerical and computer algebraic methods. In particular, if the problem (6.26) has some equivariant properties, the computational work of solving (6.30) can be greatly reduced by a systematic application of group-theoretic methods, see e.g Ashwin/Böhmer/Mei [21, 22].

If the problem (6.26) is finitely determined and the condition (6.31) holds, then a solution curve $(u(s), \lambda(s))$ of (6.26) satisfies (6.33) for appropriate k. We recall that the solution curve $(u(s), \lambda(s))$ is represented via $w(s), \alpha(s), \beta(s)$ which are in turn nonsingular solutions of the enlarged system (6.33). Numerical path following can be done directly on (6.33). Switching branch from one to another corresponds to starting from different solutions of (6.33) at $s = 0$ (cf. Mei [213, 217]). Our numerical experience on reaction-diffusion problems indicates that solutions of (6.33) can be continued for fairly large ranges of s.

Remark 6.7.2. If it is known in advance from the symmetries or the nonlinearities of a problem that a determinacy is possible only for k greater than certain minimum order k_0, we start the iteration directly with an appropriate large value k in the ansatz (6.29) and consider the equations (6.30) for $i = 1, \ldots, k$ and the enlarged system (6.33) in one step. In such a way the iteration is done simultaneously for a few steps before increasing the orders of k, see Section 6.8.2.

6.8 Corank-2 Problems with D_m-symmetry

Assume that the kernel $\mathrm{Ker}(D_u G_0)$ is two dimensional, D_m-invariant and irreducible. The reduced equations (6.30c) with D_m-symmetry are of the following form

$$g(z, \beta) = p(u, v, \beta)z + q(u, v, \beta)\bar{z}^{m-1} = 0 \tag{6.35}$$

with $u = z\bar{z}$, $v = z^m + \bar{z}^m$. Here we have taken complex coordinates $z = \alpha_1 + i\alpha_2$ in $\mathrm{Ker}(D_u G_0) \sim \mathbf{R}^2 \sim \mathbf{C}$, see e.g. [129]. The k-jet of this system is not determined until $k \geq m - 1$ in (6.30). In fact, for $k < m - 1$, by (6.35) the system (6.30c) is of the form

$$g(z, \beta) = p(u, \beta)z = p(z\bar{z}, \beta)z = 0, \tag{6.36}$$

which is $O(2)$-equivariant and has continuous $O(2)$-orbit in its solution set. This makes the system (6.30c) degenerate, i.e., the Jacobi-matrix of the left hand side of (6.30c) is singular at all solution points. The continuous symmetry $O(2)$ is broken only if higher order terms ($k \geq m - 1$) are considered. In that case, the system (6.30c) is generically of the form (6.35) and its solutions are discrete. If the Jacobian of (6.30c) has maximal rank at all solution points, the bifurcation scenario is determined by (6.33) and (6.30). In other words, for the determination of bifurcation scenario of a D_m-equivariant problem, the power k of s in (6.33) should be chosen greater than or equal to $m-1$. As examples, we consider in the following subsections D_4- and D_6-equivariant bifurcations.

6.8.1 Semilinear elliptic PDEs on a square

Consider the model equation

$$\Delta u + \lambda f(u) = 0 \qquad \text{in } \Omega := [0, 1] \times [0, 1], \tag{6.37}$$

and the homogeneous Neumann boundary conditions (cf. Mei [223])

$$\frac{\partial u}{\partial n} = 0 \quad \text{on } \partial\Omega.$$

Here f is assumed to be smooth and $f(0) = 0$. If f is an odd function with $f'(0) \cdot f'''(0) \neq 0$, the complete bifurcation scenario along the trivial solution branch $\{(0,\lambda);\ \lambda \in \mathbf{R}\}$ is determined by the 3-jet of the reduced bifurcation equations, which is equal to (6.30c) with $k = 3$ (cf. Allgower/Böhmer/Mei [9]).

Suppose

$$\text{a)}\quad f'(0) = 1, \qquad\qquad \text{b)}\quad f''(0) \neq 0. \tag{6.38}$$

This problem is D_4-equivariant, derived from the D_4-symmetry of the square domain, the boundary condition and Euclidean symmetry of the Laplacian. The equivariant branching lemma, see Section 5.3 and [9, 67, 131, 295], can be used to determine solution branches with maximal symmetries. To seek all bifurcating solution branches, we choose the enlarged system (6.33) for an analysis of bifurcation scenario, and simultaneously, for branch switching.

Let us consider the weak form of the problem (6.37)

$$\textit{Find}\quad (u,\lambda) \in X \times \mathbf{R} \qquad \textit{such that}\quad G(u,\lambda) := u + \lambda T f(u) = 0, \tag{6.39}$$

where $X = H^1(\Omega)$ and $T:\ g \in L^2(\Omega) \mapsto Tg \in X$ is defined implicitly by the equation

$$a(Tg, v) = -(g, v), \qquad \text{for all}\ \ v \in X, \quad g \in L^2(\Omega)$$

with $a(u,v) := \displaystyle\int_\Omega \nabla u \nabla v \, dxdy$ for $u, v \in X$ (cf. Mei [223]).

Remark 6.8.1. Owing to the appropriate regularity of solutions, the whole discussion can be done on (6.37) directly in the Hölder space $C^{2,s}$ as well (cf. Healey/Kielhöfer [155], Mei [223]).

A corank-2 bifurcation point $(0,\lambda_0)$ on the trivial solution curve $\{(0,\lambda);\ \lambda \in \mathbf{R}\}$ is characterized via

$$\lambda_0 = (m^2+n^2)\pi^2, \qquad m \neq n, \qquad \mathrm{Ker}(D_uG_0) = \mathrm{span}[\phi_1, \phi_2], \tag{6.40a}$$

$$\phi_1 := 2\cos m\pi x \cos n\pi y, \qquad \phi_2 := 2\cos n\pi x \cos m\pi y. \tag{6.40b}$$

Here ϕ_1, ϕ_2 are eigenfunctions of the Laplacian, corresponding to the eigenvalue $-\lambda_0$. They are often called the *modes.* The integers m, n are called *wavenumbers.* They describes numbers of waves of the mode in the interval $[0,1]$ on the x and y axis, respectively.

Since $D_uG_0 = I + \lambda_0 T$ is self-adjoint, we have

$$X = \mathrm{Ker}(D_uG_0) \oplus \mathrm{Im}(D_uG_0), \tag{6.41}$$

where $\oplus$ is the orthogonal sum under the inner product of $L^2(\Omega)$. The operator $Q := I - \langle\phi_1,\ \cdot\rangle\phi_1 - \langle\phi_2,\ \cdot\rangle\phi_2$ is an orthogonal projection from X onto $\mathrm{Im}(D_uG_0)$.

For $k = 3$ in (6.33), Mei [223] shows that on a nontrivial solution branch the following equalities hold

$$\beta_1 = 0, \qquad w_1 = 0, \qquad w_2 = \lambda_0 f''(0) \sum_{i+j=2} \alpha_i \alpha_j w_2[i,j], \tag{6.42}$$

where $w_2[i,j] \in \mathrm{Im}(D_u G_0)$ with $i+j=2$ are solutions of

$$D_u G_0 w_2[i,j] = -\frac{1}{2} T \phi_i \phi_j.$$

For $D_u G_0 = I + \lambda_0 T$ and $\phi_1,\ \phi_2$ in (6.40) these equations can be solved analytically. The unknown β_2 is determined by the following system which corresponds to (6.30c) for $k = 3$

$$\alpha_1(c_1\alpha_1^2 + c_2\alpha_2^2 + \frac{6\beta_2}{\lambda_0}) = 0, \tag{6.43a}$$

$$\alpha_2(c_2\alpha_1^2 + c_1\alpha_2^2 + \frac{6\beta_2}{\lambda_0}) = 0 \tag{6.43b}$$

with

$$c_1 := \frac{9}{4} f'''(0) + 6\lambda_0 f''(0)^2 \langle \phi_i^2,\ w_2[1,1]\rangle, \quad i = 1,2$$

$$c_2 := 3f'''(0) + 6\lambda_0 f''(0)^2 \Big[2\langle \phi_i\phi_j,\ w_2[i,j]\rangle + \langle \phi_i^2,\ w_2[i,j]\rangle \Big] \text{ for all } \ i \neq j.$$

If $c_1 \neq c_2$, the 3-jet (6.43) of the reduced bifurcation equations is determined and has eight nontrivial solutions for $c_1 \neq 0$ and four for $c_1 = 0$, respectively. Correspondingly, the original problem (6.37) has, four for $c_1 \neq 0$ and two for $c_1 = 0$, nontrivial solution branches bifurcating at $(0, \lambda_0)$, see Mei [223].

If $c_1 = c_2$, the system (6.43) is degenerated and contains the factor

$$\beta_2 = -\frac{\lambda_0 c_1}{6}(\alpha_1^2 + \alpha_2^2), \tag{6.44}$$

which has the S^1-symmetry ($S^1 = SO(2)$) of the circle in the (α_1, α_2)-plane. This S^1-group contains the D_4-group, the symmetry of the original problem, as a proper subgroup. The continuous symmetry indicates the problem (6.37) is no more 3-determined. However, we expect this S^1-symmetry and the degeneracy will be broken at higher orders. Let us consider the next order $k = 4$ and take the ansatz (6.29) with $k = 3$

$$\begin{aligned} u &= s(\alpha_1\phi_1 + \alpha_2\phi_2) + s^2 w_2 + s^3 w_3 + s^4 w, \\ \lambda &= \lambda_0 + \beta_2 s^2 + s^3 \beta. \end{aligned} \tag{6.45}$$

Here we have used the fact $\beta_1 = 0$, $w_1 = 0$ and β_2 , w_2 in (6.42) and (6.44), respectively. The coefficients $\alpha_1, \alpha_2, \beta \in \mathbf{R}$ and the elements $w_3, w \in \mathrm{Im}(D_u G_0)$ are to be determined. We examine the Taylor expansion $G(u, \lambda)$ at $(0, \lambda_0)$ with respect to the variables u, λ and collect the s^i components

$$G(u,\lambda) = \sum_{i=1}^{4} s^i (D_u G_0 w_i + r_i) + O(s^5),$$

where $r_1 = 0$ and

$$
\begin{aligned}
r_2 &= \frac{1}{2} D_{uu} G_0 (\alpha_1\phi_1 + \alpha_2\phi_2)^2, \\
r_3 &= D_{uu} G_0 (\alpha_1\phi_1 + \alpha_2\phi_2) w_2 + \beta_2 D_{u\lambda} G_0 (\alpha_1\phi_1 + \alpha_2\phi_2) \\
&\quad + \frac{1}{6} D_{uuu} G_0 (\alpha_1\phi_1 + \alpha_2\phi_2)^3, \\
r_4 &= \frac{1}{2} D_{uu} G_0 \Big(2(\alpha_1\phi_1 + \alpha_2\phi_2) w_3 + w_2^2 \Big) + D_{u\lambda} G_0 \Big(\beta_2 w_2 + (\alpha_1\phi_1 + \alpha_2\phi_2)\beta \Big) \\
&\quad + \frac{1}{2} \Big[\beta_2 D_{uu\lambda} G_0 (\alpha_1\phi_1 + \alpha_2\phi_2)^2 + D_{uuu} G_0 (\alpha_1\phi_1 + \alpha_2\phi_2)^2 w_2 \Big] \\
&\quad + \frac{1}{24} D_{uuuu} G_0 (\alpha_1\phi_1 + \alpha_2\phi_2)^4.
\end{aligned}
$$

The statement $r_2 \in \mathrm{Im}(D_u G_0)$ can be verified directly. Similarly, the relation $r_3 \in \mathrm{Im}(D_u G_0)$ is assured by (6.43) which is equivalent to $\langle \phi_1, r_3 \rangle = \langle \phi_2, r_3 \rangle = 0$. For $c_1 = c_2$ and β_2 in (6.44), we solve w_3 as a homogeneous cubic polynomial of $\alpha_1,\ \alpha_2$ from

$$
D_u G_0 w_3 = -r_3, \qquad \langle \phi_i,\ w_3 \rangle = 0, \quad i = 1, 2. \tag{6.46}
$$

This is a functional differential equation, because the right hand side is a cubic polynomial in $\alpha_1,\ \alpha_2$ with coefficients as functions in the spatial variables x, y.

Note that the equation for all w_i is linear and of the same structure. We exploit the D_4-symmetry in computation of w_i. Let us write the right hand side of (6.46) and the unknown w_3 as

$$
r_3 = \sum_{i+j=3} r_3[i,j] \alpha_1^i \alpha_2^j, \qquad w_3 = \sum_{i+j=3} w_3[i,j] \alpha_1^i \alpha_2^j
$$

with $r_3[i,j] \in C^2(\Omega)$ and

$$
\begin{aligned}
r_3[3,0] &= D_{uu} G_0 \phi_1 w_2[2,0] - \frac{1}{6} c_1 \lambda_0 D_{u\lambda} G_0 \phi_1 + \frac{1}{6} D_{uuu} G_0 \phi_1^3, \\
r_3[2,1] &= D_{uu} G_0 (\phi_2 w_2[2,0] + 2\phi_1 w_2[1,1]) - \frac{1}{6} c_1 \lambda_0 D_{u\lambda} G_0 \phi_2 + \frac{1}{2} D_{uuu} G_0 \phi_1^2 \phi_2.
\end{aligned}
$$

By the reflection $(x, y) \to (y, x)$ the terms $r_3[1,2]$ and $r_3[0,3]$ follow from $r_3[2,1]$ and $r_3[3,0]$, respectively. The unknowns $w_3[i,j]$ are solved via the equations

$$
D_u G_0 w_3[i,j] = -r_3[i,j], \quad i + j = 3.
$$

In particular, $w_3[1,2]$ and $w_3[0,3]$ are derived from $w_3[2,1]$, $w_3[3,0]$ by exchanging x and y. Therefore, we need only to solve two equations, i.e., those for $(i,j) = (3,0),\ (2,1)$. Rewriting these two equations back into differential form, we obtain

$$(\Delta+\lambda_0 f'(0))w_3[3,0] = \lambda_0\Big(f''(0)\phi_1 w_2[2,0] - \frac{1}{6}c_1 f'(0)\phi_1 + \frac{1}{6}f'''(0)\phi_1^3\Big);$$
$$(\Delta+\lambda_0 f'(0))w_3[2,1] = \lambda_0\Big[f''(0)\Big(\phi_2 w_2[2,0] + 2\phi_1 w_2[1,1]\Big) - \frac{1}{6}c_1 f'(0)\phi_2 + \frac{1}{2}f'''(0)\phi_1^2\phi_2\Big].$$

Since these equations are singular, we first regularize them as (6.13) and then solve the extended systems by finite difference method on a uniform mesh in x and y directions, equivalently to the finite element method with Courant's triangles. Once w_3 is available, coefficients of the equations $\langle\phi_1, r_4\rangle = 0$ are calculated with the adapted trapezoidal integrations. The equation $\langle\phi_2,\ r_4\rangle = 0$ is derived from $\langle\phi_1, r_4\rangle = 0$ by exchanging α_1 and α_2. Thereafter, isolated solutions of (6.30c) with $i = 4$ lead to solution branches of (6.37) bifurcating at (u_0, λ_0). If the equation (6.30c) still has degenerate solutions, the fifth order $i = 5$ needs to be considered.

Example 6.8.1. We choose the following function in (6.37)

$$f(u) = u + \frac{12}{4373}u^2 - \mu u^3. \tag{6.47}$$

Consider the wavenumbers $(m, n) = (1, 2)$ in (6.40) and examine bifurcations of (6.37) at $(u_0, \lambda_0) = (0, 5\pi^2)$. The bifurcation point (u_0, λ_0) is degenerate at $\mu = 32/625339$ for which the equality $c_1 = c_2 = -2677248/2734607447$ holds. Moreover, at the fourth order $i = 4$, the equations (6.30c) reduce to

$$\alpha_l\beta_3 = 0, \quad l = 1,\ 2.$$

Since $\alpha_1 = \alpha_2 = 0$ corresponds to $\dot{u}(0) = 0$, and in turn to the trivial solution, we consider $\beta_3 = 0$, $k = 5$ in Algorithm 6.7.1 and calculate (6.30c) to the fifth order

$$\beta\begin{pmatrix}\alpha_1\\ \alpha_2\end{pmatrix} - \begin{pmatrix}(a_h\alpha_1^4 + b_h\alpha_1^2\alpha_2^2 + c_h\alpha_2^4)\alpha_1\\ (c_h\alpha_1^4 + b_h\alpha_1^2\alpha_2^2 + a_h\alpha_2^4)\alpha_2\end{pmatrix} = 0. \tag{6.48}$$

Table 6.1 shows values of the coefficients $a_h,\ b_h,\ c_h$ for different meshsizes in a finite difference approximation.

h	a_h	b_h	c_h
1/10	0.3361476624	0.2682740866	1.3771682319
1/20	0.3400957075	0.2723790511	1.3742151531
1/40	0.3413393519	0.2734032481	1.3723933709
1/80	0.3417289811	0.2735640887	1.3714283718

Table 6.1. Numerical values of the coefficients in the reduced bifurcation equations of the PDE (6.37), (6.47).

We see immediately from (6.48) that the inequality $\beta > 0$ holds for nontrivial solutions. With the normalization $\beta = 1$, the system (6.48) has 8 real and isolated solutions

$$\left\{\pm\left(\frac{1}{a_h^{1/4}},0\right), \quad \pm\left(0, \frac{1}{a_h^{1/4}}\right), \quad \pm\frac{1}{(a_h+b_h+c_h)^{1/4}}(\pm 1, \ \pm 1)\right\}.$$

These induce four solutions branches of the enlarged system (6.33). Correspondingly, we obtain four solution branches of the problem (6.37) at (u_0, λ_0). Note that two of these solutions are on the same orbit of D_4-symmetry, while orbits of the other two are different.

6.8.2 A semilinear elliptic PDE on a hexagon

In this example we study a corank-2 bifurcation of the equation

$$G(u, \lambda) = \Delta u + \lambda(u + u^3) = 0 \tag{6.49}$$

on the Hexagon Ω with the unit sidelength and midpoint $(0,0)$, imposed with the homogeneous Dirichlet boundary conditions $u = 0$ on $\partial\Omega$. This problem is $D_6 \times Z_2$-equivariant, namely the D_6-symmetry of the domain and the Z_2-symmetry (the oddness in u) of the nonlinear term.

Similarly as for (6.37), $\{(0, \lambda); \quad \lambda \in \mathbf{R}\}$ is a trivial solution curve of (6.49). A bifurcation point on this curve is of the form $(0, \lambda_0)$ with λ_0 as an eigenvalue of $-\Delta$. The eigenvalues and eigenvectors of the Laplacian Δ on the hexagon are not analytically available. Numerical approximations for the first three eigenvalues are as follows

$$\begin{array}{lll} \lambda_1 \approx 7.1425, & & \text{single}, \\ \lambda_2 \approx 17.98, & & \text{double}, \\ \lambda_i \geq 26.37, & i \geq 3, & \text{multiplicity greater than one}, \end{array}$$

see e.g. Hersch [158] and Schwarzer [271].

We are interested in the behavior of (6.49) at the corank-2 bifurcation point $(u_0, \lambda_0) = (0, 17.98)$. Since the Laplacian on the hexagon with homogeneous Dirichlet boundary conditions is self-adjoint under the L^2-product, the decomposition (6.41) holds as in the previous example except that the basis of $\text{Ker}(D_u G_0) = \text{span}[\phi_1, \phi_2]$ is calculated numerically. By (6.35) the system (6.30c) for the D_6-equivariant problem is *not* determined until $k \geq 5$. With this in mind we take the following ansatz for u and λ

$$\begin{aligned} u &= u_0 + sv + sw_1 + s^2 w_2 + s^3 w_3 + s^4 w_4 + s^5 w + O(s^6) \\ \lambda &= \lambda_0 + t\beta_1 + s^2\beta_2 + s^3\beta_3 + s^4\beta + O(s^5), \end{aligned}$$

where $v = \alpha_1\phi_1 + \alpha_2\phi_2$, and $w_1 = 0$ for $D_\lambda G_0 = 0$. In this way, we start Algorithm 6.7.1 directly with $k = 5$. Expand $G(u, \lambda)$ into Taylor series at $(0, \lambda_0)$ up to the fifth order

$$
\begin{aligned}
G(u,\lambda) &= \sum_{i=1}^{5}(D_uG_0w_i + r_i)s^i + O(s^6)\\
&= s^2\left[(\Delta+\lambda_0 I)w_2 + \beta_1 v\right] + s^3\left[(\Delta+\lambda_0 I)w_3 + \beta_1 w_2 + \beta_2 v + \lambda_0 v^3\right]\\
&\quad + s^4\left[(\Delta+\lambda_0 I)w_4 + \beta_1 w_3 + \beta_2 w_2 + \beta_3 v + 3\lambda_0 v^2 w_2 + \beta_1 v^3\right]\\
&\quad + s^5\left[(\Delta+\lambda_0 I)w_5 + \beta_1 w_4 + \beta_2 w_3 + \beta_3 w_2 + \beta v\right.\\
&\quad + \left. 3\lambda_0 v^2 w_3 + 3\lambda_0 v w_2^2 + \beta_2 v^3 + 3\beta_1 v^2 w_2\right] + O(s^6).
\end{aligned}
$$

We consider successively the equations (6.30) for $i = 2,3,4,5$. For $i = 2$ the equation (6.30c) reduces to

$$\beta_1\alpha_j = 0, \quad j = 1,2.$$

The solution $\alpha_1 = \alpha_2 = 0$ corresponds to the trivial solution branch. We consider the case $\beta_1 = 0$ and obtain $w_2 = 0$ from the equations (6.30a)-(6.30b). Taking this into account, we derive from (6.30c) for $i = 3$

$$\left(\beta_2 + \lambda_0\kappa_1(\alpha_1^2+\alpha_2^2)\right)\begin{pmatrix}\alpha_1\\ \alpha_2\end{pmatrix} = 0,$$

where κ_1 is a constant. This system is $O(2)$-equivariant and includes a S^1-invariant factor. For nontrivial solution branches, we have

$$\beta_2 = -\lambda_0\kappa_1(\alpha_1^2+\alpha_2^2) \tag{6.50}$$

and the equation (6.30a) becomes

$$D_uG_0w_3 = \lambda_0[\kappa_1(\alpha_1\phi_1+\alpha_2\phi_2) + (\alpha_1\phi_1+\alpha_2\phi_2)^3].$$

In conjunction with (6.30b) w_3 is solved uniquely as a homogeneous cubic polynomial in α_1, α_2. Substituting w_2, w_3 and β_1, β_2 into (6.30c) for $i = 4$ yields

$$\beta_3\alpha_j = 0, \quad j = 1,2,$$

in which, again, only the case $\beta_3 = 0$ is relevant for nontrivial solutions. Correspondingly, $w_4 = 0$ is the unique solution for (6.30a)-(6.30b). Thereafter, the equation (6.30c) for $i = 5$ reduces to

$$\langle\phi_j,\ \beta v + \beta_2(w_3+v^3) + 3\lambda_0 v^2 w_3\rangle = 0, \quad j = 1,2. \tag{6.51}$$

Complexifying the variables α_1, α_2 with $z = \alpha_1 + i\alpha_2$, we write (6.51) as

$$\beta z - \lambda_0\kappa_1^2(z\bar{z})^2 z - 3\lambda_0^2(\kappa_2(z\bar{z})^2 z + \vartheta\bar{z}^5) = 0. \tag{6.52}$$

Here κ_1, κ_2, ϑ are constants and calculated numerically, for examples, using a uniform triangular mesh of the hexagon and taking piecewise linear approximation in the finite elements, we obtain approximations for κ_1, κ_2, θ in Table 6.2, where h represents the length of the equilateral triangle.

h	κ_1	κ_2	ϑ
1/6	0.90984382	−0.00436082	0.00023493
1/12	0.91060974	−0.00385766	0.00021589
1/24	0.91070398	−0.00374470	0.00021174
1/48	0.91071580	−0.00371718	0.00021076

Table 6.2. Numerical values of the coefficients in the reduced bifurcation equations (6.52) of the PDE (6.49).

The convergent rate of the coefficients should be in consistent with the error estimates of the order $O(h^2)$, the discretization error of the Laplacian. We see this in Table 6.2 in which the meshsize is half of the previous step.

Taking $h = 1/12$, we have $\kappa_1 = 0.9106$, $\kappa_2 = -0.00385$, $\vartheta = 0.000215$. The system (6.52) with $\beta^2 = 1$ has 25 isolated solutions. Besides the trivial solution $(0,0)$ and six pairs of complex solutions which belong to $\beta = -1$, the other 12 real solutions $(\alpha_1, \alpha_2, \beta)$ are

$$\{(0,\ 0.5495,\ 1),\ (0.5445,\ 0,\ 1)\}$$

and elements on their $D_6 \times Z_2$-orbits. The corresponding values for β_2 in (6.50) which determine bifurcation directions are $\beta_2 = -4.94$ and -4.85, respectively.

Since the solutions of (6.52) or (6.49) on the same group orbit can be derived from one of them simply by group actions, we restrict the discussion to the representatives $(0,\ 0.5495,\ 1)$ and $(0.5445,\ 0,\ 1)$. Substituting $(\alpha_1^0, \alpha_2^0, \beta^0) = (0,\ 0.5495,\ 1)$ into (6.30a)-(6.30b), we obtain a unique solution $w^0 \in \mathrm{Im}(D_u G_0)$.

For $k = 5$ and β_i, w_{i+1}, $i = 1, 2, 3$ as above, the enlarged system (6.33) is

$$F(w, \alpha, \beta, s) \tag{6.53}$$
$$:= \begin{pmatrix} G(s(\alpha_1\phi_1 + \alpha_2\phi_2) + s^3 w_3 + s^5 w,\ \lambda_0 + s^2\beta_2 + s^4\beta)/s^5 \\ \langle \phi_1, w \rangle \\ \langle \phi_2, w \rangle \\ \beta^2 - 1 \end{pmatrix} = 0.$$

At $t = 0$ we define

$$F(w, \alpha, \beta, 0) := \begin{pmatrix} D_u G_0 w + r_5 \\ \langle \phi_1, w \rangle \\ \langle \phi_2, w \rangle \\ \beta^2 - 1 \end{pmatrix} = 0. \tag{6.54}$$

Since $\beta_1 = \beta_3 = 0$ and $w_1 = w_2 = w_4 = 0$, we have

$$r_5 = \beta v + 3\lambda_0 v^2 w_3 + \beta_2 v^3, \qquad v = \alpha_1\phi_1 + \alpha_2\phi_2.$$

At the nonsingular solution point $(w^0, \alpha^0, \beta^0, 0)$ of (6.54), the implicit function theorem is applicable to F and yields a unique solution curve, which corresponds to a bifurcating solution branch of (6.49) at $(u_0, \lambda_0) = (0, 17.98)$ with the tangent $(\dot{u}(0), \dot{\lambda}(0)) = (\alpha_1\phi_1 + \alpha_2\phi_2, 0)$ and $(\alpha_1, \alpha_2) = (0,\ 0.5495)$, while $\ddot{\lambda}(0) = 2\beta_2 = -9.88$ and $\lambda^{(4)}(0) = 24$. Owing to the oddness of the nonlinearity in (6.49), if u is a solution, so is $-u$. We count $\pm u$ as one solution. The D_6-action on this solution branch gives five other solution branches of (6.49).

Similar discussion holds for $(\alpha_1^0, \alpha_2^0, \beta^0) = (0.5445,\ 0,\ 1)$. Hence, the problem (6.49) has altogether 12 different nontrivial solution branches passing through $(u_0, \lambda_0) = (0, 17.98)$. They lie on two different D_6-orbits. The structure of the solution branches is

$$\begin{aligned} u &= u_0 + s\alpha_1\phi_1 + s\alpha_2\phi_2 + s^3 w_3 + s^5 w_5 + O(s^6), \\ \lambda &= \lambda_0 + s^2\beta_2 + s^4\beta + O(s^5) \end{aligned}$$

with $\beta_2 < 0$, indicating that the bifurcations are subcritical and pitchfork. Figure 6.2 shows representatives on the orbits of two different solution branches of (6.49).

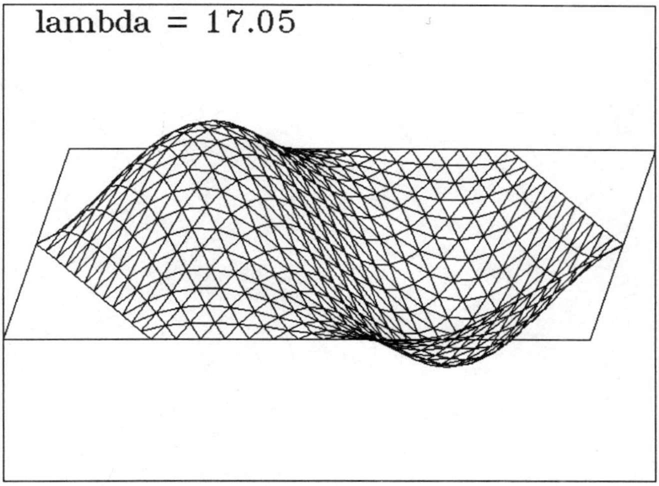

a) Solution branches corresponding to $(\alpha^0, \beta^0) = (0.5445, 0, 1)$ at $\lambda = 17.05$;

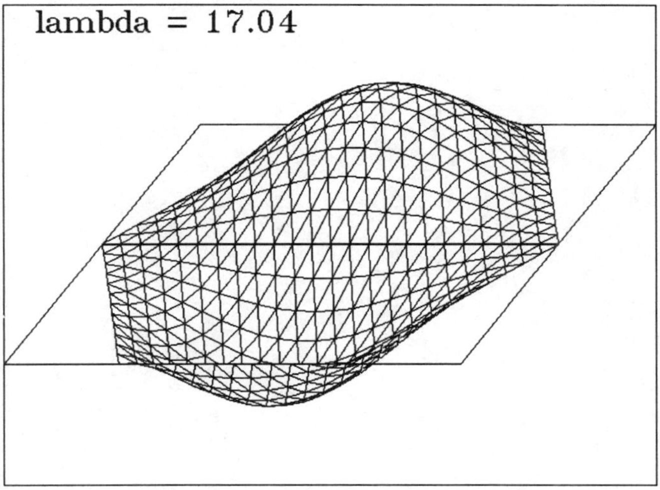

b) Solution branches with $(\alpha^0, \beta^0) = (0, 0.5495, 1)$ at $\lambda = 17.04$.

Fig. 6.2. Bifurcating solution branches.

7. Center Manifold Theory

Center manifold theory is essential for analyzing local bifurcations. As the Liapunov-Schmidt reduction for stationary and Hopf bifurcations, center manifold theory is used to reduce a dynamical system near a nonhyperbolic equilibrium or a periodic solution to a low-dimensional system with the vector field as functions of the critical modes. Furthermore, stability of solutions and local dynamics of the system can be derived from the low-dimensional system. The center manifold theorem was introduced in the sixties by Pliss [243] and Kelley [182]. Owing to the Lanford's contribution [198] this theory has been applied extensively to the study of bifurcation problems and dynamical systems, in particular, in connection with the normal form theory.

In this chapter we discuss some fundamental results of the center manifold theory, its relation to the Liapunov-Schmidt method and combination with normal form theory. We are mainly interested in applying the techniques of center manifold theory to the reaction-diffusion equations and in deriving the underlying lower dimensional systems. We refer to Carr [52], Iooss/Adelmeyer [166], Vanderbauwhede [296] for detailed discussions on center manifold theory and normal form theory for ordinary differential equations in finite dimensional spaces, and Chow/Lu [64], Henry [157], Vanderbauwhede/Iooss [298] for infinite dimensional systems.

7.1 Center Manifolds and Their Properties

Let X be a Banach space with the norm $||\cdot||$. We consider a dynamical system

$$\frac{\partial u}{\partial t} = L(\lambda)u + N(u,\lambda) =: G(u,\lambda) \tag{7.1}$$

with $u(0) = u_0 \in X$ and $L(\lambda)$ a linear operator from $D(L(\lambda)) \subset X$ into X. We assume that $L(\lambda)$ is the generator of a strongly continuous semigroup $S(t)$ on X and $N: \ X \times \mathbf{R} \to X$ is a smooth nonlinear mapping with

$$N(0,0) = 0, \qquad D_u N(0,0) = 0.$$

We follow Carr [52] for a brief description of the center manifold theory in abstract spaces, see also Henry [157] and Vanderbauwhede/Iooss [298] for detailed discussions.

Definition 7.1.1. *A function $u \in C([0,T],X)$ is said to be a weak solution of (7.1) if $u(0) = u_0 \in X$, $N(u(\cdot),\lambda) \in L^1([0,T],X)$ and for all $v \in D(L(\lambda)^*) \subset X$ the function $\langle u(t), v\rangle$ is absolutely continuous on $[0,T]$ and satisfies*

$$\frac{d}{dt}\langle u(t),v\rangle = \langle u(t), L(\lambda)^*v\rangle + \langle N(u(t),\lambda),v\rangle \quad \text{for all } v \in D(L(\lambda)^*).$$

Here $L(\lambda)^*$ is the adjoint operator of $L(\lambda)$ and $\langle\cdot,\ \cdot\rangle$ is the dual product between X and X'.

Existence of weak solutions is known as follows.

Theorem 7.1.1. *(Carr [52, pp.116] If the mapping $N: \ X \times \mathbf{R} \to X$ is local Lipschitz, then the problem (7.1) has a unique maximally defined weak solution $u \in C([0,T);X)$ given by*

$$u(t) = S(t)u_0 + \int_0^t S(t-s)N(u(s),\lambda)\, ds,$$

where $S(t)$ is the semigroup generated by $L(\lambda)$. Moreover, if $T < \infty$, then

$$\|u(t)\| \to \infty \quad as \quad t \to T^-.$$

Assumptions: At a critical value λ_0 of the parameter λ, we assume

H1) $X = X_0 \oplus X_h$, where X_0 is finite dimensional and X_h is closed;
H2) X_0 is $L(\lambda_0)$-invariant. Furthermore, let $L_0 := L(\lambda_0)\big|_{X_0}$ and the real part of all eigenvalues of L_0 be zero;
H3) If $W(t)$ is the restriction of $S(t)$ to X_h, then X_h is $W(t)$-invariant and there are positive constants $a,\ b > 0$, such that

$$\|W(t)\| \le ae^{-bt}, \quad t \ge 0.$$

Remark 7.1.1. The assumption **H3)** corresponds to the case that spectrum of the operator $L_h := L(\lambda_0)\big|_{X_h}$ is in $\mathbf{C}_- := \{\sigma \in \mathbf{C} \mid \mathrm{Re}(\sigma) \le -b\}$. We refer to Iooss/Adelmayer [166] for discussions on L_h which has both stable and unstable modes.

For any $u \in X$ and $\lambda \in \mathbf{R}$ there is a unique decomposition

$$\begin{aligned} u &= v + w, \quad && v \in X_0,\ w \in X_h, \\ \lambda &= \lambda_0 + \beta, && \beta \in \mathbf{R}. \end{aligned}$$

Let $Q:\ X \to X_h$ be a projection. We write the equation (7.1) into a canonical form

$$\begin{aligned}\frac{\partial w}{\partial t} &= L_h w + f(v+w,\beta),\\ \frac{\partial v}{\partial t} &= L_0 v + g(v+w,\beta),\\ \frac{\partial \beta}{\partial t} &= 0,\end{aligned} \tag{7.2}$$

where

$$\begin{aligned} L_h &:= QL(\lambda_0), \qquad L_0 := (I-Q)L(\lambda_0),\\ f(v+w,\beta) &:= Q[(L(\lambda_0+\beta)-L(\lambda_0))(v+w)+N(v+w,\lambda_0+\beta)],\\ g(v+w,\beta) &:= (I-Q)[(L(\lambda_0+\beta)-L(\lambda_0))(v+w)+N(v+w,\lambda_0+\beta)].\end{aligned}$$

A local invariant manifold of (7.1) is a subset M of X in which the solution $u(t)$ of (7.1) with $u(0) \in M$ stays for $0 < |t| \leq T$. If T can be chosen as infinite, then M is called invariant manifold. *A center manifold* is an invariant manifold which is tangent to X_0 at $(0, \lambda_0)$.

Remark 7.1.2. Inclusion of the equation $\frac{\partial \beta}{\partial t} = 0$ in (7.2) makes, e.g. a term $\beta(v + w)$ in (7.2) to be nonlinear, while for fixed β such a term is linear. This allows the center manifold to capture dynamics of the original problem in neighborhoods of the parameter $\lambda = \lambda_0$. The system (7.2) is also called a *suspended system.* The techniques for the computation of center manifolds of the suspended system and the original problem are identical.

Theorem 7.1.2. *There exists a center manifold for (7.1), given by*

$$M_0 := \{v + h(v,\beta) \quad | \quad (v,\beta) \in X_0 \times \mathbf{R}, \ \|(v,\beta)\| < \delta\}.$$

Furthermore, $h(0) = 0$, $D_v h(0) = 0$. *If* $N(u, \lambda)$ *is* C^k*-continuous, so is* $h(v, \beta)$ *as a mapping from* $X_0 \times \mathbf{R}$ *into* X_h. *The center manifold* M_0 *is locally attractive, i.e., all solutions of (7.1) staying in a neighborhood* U *of* M_0 *for all* $t > 0$ *tend exponentially to some solutions of (7.1) on* M_0. *The reduced equation on the center manifold is*

$$\frac{\partial v}{\partial t} = L_0 v + g(v + h(v,\beta),\beta). \tag{7.3}$$

Remark 7.1.3. The center manifold is not necessarily unique, see e.g. Iooss/Adelmayer [166, pp.11]. However, the difference is exponentially infinitesimal, i.e., the Taylor expansions of all center manifolds of a given problem are identical to each other (cf. Wan [299]).

Stabilities of the dynamical system (7.1) can be derived from those of the reduced equation (7.3) as follows.

Theorem 7.1.3. *1. If $(0,0)$ is a stable (resp. asymptotically stable, unstable) solution of (7.3), then the solution $(0,\lambda_0)$ of (7.1) is stable (resp. asymptotically stable, unstable).*

2. Let the zero solution of (7.3) be stable and $(v(t),w(t))$ be a solution of (7.2) with $\|(v(0),w(0))\|$ sufficiently small. Then there exists a solution $x(t)$ of (7.3) and a constant $\gamma>0$, such that

$$\begin{aligned} v(t) &= x(t)+O(e^{-\gamma t}),\\ w(t) &= h(x(t),\beta)+O(e^{-\gamma t}), \quad \textit{for } t\to\infty. \end{aligned} \tag{7.4a}$$

By $w=h(v,\beta)$ and (7.2) the center manifold can be formulated as the solution of the equation

$$\frac{\partial h(v,\beta)}{\partial v}\Big[L_0 v+g(v+h(v,\beta),\beta)\Big]-L_h h(v,\beta)-f(v,\beta,h(v,\beta))=0 \tag{7.5}$$

and locally approximated successively.

Theorem 7.1.4. *Let ϕ be a C^1-mapping from a neighborhood of the origin in $X_0\times\mathbf{R}$ into X_h with $\phi(0)=0$, $\phi'(0)=0$ and $\phi(v,\beta)\in D(L_0)$. If*

$$\frac{\partial \phi(v,\beta)}{\partial v}\Big[L_0 v+g(v+\phi(v,\beta),\beta)\Big]-L_h\phi(v,\beta)-f(v+\phi(v,\beta),\beta)=O(\|(v,\beta)\|^q),$$

we have

$$\|h(v,\beta)-\phi(v,\beta)\|=O(\|(v,\beta)\|^q) \quad \textit{for } \|(v,\beta)\|\ll 1. \tag{7.6}$$

7.2 Approximation of Center Manifolds and Reduced Equations

Similarly as in the Liapunov-Schmidt method, the center manifold $h(v,\beta)$ in (7.5) and the reduced equation (7.3) are explicitly available only in exceptional cases. For general problems the Taylor expansion of $h(v,\beta)$ at $(v,\beta)=(0,0)$ will be approximated successively. To simplify the notations we adapt the representation for Taylor series of mappings in Banach spaces in Iooss/Adelmeyer [166]. For examples, we write Taylor series of a mapping $G:\ X_0\times\mathbf{R}\to X_h$ as

$$G(v,\beta)=G_0+L(v,\beta)+\sum_{q=2}^{k}G_q[(v,\beta)^{(q)}]+o(\|(v,\beta)\|^k),$$

where $L:=DG(u_0,\lambda_0)$, and $(v,\beta)^{(q)}$ denotes the collection of $(v^{p_1},\ \beta^{q_1})$ with $p_1+q_1=q$ and G_q is the multilinear operator

$$G_q[(v,\beta)^{(q)}] = \frac{1}{q!} D^q G_0(v,\beta)^q = \frac{1}{q!} \sum_{p_1+q_1=q} \frac{\partial^q G_0}{\partial v^{p_1} \partial \beta^{q_1}} v^{p_1} \beta^{q_1},$$

in particular,

$$G_2[(v,\beta)^{(2)}] = \frac{1}{2} \left[D_{vv} G_0 v^2 + 2 D_{v\beta} G_0 v \beta + D_{\beta\beta} G_0 \beta^2 \right].$$

Denote $z_i = (v_i, \beta_i), i = 1, \ldots, k$ and $q = p_1 + \cdots + p_k + q_1 + \cdots + q_k$. We define

$$G_q[z_1^{(p_1,q_1)}, \ldots, z_k^{(p_k,q_k)}] = \frac{1}{q!} \frac{\partial^q G_0}{\partial v^{p_1+\cdots+p_k} \partial \beta^{q_1+\cdots+q_k}} v_1^{p_1} \cdots v_k^{p_k} \beta_1^{q_1} \cdots \beta_k^{q_k}.$$

For example, for $z = (v, \beta)$ we have

$$G_2[z^{(1,1)}, (h_2[z^{(2)}], 1)^{(1,0)}] = \frac{1}{2} \left[D_{vv} G_0 v h_2[z^{(2)}] + D_{v\beta} G_0 h_2[z^{(2)}] \beta \right],$$

which is a homogeneous cubic polynomial in z. Extensions to chain functions can be defined similarly. Therefore, with $h(0) = 0$ and $Dh(0) = 0$ we rewrite the equation (7.5) as

$$\begin{aligned} &\Big(\sum_{q=2}^{k} \frac{\partial}{\partial v} h_q[z^{(q)}] \Big) \Big(L_0 v + \sum_{q=2}^{k} g_q[(v + \sum_{p=2}^{k} h_p[z^{(p)}],\ \beta)^{(q)}] \Big) \\ &= \sum_{q=2}^{k} L_h h_q[z^{(q)}] + \sum_{q=2}^{k} f_q[(v + \sum_{p=2}^{k} h_p[z^p], \beta)^{(q)}] + o(\|z\|^k). \end{aligned}$$

This is a polynomial equation starting with the order 2. Collecting homogeneous terms yields equations for $h_q, q = 1, \ldots$. At the order 2 we obtain

$$L_h h_2[z^{(2)}] - \frac{\partial}{\partial v} h_2[z^{(2)}] L_0 v = -f_2[z^{(2)}].$$

At the order 3, the equation for $h_3[z^{(3)}]$ is

$$L_h h_3[z^{(3)}] - \frac{\partial}{\partial v} h_3[z^{(3)}] L_0 v = \frac{\partial}{\partial v} h_2[z^{(2)}] g_2[z^{(2)}] - f_3[z^{(2)}] - 2 f_2[z, h_2[z^{(2)}]].$$

Generally at the order m, the equation for $h_m[z^{(m)}]$ is of the form

$$L_h h_m[z^{(m)}] - \frac{\partial}{\partial v} h_m[z^{(m)}] L_0 v = -R_m[z^{(m)}], \tag{7.7}$$

where R_m involves the mapping f, g and the homogeneous polynomials h_q, for $q = 2, \ldots, m-1$ at previous steps. Since the right hand side is in the space X_h, this linear equation for $h_m[z^{(m)}]$ is solvable. For example, if the real part of spectrum of L_h is negative, we have formally

$$h_m[z^{(m)}] = \int_0^\infty e^{tL_h} R_m[(e^{-tL_h} z)^{(m)}] dt.$$

Example 7.2.1. Consider a semi-linear parabolic partial differential equation

$$\frac{\partial u}{\partial t} = \Delta u + \lambda(u + u^2 + au^3) \quad \text{in} \quad \Omega := [0,1] \times [0,1], \tag{7.8a}$$
$$\frac{\partial u}{\partial n} = 0 \quad \text{on} \quad \partial\Omega. \tag{7.8b}$$

Let $X := L^2(\Omega)$. The Laplacian induces a bilinear form on X and an isomorphism from $H^1(\Omega)$ onto its dual space $H^{-1}(\Omega)$ (cf. Temam [290, pp.52]). We consider a weak form of (7.8)

$$\frac{\partial u}{\partial t} = \Delta u + \lambda N(u),$$

where $D(\Delta) = X \cap H^2(\Omega)$ and $N(u) = u + u^2 + au^3$. On the trivial solution curve $\{(0,\lambda),\ \lambda \in \mathbf{R}\}$, a corank-2 bifurcation point is of the form $(u_0, \lambda_0) = (0, (m^2+n^2)\pi^2)$, $m \neq n$. The assumptions of center manifold theory are satisfied with

$$X_0 = \text{span}[\phi_1, \phi_2], \qquad X_h = X_0^{\perp},$$

and

$$\phi_1 := 2\cos mx \cos ny, \qquad \phi_2 := 2\cos nx \cos my.$$

To study dynamics near (u_0, λ_0) we decompose the unknown (u, λ) into

$$u = \alpha_1\phi_1 + \alpha_2\phi_2 + w, \quad w \in X_h,$$
$$\lambda = \lambda_0 + \beta$$

and the equation (7.8) into

$$\frac{d\alpha_1}{dt} = \alpha_1\beta + (\lambda_0+\beta)\langle\phi_1,\ N(\alpha_1\phi_1 + \alpha_2\phi_2 + w)\rangle,$$
$$\frac{d\alpha_2}{dt} = \alpha_2\beta + (\lambda_0+\beta)\langle\phi_2,\ N(\alpha_1\phi_1 + \alpha_2\phi_2 + w)\rangle,$$
$$\frac{\partial w}{\partial t} = \Delta w + (\lambda_0+\beta)w$$
$$+\ (\lambda_0+\beta)[I - \langle\phi_1,\cdot\rangle\phi_1 - \langle\phi_2,\cdot\rangle\phi_2]N(\alpha_1\phi_1 + \alpha_2\phi_2 + w).$$

Formulating the center manifold as a function $h(\alpha,\beta)$ of the coordinates $\alpha := (\alpha_1,\ \alpha_2)$ and β, we derive from (7.5) the following equation for h

$$\sum_{i=1}^{2} \frac{\partial h}{\partial \alpha_i}\Big[\alpha_i\beta + (\lambda_0+\beta)\Big\langle\phi_i,\ N(\alpha_1\phi_1 + \alpha_2\phi_2 + h(\alpha,\beta))\Big\rangle\Big]$$
$$= \Big[\Delta + (\lambda_0+\beta)\Big]h(\alpha,\beta)$$
$$+\ (\lambda_0+\beta)\Big[I - \sum_{i=1}^{2}\langle\phi_i,\cdot\rangle\phi_i\Big]N(\alpha_1\phi_1 + \alpha_2\phi_2 + h(\alpha,\beta)).$$

Since $u \equiv 0$ is a trivial solution for all $\lambda \in \mathbf{R}$, we have

$$h(0,\beta) = 0, \quad \beta \in \mathbf{R}.$$

On the other hand, $D_\alpha h(0,0) = 0$. The quadratical terms of

$$h(\alpha,\beta) = h_{200}\alpha_1^2 + h_{110}\alpha_1\alpha_2 + h_{020}\alpha_2^2 + h_{101}\alpha_1\beta + h_{011}\alpha_2\beta + O(\|(\alpha,\beta)\|^3)$$

with $h_{ijk} \in X_h$ are determined by the equations

$$\begin{aligned}
(\Delta + \lambda_0)h_{200} &= -\frac{1}{2}\lambda_0 D_{uu}N_0\phi_1^2, \\
(\Delta + \lambda_0)h_{110} &= -\lambda_0 D_{uu}N_0\phi_1\phi_2, \\
(\Delta + \lambda_0)h_{002} &= -\frac{1}{2}\lambda_0 D_{uu}N_0\phi_2^2, \\
(\Delta + \lambda_0)h_{101} &= 0, \\
(\Delta + \lambda_0)h_{011} &= 0.
\end{aligned}$$

Hence, the reduced equations on the center manifold are

$$\begin{aligned}
\frac{d\alpha_i}{dt} &= \alpha_i\beta + (\lambda_0 + \beta)\Big\langle \phi_i, \frac{1}{6}D_{uuu}N_0(\alpha_1\phi_1 + \alpha_2\phi_2)^3 \\
&+ D_{uu}N_0(\alpha_1\phi_1 + \alpha_2\phi_2)(h_{200}\alpha_1^2 + h_{110}\alpha_1\alpha_2 + h_{020}\alpha_2^2)\Big\rangle \qquad (7.9) \\
&+ O(\|(\alpha,\beta)\|^4), \quad i = 1,2.
\end{aligned}$$

For $m = 1,\ n = 2$ we have

$$\begin{aligned}
\frac{d\alpha_1}{dt} &= \alpha_1\Big\{\beta + \frac{5}{\pi^2}\Big[(\frac{9}{4}a - \frac{421}{330})\alpha_1^2 + (3a - \frac{10}{39})\alpha_2^2\Big]\Big\} + h.o.t.\ , \\
\frac{d\alpha_2}{dt} &= \alpha_2\Big\{\beta + \frac{5}{\pi^2}\Big[(3a - \frac{10}{39})\alpha_1^2 + (\frac{9}{4}a - \frac{421}{330})\alpha_2^2\Big]\Big\} + h.o.t.\ .
\end{aligned}$$

This system of the reduced equations is 3-determined except at $a = -9002/4485$ for which the 3-jet contains a factor with the continuous $O(2)$ symmetry. Thus we consider the truncated system

$$\begin{aligned}
\dot\alpha_1 &= \tilde g_1(\alpha_1,\alpha_2,\beta), \\
\dot\alpha_2 &= \tilde g_2(\alpha_1,\alpha_2,\beta)
\end{aligned} \qquad (7.10)$$

with the vector field

$$\begin{aligned}
\tilde g_1(\alpha_1,\alpha_2,\beta) &= \alpha_1\Big\{\beta + \frac{5}{\pi^2}\Big[(\frac{9}{4}a - \frac{421}{330})\alpha_1^2 + (3a - \frac{10}{39})\alpha_2^2\Big]\Big\}, \\
\tilde g_2(\alpha_1,\alpha_2,\beta) &= \alpha_2\Big\{\beta + \frac{5}{\pi^2}\Big[(3a - \frac{10}{39})\alpha_1^2 + (\frac{9}{4}a - \frac{421}{330})\alpha_2^2\Big]\Big\}.
\end{aligned}$$

This system has the following equilibria (α_1, α_2)

$$(\pm a_1,\ 0),\quad (0,\ \pm a_1)\quad \text{with}\quad a_1 := \pi\sqrt{-\frac{\beta}{5}\Big(\frac{9}{4}a - \frac{421}{330}\Big)^{-1}}$$

$$\Big(\pm a_2, \pm a_2)\quad \text{with}\quad a_2 := \pi\sqrt{-\frac{\beta}{5}\Big(\frac{12}{4}a - \frac{2191}{1430}\Big)^{-1}}\,.$$

We see that if $a < 626/2145$, all solutions are real for $\beta > 0$. If $a > 842/1485$, then $\beta < 0$ ensures all solutions in real. For a in $(626/2145,\ 842/1485)$ we need to adjust the sign of β to obtain real solutions. Stability of these stationary solution branches are determined by eigenvalues of the matrix

$$\begin{pmatrix} m_{11} & m_{12} \\ m_{21} & m_{22} \end{pmatrix}$$

with

$$\begin{aligned} m_{11} &:= \beta + \frac{5}{\pi^2}\Big(3(\frac{9a}{4} - \frac{421}{310})\alpha_1^2 + (3a - \frac{10}{39})\alpha_2^2\Big) \\ m_{12} &= m_{21} = \frac{10}{\pi^2}(3a - \frac{10}{3a})\alpha_1\alpha_2 \\ m_{22} &:= \beta + \frac{5}{\pi^2}\Big((3a - \frac{10}{39})\alpha_1^2 + 3(\frac{9a}{4} - \frac{421}{310})\alpha_2^2\Big) \end{aligned}$$

on the corresponding solution curves. For example, on the branch $(\alpha_1, \alpha_2) = (\pm a_1, 0)$ the above matrix reduces to

$$\begin{pmatrix} -2\beta & 0 \\ 0 & -(96525a + 34984)\beta/(495a - 3368) \end{pmatrix}.$$

Thus stability of the solution curve follows easily from signs of the two diagonal elements. Namely, if there are negative then the solution is stable, otherwise unstable.

7.3 Center Manifold Reduction versus Liapunov-Schmidt Method

If (u_0, λ_0) is a stationary bifurcation point of (7.1) and zero is a semi-simple eigenvalue of $L(\lambda_0)$, then $L_0 = 0$ and the equation (7.7) reduces to

$$L_h h_m[z^{(m)}] = -R_m[z^{(m)}].$$

For $m = 2$ this equation is identical to that in calculations of cubic terms of the reduced bifurcation equations in the Liapunov-Schmidt reduction. Namely

$$L_h[z^{(2)}] = -\frac{1}{2}\frac{\partial^2 f_0}{\partial^2 z}z^2.$$

The reduced equation on the center manifold is

$$\frac{dv}{dt} = g_2[z^{(2)}] + g_2[z, h_2[z^{(2)}]] + g_3[z^{(3)}] + O(\|z\|^4). \tag{7.11}$$

We claim that the right hand side coincides with the Taylor expansion of reduced bifurcation equations in the Liapunov-Schmidt reduction up to the cubic orders. To see these more precisely, let us consider the equation

$$\frac{\partial u}{\partial t} = G(u, \lambda), \tag{7.12}$$

where G is a smooth mapping from the Hilbert space $X \times \mathbf{R}$ into X. We assume that (u_0, λ_0) is a bifurcation point ($D_\lambda G_0 \in \mathrm{Im}(D_u G_0)$) and $D_u G(u_0, \lambda_0)$ is a Fredholm operator of index 0. Decompose

$$X = X_0 \oplus X_h = \tilde{X}_0 \oplus \tilde{X}_h$$

with

$$X_0 = \mathrm{Ker}(D_u G_0) = \mathrm{span}[\phi_1, \ldots, \phi_n], \qquad X_h = X_0^\perp,$$
$$\tilde{X}_0 = \mathrm{Ker}(D_u G_0^*) = \mathrm{span}[\phi_1^*, \ldots, \phi_n^*], \qquad \tilde{X}_h = \tilde{X}_0^\perp, \quad \langle \phi_i^*,\ \phi_j \rangle = \delta_{ij}.$$

We write $(u, \lambda) \in X \times \mathbf{R}$ as

$$u = u_0 + \sum_{i=1}^n \alpha_i \phi_i + w, \quad w \in X_h$$
$$\lambda = \lambda_0 + \beta$$

and choose the following projection from X onto X_h

$$Q := I - \sum_{i=1}^n \langle \phi_i^*,\ \cdot \rangle \phi_i.$$

By identifying $v = \sum_{i=1}^n \alpha_i \phi_i \in X_0$ with its coordinates $\alpha := (\alpha_1, \ldots, \alpha_n)$, the defining equation for center manifold becomes

$$\sum_{i=1}^n \frac{\partial h}{\partial \alpha_i} \Big\langle \phi_i^*,\ G(u_0 + \sum_{i=1}^n \alpha_i \phi_i + h(\alpha, \beta),\ \lambda_0 + \beta) \Big\rangle$$
$$= D_u G_0 h(\alpha, \beta) + QG(u_0 + \sum_{i=1}^n \alpha_i \phi_i + h(\alpha, \beta), \lambda_0 + \beta).$$

Since $h(0,0) = 0$ and $D_\alpha h(0,0) = 0$, we have

$$h(\alpha, \beta) = h_{01}^{(1)} \beta + \sum_{i,j=1}^n h_{ij0}^{(2)} \alpha_i \alpha_j + \sum_{i=1}^n h_{0i1}^{(2)} \alpha_i \beta + h_{002}^{(2)} \beta^2$$
$$+ O(\|(\alpha, \beta)\|^3), \quad h_{ijk}^{(2)} \in X_h,$$

where $h_{01}^{(1)} \in X_h$ is the solution of $D_uG_0w = -D_\lambda G_0$. Substituting this into the equation for $h(\alpha, \beta)$ and comparing coefficients yield the following equations for $h_{ijk} \in X_h$,

$$\begin{aligned}
D_uG_0h_{ii0}^{(2)} &= -\frac{1}{2}QD_{uu}G_0\phi_i^2, \\
D_uG_0h_{ij0}^{(2)} &= -QD_{uu}G_0\phi_i\phi_j, \quad i \neq j \\
D_uG_0h_{0i1}^{(2)} &= -QD_{u\lambda}G_0\phi_i, \\
D_uG_0h_{002}^{(2)} &= -\frac{1}{2}QD_{\lambda\lambda}G_0.
\end{aligned}$$

The reduced equations on the center manifold up to the cubic terms are

$$\begin{aligned}
\frac{d\alpha_k}{dt} &= \frac{1}{2}\Big\langle \phi_k^*,\ D^2G_0(\sum_{i=1}^n \alpha_i\phi_i + h_{01}^{(1)}\beta, \beta)^2\Big\rangle \\
&+ \frac{1}{6}\Big\langle \phi_k^*,\ D^3G_0(\sum_{i=1}^n \alpha_i\phi_i + h_{01}^{(1)}\beta, \beta)^3\Big\rangle \\
&+ \Big\langle \phi_k^*,\ D_uDG_0(\sum_{i=1}^n \alpha_i\phi_i + h_{01}^{(1)}\beta, \beta) \sum_{i,j=1}^n h_{ij0}^{(2)}\alpha_i\alpha_j \\
&+ \sum_{i=1}^n h_{0i1}^{(2)}\alpha_i\beta + h_{002}^{(2)}\beta^2)\Big\rangle \\
&+ O(\|(\alpha, \beta)\|^4), \quad k = 1, \ldots, n.
\end{aligned}$$

The right hand side coincides with the reduced bifurcation equations in Liapunov-Schmidt reduction up to the third orders, see (6.25). This has been shown for Hopf bifurcations with symmetries by Chossat and Golubitsky in [63]. In conclusion at a stationary bifurcation point (u_0, λ_0), if zero is a semi-simple eigenvalue of D_uG_0 and the problem is 3-determined, Liapunov-Schmidt method with the reduced stability analysis yields the same results as the center manifold theory. At higher orders of the reduced equations, the center manifold and the Liapunov-Schmidt reduction will be different, see also Chossat/Golubitsky [63]. We illustrate this with a simple example.

Example 7.3.1. Consider the system

$$\begin{aligned}
\dot{x} &= \lambda x + xy, \\
\dot{y} &= -y + \alpha x^2 + x^4.
\end{aligned}$$

Liapunov-Schmidt method at the bifurcation point $x = y = 0,\ \lambda = 0$ yields immediately the reduced bifurcation equation

$$\lambda x + \alpha x^3 + x^5 = 0. \tag{7.13}$$

On the other hand, taking the following ansatz for the center manifold $h(x, \lambda)$

$$h(x,\lambda) = \sum_{i+j=2}^{4} h_{ij} x^i \lambda^j + O(\|(x,\lambda\|^5),$$

we derive from the equation

$$h'(x,\lambda)(\lambda x + x h(x,\lambda)) = -h(x,\lambda) + \alpha x^2 + x^4.$$

that

$$h_{20} = \alpha, \quad h_{21} = -2\alpha, \quad h_{22} = 4\alpha, \quad h_{40} = 1 - 2\alpha^2$$

and

$$h_{11} = h_{02} = h_{12} = h_{30} = h_{31} = h_{13} = h_{04} = 0.$$

The reduced equation on center manifold is

$$\dot{x} = \lambda x + \alpha x^3 - x[2\alpha\lambda x^2 - (1-2\alpha^2)x^4 - 4\alpha\lambda^2 x^2] + o(\|(x,\lambda)\|^5). \quad (7.14)$$

It is evident that the right hand side of this equation differs from the reduced bifurcation equation (7.13) from the fourth order on.

Center manifold reduction provides a natural approach for stability analysis of bifurcating solution branches, especially if the nonlinear stability of equilibria and periodic solutions is determined by high order terms in Taylor expansion of the reduced equations. Nevertheless, often concepts of the reduced stability in McLeod/Sattinger [209] can be applied directly to the bifurcation equations in the Liapunov-Schmidt method.

7.4 Symmetry and Normal Form

Let Γ be a compact group and the vector field $G(u, \lambda)$ be Γ-equivariant

$$G(\gamma u, \lambda) = \gamma G(u, \lambda) \quad \text{for all } (u,\lambda) \in X \times \mathbf{R}, \quad \gamma \in \Gamma.$$

The reduced equations on the center manifold inherit the Γ-equivariance with respect to the induced action of Γ in the critical eigenspace. This property can be utilized to derive structure of the reduced equations and center manifold before starting the reduction process.

It has been known for long time that the reduced equations (7.3) can be simplified further by a series of near identity transformations. Such techniques have been used by Poincarè and Birkhoff (thus the Poincarè/Birkhoff normal form). Coullet/Spiegel [70] and Elphick *et al* [99] unify the center manifold reduction with the coordinate transformations such that the two-step process—computation of the Taylor expansion of the center manifold and simplification of the reduced equations, is realized simultaneously in a combination of normal form theory and center manifold reduction. In other words, the normal form of the reduced bifurcation equations is derived directly from the original equation, see e.g. Iooss/Adelmayer [166, pp.18].

Theorem 7.4.1. *Suppose that $G(u,\lambda)$ is Γ-equivariant and the assumptions* **H1)-H3)** *are satisfied. Then the center manifold*

$$M_0 = \{v + h(v,\beta) \quad | \quad v \in X_0, \quad \beta \in \mathbf{R}^p\}$$

has the property

$$h(\gamma v, \beta) = \gamma h(v,\beta) \quad \text{for all} \quad (v,\beta) \in X_0 \times \mathbf{R}^p,\ \gamma \in \Gamma,\ \|(v,\beta)\| \le \delta. \tag{7.15}$$

Furthermore, there are Γ-equivariant polynomials $P_i : X_0 \times \mathbf{R}^p \to X_0$ of the degree $\le k$ with $P_i(0,0) = 0$, $D_v P_i(0,0) = 0$, $i = 0,1$, such that for $v = \tilde{v} + P_1(\tilde{v},\beta)$ the reduced equations (7.3) is of the normal form

$$\frac{d\tilde{v}}{dt} = L_0\tilde{v} + P_0(\tilde{v},\beta) + O((\|\tilde{v}\| + \|\beta\|)^{k+1}). \tag{7.16}$$

Here P_0 has the property

$$P_0(e^{tL_0^*}\tilde{v},\beta) = e^{tL_0^*}P_0(\tilde{v},\beta) \quad \text{for all} \quad (\tilde{v},\beta) \in X_0 \times \mathbf{R}^p,\ t \in \mathbf{R}. \tag{7.17}$$

In computation of center manifold and normal form of the reduced equations we derive first the structure of the polynomial

$$P_0(v,\beta) = \sum_{p,q=0}^{k} P_{p,q}[v^{(p)},\beta^{(p)}]$$

with the symmetry (7.17). Similarly, the Γ-equivariance is used to determine the structure of the Taylor expansion of the mapping

$$h(v,\beta) = \sum_{p,q=0}^{k} h_{p,q}[v^{(p)},\beta^{(p)}] + h.o.t.$$

Thereafter, normal form of the reduced equations and Taylor expansion of the center manifold are calculated simultaneously via the equation

$$\frac{\partial}{\partial v}\Big(v + h(v,\beta)\Big)\Big(L_0 v + \sum_{p,q=0}^{k} P_{p,q}[v^{(p)},\beta^{(p)}]\Big) = L_h h(v,\beta) + G(v + h(v,\beta),\beta). \tag{7.18}$$

7.4.1 Simple bifurcation points

Assume that X is a Hilbert space and $(0,0)$ is a stationary solution of

$$\frac{\partial u}{\partial t} = G(u,\lambda).$$

Let $S \ne I$ be a reflection in X and

$$G(Su,\lambda)=SG(u,\lambda) \quad \text{for all} \quad (u,\lambda)\in X\times\mathbf{R}.$$

We assume that $L=D_uG(0,0)$ is a Fredholm operator of index zero. If zero is a simple eigenvalue of L and L has no other eigenvalues on the imaginary axis, we have

$$X=X_0\oplus X_h=X_0^*\oplus X_h^*$$

with

$$X_0:=\mathrm{Ker}(L)=\mathrm{span}[\phi], \qquad X_h=\mathrm{Im}(L^*)$$

and

$$X_0^*:=\mathrm{Ker}(L^*)=\mathrm{span}[\phi^*], \qquad X_h^*=\mathrm{Im}(L), \qquad \langle\phi^*,\ \phi\rangle=1.$$

The induced action of S in X_0 is given by $S\phi=\pm\phi$ or equivalently $S\alpha=\pm\alpha$ for all $v=\alpha\phi\in X_0$. The reduced equation is of the form

$$\frac{d\alpha}{dt}=g(\alpha,\beta)$$

where $g(0,0)=0$, $\dfrac{\partial g}{\partial\alpha}(0,0)=0$. Furthermore,

$$g(S\alpha,\beta)=Sg(\alpha,\beta), \qquad h(S\alpha,\beta)=Sh(\alpha,\beta) \quad \text{for all} \quad \alpha,\ \beta\in\mathbf{R}.$$

Since $L_0=L\big|_{X_0}=0$, the statement (7.17) is trivial.

Case I). $S\alpha=\alpha$: In this case we have

$$g(\alpha,\beta)=P(\alpha,\beta)+O((|\alpha|+|\beta|)^3)$$

with $P(\alpha,\beta)=g_{01}\beta+g_{20}\alpha^2+g_{11}\alpha\beta+g_{02}\beta^2$. Similarly,

$$h(\alpha,\beta)=h_{01}\beta+h_{20}\alpha^2+h_{11}\alpha\beta+h_{02}\beta^2+h.o.t.$$

We write the equation (7.18) as

$$\begin{aligned}&(\phi+2h_{20}\alpha+h_{11}\beta)(g_{01}\beta+g_{20}\alpha^2+g_{11}\alpha\beta+g_{02}\beta^2)\\ &=L(h_{01}\beta+h_{20}\alpha^2+h_{11}\alpha\beta+h_{02}\beta^2)+D_\lambda G_0\beta\\ &\quad+\frac{1}{2}D^2G_0(\alpha\phi+h_{01}\beta,\beta)^2+O((|\alpha|+|\beta|)^3).\end{aligned}$$

Comparing the coefficients of different orders of $\alpha^i\beta^j$ yields

terms	equations
β	$Lh_{01}=-D_\lambda G_0+g_{01}\phi$
α^2	$Lh_{20}=-\frac{1}{2}D_{uu}G_0\phi^2+g_{20}\phi$
$\alpha\beta$	$Lh_{11}=2h_{20}g_{01}-D_{u\lambda}G_0\phi-D_{uu}G_0\phi h_{01}+g_{11}\phi$
β^2	$Lh_{02}=h_{11}g_{01}-\frac{1}{2}D^2G_0(h_{01},1)^2+g_{02}\phi$

Solvability conditions of these equations yield

$$\begin{aligned} g_{01} &= \langle \phi^*,\ D_\lambda G_0 \rangle, \\ g_{20} &= \frac{1}{2}\langle \phi^*,\ D_{uu}G_0\phi\phi \rangle, \\ g_{11} &= \langle \phi^*,\ D_{u\lambda}G_0\phi + D_{uu}G_0\phi h_{01} \rangle, \\ g_{01} &= \langle \phi^*,\ D_{\lambda\lambda}G_0 + D_{uu}G_0 h_{01}^2 \rangle. \end{aligned} \tag{7.19}$$

If $D_\lambda G_0 \in \operatorname{Im}(L)$, these quantities coincide with those in the Liapunov-Schmidt reduction, particularly, $g_{01} = 0$. Correspondingly, the reduced equation is

$$\frac{d\alpha}{dt} = g_{20}\alpha^2 + g_{11}\alpha\beta + g_{02}\beta^2 + O((|\alpha| + |\beta|)^3).$$

Case II). $S\alpha = -\alpha$: In this situation by equivariance we know that $g(\alpha, \beta)$ is odd in α. Therefore,

$$g(\alpha, \beta) = g_{11}\alpha\beta + g_{30}\alpha^3 + O(|\beta|(|\alpha| + |\beta|)^2).$$

On the other hand, taking the general form

$$h(\alpha, \beta) = h_{01}\beta + h_{20}\alpha^2 + h_{11}\alpha\beta + h_{02}\beta^2 + h_{30}\alpha^3 + O(|\beta|(|\alpha| + |\beta|)^2),$$

we see from the equality $h(\alpha, \beta) = Sh(S\alpha, \beta) = Sh(-\alpha, \beta)$ the properties

$$Sh_{0i} = h_{0i},\ i = 1, 2, \quad Sh_{20} = h_{20}, \quad Sh_{11} = h_{11}, \quad Sh_{30} = -h_{30}.$$

Substituting these into the equation (7.18)

$$\begin{aligned} &(\phi + 2h_{20}\alpha + h_{11}\beta + 3h_{30}\alpha^2)(g_{11}\alpha\beta + g_{30}\alpha^2) \\ &= L(h_{01}\beta + h_{20}\alpha^2 + h_{11}\alpha\beta + h_{02}\beta^2 + h_{30}\alpha^3) + D_\lambda G_0\beta \\ &\quad + \frac{1}{2}D^2G_0(\alpha\phi + h_{01}\beta, \beta)^2 + \frac{1}{6}(D_{uuu}G_0(\phi^3 + 3\phi h_{20})\alpha^3 + O(|\beta|(|\alpha| + |\beta|)^2), \end{aligned}$$

we derive the following equations for g_{ij} and h_{ij}.

terms	equations
β	$Lh_{01} = -D_\lambda G_0 + g_{01}\phi$
α^2	$Lh_{20} = -\frac{1}{2}D_{uu}G_0\phi^2$
$\alpha\beta$	$Lh_{11} = -D_{u\lambda}G_0\phi - D_{uu}G_0\phi h_{01} + g_{11}\phi$
β^2	$Lh_{02} = -\frac{1}{2}D^2G_0(h_{01}, 1)^2$
α^3	$Lh_{03} = -\frac{1}{6}D_{uuu}G_0(\phi^3 + 3\phi h_{20}) + g_{30}\phi$

Solvability conditions of these equations correspond to

$$\begin{aligned} \langle \phi^*,\ D_\lambda G_0 \rangle = 0 &\Longleftrightarrow D_\lambda G_0 \in \operatorname{Im}(L), \\ \langle \phi^*,\ D_{uu}G_0\phi^2 \rangle = 0 &\Longleftrightarrow D_{uu}G_0\phi^2 \in \operatorname{Im}(L), \\ g_{11} = \langle \phi^*,\ D_{u\lambda}G_0\phi + D_{uu}G_0\phi h_{01} \rangle, &\qquad h_{01} = -L_h^{-1}D_\lambda G_0 \in X_h \\ \langle \phi^*,\ D^2G_0(h_{01}, 1)^2 \rangle = 0 &\Longleftrightarrow D^2G_0(h_{01}, 1)^2 \in \operatorname{Im}(L), \\ g_{30} = \frac{1}{6}\langle \phi^*,\ D_{uuu}G_0(\phi^3 + 3\phi h_{20}) \rangle, &\qquad h_{20} = -\frac{1}{2}L_h^{-1}D_{uu}G_0\phi^2 \in X_h. \end{aligned} \tag{7.20}$$

The equalities $\langle \phi^*,\ D_\lambda G_0\rangle = \langle \phi^*,\ D_{uu}G_0\phi^2\rangle = \langle \phi^*,\ D^2G_0(h_{01},1)^2\rangle = 0$ are necessary conditions of a pitchfork bifurcation with a nontrivial reflection symmetry. The reduced equation is

$$\frac{d\alpha}{dt} = g_{11}\alpha\beta + g_{30}\alpha^3 + h.o.t.$$

Here the coefficients $g_{11},\ g_{30}$ in (7.20) are consistent with those derived from the Liapunov-Schmidt reduction.

7.4.2 Hopf bifurcations

Let $(0,0)$ be a stationary solution of

$$\frac{\partial u}{\partial t} = G(u,\lambda).$$

Assume that $L = D_uG(0,0)$ has a pair simple eigenvalues $\pm\omega_0 i,\ \omega_0 > 0$ and no other eigenvalues on the imaginary axis. Denote $\phi,\ \bar\phi$ the eigenvectors of L corresponding to $\pm\omega_0 i$,

$$L\phi = i\omega_0\phi, \qquad L\bar\phi = -i\omega_0\bar\phi.$$

The critical eigenspace X_0 is spanned by the real and imaginary part of ϕ, i.e.,

$$X_0 = \mathrm{span}[\mathrm{Re}(\phi),\ \mathrm{Im}(\phi)].$$

We identify this two dimensional space with the subspace $\{(z,\ \bar z) \mid z \in \mathbf{C}\} \subset \mathbf{C}^2$ via

$$v = z\phi + \overline{z\phi} \in X_0 \longleftrightarrow (z,\ \bar z) \in \mathbf{C}^2.$$

With respect to the coordinates $z,\ \bar z$ the restriction L_0 of L in X_0 is a 2×2 matrix

$$L_0 = \begin{pmatrix} i\omega_0 & 0 \\ 0 & -i\omega_0 \end{pmatrix}.$$

Let the assumption **H3)** be satisfied. The reduced equations on the center manifold are of the form

$$\frac{d}{dt}\begin{pmatrix} z \\ \bar z \end{pmatrix} = L_0\begin{pmatrix} z \\ \bar z \end{pmatrix} + \begin{pmatrix} P(z,\bar z,\beta) \\ \overline{P(z,\bar z,\beta)} \end{pmatrix} + h.o.t.$$

Here $\overline{P(z,\bar z,\beta)}$ is the complex conjugate of $P(z,\bar z,\beta)$. The center manifold is given by $M_0 := \{z\phi + \overline{z\phi} + h(z,\bar z,\beta)\}$. Note that

$$e^{tL_0^*} = \begin{pmatrix} e^{-i\omega_0 t} & 0 \\ 0 & e^{i\omega_0 t} \end{pmatrix}$$

and the property (7.17) of the normal form, i.e., $e^{tL_0^*}$ commutes with the mapping $(P(z,\bar z,\beta),\ \overline{P(z,\bar z,\beta)})^T$. We have

$$P(e^{-i\omega_0 t}z,\ e^{i\omega_0 t}\bar{z},\ \beta) = e^{-i\omega_0 t}P(z, \bar{z}, \beta).$$

Let $t = \arg(z)/\omega_0$, then

$$P(z, \bar{z}, \beta) = e^{\arg(z)i}P(|z|, |z|, \beta) \quad \text{for all } z \in \mathbf{C},\ \beta \in \mathbf{R}.$$

On the other hand, choosing $t = \pi/\omega_0$, we derive

$$P(z, \bar{z}, \beta) = -P(-z, -\bar{z}, \beta).$$

Therefore, $P(|z|, |z|, \beta)$ is an odd polynomial in $|z|$ and of the form

$$P(|z|, |z|, \beta) = |z|P_0(|z|^2, \beta).$$

This in turn implies

$$P(z, \bar{z}, \beta) = zP_0(|z|^2, \beta).$$

Here P_0 is a complex polynomial of the degree $\leq k$ in $|z|^2$. Its coefficients are polynomials in β and $P_0(0,0) = 0$. Now we can write the reduced equations on the center manifold as

$$\frac{dz}{dt} = \omega_0 iz + zP_0(|z|^2, \beta) + O(|z|^{2k+2}). \tag{7.21}$$

The equation for $\bar{z}$ is the complex conjugate of (7.21). For $k = 1$ we have

$$\frac{dz}{dt} = \omega_0 iz + z(a|z|^2 + b\beta) + O(|z|^4 + \beta^2) \quad \text{for } a, b \in \mathbf{C}. \tag{7.22}$$

The coefficients $a,\ b \in \mathbf{C}$ are determined by identifying the lower order terms in the defining equation

$$\begin{aligned} &(\phi + D_z h(z, \bar{z}, \beta))(\omega_0 iz + z(a|z|^2 + b\beta)) \\ &\quad + (\bar{\phi} + D_{\bar{z}} h(z, \bar{z}, \beta))(-\omega_0 i\bar{z} + \bar{z}(\bar{a}|z|^2 + \bar{b}\beta)) \\ &= G(z\phi + \bar{z}\bar{\phi} + h(z, \bar{z}, \beta),\ \beta) + O(|z|^4 + \beta^2), \end{aligned} \tag{7.23}$$

where $h(z, \bar{z}, \beta)$ satisfies $h(0,0,0) = 0,\ D_z h(0,0,0) = 0,\ D_{\bar{z}} h(0,0,0) = 0$. More precisely, with the Taylor expansion of $h(z, \bar{z}, \beta)$,

$$\begin{aligned} h(z, \bar{z}, \beta) &= h_{001}\beta + h_{200}z^2 + h_{110}z\bar{z} + h_{020}\bar{z}^2 + h_{101}z\beta + h_{011}\bar{z}\beta + h_{002}\beta^2 \\ &\quad + h_{300}z^3 + h_{210}z|z|^2 + h_{120}\bar{z}|z|^2 + h_{030}\bar{z}^3 + \cdots, \end{aligned}$$

we derive from (7.23) the following equations for the coefficients h_{ijk} and $a,\ b$:

terms	equations
β	$Lh_{001} = -D_\lambda G_0$
z^2	$(L - 2\omega_0 i)h_{200} = -\frac{1}{2}D_{uu}G_0\phi^2$
$z\bar{z}$	$Lh_{110} = -D_{uu}G_0\phi\bar{\phi}$
$z\beta$	$(L - \omega_0 i)h_{101} = b\phi - D_{u\lambda}G_0\phi - D_{uu}G_0\phi h_{001}$
$z^2\bar{z}$	$(L - \omega_0 i)h_{210} = a\phi - D_{uu}G_0(\phi h_{110} + \bar{\phi}h_{200} - \frac{1}{2}D_{uu}$ [illegible]
z^3	$(L - 3\omega_0 i)h_{300} = -\frac{1}{6}D_{uuu}G_0\phi^3 - D_{uu}G_0\phi h_{200})$

Since the operators $L - k\omega_0 i,\ k \neq \pm 1$ are invertible, the equations for $h_{001}, h_{200}, h_{110}$ and h_{300} can be solved directly. Solvability conditions for the coefficients h_{101} and h_{210} yield

$$\begin{aligned} a &= \langle \phi^*,\ D_{uu}G_0(\phi h_{110} + \bar{\phi} h_{200}) + \tfrac{1}{2} D_{uuu}G_0 \phi^2 \bar{\phi} \rangle, \\ b &= \langle \phi^*,\ D_{u\lambda}G_0 \phi + D_{uu}G_0 \phi h_{001} \rangle. \end{aligned} \tag{7.24}$$

Other coefficients in the Taylor expansion of $h(z, \bar{z}, \beta)$ can be determined similarly, in particular, we have

$$h_{020} = \bar{h}_{200}, \quad h_{011} = \bar{h}_{101}, \quad h_{120} = \bar{h}_{210}, \quad h_{030} = \bar{h}_{300}.$$

The coefficient h_{002} is determined by the equation

$$Lh_{002} = -\frac{1}{2} D^2 G_0 (h_{001}, 1)^2.$$

7.5 Waves in Reaction-Diffusion Equations

Generally an unstirred chemical reaction process is described as a reaction-diffusion problem

$$\frac{\partial u}{\partial t} = D \Delta u + G(u, \lambda) \tag{7.25}$$

depending on the control parameter λ. Here u represents concentrations of some involved chemical species. The matrix D consists of diffusion constants. It is symmetric, positive semi-definite and often diagonal. The case $D = 0$ corresponds to dynamics of reactions which is either local and uncoupled with the surroundings or a continuously and uniformly stirred entire system. The function $G : \mathbf{R}^m \times \mathbf{R}^p \to \mathbf{R}^m$ represents the reaction term and is sufficiently smooth. Furthermore, we assume that for any fixed parameter λ there are constants $c_1,\ c_2 > 0$ such that

$$||G(u, \lambda)|| \leq c_1 + c_2 ||u||^s, \tag{7.26}$$

where $1 < s < (n+2)/(n-2)$ for $n \geq 2$. We consider bounded domains $\Omega \in \mathbf{R}^n (n \leq 3)$ with $C^{k,1}$-boundaries $(k \geq 0)$. The boundary conditions is chosen to be of the Robin type

$$\mu u + (1 - \mu) \frac{\partial u}{\partial n} = 0 \quad \text{on } \partial \Omega. \tag{7.27}$$

Stationary solutions of (7.25) are described by the equation

$$D \Delta u + G(u, \lambda) = 0.$$

A uniform steady solution, i.e., a stationary solution which is independent of the space variables, satisfies the equation

$$G(u, \lambda) = 0.$$

A steady state solution with spatial variations depends smoothly on the parameters λ and μ.

Let u_0 be a uniform steady solution. Stability of this solution is described locally by the spectrum of the linearized problem

$$\frac{\partial u}{\partial t} = (D\Delta + L(\lambda))u, \qquad L(\lambda) := D_u G(u_0, \lambda) \tag{7.28}$$

with the boundary condition (7.27). As we will see in Chapter 14, the Laplacian $-\Delta$ with Robin boundary condition (7.27) is self-adjoint, positive for $\mu \in (0,1]$ and semi-positive definite for $\mu = 0$ respectively. Its spectrum consists of discrete eigenvalues ordered as

$$0 \leq \kappa_1(\mu) < \kappa_2(\mu) < \cdots \to \infty. \tag{7.29}$$

Denote $\phi_i(\mu)$, $i = 1, 2, \ldots$ the corresponding eigenfunctions. Since the operator $D\Delta + L(\lambda)$ maps the subspace $\mathbf{R}^m \phi_i$ into itself for all $i = 1, 2, \ldots$, a solution of (7.28) is a linear combination of the basis solutions

$$\mathbf{V}_l e^{\sigma t} \phi_l, \quad l \in N. \tag{7.30}$$

Here $(\sigma,\ \mathbf{V}_l)$ is an eigenpair of the $m \times m$ matrix $\mathcal{L}_l(\lambda) := L(\lambda) - \kappa_l D$, i.e.,

$$(L(\lambda) - \kappa_l D)\mathbf{V}_l = \sigma \mathbf{V}_l. \tag{7.31}$$

If symmetry is involved, eigenvalues κ_l of the Laplacian can be multiple, so are the corresponding eigenvalues σ of $\mathcal{L}_l(\lambda)$. The multiplicity of σ is greater than or equal to that of κ_l. The multiplicity $multi(\kappa_l)$ reflects spatial and structural symmetries in the system, while $multi(\sigma) - multi(\kappa_l)$ describes interactions of different species. The behavior of (7.25) at mode interactions caused by symmetries of the system differs from those resulted from interaction of the species. This is illustrated in the following example.

Example 7.5.1. Consider the problem

$$\begin{aligned}
\frac{\partial u_1}{\partial t} &= \Delta u_1 + \lambda(u_1 + u_2) - 2u_2 + f_1(u_1, u_2), \\
\frac{\partial u_2}{\partial t} &= \Delta u_2 + \lambda(u_1 + u_2) - 2u_1 + f_2(u_1, u_2)
\end{aligned}$$

with the homogeneous Dirichlet boundary condition on the square domain $\Omega := [0, \pi] \times [0, \pi]$ and $f_i(0) = 0,\ Df_i(0) = 0,\ i = 1, 2$. Since

$$\mathcal{L}_l(\lambda) = \begin{pmatrix} -\kappa_l + \lambda & \lambda - 2 \\ \lambda - 2 & -\kappa_l + \lambda \end{pmatrix},$$

zero is an eigenvalue of this matrix if and only if

$$\lambda = \frac{\kappa_l + 2}{2} \quad \text{or} \quad \kappa_l = 2, \ \lambda \in \mathbf{R}.$$

Let $\kappa = 2m^2\pi^2$, $m \in \mathbf{N}$ be a simple eigenvalue of the Laplacian $-\Delta$ with the corresponding eigenfunction $\phi_1 = 2 \sin mx \sin my$. For $m = 1$ and $\lambda = \kappa_1 = 2$ the matrix

$$\mathcal{L}_1(2) = \begin{pmatrix} 0 & 0 \\ 0 & 0 \end{pmatrix}$$

has a double eigenvalue 0 with the associated eigenvectors $(1,0)^T$ and $(0,1)^T$. This leads to two different critical modes

$$(1,0)^T e^{\sigma t}\phi_1, \qquad (0,1)^T e^{\sigma t}\phi_1 \tag{7.32}$$

of (7.28). These modes have identical spatial distributions.

Eigenvalues of $-\Delta$ in the square with the homogeneous Dirichlet boundary conditions are of the form $\kappa_i = k^2 + l^2$ with $k,\ l \in \mathbf{N}$. If $k \neq l$, κ_i is generically of the multiplicity two. The associated eigenfunctions are

$$\phi_i = 2 \sin kx \sin ly, \qquad \tilde{\phi}_i = 2 \sin lx \sin ky.$$

If $\lambda = (\kappa_i + 2)/2 \neq 2$, zero is a simple eigenvalue of $\mathcal{L}_i(\lambda)$ and the associated eigenvector is $(1,1)^T$. Thereafter,

$$(1,1)^T e^{\sigma t}\phi_i, \qquad (1,1)^T e^{\sigma t}\tilde{\phi}_i \tag{7.33}$$

are two critical modes of (7.28) and they have different spatial distributions. This is a consequence of D_4-symmetry in the problem.

The bifurcation scenario at $\lambda = 2$ with the critical modes (7.32) is obviously different from those at $\lambda = (\kappa_i + 2)/2,\ i = 2, 3\ldots$ with the critical modes (7.33), though in both cases the dimension of critical eigenspaces is two.

Corresponding to the behavior (7.29) of eigenvalues κ_l of the Laplacian, we have qualitatively three different situations for λ in a bounded range

$$a)\ \kappa_l \ll O(\|L(\lambda)\|), \quad b)\ \kappa_l \equiv O(\|L(\lambda)\|), \quad c)\ \kappa_l \gg O(\|L(\lambda)\|). \tag{7.34}$$

This happens typically if we vary other parameters, e.g. size of the domain or the homotopy parameter μ in the boundary condition (7.27), which causes the eigenvalue κ_l to vary continuously. Perturbation theory, see e.g. Kato [178], shows that for (7.34c) eigenvalues of $\mathcal{L}_l(\lambda)$ behave approximately like $-\kappa_l$ and are far in the left half of the complex plan, while for (7.34a) eigenvalues of $\mathcal{L}_l(\lambda)$ are dominated by those of $L(\lambda)$ of the uniform system. In (7.34b) the effects of diffusion and reaction are of the same magnitude and induce sophisticated nonlinear scenarios.

7.5.1 Oscillating waves

Suppose that (u_0, λ_0) is a uniform steady solution and the matrix $L(\lambda)$ has one pair complex eigenvalues $\sigma(\lambda)$ satisfying

$$\sigma(\lambda_0) = \pm\omega_0 i, \qquad \frac{d}{d\lambda}\mathrm{Re}(\sigma(\lambda))\Big|_{\lambda=\lambda_0} \neq 0. \tag{7.35}$$

Furthermore, let all other eigenvalues of matrices $L(\lambda_0)$ and $\mathcal{L}_i(\lambda_0)$ have strictly negative real parts, i.e., (u_0, λ_0) is a Hopf bifurcation point of the ordinary differential equation (ODE)

$$\dot{u} = G(u, \lambda). \tag{7.36}$$

There is a unique and uniform steady solution $(u(\lambda), \lambda)$ passing through (u_0, λ_0). Under nondegeneracy conditions there is a unique periodic solution of (7.36) bifurcating at (u_0, λ_0). Thus a timely periodic and spatially uniform solution of the reaction-diffusion problem (7.25) emanates from (u_0, λ_0).

7.5.2 Long waves

It is known that diffusion is responsible for the spatial variations of the substances. We study impact of diffusions on the solutions which are uniform in space and periodic in time, particularly for long spatial waves of (7.25) in the neighborhoods of (u_0, λ_0). To this end, we take a transformation of the spatial coordinates

$$(\tilde{x},\ \tilde{y}) = (\epsilon x,\ \epsilon y), \quad \epsilon > 0.$$

This scales the domain $\Omega \to \tilde{\Omega}$. Correspondingly the Laplacian operator becomes

$$\Delta = \epsilon^2 \tilde{\Delta}.$$

In other words, long waves correspond to slow spatial variations. We consider them as small perturbations of the temporal oscillations. Taking these factors into account we write the equation (7.25) as

$$\frac{\partial u}{\partial t} = \epsilon^2 D \Delta u + G(u, \lambda) \tag{7.37}$$

with ϵ as an additional parameter. Evidently (u_0, λ_0) is a Hopf bifurcation point of (7.37). Near the critical value λ_0 we set

$$u = u_0 + v, \quad \lambda = \lambda_0 + \beta$$

and consider the following perturbation of (7.36)

$$\begin{aligned} \frac{\partial v}{\partial t} &= L_0 v + \tilde{G}(v, \beta) + D\epsilon^2 \Delta v \\ \frac{\partial \beta}{\partial t} &= 0, \\ \frac{\partial \epsilon}{\partial t} &= 0, \end{aligned} \tag{7.38}$$

where $\tilde{G}(v,\beta) = G(u_0 + v, \lambda_0 + \beta) - L_0 v$. Let ϕ be the eigenfunction of $L_0 = D_u G(u_0, \lambda_0)$, associated to the eigenvalue $\sigma_0 := \sigma(\lambda_0) = i\omega_0$,

$$L_0\phi = \sigma_0\phi, \qquad L_0\bar{\phi} = \bar{\sigma}_0\bar{\phi}.$$

Correspondingly, let ϕ^* be eigenfunction of the adjoint operator L_0^*, associated to σ_0

$$L_0^*\phi^* = \sigma_0\phi^*, \qquad L_0^*\bar{\phi}^* = \bar{\sigma}_0\bar{\phi}^*.$$

The statement $\langle\phi^*,\ \bar{\phi}\rangle = \langle\bar{\phi}^*,\ \phi\rangle = 0$ holds for the inner product in $\mathbf{C}^n$. Furthermore, these eigenfunctions can be normalized as

$$\langle\phi^*,\ \phi\rangle = \langle\bar{\phi}^*,\ \bar{\phi}\rangle = 1.$$

Then $\sigma_0 = \langle\phi^*,\ L_0\phi\rangle$. By Iooss and Mielke [167, 228] the system (7.38) has an infinite dimensional center manifold of the form

$$M := \{A\phi + \bar{A}\bar{\phi} + h(A,\bar{A},\beta,\epsilon^2) \ \mid \ A \in \mathbf{C} \text{ (depending on } x,y,t)\}. \tag{7.39}$$

Here $h(0,0,0,\epsilon^2) \equiv 0$, $D_A h(0) = 0$ and $D_{\bar{A}} h(0) = 0$. The amplitude A is a complex function of the temporal variable t, and the spatial variables x, y which we treat simply as parameters for the moment. Furthermore, from the normal form theory at a Hopf bifurcation for ODE we know that the reduced equation on the center manifold is of the form

$$\frac{\partial A}{\partial t} = \omega_0 i A + aA|A|^2 + b\beta A + \epsilon^2 cA + h.o.t.$$

The coefficients a, b and c are linear operators. With this in mind we substitute $u = A\phi + \bar{A}\bar{\phi} + h(A,\bar{A},\beta,\epsilon)$ into (7.38)

$$\begin{aligned}
&\Big(\phi + \frac{\partial}{\partial A}h(A,\bar{A},\beta,\epsilon^2)\Big)\Big(\omega_0 iA + aA|A|^2 + b\beta A + c\epsilon^2 A\Big) \\
&\quad + \ \Big(\bar{\phi} + \frac{\partial}{\partial \bar{A}}h(A,\bar{A},\beta,\epsilon^2)\Big)\Big(- \omega_0 i\bar{A} + \bar{a}\bar{A}|A|^2 + \bar{b}\beta\bar{A} + \bar{c}\epsilon^2\bar{A}\Big) \\
&= \epsilon^2 D\Delta(A\phi + \bar{A}\bar{\phi} + h(A,\bar{A},\beta,\epsilon^2)) \\
&\quad + \ G(u_0 + A\phi + \bar{A}\bar{\phi} + h(A,\bar{A},\beta,\epsilon^2),\ \lambda_0 + \beta).
\end{aligned}$$

Taking the Taylor expansion of h at $(A,\bar{A},\beta,\epsilon) = (0,0,0,0)$

$$\begin{aligned}
h(A,\bar{A},\beta,\epsilon) &= h_{0010}\beta + h_{2000}A^2 + h_{1100}A\bar{A} + h_{0200}\bar{A}^2 \\
&\quad + h_{1010}A\beta + h_{1001}A\epsilon^2 + h_{0110}\bar{A}\beta + h_{0101}\bar{A}\epsilon^2 \\
&\quad + h_{0011}\beta\epsilon^2 + h_{0020}\beta^2 + h_{3000}A^3 + h_{2100}A|A|^2 + h_{0300}\bar{A}^3 \\
&\quad + h_{1200}\bar{A}|A|^2 + h.o.t.
\end{aligned}$$

we derive the following equations for h_{ijkl} and a, b, c.

term	equation	solvability
β	$L_0 h_{0010} = -D_\lambda G_0$	solvable; h_{0010} is uniform $x,\ y$.
$\beta\epsilon^2$	$Lh_{0011} + D\Delta h_{0010} = 0$	$h_{0011} = 0$ owing to uniform h_{0010}.
A^2	$(L_0 - 2\omega_0 i)h_{2000} = -\frac{1}{2}D_{uu}G_0\phi^2$	solvable.
$A\epsilon^2$	$(L_0 - \omega_0 i)h_{1001}A = (cA - D\Delta A)\phi$	$c = D\Delta$
$A\beta$	$(L_0 - \omega_0 i)h_{1010} = b\phi - D_{uu}G_0\phi h_{0010}$ $-D_{u\lambda}G_0\phi$	$b = \langle\phi^*,\ D_{uu}G_0\phi h_{0010} + D_{u\lambda}G_0\phi\rangle$.
$A^2\bar{A}$	$(L_0 - \omega_0 i)h_{2100} = a\phi - D_{uu}G_0\phi h_{1100}$ $-D_{uu}G_0\bar{\phi}h_{2000} - \frac{1}{2}D_{uuu}\phi^2\bar{\phi}$	$a = \langle\phi^*,\ D_{uu}G_0\phi h_{1100}$ $+D_{uu}G_0\bar{\phi}h_{2000} + \frac{1}{2}D_{uuu}\phi^2\bar{\phi}\rangle$.
A^3	$(L_0 - 3\omega_0 i)h_{3000} = -\frac{1}{6}D_{uuu}G_0\phi^3$ $-D_{uu}G_0\phi h_{2000}$	solvable.

We see from the table above that $a,\ b \in \mathbf{C}$ are complex constants, while c is a linear differential operator. The reduced equation is of the form

$$\frac{\partial A}{\partial t} = \omega_0 iA + aA|A|^2 + b\beta A + \epsilon^2 D\Delta A + h.o.t. \tag{7.40}$$

The equation for $\bar{A}$ is the complex conjugate of it. Consequently, the system of equations for A and $\bar{A}$ describes approximately perturbation of a uniform oscillation via diffusion.

7.5.3 Long time and large spatial behavior

To trace slow temporal variations, we scale the time variable

$$\tau = \epsilon^2 t, \qquad \frac{\partial}{\partial t} = \epsilon^2\frac{\partial}{\partial\tau}$$

and consider the special case $\beta = \epsilon^2$. Let

$$\tilde{A} = \epsilon A e^{-\omega_0 t\ i}$$

and truncate the high order terms in (7.40). We obtain an equation for $\tilde{A}$:

$$\frac{\partial\tilde{A}}{\partial\tau} = a\tilde{A}|\tilde{A}|^2 + b\tilde{A} + D\Delta\tilde{A}. \tag{7.41}$$

This is the well-known Ginzburg-Landau equation. The discussion above shows that the Ginzburg-Landau equation describes long time and large spatial behavior of reaction-diffusion equations near a uniform oscillating wave. We mention that this equation can be also derived directly by the perturbation theory.

8. A Numerical Bifurcation Function for Homoclinic Orbits

In this chapter we adapt the discussion in Ashwin/Mei [26] and consider a numerical method for studying (quasi-) periodic solutions near a homoclinic orbit of the parameter-dependent autonomous differential equation

$$\dot{x}(t) = f(x(t), \lambda), \tag{8.1}$$

where $f : \mathbf{R}^n \times \mathbf{R}^p \to \mathbf{R}^n$ is a C^k-continuous ($k \geq 2$) mapping and $\lambda \in \mathbf{R}^p$ represents the control parameters. This ODE may be thought of as a large system derived from a spatial discretization or from ansatz for traveling waves of reaction-diffusion equations. In one-parameter problems homoclinic orbits arise typically as limits of a branch of periodic solutions when the periods tend to infinity and the solutions stay bounded in phase space (cf. Guckenheimer/Holmes [144], Glendinning [126], Doedel/Kernevez [94] and Rinzel [251]). Dynamics near a homoclinic orbit reveals long time behavior of a system. It gives also hints on global bifurcations, namely bifurcation of homoclinic orbits. This is a complementary to the local bifurcations which we have studied with the Liapunov-Schmidt method and the center manifold theory.

It is well-known that a homoclinic orbit can induce rich dynamics in the system (8.1) (cf. Guckenheimer/Holmes [144], Kuznetsov [197] and Lin [205]). Extensions of Liapunov-Schmidt reduction (cf. Hale [148] and Beyn [39]) and Melnikov function (cf. Guckenheimer/Holmes [144], Lin [205]) have been used to investigate homoclinic bifurcation near TB-points. Under the hypothesis that the original system (*2 dimensional or Halmitonian*) has an analytically available homoclinic orbit, persistence of this homoclinic orbit is assured for perturbed problems under appropriate geometric conditions (cf. Guckenheimer/Holmes [144], Lin [205]) and algebraic conditions (cf. Hale [148] and Beyn [39]). We are interested in bifurcation of periodic orbits near a homoclinic orbit and continuation of the homoclinic orbit with respect to control parameters in λ. In [205] Lin introduces a reduction method for locating (quasi-) periodic orbits near homoclinic orbits. Similarly to the Liapunov-Schmidt reduction, this method reduces dynamical system to a bifurcation function for periodic orbits whose period is asymptotic to infinity on approaching a homoclinic orbit.

We want to approximate the bifurcation function and to investigate bifurcation of periodic orbits from a homoclinic orbit. Calculation of the bifurcation function involves numerical approximation of the homoclinic orbit and numerical solution of the adjoint variational equation (AVE). We discuss a class of methods for solving the adjoint variational equation. In addition we consider effects of continuous symmetries of equations.

After introducing a bifurcation function in Section 8.1, we discuss in Section 8.2 briefly methods for computing homoclinic orbits and then move on in Section 8.3 to the problem of solving the adjoint variational equation. The structure of the AVE means that there is a conservation property that must be preserved in the numerical approximations (see (8.14) below). In Subsection 8.3.2 we consider implications of continuous symmetries and discuss how to handle bifurcations of relative periodic orbits from relative homoclinic orbits. Section 8.4 is devoted to numerical approximation of the bifurcation function (8.4). We show that a predictor for continuation of the homoclinic orbit is easily available. Finally in Section 8.5 we examine the equations for an electronic circuit of Freire *et al.* [254] and symmetry breaking perturbations of the Kuramoto-Sivashinsky equation [62].

8.1 A Bifurcation Function

Suppose that at $\lambda = 0$ the flow has a homoclinic orbit $q(t)$, a solution of the system (8.1), which converges to an equilibrium p as $t \to \pm\infty$, i.e.,

$$p = \lim_{t \to \pm\infty} q(t). \tag{8.2}$$

Furthermore, we assume that the equilibrium p is hyperbolic, i.e., the Jacobian matrix $D_x f(p, 0)$ has no eigenvalues on the imaginary axis.

In order to characterize Silnikov (and many other) phenomena for a generic set of such systems, notably including those which have one-dimensional intersection of tangent spaces of the stable and unstable manifolds of p along the homoclinic orbit $q(t)$, Lin introduces in [205] a bifurcation function $G(\omega, \lambda)$ describing the jump in piecewise smooth solutions of (8.1) on varying a time interval $[0, 2\omega]$, $\omega \in \mathbf{R}^+$ and parameter λ. More precisely, Lin shows that for appropriate vectors Δ and $\bar{\psi}$, for large enough ω and small enough λ there is a unique solution $(x_\omega(t), G(\omega, \lambda))$ of the boundary value problem

$$\begin{aligned} \dot{x} - f(x, \lambda) &= 0 \quad t \in [0, 2\omega], \\ \langle \bar{\psi},\ x(0) - q(0) \rangle &= 0, \\ x(2\omega) - x(0) &= G\Delta. \end{aligned} \tag{8.3}$$

Here and in the sequel, $\langle \cdot,\ \cdot \rangle$ is the inner product in $\mathbf{R}^n$. Asymptotically as $\omega \to \infty$ and $|\lambda| \to 0$, the jump $G(\omega, \lambda)$, known as the *bifurcation function*, can be approximated as

$$G(\omega,\lambda) = \langle\psi(\omega),\ q(-\omega)\rangle - \langle\psi(-\omega),\ q(\omega)\rangle \\ + \sum_{i=1}^{d}\lambda_i\int_{-\infty}^{\infty}\langle\psi(s),\ \frac{\partial f}{\partial\lambda_i}(q(s),0)\rangle\, ds + \text{h.o.t.}, \tag{8.4}$$

where $\psi(t)$ is a nontrivial bounded solution of the so-called *adjoint variational equation* (or AVE)

$$\dot{\psi} = -(D_x f(q(t),0))^T\psi. \tag{8.5}$$

Zeros of the bifurcation function $G(\omega,\lambda) = 0$ at finite ω and small $|\lambda|$ are in one-to-one correspondence with period-2ω solutions of (8.1) near the homoclinic orbit $q(t)$.

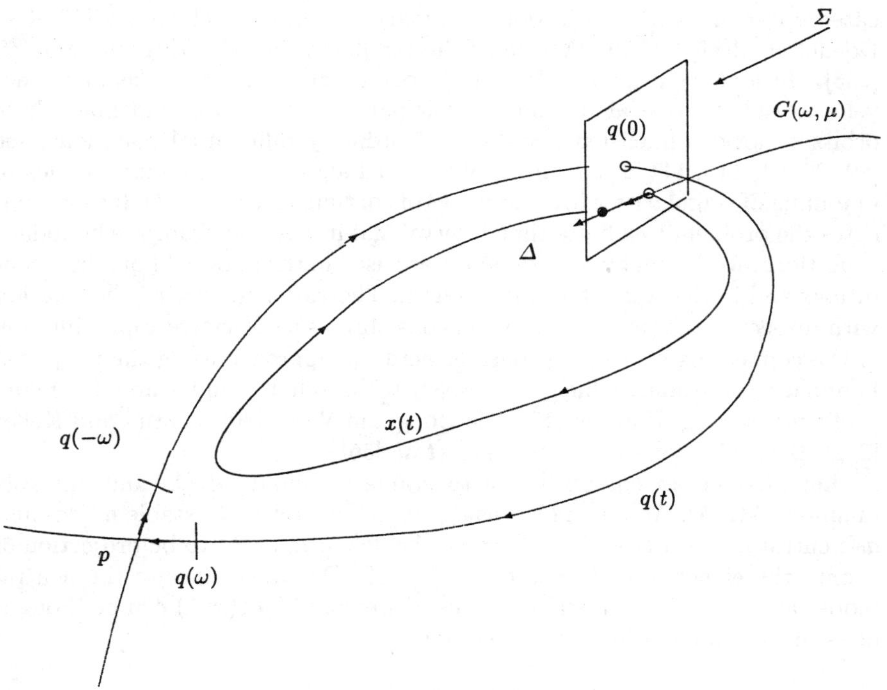

Fig. 8.1. A bifurcation function for homoclinic orbit.

Moreover, solutions of

$$G(\infty,\lambda) = \lim_{\omega\to\infty} G(\omega,\lambda) = 0$$

give values of λ for which the system (8.1) has homoclinic orbits (see also [297]). Thus this function can be used for continuation of a homoclinic orbit with respect to the parameter λ. Using a set of these jumping (bifurcation) functions of the corresponding return maps, Sandstede [264, 265] has been able to analyze bifurcations and period-doubling of homoclinic orbits. Thus the bifurcation function plays the same role as the reduced bifurcation equations in the Liapunov-Schmidt method.

8.2 Approximation of Homoclinic Orbits

Several methods are available for numerical approximation of homoclinic orbits, e.g., periodic orbits with large period (cf. Doedel/Kernevez [94]), collocation methods (cf. Moore [233]), boundary value problems on truncated intervals with projection boundary conditions (cf. Beyn [37, 38], Doedel et al[87, 93, 89, 111] and Schecter [269]) and shooting method (cf. [253]). In a series of papers Doedel, Friedman and their coworkers have developed an effective algorithm for calculating heteroclinic and homoclinic orbits of lower dimensional systems of ordinary differential equations, see [89, 90, 91, 110, 111, 112]. This sophisticated algorithm combines the idea of unfolding of singularity with techniques of continuation methods. It first truncates the problem to a finite time interval and imposes appropriate boundary conditions. Additional free parameters are used in the imposed boundary conditions and in the truncated time interval. Thereafter by tracing the solution with respect to the various parameters one obtains a successive approximation of the connecting orbits. We consider here the approach using the projected boundary conditions, which has proven to be reliable and stable for many problems, see e.g. Hassard [151], de Hoog and Weiss [80], Lentini and Keller [202], Beyn [37, 38] and Champneys *et al.* [56].

Let $p(\lambda)$ be an equilibrium with stable manifold $\mathcal{M}_s(\lambda)$ and unstable manifold $\mathcal{M}_u(\lambda)$. Assume $p(\lambda)$ has d_+ unstable modes, d_- stable modes and d_0 neutral modes ($d_+ + d_- + d_0 = n$). We define $L_u(x, \lambda)$ to be projection of x onto the eigenspace of unstable modes of $p(\lambda)$, $L_n(x, \lambda)$ onto the neutral modes and $L_s(x, \lambda)$ onto stable modes. For example, $L_s(x, \lambda)$ can be thought of as linear combinations of d_- vectors

$$\{\langle \phi_i^*,\ x\rangle \phi_i \ :\ i = 1, \ldots, d_-\},$$

where $\{\phi_i\}$ (resp. $\{\phi_i^*\}$) are the d_- right (resp. left) normalized eigenvectors of $D_x f(p(\lambda), \lambda)$ corresponding to the stable eigenvalues. The projections $L_n(x, \lambda)$ and $L_u(x, \lambda)$ are defined in a similar manner.

For one-parameter problems, an orbit $q(t)$ for $\lambda = 0$ that is homoclinic to p is said to be *nondegenerate* (cf. Beyn [37]), if p is hyperbolic (i.e. $d_0 = 0$), the intersection of the manifolds $\mathcal{M}_s$ and $\mathcal{M}_u$ is one dimensional and the linearized problem

$$\dot{z} = D_x f(q(t), 0)z, \quad \lim_{t \to \pm\infty} z(t) = 0 \tag{8.6}$$

has only one bounded solution given by $c\dot{q}(t)$, $c \in \mathbf{R}$. A sufficient condition for $q(t)$ to be nondegenerate is that solution spaces of the problems (8.5) and (8.6) are exactly one dimensional, respectively, and

$$\int_{-\infty}^{\infty} \psi(s)^T \frac{\partial f}{\partial \lambda}(q(s), 0)\, ds \neq 0. \tag{8.7}$$

Assume that we can find $p(\lambda)$ and the projections $L_u(\cdot, \lambda)$, $L_s(\cdot, \lambda)$ for a homoclinic orbit near $\lambda = 0$. In order to remove the degeneracy of $q(t)$ (all time-shifts are also homoclinic orbits) we choose a *phase condition* $\Psi = 0$, see e.g. Beyn [37], with

$$\Psi(x) = \langle \dot{\hat{x}}(0),\; x(0) - \hat{x}(0) \rangle \qquad \text{or}$$
$$\Psi(x) = \int_{-\infty}^{\infty} \langle \dot{\hat{x}}(t),\; x(t) - \hat{x}(t) \rangle\, dt$$

for some initial approximation $\hat{x}(t)$ of $q(t)$, in fact, an arbitrary function that decays fast enough at $\pm\infty$ can be used.

Let $T > 0$ be a constant. To approximate the homoclinic orbit over the finite time interval $[-T, T]$, we choose T sufficiently large such that at $t = T$ (resp. $t = -T$) the ends of the orbits are in the tangent spaces of the stable (resp. unstable) manifold to the fixed point. We consider the following boundary value problem on $[-T, T]$

$$\begin{aligned} \dot{x}(t) - f(x(t), \lambda) &= 0, \quad t \in [-T, T], \\ L_s(x(-T) - p(\lambda), \lambda) &= 0, \\ L_u(x(T) - p(\lambda), \lambda) &= 0, \\ \Psi(x, \lambda) &= 0. \end{aligned} \tag{8.8}$$

In this way, we truncate the dynamic system (8.1) into the finite interval $[-T, T]$ and then imposed with the *projection boundary conditions* and a phase condition. As demonstrated by Beyn [38], solutions of this problem converge exponentially to a given nondegenerate homoclinic orbit in phase space and parameter space as $T \to \infty$. Doedel and his coworkers have used various linear and higher order approximations of stable and unstable manifolds near the equilibrium p to reduce the problem (8.1) to a robust boundary value problem on a finite interval, see [88, 89, 90, 91, 93, 111, 112]. Moreover, they have developed very effective continuation and successive continuation methods for locating connecting orbits.

There are many ways to solve (8.8) numerically. For smaller systems one can use Newton iteration to solve it directly with a collocation or finite difference discretization. For larger systems this becomes computationally much more intensive. Iterative methods are preferred, see Ashwin/Mei [27] and

Garbotz [116]. On the other hand, since the homoclinic orbit converges exponentially to the equilibrium as $t \to \infty$, the grid points in the interval $[-T, T]$ can be distributed exponentially to achieve uniform accuracy. Shooting methods with appropriate projections may be another alternative for discretizing the truncated problem (8.8), at least for cases where d_+ is small (cf. Canale [51] and Kubíček/Schreiber [192]).

8.3 Solving the Adjoint Variational Problem

The linear variational problem for (8.1) at a homoclinic orbit $q(t)$ (assumed to be at $\lambda = 0$) is the non-autonomous linear equation

$$\dot{z} = D_x f(q(t), 0) z. \tag{8.9}$$

It has at least a one dimensional space of bounded solutions because $z = c\dot{q}$ satisfies the equation for all $c \in \mathbf{R}$. We shall consider a case where symmetry may cause additional dimensions of solution space in Section 8.3.2, but in the absence of symmetry it is generic that $\dot{q}(t)$ gives a basis of the solution space.

Associated with this linear equation is the adjoint variational equation

$$\dot{\psi} = -D_x f(q(t), 0)^T \psi \tag{8.10}$$

which, if the homoclinic orbit is nondegenerate, will also have precisely one bounded solution, i.e.,

$$\mathrm{Ker}\left(\frac{d}{dt} + (D_x f(q(t), \lambda))^T\right) = \mathrm{span}[\psi(t)]. \tag{8.11}$$

If a solution $\psi(t)$ of (8.10) is bounded and the fixed point is hyperbolic, it must decay exponentially to zero as $t \to \pm\infty$, i.e.,

$$\lim_{t \to \pm\infty} \psi(t) = 0. \tag{8.12}$$

Lemma 8.3.1. (Owing to Chossat [62]) *If $A(t) \in \mathbf{R}^{n \times n}$ is a continuous and bounded matrix function for all $t \in \mathbf{R}$, then the conservation property*

$$\langle \psi(t), z(t) \rangle \equiv \textit{constant (independent of } t) \tag{8.13}$$

holds between any solutions $z(t)$ of the equation

$$\dot{z} = A(t) z$$

and $\psi(t)$ of

$$\dot{\psi} = -A(t)^T \psi.$$

Proof: By taking time derivatives and substituting in the equations we derive directly

$$\frac{d}{dt}\langle\psi,\ z\rangle = \langle\dot{\psi},\ z\rangle + \langle\psi,\ \dot{z}\rangle = \langle -A(t)^T\psi,\ z\rangle + \langle\psi,\ A(t)z\rangle = 0. \quad \blacksquare$$

Evidently, the conservation property (8.13) holds for any solutions $z(t)$ of (8.9) and $\psi(t)$ of (8.10) even if the homoclinic orbit is degenerate. In particular, if we wish to compute a bounded solution of (8.10), the fact that $\dot{q}(t)$ decays to zero at both ends implies that any such solution must obey

$$\langle\psi(t),\ \dot{q}(t)\rangle = 0 \quad \text{for all} \quad t \in \mathbf{R}. \tag{8.14}$$

In other words, bounded solutions $\psi(t)$ of (8.10) are orthogonal to $\dot{q}(t)$ in the space $\mathbf{R}^n$ for all points $t \in \mathbf{R}$. In addition, we claim that

$$\dot{q}(t) \notin \operatorname{Im}\Big(\frac{d}{dt} + (D_x f(q(t),0))^T\Big),$$

where

$$\begin{aligned}&\operatorname{Im}\Big(\frac{d}{dt} + (D_x f(q(t),0))^T\Big)\\ &:= \Big\{\frac{d\psi}{dt} + (D_x f(q(t),0))^T\psi \mid \psi \in C^1(\mathbf{R}) \ \text{ and bounded }\Big\}.\end{aligned}$$

In fact, if the equality

$$\frac{d\psi}{dt} + (D_x f(q(t),0))^T\psi = c\dot{q}(t)$$

holds for some bounded function $\psi \in C^1(\mathbf{R})$ and finite constant c, we derive

$$\begin{aligned}c\int_{-\infty}^{\infty}\langle\dot{q}(t),\ \dot{q}(t)\rangle &= c\int_{-\infty}^{\infty}\Big\langle\dot{q}(t),\ \frac{d\psi}{dt} + (D_x f(q(t),0))^T\psi\Big\rangle\\ &= c\langle -\ddot{q}(t) + (D_x f(q(t),0))\dot{q}(t),\ \psi(t)\rangle + c\langle\dot{q}(t),\ \psi(t)\rangle\Big|_{-\infty}^{\infty}\\ &= 0.\end{aligned}$$

The statement $c = 0$ follows from the fact $\dot{q}(t) \not\equiv 0$, leading to a contradiction. Consequently, by $\dot{q}(t) = f(q(t),0)$ a bounded solution ψ of the adjoint variational equation (8.10) can be described equivalently with the equation

$$\frac{d\psi}{dt} + (D_x f(q(t),0))^T\psi + \epsilon f(q(t),0) = 0 \tag{8.15}$$

for the unknowns ψ and $\epsilon \in \mathbf{R}$, see also [56]. We emphasize that ϵ is an artificially introduced parameter and $\epsilon = 0$ holds for all bounded solutions of (8.15). We may utilize these properties and determine bounded solutions of (8.10) with the equation

$$\frac{d\psi}{dt} + (D_x f(q(t),0))^T \psi + \Phi(\psi) f(q(t),0) = 0. \tag{8.16}$$

Here (implicitly) $\Phi(x) = 0$ is a linear normalization condition that excludes the trivial solution and selects a bounded solution. Because kernel of the operator $\frac{d}{dt} + (D_x f(q(t),0))^T$ for a nondegenerate homoclinic orbit $q(t)$ is one-dimensional, we may, for example, choose $\Phi(\psi) = \langle \psi(0),\ \Delta \rangle - 1$ for some non-zero vector Δ such that $\langle \psi(0),\ \Delta \rangle \neq 0$. The operator

$$\frac{d}{dt} + (D_x f(q(t),0))^T + D_\psi \Phi$$

is generically nonsingular for a nondegenerate homoclinic orbit.

For one parameter problems the nondegeneracy condition (8.7) of the homoclinic orbit yields another choice for the normalization of $\psi(t)$

$$\Phi(\psi) := \int_{-\infty}^{\infty} \psi(s)^T \frac{\partial f}{\partial \lambda}(q(s),0)\ ds - 1 = 0.$$

Moreover, the bifurcation function (8.4) reduces to

$$G(\omega,\lambda) = \langle \psi(\omega),\ q(-\omega) \rangle - \langle \psi(-\omega),\ q(\omega) \rangle + \lambda + \text{h.o.t.}.$$

Based on an approximation $(x(t),\lambda)$ of the homoclinic orbit $(q(t),0)$ in the interval $[-T,T]$, we may approximate $\psi(t)$ by numerically solving the linear boundary value problem

$$\begin{aligned} \dot{\psi}(t) + (D_x f(x(t),\lambda))^T \psi(t) + \Phi(\psi) f(x(t),\lambda) &= 0, \quad t \in [-T,T], \\ L_s^*(\psi(T),\lambda) &= 0, \\ L_u^*(\psi(-T),\lambda) &= 0. \end{aligned} \tag{8.17}$$

Here the projections $L_s^*(x,\lambda)$, $L_u^*(x,\lambda)$ are defined similarly as $L_s(x,\lambda)$, $L_u(x,\lambda)$ respectively, by taking the right eigenvectors instead of the left eigenvectors of $D_x f(p(\lambda),\lambda)$. Note that we let the parameter λ vary to adjust any shift of the homoclinic orbit and the bounded solution of AVE in the parameter space. We may use continuation method to drive the solution toward $\lambda = 0$.

Suppose that we have an approximation $x(t)$ given by a discretization of the finite interval boundary value problem (8.8) and wish to approximate $\psi(t)$ using (8.17). Since $x(t)$ is not an exact solution, we cannot expect *a priori* that a numerical method leaves $\langle \psi(t), \dot{x}(t) \rangle$ independent of t; if it tends to a non-zero value at either end of the interval $[-T,T]$ then $\psi(t)$ must blow up since $\dot{x}(t)$ decays. However, as we are interested in bounded solutions of the adjoint variational equation, the conservation property (8.14) is an important criterion for numerical solutions.

One way to get around this, used by Champneys and Kuznetsov [55], is to replace (8.10) by a normalized adjoint variational equation for $\hat{\psi}(t) := \psi(t)/||\psi(t)||$, the unit normal in the direction of $\psi(t)$. This has a disadvantage, noted by Champneys *et al.* [56], that it turns a linear equation into a nonlinear one.

Another robust approach has been developed by Sandstede and used in [56] is to modify (8.10) to (8.15). Treating both ψ and ϵ as the unknowns makes (8.15) to be a nonsingular system. The component ϵ in the solution should be small. It is a measure for the error from time-discretization and truncation of the problems to the finite interval. By using effective methods for boundary value problem, which is possible, in particular, for small systems, the truncation error will certainly dominate. In fact, ϵ satisfies

$$\frac{d}{dt}\langle \dot{x}, \psi \rangle + \epsilon \langle \dot{x}, \dot{x} \rangle = 0 \quad \text{for all } t \in [-T, T].$$

This method will produce very reasonable approximations ψ such that the bifurcation function (8.4) can be approximated.

We propose here an alternative class of methods that approximate ψ while preserving (8.14) and the linearity exactly.

8.3.1 Preserving the inner product

Let

$$\{\hat{q}_i(t) \in \mathbf{R}^n \;:\; i = 1, \ldots, n-1\}$$

be a set of linearly independent vectors, piecewise continuous in time, that span a space complementary to span$[\dot{q}(t)]$ in $\mathbf{R}^n$ for each $t \in \mathbf{R}$.

Theorem 8.3.1. *A bounded solution of the equation (8.10) satisfies*

$$\begin{aligned} \langle \dot{\psi}(t) + (D_x f(q(t),0))^T \psi(t),\; \hat{q}_i(t) \rangle &= 0, \quad i = 1, \ldots, n-1, \\ \langle \psi(t),\; \dot{q}(t) \rangle &= 0, \end{aligned} \tag{8.18}$$

where $\langle \cdot, \cdot \rangle$ is the scalar product in $\mathbf{R}^n$. Conversely, any solution of (8.18) is also a solution of (8.10).

Proof: From the above discussion we see that the first part of the conclusion is the representation of (8.14) with respect to the basis $\{\dot{q}(t),\; \hat{q}_i(t); i = 1, \ldots, n-1\}$ of $\mathbf{R}^n$. Suppose $\psi(t)$ is a solution of (8.18). The spanning property of $\{\hat{q}_i(t)\}$ implies that

$$\dot{\psi}(t) + (D_x f(q(t),0))^T \psi(t) = c\dot{q}(t)$$

for some scalar $c \in \mathbf{R}$. On the other hand, from the last equation of (8.18) we derive

$$\begin{aligned} c\langle \dot{q}(t),\ \dot{q}(t)\rangle &= c\langle \dot{\psi}(t) + (D_x f(q(t),0))^T \psi(t),\ \dot{q}(t)\rangle \\ &= c[\langle \dot{\psi}(t),\ \dot{q}(t)\rangle + \langle \psi(t),\ \ddot{q}(t)\rangle] \\ &= c\frac{d}{dt}\langle \psi(t),\ \dot{q}(t)\rangle \\ &= 0. \end{aligned}$$

Hence, $c = 0$ and $\psi(t)$ satisfies (8.10). ■

For numerical approximation of the bounded solution ψ of (8.10) and the preservation of the inner product (8.14), we will use the system (8.18) with a shift condition. Suppose we have an approximation $(x(t), \lambda)$ of $(q(t), 0)$, and correspondingly the equilibrium is shifted from the origin to $(p(\lambda), \lambda)$. Let E_s^* and E_u^* be the stable and unstable eigenspaces of $(D_x f(p(\lambda), \lambda))^T$. Taking the projections $y_s := L_s^*(\dot{x}(T), \lambda) \in E_s$ and $y_u^* := L_u^*(\dot{x}(-T), \lambda) \in E_s$, we decompose the spaces E_s^* and E_u^* as

$$E_s^* = \text{span}[y_s] \oplus \tilde{E}_s^*, \qquad E_u^* = \text{span}[y_u] \oplus \tilde{E}_u^*.$$

Define $\tilde{L}_s^*$ and $\tilde{L}_u^*$ to be the projections from R^n onto the subspaces $\tilde{E}_s^*$ and $\tilde{E}_u^*$ respectively. Let $\hat{x}_i(t)$, $i = 1, \ldots, n-1$ be an orthonormal basis for $(\text{span}[\dot{x}(t)])^\perp$ in $\mathbf{R}^n$ and

$$A(t) := (\hat{x}_1(t), \ldots, \hat{x}_{n-1})^T \in \mathbf{R}^{(n-1)\times n}.$$

We truncate the equation (8.18) in $[-T, T]$ and consider the following system of equations for $\psi(t)$

$$\begin{aligned} A(t)\dot{\psi}(t) + A(t)D_x f(x(t), \lambda))^T \psi(t) &= 0, \quad t \in [-T, T], \quad i = 1, \ldots, n-1 \\ \dot{x}(t)^T \psi(t) &= 0, \quad t \in [-T, T], \\ \tilde{L}_s^*(\psi(T), \lambda) &= 0, \\ \tilde{L}_u^*(\psi(-T), \lambda) &= 0, \\ \Phi(\psi) &= 0. \end{aligned} \tag{8.19}$$

This is a system of *differential-algebraic equations* (DAE) consisting of $n-1$ implicit differential equations and one algebraic equation. The $n-2$ projection boundary conditions and one normalization condition make the system well-posed. Recall that $(p(\lambda), \lambda)$ is a hyperbolic steady state solution. The statements $y_s \neq 0$ and $y_u \neq 0$ hold for sufficiently large T. We may define the projections $\tilde{L}_s^*$ and $\tilde{L}_u^*$ directly with the basis of E_s^* and E_u^*. Alternatively, we can incorporate the conserved equalities at the ends of the truncated intervals into the projection boundary conditions. Rewrite the equations $\langle \psi(t),\ \dot{x}(t)\rangle = 0$ at $t = \pm T$ as

$$\langle \psi(T),\ \dot{x}(T)\rangle y_s = 0, \qquad \langle \psi(-T),\ \dot{x}(-T)\rangle y_u = 0.$$

Note that $\dot{x}(T) = y_s + (I - L_s^*)\dot{x}(T)$ and $\dot{x}(-T) = y_u + (I - L_u^*)\dot{x}(-T)$. We have

$$\langle\psi(T),\ y_s\rangle y_s = -\langle\psi(T),\ (I - L_s^*)\dot{x}(T)\rangle y_s,$$
$$\langle\psi(-T),\ y_u\rangle y_u = -\langle\psi(-T),\ (I - L_u^*)\dot{x}(-T)\rangle y_u.$$

Adding these equations to the boundary conditions in (8.19) leads to

$$\begin{aligned} A(t)\dot{\psi}(t) + A(t)D_x f(x(t),\lambda))^T\psi(t) &= 0, \quad t \in [-T,T], \quad i = 1,\dots,n-1 \\ \dot{x}(t)^T\psi(t) &= 0, \quad t \in (-T,T), \\ L_s^*(\psi(T),\lambda) &= -\langle\psi(T),(I - L_s^*)\dot{x}(T)\rangle y_s, \\ L_u^*(\psi(-T),\lambda) &= -\langle\psi(-T),(I - L_u^*)\dot{x}(-T)\rangle y_u, \\ \Phi(\psi) &= 0. \end{aligned}$$

Since the matrix

$$\begin{pmatrix} A(t) \\ \dot{x}(t)^T \end{pmatrix}$$

is nonsingular, the system (8.19) of differential-algebraic equations is of index one (cf. Bai [29], Gear [124]). In particular, we can approximate $\psi(t)$ by exploiting effective numerical methods for boundary value problems of DAEs, see e.g. Ascher [18], Bai [29, 30] and Griepentrog/März [139] for more details. To illustrate the main ideas, we choose a grid $\{t_1 = -T < t_2 < \cdots < t_m = T\}$ and discretize (8.19) with a finite difference approximation. Thus we obtain the following linear system of mn equations for the mn unknowns $\{\psi_i(t_j)\ :\ i = 1,\dots,n,\ j = 1,\dots,m\}$

$$\begin{aligned} \Big\langle \frac{\psi(t_{j+1}) - \psi(t_j)}{t_{j+1} - t_j} + (D_x f(x(t_j),\lambda))^T\psi(t_j),\ \hat{x}_i(t) \Big\rangle &= 0,\ j = 1,\dots,m-1, \\ \langle\psi(t_j),\ \dot{x}(t_j)\rangle &= 0,\ j = 1,\dots,m, \\ \tilde{L}_s^*(\psi(t_m),\lambda) &= 0, \\ \tilde{L}_u^*(\psi(t_1),\lambda) &= 0, \\ \Phi(\psi) &= 0. \end{aligned} \tag{8.20}$$

Note that $\{\hat{x}_i(t) \in \mathbf{R}^n;\ i = 1,\dots,n-1\}$ can be defined as the standard basis unit vectors excluding one which attains largest modulus inner product with $\dot{x}(t)$. This system can be also considered as a direct discretization of (8.18) in which we approximate simultaneously $(q(t),0)$ with $(x(t),\lambda)$. Thus the error caused by approximating the homoclinic orbit is treated as a discretization error.

Alternatively, instead of projecting the boundary conditions we can omit the equations $\langle\psi(\pm T),\dot{x}(\pm T)\rangle = 0$, see also the property (8.12). In this case the terms $\langle\psi(\pm T),\dot{x}(\pm T)\rangle$ give a measure of the errors in the discretization.

This method inherits the $O(h)$ errors of the forward difference method and will not provide an accurate solution unless the number of gridpoints is taken large enough, which in turn leads to large scale systems and involves high computational effort, especially if the real part of a leading eigenvalues (i.e.

those closest to the imaginary axis, [55]) is of small magnitude. To handle this kind of problems we need appropriate strategy for adaptive mesh refinement, e.g. incorporation of decaying rate into discretizations.

One possible higher order method is to approximate (8.19) by multistep shooting. In the case that $d_+ = 1$, this means solving a zero-parameter shooting problem. For example, using Gear's method as time integrator between two consecutive gridpoints, we solve (8.19) first without respecting the conservation equation, and then impose $\langle\psi(t), \dot{q}(t)\rangle = 0$ by projecting orthogonally onto the codimension-one surface with (8.14). The sum of the squares of these corrections gives a measure of the error in ψ. It is this projection that forces the solution to be bounded. Results of using this method are presented in examples in Section 8.5 and in [23].

8.3.2 Systems with continuous symmetries

It often happens that the system (8.1) commutes with action of a k-dimensional compact Lie group Γ. Let Γ_0 be the connected component of the identity in Γ. A *relative equilibrium* is defined to be a dynamically invariant group orbit of a single steady state solution.

Assume that we have an trajectory $q(t)$ which limits to a relative equilibrium p both forwards and backwards in time. Assume that the end points and the connection have trivial isotropy and so in particular, for each finite value of t, $\Gamma_0 q(t)$ is a k-dimensional manifold. We cover a neighborhood of $q(t)$ by a global product $\Gamma_0 \times X$, where X is a domain in a linear subspace of $\mathbf{R}^n$ (this may cover the end points twice). We investigate the relative homoclinic orbit by lifting it to X; in this subspace it is a connecting orbit. The existence of a group action can cause there to be such orbits in a structurally stable manner, see [184] and the example in Section 8.5.2. This is typically due to the fact that the group action leaves invariant subspaces where the connection is from a saddle (relative equilibrium) to a sink. Numerical confirmation is done on Kuramoto-Sivashinsky equation with spectral and finite difference method, see Section 8.5.2 and Garbotz [116].

The linear variational problem of (8.1) is then

$$\frac{dz}{dt} = D_x f(q(t), 0)z. \tag{8.21}$$

In the least degenerate case it has a $k+1$-dimensional solution space

$$\operatorname{Ker}\Big(\frac{d}{dt} - D_x f(q(t), 0)\Big) = \operatorname{span}[z_1, z_2, \dots, z_{k+1}].$$

We set $z_{k+1}(t) = c\dot{q}(t), c \in \mathbf{R}$ and $z_i(t) = \gamma_i q(t), i = 1, \dots, k$ where $\{\gamma_i\}$ generates the Lie algebra of Γ. In other words, once the homoclinic orbit $q(t)$ is known, we derive z_{k+1} by differentiation with respect to t and z_i, $i = 1, \dots, k$ by the infinitesimal group action. Similarly, the kernel

$$\operatorname{Ker}\Big(\frac{d}{dt} + D_x f(q(t),0)^T\Big) = \operatorname{span}[\psi_1, \psi_2, \dots, \psi_{k+1}]$$

is also $k+1$ dimensional. The equality

$$\langle \psi(t),\ z(t)\rangle \equiv \ \text{constant}$$

holds for a given solution $z(t)$ of the adjoint problem. However because of the continuous group orbit, the relative equilibrium p is non-hyperbolic. The constant above needs not to vanish, i.e., ψ_i and z_i do not necessarily decay at infinity. For this reason we calculate a basis $\{\psi_i\}$ from following system

$$\begin{aligned} \langle \psi_i, z_j\rangle &= \delta_{ij}, \quad (i,j) = 1,\dots,k,\\ \langle \psi_i, z_{k+1}\rangle &= 0, \quad\ \ i = 1,\dots,k+1,\\ (I - \sum_{j=1}^{k+1} z_j z_j^T)\Big[\frac{\partial}{\partial t} + (D_x f(q(t),0))^T\Big]\psi_i &= 0,\\ \tilde{L}_s^* \psi_i(T) &= 0,\\ \tilde{L}_u^* \psi_i(-T) &= 0,\\ \Phi(\psi_i) &= 0. \end{aligned} \tag{8.22}$$

The unique decaying bounded solution of this system will be ψ_{k+1}.

8.4 The Approximate Bifurcation Function

Given $q(t)$ and $\psi(t)$, the leading terms of the bifurcation function by [205, eq. (4.7)] follow directly as

$$\begin{aligned} G(\omega,\lambda) = \langle \psi(\omega),\ q(-\omega)\rangle - \langle \psi(-\omega),\ q(\omega)\rangle + \lambda \int_{-\infty}^{\infty} \langle \psi(s),\ D_\lambda f(q(s),0)\, ds \\ +\ o(|\lambda| + e^{-\delta\omega}) \end{aligned}$$

for $\omega \in [0,T]$, where $\delta > 2\rho$ and ρ is real part of the leading eigenvalue $\rho \pm i\beta$ of $D_x f$ at the fixed point. In particular, for generic situations we have [205, thm 4.6]

$$G(\omega,0) = e^{-2\rho\omega}\Theta(\omega) + \text{h.o.t.} \tag{8.23}$$

where Θ is a quasiperiodic function of ω, a sum of trigonometric functions of period π/β and h.o.t. are higher order terms that decay at an exponentially faster rate.

We can fit $q(t)$ and $\psi(t)$ with their approximate values at the points $t = \pm T$. Suppose that $\{\sigma_i,\ v_i\}$ are the leading eigenpairs of $D_x f(p,0)$ and $\mathrm{Re}(\sigma_i) = \rho > 0$. Similarly, let $\{\sigma_i,\ v_i^*\}$ be the leading eigenpairs of $(D_x f(0,0))^T$. We make the generic assumption that $D_x f(p,0)$ is semi-simple, i.e., complex diagonalizable. There is a $\delta > 0$ and real quantities A_i, B_i such that

$$q(t) = \sum_i A_i v_i e^{\sigma_i t} + o(e^{(\rho+\delta)t})$$

for $t \to -\infty$ whereas for $t \to \infty$ we have

$$\psi(t) = \sum_i B_i v_i^* e^{-\sigma_i t} + o(e^{-(\rho+\delta)t}).$$

Since we know approximations of $q(t)$ and $\psi(t)$ at $t = \pm T$, we can estimate the coefficients A_i and B_i. Namely, fixing t to be one of the endpoints, $\pm T$, taking inner products of these equations with $\{v_j^*\}$, resp. $\{v_j\}$ we obtain linear systems for A_i, B_i, respectively. Solving them yields approximations of A_i, B_i, which in turn give us the asymptotics of $q(t)$ and $\psi(t)$ at infinity.

In this way we also obtain the leading order behavior of $G(\omega, 0)$. In the case described above (which includes both examples discussed in Section 8.5) we write

$$\begin{aligned} G(\omega,0) &= \langle \psi(\omega),\ q(-\omega)\rangle - \langle \psi(-\omega),\ q(\omega)\rangle + \text{h.o.t.} \\ &= -\langle \psi(-\omega),\ q(\omega)\rangle + \text{h.o.t.} \\ &= -\sum_{i,j} A_i B_j e^{-(\sigma_i+\sigma_j)\omega} \langle v_j^*, v_i\rangle + \text{h.o.t.} \end{aligned}$$

The integral term in (8.4) is approximated by numerical integration, e.g. by Simpson's rule. Thereafter we obtain an approximate bifurcation function and can extract information about a continuum of periodic orbits with asymptotically large periods. This is analogous to the prediction of branches of stationary solutions near a steady state bifurcation point with the reduced bifurcation equations in the Liapunov-Schmidt reduction, see Chapter 6.

Since approximation of $q(t)$ and $\psi(t)$ is also necessary for detection of degenerate homoclinic orbits in continuation problems [56], we remark here briefly some properties in continuation of homoclinic orbits.

Continuation of homoclinic orbits with respect to parameters: Once the adjoint variational problem is solved, we can make a predictor for the continuation of homoclinic orbits.

Consider the linearization of (8.1) with respect to λ at $(q(t), 0)$

$$\dot{u}(t) - D_x f(q(t), 0) u(t) = D_\lambda f(q(t), 0)\nu. \tag{8.24}$$

A necessary condition for (8.24) to have a bounded solution is that the perturbation ν in parameter space $\mathbf{R}^p$ should be in the hyperplane

$$\left\{ \nu \in \mathbf{R}^p \;\middle|\; \sum_{i=1}^{p} \nu_i \int_{-\infty}^{\infty} \langle \psi(t),\ \frac{\partial f}{\partial \nu_i}(q(t), 0)\rangle \, dt = 0 \right\}.$$

Here ψ with the property (8.12) is a bounded solution of the adjoint variational problem. In other words, the "Melnikov" integral equation

$$\int_{-\infty}^{\infty} \langle \psi(t),\ D_\lambda f(q(t),0)\nu \rangle\ dt = 0$$

is a solvability condition for (8.24). For one-parameter problems the nondegeneracy condition (8.7) implies that (8.24) has no solution for $\nu \in \mathbf{R}$. Thus the homoclinic orbit is isolated.

Knowing that the linear operator $\frac{d}{dt} - D_x f(q(t),0)$ has a one dimensional kernel span$[\dot{q}(t)]$, we are at liberty to select solutions of this equation such that

$$\langle \dot{q}(t),\ u(t) \rangle = 0.$$

Thereafter, for given ν and solution $u(t)$ of the equation (8.24), we choose a small constant $\epsilon \in \mathbf{R}$ and

$$(q(t),0) \mapsto (q(t) + \epsilon u(t), \epsilon\nu)$$

as a linear predictor for continuation of the homoclinic orbit.

8.5 Examples

8.5.1 Freire *et al.*'s circuit

As an example of ordinary differential equations with interesting periodic orbit structure near a homoclinic orbit, we consider the following equations of Freire *et al.* [254], also used as a test system by Champneys and Kuznetsov [55]

$$\begin{aligned} r\dot{x} &= -\nu x + \beta(y-x) - A_3 x^3 + B_3(y-x)^3 \\ \dot{y} &= -\beta(y-x) - z - B_3(y-x)^3 \\ \dot{z} &= y \end{aligned}$$

where $(x,y,z) \in \mathbf{R}^3$ are unknowns, $r = 0.6$, $A_3 = 0.328578$, $B_3 = 0.933578$, ν and β are treated as free parameters. The origin $(x,y,z) = (0,0,0)$ is an equilibrium of this system. Furthermore, there is an orbit homoclinic to the origin. We approximate the homoclinic orbit and the solution to the adjoint variational equation as described in Sections 8.2 and 8.3. We use a collocation in the time interval $[-17,17]$ for approximation of the homoclinic orbit and Gear's method for shooting in conjunction with the projection correction for solution of the AVE, see Figure 8.2, where ψ is shown at the grid points with linear interpolation.

The corresponding bifurcation function $G(\omega,\lambda)$ is calculated as in Section 8.4. A comparison with a bifurcation diagram computed by AUTO86 [94] is shown in Figure 8.3. The curve labeled "AUTO" is the result of path following , and the prediction from truncating (8.4) is labeled "$G = 0$ prediction".

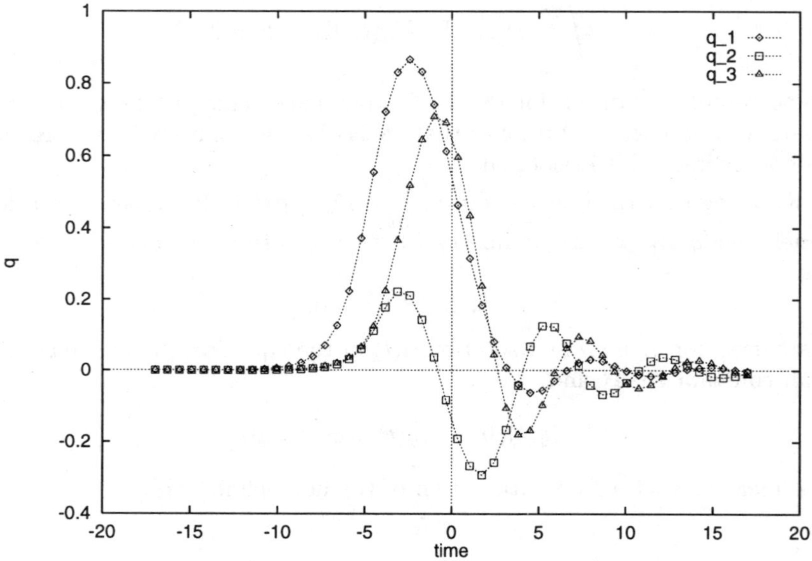

(i)

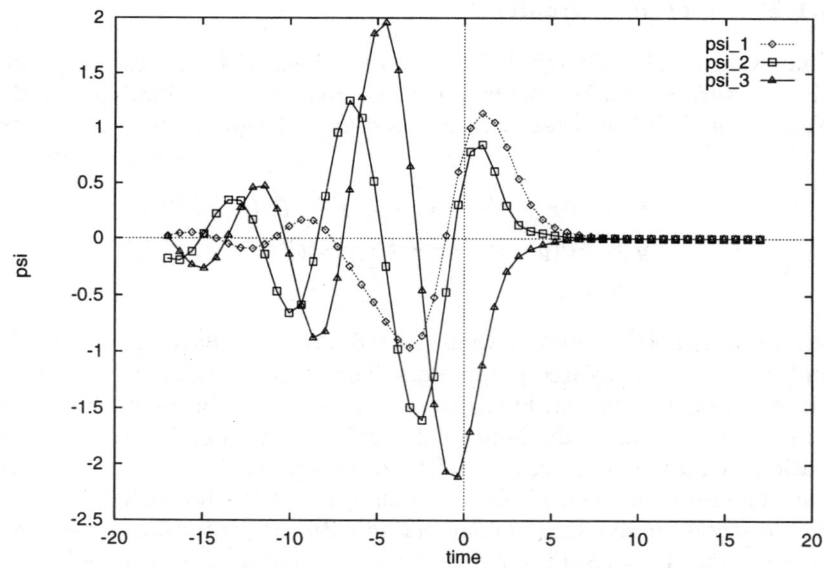

(ii)

Fig. 8.2. Approximations of the three components of the homoclinic orbit $q(t)$ and the solution of the adjoint variational equations $\psi(t)$ for Freire *et al.*'s circuit equations on the truncated time interval $[-17, 17]$. Note that the plotted points are calculated to higher accuracy than the piecewise linear interpolation suggests. The homoclinic orbit is located at $(\nu, \beta) = (-0.775883, 0.3)$.

The other curve is obtained by using an exponential fit and (8.23). As can be seen, all three curves are in excellent agreement, even right down to near the Hopf bifurcation at $(\nu, \beta) = (0.55, 0.3)$.

As expected, for small period the exponential fit has larger errors and moves away from the true curve before the "$G = 0$ prediction" curve does, but has the advantage that it provides an extrapolation of the bifurcation function to periods longer than the interval in which $q(t)$ is calculated.

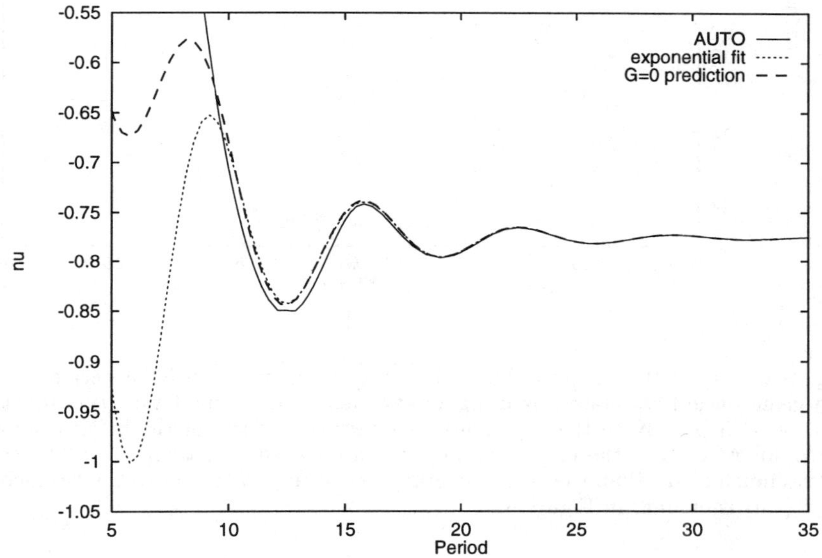

Fig. 8.3. Computed bifurcation diagram at $\beta = 0.3$ in the (P, ν) plane. Here $P = 2\omega$ is the period of the solution. These three curves correspond to a path following from AUTO86; the approximate zeros of the bifurcation function with $G(\omega, 0)$ calculated by exponential fit and by direct numerical calculations, respectively. Note that all three curves remain close right down to a neighborhood of the Hopf bifurcation at $\nu = -0.55$ that creates the periodic orbit.

Figure 8.4 shows the error of the leading exponential fit to the AUTO prediction of the value of ν necessary for the appearance of a periodic orbit of a given period. Note that the error decays exponentially fast up to approximately $\omega = 40$ where it levels off at the error tolerance used for the AUTO calculation.

8.5.2 Kuramoto-Sivashinsky equation

Consider the perturbed Kuramoto-Sivashinsky (K-S) equation on the interval $[0, 2\pi)$

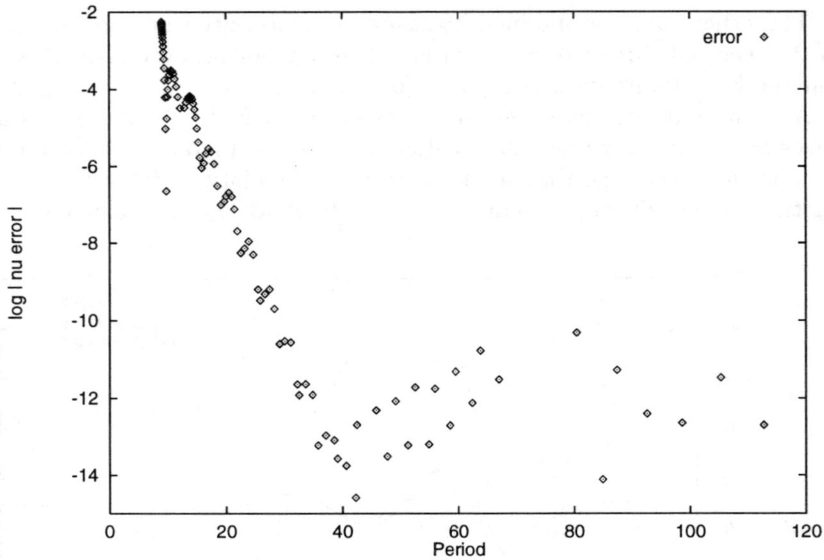

Fig. 8.4. Log of the absolute difference between the values ν for a periodic orbit of a given period calculated by using an exponential fit to the bifurcation function and by AUTO86. Note the exponential convergence of the methods with periods up to 40. After that the error saturates at an error smaller than e^{-10}. The error decays much faster than the rate of convergence of the parameter to the homoclinic orbit (this is evident in Figure 8.3).

$$\frac{\partial v}{\partial t} + 4\frac{\partial^4 v}{\partial x^4} + \alpha\left(\frac{\partial^2 v}{\partial x^2} + v\frac{\partial v}{\partial x}\right) - \beta\frac{\partial v}{\partial x} - \gamma\frac{\partial^3 v}{\partial x^3} = 0, \tag{8.25}$$

where $\alpha,\ \beta,\ \gamma \in \mathbf{R}$ are parameters and $v(x,t)$ is the unknown, imposed with an initial value

$$v(x,0) = v_0$$

and the periodic boundary condition

$$v(0,t) = v(2\pi,t).$$

The parameters β and γ are used to break the x-reflection symmetry. Garbotz discusses in his thesis [116] some other types of boundary conditions, e.g.,

$$v(x,t) = v_{xx} = 0 \quad \text{for } x = 0,\ 2\pi$$

or

$$v(x,t) + \frac{\partial v}{\partial n}(x,t) = 0, \quad v_x(x,t) + \frac{\partial v_{xx}}{\partial n}(x,t) = 0 \quad \text{for } x = 0,\ 2\pi.$$

Define

$$L_0^2[0,2\pi] := \{v \in L^2[0,2\pi] \;\Big|\; v(x,t) = v(x+2\pi,t)\}$$

and

$$\mathcal{X} := \left\{ v \in L_0^2[0,2\pi] \;\Big|\; \frac{\partial^i v}{\partial x^i} \in L_0^2[0,2\pi], i = 1,2, \int_0^{2\pi} v(x)\,dx = 0 \right\}$$

i.e., 2π-periodic and two times weak differentiable functions with zero mean over the interval $[0,2\pi]$. The inner product is defined as

$$\langle u,\ v\rangle_{\mathcal{X}} = \int_0^{2\pi} u_{xx} v_{xx}\,dx.$$

The restriction to zero mean comes from the fact that $v = u_x$ for the original form of the K-S equation, with u periodic in x. We consider the dual product

$$\langle u,\ v\rangle = \frac{1}{2\pi}\int_0^{2\pi} u(x)v(x)\,dx.$$

and perform the analysis in the space $L^2[0,2\pi]$.

The K-S equation has been shown to have an inertial manifold which is forward time invariant and contains the global attractor, in particular, we expect that the homoclinic orbits lie on the inertial manifold. Foias *et al.* [107] show that solutions of K-S equation tend rapidly to the inertial manifold and discuss estimations of the dimension of the inertial manifold.

For $\beta = \gamma = 0$ this equation has $\mathbf{O}(2)$ symmetry acting by

$$R_\theta u(x) = u(x+\theta), \quad \kappa u(x) = -u(2\pi - x), \tag{8.26}$$

where R_θ is a rotation and κ a reflection. Since the introduced flux-like terms destroy the reflection symmetry, the perturbed equation has $\mathbf{SO}(2)$-symmetry for nonvanishing β and γ.

The K-S equation is an evolution on suitable subspaces of functions $\mathcal{X}$, see Kevrekedis [184]. The $\mathbf{O}(2)$ symmetry (8.26) forces the existence of a hierarchy of subspaces of $\mathcal{X}$ invariant under the dynamics. We consider the following two subspaces

$$\begin{aligned}
\mathcal{Y} &= \{u \in \mathcal{X} \quad | \quad u(-x) = -u(x)\}, \\
\mathcal{Z} &= \{u \in \mathcal{X} \quad | \quad u(-x) = -u(x) \text{ and } u(x) = u(x+\pi)\}.
\end{aligned}$$

Note that

$$\mathcal{Z} \subset \mathcal{Y} \subset \mathcal{X}$$

and all three spaces are linear. In addition, we have $R_{\pi/2}\mathcal{Z} = \mathcal{Z}$; this rotation is a "hidden symmetry" of $\mathcal{Z}$.

Kevrekedis *et al.* [184] found that there is a structurally stable homoclinic cycle for $\beta = \gamma = 0$ and α in the range $(16.1399, 22.557)$. This orbit is different

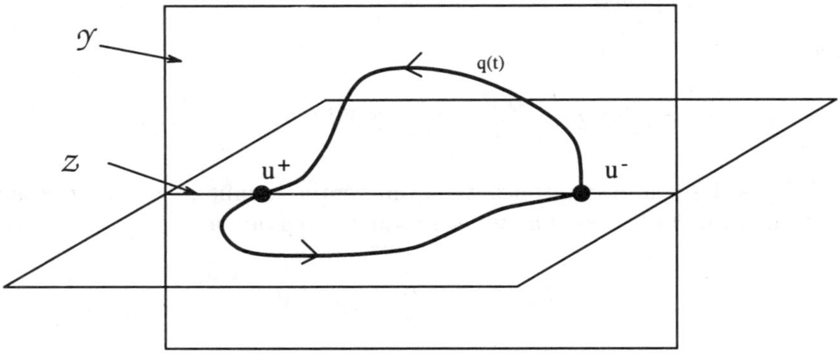

Fig. 8.5. A schematic representation of the invariant subspaces for the dynamics and the location of a structurally stable homoclinic cycle (cf. Kevrekedis *et al.* [184]). There is a one-parameter family of such cycles gained by translating in the x direction.

from the homoclinic orbits derived by the ansatz of traveling wave, which reduces the problem to ODEs.

The homoclinic cycle is found to be attracting for $\alpha \in (16.8, 22.557)$. We show the cycle schematically in Fig. 8.5; a representative cycle connects a point $u^-(x) \in \mathcal{Z}$ and $u^+(x) = u^-(x + \pi/2) \in \mathcal{Z}$ by a trajectory that remains entirely in $\mathcal{Y}$ and back again by a trajectory that remains entirely in $R_{\pi/2}\mathcal{Y}$. Within these subspaces, the connecting orbits pass from a saddle point with one positive eigenvalue to a linearly stable sink. By considering the group orbit of one such cycle we generate a one parameter family of them. These homoclinic cycles bifurcate to two-frequency quasiperiodic attractors for $\gamma \neq 0$, see e.g. [23] and [62].

To investigate this cycle in more detail we perform a spatial discretisation of the PDE via a Galerkin projection onto $2N$ Fourier modes to obtain an ODE in $\mathbf{R}^{2N}$. We write

$$v(x,t) = \sum_{k=1}^{N} \Big(y_k(t) \cos kx + y_{N+k}(t) \sin kx \Big).$$

Since for any N the Fourier basis contains its group orbit under $\mathbf{O}(2)$, we do not expect the continuous symmetry to be broken except by round-off errors. Projecting the equation (8.25) onto these $2N$ Fourier modes we obtain a spatially discrete, timely continuous system

$$\dot{y}_k = F_k(y_1, \ldots, y_{2N}; \alpha, \beta, \gamma), \quad k = 1, \ldots, 2N. \tag{8.27}$$

This system describes an evolution in a linear subspace of $\mathcal{X}$ and approximates evolution of the original equation.

For $\alpha = 22$ as in [23], we calculate the homoclinic orbit $q(t)$ between two fixed points $u^{\pm}$ on the same group orbit such that

$$q(\pm t) \to u^{\pm}$$

as $t \to \infty$. The single unstable eigenvalue is $\sigma = 8.269$, while the leading stable eigenvalue is -50.88. There is also a neutral eigenvalue given by the continuous group action. The eigenvector of the linearization at $u^{\pm}$ corresponding to σ is denoted as as $\phi_u^{\pm}$ while that of the adjoint is $\phi_u^{*\pm}$. The homoclinic orbit and solution of the AVE are calculated on the time-interval $[-1.5, 1.5]$. The leading terms in $G(\omega, 0)$ are of the form

$$G(\omega, 0) = Ce^{-2\sigma\omega} + I\gamma + \text{h.o.t.},$$

where the constant C is calculated from fitting exponentials to the endpoints as described in Section 8.4. The integral term I is calculated by the Simpson's rule as in [23]. Table 8.1 shows C and I calculated as functions of a spatial

N	δ	C	I
8	0.4	−0.3054	0.4698
8	0.2	−0.3231	0.4747
16	0.2	−0.3228	0.4747
16	0.1	−0.3308	0.4753

Table 8.1. Constants for the approximation of the bifurcation function for the perturbed Kuramoto-Sivashinsky equation. N and δ are the spatial and temporal discretization parameters, respectively.

discretization using $2N$ Galerkin modes and maximum stepsize δ for shooting with the Gear's method. Thus we get the approximation

$$G(\omega, \gamma) = 0.3231 \exp(-16.538\omega) + 0.475\gamma + o\left(\gamma + \exp(-16.538\omega)\right). \quad (8.28)$$

The signs of the constants imply that we should see quasiperiodicity in the original equations for small $\gamma < 0$, which remains close to the group orbit of the original homoclinic cycle, see Figure 8.7 and [23] for more details. Figure 8.6 shows good agreement between the measured period from direct simulation of 16 mode truncated equations (calculated by using a variable-step Runge-Kutta integrator) and the prediction from (8.28). We also get quasiperiodicity for small $\gamma > 0$ as can be seen by considering the symmetry $(x, \gamma) \to (2\pi - x, -\gamma)$.

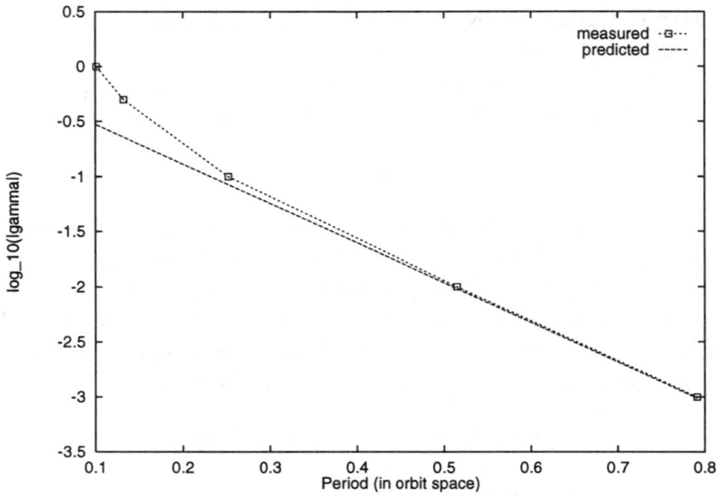

Fig. 8.6. Periods of (relative) periodic solutions in the orbit space with respect to the $O(2)$-symmetry for the perturbed Kuramoto-Sivashinsky equation as a function of $\log_{10}(|\gamma|)$. The points indicate measurements from simulations of 16 mode truncated equations. The prediction is from (8.28).

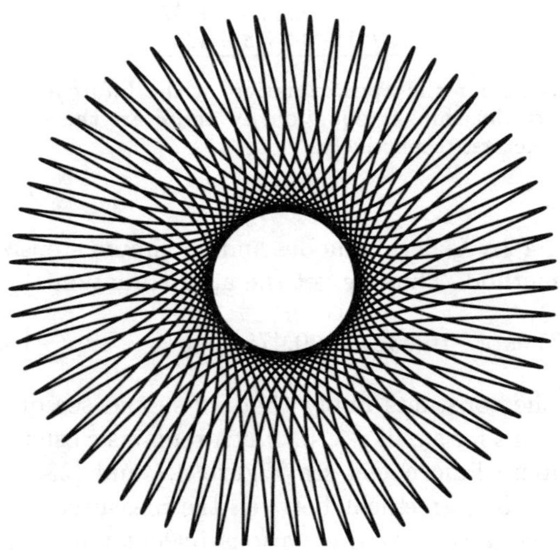

Fig. 8.7. A quasiperiodic solution depicted in the Fourier modes $\cos(x)$ and $\sin(x)$ at $\alpha = 22$, $\beta = \gamma = 1$.

9. One-Dimensional Reaction-Diffusion Equations

We study double bifurcations in a system of one-dimensional reaction-diffusion equations, and numerical continuation of bifurcating solution branches. To ensure a correct reflection of bifurcation scenario in discretizations and to reduce imperfection of singularities, we consider a preservation of multiplicities of the bifurcation points in the discrete problems. A continuation-Arnoldi algorithm is exploited to trace the solution branches and to detect secondary bifurcations.

9.1 Introduction

We consider a system of one-dimensional reaction-diffusion equations

$$\left.\begin{aligned} \frac{\partial u}{\partial t} &= d_1 \frac{\partial^2 u}{\partial x^2} + f(u, v, \lambda) \\ \frac{\partial v}{\partial t} &= d_2 \frac{\partial^2 v}{\partial x^2} + g(u, v, \lambda) \end{aligned}\right. \quad \text{in } \Omega := [0, \ell], \tag{9.1}$$

subject to the homogeneous Dirichlet boundary conditions

$$u(0,t) = u(\ell,t) = u_0, \quad v(0,t) = v(\ell,t) = v_0. \tag{9.2}$$

Here the unknowns u, v are state variables; d_1 and d_2 are diffusion rates; λ is one of the control parameters in the system; and (u_0, v_0) is a uniform steady state solution, i.e., $u_0,\ v_0$ are independent of the variables $t,\ x$, and satisfy

$$\begin{aligned} f(u_0, v_0, \lambda) &= 0, \\ g(u_0, v_0, \lambda) &= 0. \end{aligned}$$

If we choose

$$\begin{aligned} f(u, v, \lambda) &= -(\lambda + 1)u + \alpha + u^2 v, \\ g(u, v, \lambda) &= \lambda u - u^2 v \end{aligned} \tag{9.3}$$

and the boundary conditions

$$u(0,t) = u(\ell,t) \ = \ \alpha, \quad v(0,t) = v(\ell,t) \ = \ \frac{\lambda}{\alpha}, \tag{9.4}$$

we obtain the well-known "Brusselator equations", a mathematically simplified model of chemical reactions involving λ, α as initial and final products, and u, v as intermediate substances, respectively (cf. Nicolis/Prigogine [237], Schaeffer/Golubitsky [268], Dangelmayr [79]).

We are interested in the steady state solutions of (9.1) at double bifurcation points and preservation of the bifurcation scenario in discretizations as well as numerical continuation of the bifurcating solution branches. We show that for the multiple parameter problems

$$\begin{aligned} d_1 \frac{\partial^2 u}{\partial x^2} + f(u,v,\lambda) &= 0 \\ d_2 \frac{\partial^2 v}{\partial x^2} + g(u,v,\lambda) &= 0 \end{aligned} \qquad \text{in } \Omega := [0,\ell], \tag{9.5}$$

a double bifurcation can persist in discretizations if one adjusts parameters properly. That is, though discretizations may split a double bifurcation point into two simple ones, by varying other parameters we can force these two simple bifurcation points to merge in a neighborhood of the original double bifurcation point. Preserving the multiplicity of a bifurcation point in discretizations reduces imperfection of the bifurcations. It is crucial for a correct reflection of bifurcation scenario in the discrete problem. With λ and ℓ as bifurcation parameters we show that Hopf bifurcations occur merely for ℓ large enough, in other words, bifurcation of timely periodic solutions from the stationary solution curve or onset of chemical oscillations takes place only if the domain is properly large, which was noted already by Turing in [293].

In Section 9.2 we consider linear stability of the system (9.1) at the uniform steady state solution and discuss conditions for mode interactions, in particular, for the Brusselator equations. Section 9.3 is devoted to existence and parameterization of the bifurcating solutions. To this end, we transform bifurcating solutions of the original problem into nonsingular solutions of an enlarged system. This system is also appropriate for branch-switching and path-following at bifurcation points. In Section 9.4 we discretize (9.5) with finite difference methods and examine the influence of discretizations on bifurcations. In order to achieve a correct approximation of bifurcation scenario of the original problem, especially to avoid secondary bifurcations which are induced by perturbations and discretizations, we study possibilities to preserve the multiplicity of bifurcation points in discretizations. The continuation-Arnoldi algorithm is used to trace the bifurcating solution branches. Numerical results on the one-dimensional Brusselator equations are given in Section 9.5, which shows that mode interactions can yield both transcritical and pitchfork solution branches.

9.2 Linear Stability Analysis

In this section we study linear stabilities of the uniform steady state solutions (u_0, v_0) by varying the parameters λ and ℓ. To simplify the discussion and the numerical approximation in Section 9.4, we shift the homogeneous state to $(0,0)$ by the transformation $(u,v) = (u_0 + \tilde{u}, v_0 + \tilde{v})$. Furthermore, we incorporate explicitly the length ℓ into the equations by the transformation $x = \ell\tilde{x}$, which changes the domain $\Omega = [0,\ell]$ to the unit interval $\tilde{\Omega} := [0,1]$ and the equations (9.1) into

$$\left.\begin{aligned}\frac{\partial \tilde{u}}{\partial t} &= \frac{d_1}{\ell^2}\frac{\partial^2 \tilde{u}}{\partial \tilde{x}^2} + f(u_0 + \tilde{u},\ v_0 + \tilde{v}, \lambda)\\ \frac{\partial \tilde{v}}{\partial t} &= \frac{d_2}{\ell^2}\frac{\partial^2 \tilde{v}}{\partial \tilde{x}^2} + g(u_0 + \tilde{u},\ v_0 + \tilde{v}, \lambda)\end{aligned}\right\} \quad \text{in } \tilde{\Omega} = [0,1]. \tag{9.6}$$

For simplicity we denote $\tilde{\Omega}$, $\tilde{x}$, $f(u_0+\tilde{u}, v_0+\tilde{v}, \lambda)$, $g(u_0+\tilde{u}, v_0+\tilde{v}, \lambda)$ again by Ω, x and $f(u,v,\lambda)$, $g(u,v,\lambda)$, respectively, when no confusion arises.

9.2.1 The general system

Let

$$C_0^2(\Omega) := \{u \in C^2(\Omega);\ u(0) = u(1) = 0\}$$

and

$$X := \left(C_0^2(\Omega)\right)^2, \qquad Y := \left(C(\Omega)\right)^2. \tag{9.7}$$

We rewrite (9.6) as an operator equation

$$\frac{\partial \mathbf{u}}{\partial t} = G(\mathbf{u}, \lambda), \tag{9.8a}$$

where $\mathbf{u} := (u,v)$, and the mapping $G:\ X \times \mathbf{R} \to Y$ is defined by

$$G(\mathbf{u}, \lambda) := \begin{pmatrix} \dfrac{d_1}{\ell^2}\dfrac{\partial^2 u}{\partial x^2} \\ \dfrac{d_2}{\ell^2}\dfrac{\partial^2 v}{\partial x^2} \end{pmatrix} + \begin{pmatrix} f(u,v,\lambda) \\ g(u,v,\lambda) \end{pmatrix}. \tag{9.8b}$$

Differentiating G with respect to $\mathbf{u}$ at the homogeneous equilibrium $\mathbf{u}_0 \equiv (0,0)$, we obtain the linearization L of G,

$$L := D_{\mathbf{u}}G(\mathbf{u}_0, \lambda) = \begin{pmatrix} \dfrac{d_1}{\ell^2}\dfrac{\partial^2}{\partial x^2} & 0 \\ 0 & \dfrac{d_2}{\ell^2}\dfrac{\partial^2}{\partial x^2} \end{pmatrix} + \begin{pmatrix} \dfrac{\partial f}{\partial u}(0,0,\lambda) & \dfrac{\partial f}{\partial v}(0,0,\lambda) \\ \dfrac{\partial g}{\partial u}(0,0,\lambda) & \dfrac{\partial g}{\partial v}(0,0,\lambda) \end{pmatrix}.$$

Thereafter, stabilities of $\mathbf{u}_0$ can be analyzed via solutions of the variational problem

$$\frac{\partial \mathbf{u}}{\partial t} = L\mathbf{u}.$$

Stationary or Hopf bifurcations occur when the operator $L : X \times \mathbf{R} \to Y$ has zero or pure imaginary eigenvalues. To examine the spectrum of L, we note that the direct sum

$$X = \sum_{k=1}^{\infty} \oplus X_k, \quad X_k := \left\{ \begin{pmatrix} c_1 \\ c_2 \end{pmatrix} \sin k\pi x;\ c_1, c_2 \in \mathbf{R} \right\}, \quad k \in \mathbf{N} \tag{9.9}$$

holds under the $L^2(\Omega)$-product

$$\langle \mathbf{u},\ \mathbf{v} \rangle := \int_{\Omega} (u_1 v_1 + u_2 v_2)\, dx \quad \text{for all } \mathbf{u} = \begin{pmatrix} u_1 \\ u_2 \end{pmatrix},\ \mathbf{v} = \begin{pmatrix} v_1 \\ v_2 \end{pmatrix} \in Y. \tag{9.10}$$

Moreover, L maps X_k into itself and

$$\begin{aligned} & L \begin{pmatrix} c_1 \\ c_2 \end{pmatrix} \sin k\pi x \\ & = \begin{pmatrix} -\frac{\pi^2 k^2 d_1}{\ell^2} + f_u(0,0,\lambda) & f_v(0,0,\lambda) \\ g_u(0,0,\lambda) & -\frac{\pi^2 k^2 d_2}{\ell^2} + g_v(0,0,\lambda) \end{pmatrix} \begin{pmatrix} c_1 \\ c_2 \end{pmatrix} \sin k\pi x. \end{aligned}$$

Here and in the sequel, $f_u^0,\ f_v^0, \ldots$ represent the partial derivatives of f with respect to u and v at $(0,0,\lambda)$, respectively. The restriction of L in the subspace X_k is a 2×2 matrix

$$M_k(\lambda, \ell) := L\Big|_{X_k} = \begin{pmatrix} -\frac{\pi^2 k^2 d_1}{\ell^2} + f_u^0 & f_v^0 \\ g_u^0 & -\frac{\pi^2 k^2 d_2}{\ell^2} + g_v^0 \end{pmatrix}, \quad k = 1, 2, \ldots . \tag{9.11}$$

Thus eigenvalues of L consist of those of $M_k \in \mathbf{R}^{2\times 2}$ for all $k \in N$. The eigenvalues $\nu_1,\ \nu_2$ of a 2×2 matrix $M \in \mathbf{R}^{2\times 2}$ have the properties

$$\begin{aligned} \det(M) &= \nu_1 \nu_2, \\ \operatorname{trace}(M) &= \nu_1 + \nu_2. \end{aligned} \tag{9.12}$$

If $\det(M) > 0$ and $\operatorname{trace}(M) = 0$, then ν_1 and ν_2 are purely imaginary.

A stationary bifurcation occurs on the solution curve $\mathbf{u} \equiv 0$ at $\lambda = \lambda_0$ if and only if there is at least one integer $k \in \mathbf{N}$ such that

$$\det(M_k(\lambda_0, \ell)) = \Big(-\frac{\pi^2 k^2 d_1}{\ell^2} + f_u^0 \Big)\Big(-\frac{\pi^2 k^2 d_2}{\ell^2} + g_v^0 \Big) - f_v^0 g_u^0 = 0. \tag{9.13}$$

Therefore, for $k \in \mathbf{N}$ a solution $\lambda_0 \in \mathbf{R}$ of (9.13) corresponds to a stationary bifurcation point of (9.8). Similarly, a Hopf bifurcation takes place at $\lambda = \lambda_0$ whenever for some $k \in \mathbf{N}$, the following statements hold

$$\operatorname{trace}(M_k(\lambda_0, \ell)) = -\frac{\pi^2 k^2 d_1}{\ell^2} + f_u^0 - \frac{\pi^2 k^2 d_2}{\ell^2} + g_v^0 = 0, \tag{9.14}$$

$$\det(M_k(\lambda_0, \ell)) = \Big(-\frac{\pi^2 k^2 d_1}{\ell^2} + f_u^0 \Big)\Big(-\frac{\pi^2 k^2 d_2}{\ell^2} + g_v^0 \Big) - f_v^0 g_u^0 > 0. \tag{9.15}$$

The eigenvectors associated with the simple eigenvalues of the operator L are also called *modes*. If we consider the length ℓ as the second bifurcation parameter, a linear degeneracy may occur such that two different modes become unstable simultaneously as the corresponding eigenvalues cross the imaginary axis. Owing to the nonlinearity in the problem, different modes of L can interact and yield new solutions. This kind of phenomena is called *mode interaction*. We use the reaction-diffusion equation (9.8) as an example to introduce briefly different types of mode interactions. In Chapters 12, 13 and 16 we shall discuss bifurcation scenario at mode interactions in more details. Here we refer to the monographs Golubitsky/Schaeffer [129, Chapters 9, 10] and Golubitsky/Stewart/Schaeffer [131, Chapter 19] for further discussions on the mode interactions and on the associated nondegeneracy conditions.

(i) **Steady/steady state mode interactions:**

Suppose that for certain $\lambda = \lambda_0$ and $\ell = \ell_0$, zero is a double eigenvalue of L, i.e., there are two integers $k,\ m \in \mathbf{N}$, such that the equalities

$$\det(M_k(\lambda_0, \ell_0)) = 0, \qquad \det(M_m(\lambda_0, \ell_0)) = 0 \tag{9.16}$$

hold. If $k = m$, the matrix M_k is either similar to

$$\begin{pmatrix} 0 & 1 \\ 0 & 0 \end{pmatrix},$$

which corresponds to a generic Takens-Bogdanov bifurcation (cf. Golubitsky/Stewart/Schaeffer [131, pp.414]), or M_k is equal to the zero matrix which corresponds to a nongeneric codimension four bifurcation. The latter, and the case $k \neq m$ yields a two dimensional kernel of L, respectively. Correspondingly, we call $(\mathbf{u}, \lambda, \ell) = (0, \lambda_0, \ell_0)$ a *double bifurcation point* of (9.8).

We restrict the discussion to the case $k \neq m$ and write (9.16) as

$$\begin{aligned} \pi^2 d_1 d_2 (k^2 + m^2) - \ell_0^2 (d_1 g_v^0 + d_2 f_u^0) = 0, \\ \Big(- \frac{\pi^2 k^2 d_1}{\ell_0^2} + f_u^0 \Big) \Big(- \frac{\pi^2 k^2 d_2}{\ell_0^2} + g_v^0 \Big) - f_v^0 g_u^0 = 0. \end{aligned} \tag{9.17}$$

Normally these equations are solvable for the unknowns λ, ℓ. They are generic conditions for a double bifurcation of (9.8).

Remark 9.2.1. From the first equation of (9.17) we see that the equality $\det(M_k(\lambda_0, \ell_0)) = 0$ cannot hold for three different integers $k \in \mathbf{N}$, in other words, a triple bifurcation cannot occur on the trivial solution curve of (9.8).

Let (λ_0, ℓ_0) be a solution of (9.17). Then

$$\mathrm{Ker}(L) = \mathrm{span}[\phi_1,\ \phi_2] \tag{9.18}$$

with

$$\phi_1 := \begin{pmatrix} f_v^0 \\ \frac{\pi^2 k^2 d_1}{\ell_0^2} - f_u^0 \end{pmatrix} \sin(k\pi x), \qquad \phi_2 := \begin{pmatrix} f_v^0 \\ \frac{\pi^2 m^2 d_1}{\ell_0^2} - f_u^0 \end{pmatrix} \sin(m\pi x).$$

The integers k, m are also called wavenumbers.

(ii) **Hopf/steady state mode interactions:**

If there are λ_0, ℓ_0 and k, $m \in \mathbf{N}$, $k \neq m$, such that the following statements hold

$$\operatorname{trace}(M_k(\lambda_0,\ell_0)) = -\frac{\pi^2 k^2 d_1}{\ell_0^2} + f_u^0 - \frac{\pi^2 k^2 d_2}{\ell_0^2} + g_v^0 = 0, \tag{9.19a}$$

$$\det(M_k(\lambda_0,\ell_0)) > 0,$$

$$\det(M_m(\lambda_0,\ell_0)) = \Big(\frac{\pi^2 m^2 d_1}{\ell_0^2} - f_u^0\Big)\Big(\frac{\pi^2 m^2 d_2}{\ell_0^2} - g_v^0\Big) - f_v^0 g_u^0 = 0, \tag{9.19b}$$

then a Hopf/steady state mode interaction occurs at $\mathbf{u} = 0$, $\lambda = \lambda_0$, $\ell = \ell_0$. The equations in (9.19) define a generically solvable system for λ_0, ℓ_0.

(iii) **Hopf/Hopf mode interactions:**

In this case there should be two different wavenumbers k, $m \in N$, such that

$$\operatorname{trace}(M_k) = -\frac{\pi^2 k^2 d_1}{\ell_0^2} + f_u^0 - \frac{\pi^2 k^2 d_2}{\ell_0^2} + g_v^0 = 0, \qquad \det(M_k) > 0,$$

and

$$\operatorname{trace}(M_m) = -\frac{\pi^2 m^2 d_1}{\ell_0^2} + f_u^0 - \frac{\pi^2 m^2 d_2}{\ell_0^2} + g_v^0 = 0, \qquad \det M_m > 0$$

for some λ_0, $\ell_0 > 0$. However, it is evident that this is impossible. Hence, no Hopf/Hopf mode interactions can occur along the uniform solution curve of (9.8).

9.2.2 The Brusselator equations

From (9.4) we know that the Brusselator equations have a uniform steady state solution

$$(u_0, v_0) \equiv (\alpha, \frac{\lambda}{\alpha}).$$

After shifting it to the origin $(0,0)$, the equations (9.3) can be written as

$$\begin{aligned} f(u,v,\lambda) &= (\lambda-1)u + \alpha^2 v + \Big(\frac{\lambda}{\alpha}u^2 + 2\alpha uv + u^2 v\Big), \\ g(u,v,\lambda) &= -\lambda u - \alpha^2 v - \Big(\frac{\lambda}{\alpha}u^2 + 2\alpha uv + u^2 v\Big). \end{aligned} \tag{9.20}$$

The statement (9.11) becomes

$$M_k(\lambda,\ell) := \begin{pmatrix} -\frac{\pi^2 k^2 d_1}{\ell^2} + \lambda - 1 & \alpha^2 \\ -\lambda & -\frac{\pi^2 k^2 d_2}{\ell^2} - \alpha^2 \end{pmatrix}. \tag{9.21}$$

Hence,

$$\operatorname{trace}(M_k) = -\frac{\pi^2 k^2}{\ell^2}(d_1 + d_2) - \alpha^2 - 1 + \lambda,$$
$$\det(M_k) = \frac{\pi^4 k^4}{\ell^4} d_1 d_2 + \frac{\pi^2 k^2}{\ell^2}[d_1\alpha^2 + d_2(1-\lambda)] + \alpha^2.$$

Steady state bifurcation: A stationary bifurcation corresponds to $\det(M_k) = 0$ for some $k \in \mathbf{N}$, which is satisfied by

$$\lambda_k := 1 + \frac{d_1}{d_2}\alpha^2 + \frac{\pi^2 k^2}{\ell^2} d_1 + \left(\frac{\ell}{\pi k}\right)^2 \frac{\alpha^2}{d_2} > 0. \tag{9.22}$$

Furthermore, if $\lambda > \lambda_k$, then $\det(M_k) < 0$. By (9.12) the two eigenvalues of M_k are real and have opposite signs. Consequently, the uniform steady state solution is unstable for $\lambda > \lambda_k,\ k \in \mathbf{N}$.
Hopf bifurcations: The trace of M_k changes sign at

$$\lambda = 1 + \alpha^2 + \frac{\pi^2 k^2}{\ell^2}(d_1 + d_2),\ k \in \mathbf{N}. \tag{9.23}$$

A Hopf bifurcation occurs when λ is chosen as (9.23) for some k, at the same time, the following inequality holds

$$\begin{aligned}
&\det(M_k(\lambda_0, \ell)) \\
&= -\left(\frac{\pi k}{\ell}\right)^4 d_2^2 + \left(\frac{\pi k}{\ell}\right)^2 (d_1 - d_2)\alpha^2 + \alpha^2 \\
&= -\frac{1}{d_2^2}\left[\left(\frac{\pi k}{\ell}\right)^2 d_2^2 - \frac{1}{2}\left(\alpha^2(d_1 - d_2) + \sqrt{\alpha^4(d_1 - d_2)^2 + 4\alpha^2 d_2^2}\,\right)\right] \\
&\quad \cdot \left[\left(\frac{\pi k}{\ell}\right)^2 d_2^2 - \frac{1}{2}\left(\alpha^2(d_1 - d_2) - \sqrt{\alpha^4(d_1 - d_2)^2 + 4\alpha^2 d_2^2}\,\right)\right] \\
&> 0.
\end{aligned}$$

This yields a necessary condition for Hopf bifurcation

$$0 < \frac{k^2}{\ell^2} < \frac{1}{2 d_2^2 \pi^2}\left(\alpha^2(d_1 - d_2) + \sqrt{\alpha^4(d_1 - d_2)^2 + 4\alpha^2 d_2^2}\,\right). \tag{9.24}$$

In other words, Hopf bifurcations occur only when the domain is large enough, at least

$$\ell^2 > \frac{2 d_2^2 \pi^2}{\alpha^2(d_1 - d_2) + \sqrt{\alpha^4(d_1 - d_2)^2 + 4\alpha^2 d_2^2}} > 0. \tag{9.25}$$

We summarize that a Hopf or a stationary bifurcation takes place only when

$$\lambda > \min_k\{\lambda_k, 1 + \alpha^2 + \frac{\pi^2 k^2}{\ell^2}(d_1 + d_2)\}. \tag{9.26}$$

Mode interaction: Since we are interested in the cases $\alpha,\ \lambda > 0$, by (9.16) a double bifurcation or a steady/steady mode interaction occurs for the Brusselator equations if and only if there are $k,\ m \in \mathbf{N},\ k \neq m$, such that

$$\lambda_k = \lambda_m \iff \ell^4 = \pi^4 \alpha^{-2} d_1 d_2 m^2 k^2. \tag{9.27}$$

Moreover, substituting λ_k of (9.22) into the matrix M_k in (9.21), we obtain

$$M_k(\lambda_k, \ell) = \begin{pmatrix} \frac{d_1}{d_2}\alpha^2 + (\frac{\ell}{\pi k})^2 \frac{\alpha^2}{d_2} & \alpha^2 \\ -\frac{\pi^2 k^2}{\ell^2} d_1 - \frac{d_1}{d_2}\alpha^2 - 1 - (\frac{\ell}{\pi k})^2 \frac{\alpha^2}{d_2} & -\frac{\pi^2 k^2}{\ell^2} d_2 - \alpha^2 \end{pmatrix}.$$

Thus the null space $\mathrm{Ker}(D_{\mathbf{u}}G(0, \lambda_k))$ can be given explicitly as

$$\mathrm{Ker}(D_{\mathbf{u}}G(0, \lambda_k)) = \mathrm{span}[\phi_1,\ \phi_2], \tag{9.28}$$

with

$$\phi_1 := \begin{pmatrix} -1 \\ \frac{d_1}{d_2} + (\frac{\ell}{\pi k d_2})^2 \end{pmatrix} \sin k\pi x,$$
$$\phi_2 := \begin{pmatrix} -1 \\ \frac{d_1}{d_2} + (\frac{\ell}{\pi k d_2})^2 \end{pmatrix} \sin m\pi x.$$

For Hopf/steady state mode interaction we derive from (9.19) that it takes place in Brusselator equations if and only if there are $\lambda_0,\ \ell_0 > 0$ and some $k,\ m \in \mathbf{N},\ k \neq m$, such that

$$\ell_0^4 \alpha^2 + \alpha^2 \pi^2 m^2 \ell_0^2 (d_2 - d_1) + \pi^4 m^2 d_2 (m^2 d_1 - k^2 d_1 - k^2 d_2) = 0,$$
$$\lambda_0 = 1 + \alpha^2 + \frac{\pi^2 k^2}{\ell_0^2}(d_1 + d_2),$$
$$\frac{\pi^4 k^4}{\ell_0^4} d_1 d_2 + \frac{\pi^2 k^2}{\ell_0^2}[d_1 \alpha^2 + d_2(1 - \lambda_0)] + \alpha^2 > 0.$$

9.3 Solution Branches at Double Bifurcations

In this section we study the stationary problem

$$G(\mathbf{u}, \lambda) = 0 \tag{9.29}$$

at a double bifurcation point $(\mathbf{u}_0, \lambda_0) = (0, \lambda_0)$. We integrate the algebraic bifurcation equations (ABE) of Decker/Keller [81] and Jepson/Decker [170] into a slightly enlarged system. This system regularizes (9.29) at double bifurcation points, and simultaneously parameterizes the bifurcating solution branches. Our approach is constructed for the numerical continuation of bifurcating solution branches, in particular, for the branch-switching at the bifurcation points. The analysis is based mainly on the implicit function theorem.

For a one dimensional domain Ω the operator $L = D_{\mathbf{u}}G(\mathbf{u}_0, \lambda_0)$ is Fredholm of index zero (cf. Golubitsky/Schaeffer [129, pp.334]. Furthermore, under the $L^2(\Omega)$-product (9.10) the spaces X and Y in (9.7) can be decomposed as

$$X = \mathrm{Ker}(L) \oplus M, \qquad Y = \mathrm{Ker}(L^*) \oplus \mathrm{Im}(L), \tag{9.30}$$

where $L^* : X \to Y$ is the adjoint operator of L,

$$L^* = \begin{pmatrix} \frac{d_1}{\ell^2}\frac{\partial^2}{\partial x^2} & 0 \\ 0 & \frac{d_2}{\ell^2}\frac{\partial^2}{\partial x^2} \end{pmatrix} + \begin{pmatrix} f_u^0 & g_u^0 \\ f_v^0 & g_v^0 \end{pmatrix}.$$

At the double bifurcation point $\mathbf{u}_0 = 0,\ \lambda = \lambda_0,\ \ell = \ell_0$, we have

$$\mathrm{Ker}(L) = \mathrm{span}[\phi_1,\ \phi_2], \qquad \mathrm{Ker}(L^*) = \mathrm{span}[\psi_1^*,\ \psi_2^*] \tag{9.31a}$$

with $\langle \phi_i, \psi_j^* \rangle = \langle \psi_i^*, \psi_j^* \rangle = \delta_{i,j},\ i,j = 1,2$, and

$$\phi_1 = \mathbf{c}_1 \sin k\pi x, \qquad \phi_2 = \mathbf{c}_2 \sin m\pi x, \tag{9.31b}$$

and

$$\psi_1^* = \tilde{\mathbf{c}}_1 \sin k\pi x, \qquad \psi_2^* = \tilde{\mathbf{c}}_2 \sin m\pi x \tag{9.31c}$$

for some $k, m \in \mathbf{N}$. Here, $\mathbf{c}_1, \mathbf{c}_2 \in \mathbf{R}^2$ are null vectors of $M_k,\ M_m$ and $\tilde{\mathbf{c}}_1, \tilde{\mathbf{c}}_2 \in \mathbf{R}^2$ are null vectors of M_k^T, M_m^T, respectively, see also (9.18). A solution curve $(\mathbf{u}(t), \lambda(t))$ passing through the bifurcation point $(\mathbf{u}(0), \lambda(0)) = (0, \lambda_0)$ is of the form

$$\begin{aligned} \mathbf{u}(t) &= t\alpha_1(t)\phi_1 + t\alpha_2(t)\phi_2 + tw(t),\ w(t) \in M, \\ \lambda(t) &= \lambda_0 + t\beta(t). \end{aligned}$$

Differentiating $G(\mathbf{u}(t), \lambda(t)) = 0$ with respect to t at $t = 0$ yields $w(0) = 0$, see e.g. Section 4.1 and Mei [223]. Therefore, the solution curve $(\mathbf{u}(t),\ \lambda(t))$ can be rewritten as

$$\begin{aligned} \mathbf{u}(t) &= t\alpha_1(t)\phi_1 + t\alpha_2(t)\phi_2 + t^2 w(t), \quad w(t) \in M, \\ \lambda(t) &= \lambda_0 + t\beta(t). \end{aligned} \tag{9.32}$$

By Puiseux's theorem (cf. Fischer [106, pp.104]) we can choose the parameter t such that

$$|\alpha_1(0)| + |\alpha_2(0)| + |\beta(0)| > 0. \tag{9.33}$$

We will determine $\alpha_1(t),\ \alpha_2(t),\ w(t)$ and $\beta(t)$ via a nonsingular problem.

9.3.1 The reflection symmetry and its induced action

The interval $\Omega = [0, 1]$ is invariant under the reflection

$$S : x \in \Omega \to Sx := 1 - x \in \Omega.$$

Define

$$S : \ \mathbf{u} \in X \to S\mathbf{u} := \begin{pmatrix} u(Sx) \\ v(Sx) \end{pmatrix} \in X.$$

It is easy to verify

$$G(S\mathbf{u}, \lambda) = SG(\mathbf{u}, \lambda) \quad \text{for all } \ \mathbf{u} \in X, \ \lambda \in \mathbf{R} \tag{9.34}$$

and

$$\begin{aligned} S\phi_1 &= (-1)^{k-1}\phi_1, & S\phi_2 &= (-1)^{m-1}\phi_2, \\ S\psi_1^* &= (-1)^{k-1}\psi_1^*, & S\psi_2^* &= (-1)^{m-1}\psi_2^*. \end{aligned}$$

If we identify the null space $\mathrm{Ker}(L)$ with $\mathbf{R}^2$ via

$$w = \alpha_1\phi_1 + \alpha_2\phi_2 \in \mathrm{Ker}(L) \longleftrightarrow \begin{pmatrix} \alpha_1 \\ \alpha_2 \end{pmatrix} \in \mathbf{R}^2,$$

the equality $Sw = (-1)^{k-1}\alpha_1\phi_1 + (-1)^{m-1}\alpha_2\phi_2$ induces the following action of S in $\mathbf{R}^2$

$$S\begin{pmatrix} \alpha_1 \\ \alpha_2 \end{pmatrix} = \begin{pmatrix} (-1)^{k-1}\alpha_1 \\ (-1)^{m-1}\alpha_2 \end{pmatrix} \Longleftrightarrow S = \begin{pmatrix} (-1)^{k-1} & 0 \\ 0 & (-1)^{m-1} \end{pmatrix}. \tag{9.35}$$

We see that action of the reflection S in $\mathbf{R}^2$ depends on the parity of the wave numbers k and m. Moreover, S has the same action in $\mathrm{Ker}(L^*)$ as in $\mathrm{Ker}(L)$. In the following subsections we study different combinations of the wave numbers.

9.3.2 (k, m) = (odd, odd) or (odd, even)

To determine $\alpha_1(t)$, $\alpha_2(t)$, $w(t)$ and $\beta(t)$ in (9.32), we consider the following slightly enlarged system

$$H(w, \alpha_1, \alpha_2, \beta, t) := \begin{pmatrix} \dfrac{1}{t^2}G(t\alpha_1\phi_1 + t\alpha_2\phi_2 + t^2 w, \ \lambda_0 + t\beta) \\ \langle \phi_1, \ w \rangle \\ \langle \phi_2, \ w \rangle \\ \alpha_1^2 + \alpha_2^2 + \beta^2 - 1 \end{pmatrix} = 0, \tag{9.36a}$$

where $H : \ X \times \mathbf{R}^4 \to Y \times \mathbf{R}^3$ and

$$H(w,\alpha_1,\alpha_2,\beta,0) \tag{9.36b}$$
$$:= \begin{pmatrix} D_{\mathbf{u}}G_0 w + \beta D_{\mathbf{u}\lambda}G_0(\alpha_1\phi_1+\alpha_2\phi_2)+\frac{1}{2}D_{\mathbf{uu}}G_0(\alpha_1\phi_1+\alpha_2\phi_2)^2 \\ \langle\phi_1,\ w\rangle \\ \langle\phi_2,\ w\rangle \\ \alpha_1^2+\alpha_2^2+\beta^2-1 \end{pmatrix}.$$

The system (9.36) is equivalent to (9.29), see Section 6.7 and Mei [216]. Consider the system (9.36b). It is solvable for w if and only if there exist $\alpha_1,\alpha_2,\beta\in\mathbf{R}$ such that

$$B(\alpha_1,\alpha_2,\beta) \tag{9.37}$$
$$:= \begin{pmatrix} \langle\psi_1^*,\ \beta D_{\mathbf{u}\lambda}G_0(\alpha_1\phi_1+\alpha_2\phi_2)+\frac{1}{2}D_{\mathbf{uu}}G_0(\alpha_1\phi_1+\alpha_2\phi_2)^2\rangle \\ \langle\psi_2^*,\ \beta D_{\mathbf{u}\lambda}G_0(\alpha_1\phi_1+\alpha_2\phi_2)+\frac{1}{2}D_{\mathbf{uu}}G_0(\alpha_1\phi_1+\alpha_2\phi_2)^2\rangle \\ \alpha_1^2+\alpha_2^2+\beta^2-1 \end{pmatrix} = 0.$$

On the other hand, for each isolated solution $(\alpha_1^0,\alpha_2^0,\beta^0)$ of (9.37) there is a unique $w^0\in M$ such that $(w^0,\alpha_1^0,\alpha_2^0,\beta^0,0)$ is a nonsingular solution of (9.36). Thereafter, the implicit function theorem implies that (9.36) has a unique solution curve $(w(t),\alpha_1(t),\alpha_2(t),\beta(t),t)$ in the neighborhood of $(w^0,\alpha_1^0,\alpha_2^0,\beta^0,0)$ and

$$w(0)=w^0,\quad \alpha_1(0)=\alpha_1^0,\quad \alpha_2(0)=\alpha_2^0,\quad \beta(0)=\beta^0.$$

Thus we obtain via (9.32) a solution branch of (9.29) which bifurcates from the trivial solution curve at $(0,\lambda_0)$. We conclude that the isolated solutions of the system (9.37) are important for identifying solution branches at the bifurcation point. The system (9.37) consists the algebraic bifurcation equations in Decker/Keller [81] and Jepson/Decker [170].

Based on the definitions of ϕ_i, ψ_i^*, $i=1,2$ in (9.31b) and (9.31c), respectively, and the orthogonal properties of the trigonometric functions, we rewrite (9.37) as

$$B(\alpha_1,\alpha_2,\beta) = \begin{pmatrix} a_1\alpha_1^2+a_2\alpha_1\alpha_2+a_3\alpha_2^2+a_4\beta\alpha_1 \\ b_1\alpha_1^2+b_2\alpha_1\alpha_2+b_3\alpha_2^2+b_4\beta\alpha_2 \\ \alpha_1^2+\alpha_2^2+\beta^2-1 \end{pmatrix} = 0, \tag{9.38}$$

where

$$\begin{aligned} a_1 &= \frac{1}{2}\langle\psi_1^*,\ D_{\mathbf{uu}}G_0\phi_1^2\rangle, \\ a_2 &= \langle\psi_1^*,\ D_{\mathbf{uu}}G_0\phi_1\phi_2\rangle, \\ a_3 &= \frac{1}{2}\langle\psi_1^*,\ D_{\mathbf{uu}}G_0\phi_2^2\rangle, \\ a_4 &= \langle\psi_1^*,\ D_{\mathbf{u}\lambda}G_0\phi_1\rangle. \end{aligned}$$

Replacing ψ_1^* by ψ_2^* in these statements yields the coefficients b_i, $i=1,2,3$, respectively. Similarly we have $b_4=\langle\psi_2^*,\ D_{\mathbf{u}\lambda}G_0\phi_2\rangle$. If $(\alpha_1,\alpha_2,\beta)$ satisfies (9.38), so does $-(\alpha_1,\alpha_2,\beta)$.

The case $(k, m) =$*(odd, odd).* In this situation generically the coefficients $a_i,\ b_i,\ i = 1, \ldots, 4$ do not vanish and the system (9.38) has 8 isolated solutions. Since the solutions $\pm(\alpha_1, \alpha_2, \beta)$ lead to the same bifurcating solution branch of (9.29), and $\alpha_1 = \alpha_2 = 0,\ \beta = \pm 1$ correspond to the trivial solution curve, there are generically three nontrivial solution branches of (9.29) across $(0, \lambda_0)$.

The case (k, m)=*(odd, even).* Under this assumption we can simplify the system (9.38) further. From the properties $\psi_2^*(1-x) = -\psi_2^*(x)$ and $\phi_1(1-x) = \phi_1(x)$ we derive the equality

$$\begin{aligned} b_1 &= \frac{1}{2}\langle \psi_2^*,\ D_{\mathbf{uu}} G_0 \phi_1^2 \rangle \\ &= \frac{1}{2}\int_0^1 \psi_2^*(x)^T D_{\mathbf{uu}} G_0 \phi_1^2(x)\ dx \\ (x \longrightarrow 1-y) &= \frac{1}{2}\int_0^1 \psi_2^{*T}(1-y) D_{\mathbf{uu}} G_0 \phi_1^2(1-y)\ dy \\ &= -b_1. \end{aligned}$$

Thus we have $b_1 = 0$, and similarly, $a_2 = b_3 = 0$. Therefore,

$$B(\alpha_1, \alpha_2, \beta) = \begin{pmatrix} a_1\alpha_1^2 + a_3\alpha_2^2 + a_4\beta\alpha_1 \\ b_2\alpha_1\alpha_2 + b_4\beta\alpha_2 \\ \alpha_1^2 + \alpha_2^2 + \beta^2 - 1 \end{pmatrix} = 0. \tag{9.39}$$

Its eight solutions are

$$\begin{aligned} &(0,\ 0,\ \pm 1), \quad \Big(\frac{a_4}{\sqrt{a_1^2 + a_4^2}},\ 0,\ -\frac{a_1}{\sqrt{a_1^2 + a_4^2}}\Big), \\ &\Big(-\frac{a_4}{\sqrt{a_1^2 + a_4^2}},\ 0,\ \frac{a_1}{\sqrt{a_1^2 + a_4^2}}\Big), \\ &\Big(b_4\sqrt{\frac{a_3}{c}},\ \pm\sqrt{\frac{b_4(a_4 b_2 - a_1 b_4)}{c}},\ -b_2\sqrt{\frac{a_3}{c}}\Big), \\ &\Big(-b_4\sqrt{\frac{a_3}{c}},\ \mp\sqrt{\frac{b_4(a_4 b_2 - a_1 b_4)}{c}},\ b_2\sqrt{\frac{a_3}{c}}\Big), \\ &c := a_3 b_2^2 + (a_3 - a_1) b_4^2 + a_4 b_2 b_4. \end{aligned}$$

Except $(0, 0, \pm 1)$, the other six solutions, if they are real, lead to three nontrivial solution branches of (9.29) at $(0, \lambda_0)$.

9.3.3 $(k, m) =$ (even, even)

If both k, m are even, the induced action of S in $\mathrm{Ker}(L)$ and $\mathrm{Ker}(L^*)$ is $S = -I$. This makes $a_i = b_i = 0,\ i = 1, 2, 3$ in (9.38), in particular, its solutions are no more isolated and we cannot distinguish different bifurcating solution branches. Therefore, the enlarged system (9.36) has to be modified. Recalling the statements in (9.32) and (9.33), and differentiating $G(\mathbf{u}(t), \lambda(t)) = 0$ twice with respect to t at $t = 0$, we derive that $(\mathbf{u}(t), \lambda(t))$ is of the form

$$\mathbf{u}(t) = t\alpha_1(t)\phi_1 + t\alpha_2(t)\phi_2 + t^2\Big(\alpha_1^2(t)w_1 + 2\alpha_1(t)\alpha_2(t)w_2 + \alpha_2^2(t)w_3\Big) + t^3 w(t),$$
$$\lambda(t) = \lambda_0 + t^2\beta(t).$$

Here $w(t),\ w_i \in M,\ i = 1,2,3$, and $w_1,\ w_2,\ w_3$ are defined implicitly by

$$Lw_i = -\frac{1}{2}D_{\mathbf{uu}}G_0\phi_i^2,\ i = 1,3 \quad \text{and} \quad Lw_2 = -\frac{1}{2}D_{\mathbf{uu}}G_0\phi_1\phi_2,$$

respectively. Evidently, w_i has the S-symmetry, i.e., $Sw_i = w_i,\ i = 1,2,3$.

To determine $w(t),\ \alpha_1(t),\ \alpha_2(t)$ and $\beta(t)$, we replace the system (9.36) by

$$H(w, \alpha_1, \alpha_2, \beta, t) = 0 \tag{9.40a}$$

with

$$H(w, \alpha_1, \alpha_2, \beta, t) \tag{9.40b}$$
$$:= \begin{pmatrix} \frac{1}{t^3}G(t\alpha_1\phi_1 + t\alpha_2\phi_2 + t^2(\alpha_1^2 w_1 + 2\alpha_1\alpha_2 w_2 + \alpha_2^2 w_3) + t^3 w,\ \lambda_0 + t^2\beta) \\ \langle\phi_1,\ w\rangle \\ \langle\phi_2,\ w\rangle \\ \alpha_1^2 + \alpha_2^2 + \beta^2 - 1 \end{pmatrix}$$

for $t \neq 0$ and

$$H(w, \alpha_1, \alpha_2, \beta, 0) := \begin{pmatrix} D_{\mathbf{u}}G_0 w + r(\alpha_1, \alpha_2, \beta) \\ \langle\phi_1,\ w\rangle \\ \langle\phi_2,\ w\rangle \\ \alpha_1^2 + \alpha_2^2 + \beta^2 - 1 \end{pmatrix}, \tag{9.40c}$$

for $t = 0$, where

$$\begin{aligned} r(\alpha_1, \alpha_2, \beta) := {} & \beta D_{\mathbf{u}\lambda}G_0(\alpha_1\phi_1 + \alpha_2\phi_2) \\ & + D_{\mathbf{uu}}G_0(\alpha_1\phi_1 + \alpha_2\phi_2)(\alpha_1^2 w_1 + 2\alpha_1\alpha_2 w_2 + \alpha_2^2 w_3) \\ & + \frac{1}{6}D_{\mathbf{uuu}}G_0(\alpha_1\phi_1 + \alpha_2\phi_2)^3. \end{aligned}$$

Similarly to the cases in Section 9.3.2, we obtain solution branches of (9.29) at $(0, \lambda_0)$ via the system (9.40) by examining roots of the algebraic system

$$\begin{aligned} B(\alpha_1, \alpha_2, \beta) &:= \begin{pmatrix} \langle\psi_1^*,\ r(\alpha_1, \alpha_2, \beta)\rangle \\ \langle\psi_2^*,\ r(\alpha_1, \alpha_2, \beta)\rangle \\ \alpha_1^2 + \alpha_2^2 + \beta^2 - 1 \end{pmatrix} \\ &= \begin{pmatrix} a_1\alpha_1^3 + a_2\alpha_1^2\alpha_2 + a_3\alpha_1\alpha_2^2 + a_4\alpha_2^3 + a_5\beta\alpha_1 \\ b_1\alpha_1^3 + b_2\alpha_1^2\alpha_2 + b_3\alpha_1\alpha_2^2 + b_4\alpha_2^3 + b_5\beta\alpha_2 \\ \alpha_1^2 + \alpha_2^2 + \beta^2 - 1 \end{pmatrix} \\ &= 0. \end{aligned} \tag{9.41}$$

Here we have used the orthogonality of $\sin k\pi x$, $\sin m\pi x$, $k \neq m$, the S-symmetry of ϕ_i, $\psi_i^*, i = 1, 2$, and the notations

$$\begin{aligned}
a_1 &:= \langle \psi_1^*, \ \frac{1}{6} D_{\mathbf{uuu}} G_0 \phi_1^3 + D_{\mathbf{uu}} G_0 \phi_1 w_1 \rangle, \\
a_2 &:= \langle \psi_1^*, \ \frac{1}{2} D_{\mathbf{uuu}} G_0 \phi_1^2 \phi_2 + 2 D_{\mathbf{uu}} G_0 \phi_1 w_2 + D_{\mathbf{uu}} G_0 \phi_2 w_1 \rangle, \\
a_3 &:= \langle \psi_1^*, \ \frac{1}{2} D_{\mathbf{uuu}} G_0 \phi_1 \phi_2^2 + 2 D_{\mathbf{uu}} G_0 \phi_2 w_2 + D_{\mathbf{uu}} G_0 \phi_1 w_3 \rangle, \qquad (9.42) \\
a_4 &:= \langle \psi_1^*, \ \frac{1}{6} D_{\mathbf{uuu}} G_0 \phi_2^3 + D_{\mathbf{uu}} G_0 \phi_2 w_3 \rangle, \\
a_5 &:= \langle \psi_1^*, \ D_{\mathbf{u}\lambda} G_0 \phi_1 \rangle.
\end{aligned}$$

The coefficients b_i, $i = 1, \ldots, 4$ are defined in a similar way by replacing ψ_1^* with ψ_2^* in (9.42), while $b_5 = \langle \psi_2^*, \ D_{\mathbf{u}\lambda} G_0 \phi_2 \rangle$.

Note that if $(\alpha_1, \alpha_2, \beta)$ is a solution of (9.41), so is $(-\alpha_1, -\alpha_2, \beta)$, while $(-\alpha_1, -\alpha_2, -\beta)$ is not, except that the statement $\beta = 0$ or $\alpha_1 = \alpha_2 = 0$ holds. Thus the solutions $(\alpha_1, \alpha_2, \beta)$ and $(-\alpha_1, -\alpha_2, \beta)$ lead to the same bifurcating solution branch of (9.29), which exists only for $\lambda \leq \lambda_0$ if $\beta < 0$, or only for $\lambda \geq \lambda_0$ if $\beta > 0$. In other words, all nontrivial solution branches of (9.29) at $(0, \lambda_0)$ are pitchfork. Except $(0, 0, \pm 1)$, the other ten roots of (9.41), if they are real, lead to five bifurcating solution branches.

Concluding the discussions above, we have

Theorem 9.3.1. *Assume $(0, \lambda_0)$ is a double bifurcation point with the properties in (9.31). If (k, m) =(odd, odd) or (even, odd), the equation (9.29) has up to three nontrivial and transcritical solution branches of the form (9.32) across $(0, \lambda_0)$. Furthermore, $(\alpha_1(0), \alpha_2(0), \beta(0))$ satisfies (9.38). If k, m are both even, then (9.29) has up to five pitchfork solution branches of the form (9.32), where $(\alpha_1(0), \alpha_2(0), \beta(0))$ satisfies (9.41).*

9.3.4 The Brusselator equations

If $\alpha = \sqrt{2}\pi^2/50$, $d_1 = 1$, $d_2 = 2$ and $(k, m) = (1, 2)$, then for $\ell = 10$ the first bifurcation point $(0, \lambda_1) = (0, \lambda_2)$ is double with

$$\lambda_1 = \lambda_2 = 1 + \frac{\pi^2}{20} + \frac{\pi^4}{2500}.$$

The system (9.39) for $(\alpha_1, \alpha_2, \beta)$ has eight solutions

$$\begin{aligned}
&(0, \ 0, \ \pm 1), \qquad (0.2722, \ 0, \ -0.9622), \qquad (-0.2722, \ 0, \ 0.9622), \\
&(0.1821, \ \pm 0.03648, \ 0.9645), \qquad (-0.1821, \ \pm 0.03648, \ -0.9645).
\end{aligned}$$

There are three solution branches bifurcating from the trivial solution curve, see Section 9.5 for the path-following of these branches.

If $\alpha = 3\sqrt{2}\pi^2/100$, $d_1 = 1$, $d_2 = 2$ and $(k,m) = (1,3)$, then for $\ell = 10$ the bifurcation point $(0, \lambda_1) = (0, \lambda_3)$ is double and

$$\lambda_1 = \lambda_3 = 1 + \frac{\pi^2}{10} + \frac{9\pi^4}{10000}.$$

The coefficients in the system (9.38) are

$$\begin{array}{llll} a_1 = -0.9432, & a_2 = 0.01338, & a_3 = -0.03578, & a_4 = 0.1397, \\ b_1 = 0.01957, & b_2 = -0.05355, & b_3 = -0.01604, & b_4 = 0.00087. \end{array}$$

The system (9.38) has 6 nontrivial solutions,

$$\begin{array}{l} \pm(-0.004202, \ -0.0407, \ -0.9992), \\ \pm(-0.003996, \ 0.09911, \ -0.9943), \\ \pm(-0.1465, \ -0.053536, \ -0.9878). \end{array}$$

These lead to 3 solution branches bifurcating from the trivial solution curve, see Section 9.5 for the path-following of these branches.

9.4 Central Difference Approximations

In this section we discuss discretizations of the equation (9.1) with the central difference method and bifurcations of the discrete problem.

9.4.1 General systems

Let

$$\Omega_h = \{0 < x_1 < x_2 < \cdots < x_N < 1 \ \ | \ \ x_i = i \cdot h\}$$

be a uniform grid on the unit interval with meshsize $h = 1/(N+1)$. At the discrete points the functions $u(x)$ and $v(x)$ are approximated by the net functions $U(x_i) := U_i$ and $V(x_i) := V_i$, $i = 1, \ldots, N$, respectively. Let

$$U := (U_1, \cdots, U_N)^T, \qquad V := (V_1, \cdots, V_N)^T \in \mathbf{R}^N.$$

Discretizing $\frac{\partial^2}{\partial x^2}$ with the central difference scheme, and denoting F, G the net functions associated to the functions f and g, respectively, we obtain the discretization of $G(\mathbf{u}, \lambda) = 0$ with G in (9.8b)

$$H(Z, \lambda) = \begin{pmatrix} d_1 AU \\ d_2 AV \end{pmatrix} - h^2\ell^2 \begin{pmatrix} F(Z,\lambda) \\ G(Z,\lambda) \end{pmatrix} = 0, \tag{9.43}$$

where $Z = (U, \ V)$, and

$$A = \begin{pmatrix} 2 & -1 & & & \\ -1 & 2 & -1 & & \\ & \ddots & \ddots & \ddots & \\ & & \ddots & \ddots & -1 \\ & & & -1 & 2 \end{pmatrix}$$

and

$$F(Z,\lambda) = \begin{pmatrix} f(U_1,V_1,\lambda) \\ \vdots \\ f(U_N,V_N,\lambda) \end{pmatrix}, \qquad G(Z,\lambda) = \begin{pmatrix} g(U_1,V_1,\lambda) \\ \vdots \\ g(U_N,V_N,\lambda) \end{pmatrix}.$$

Evidently,

$$H(0,\lambda) \equiv 0 \qquad \text{for all} \quad \lambda \in \mathbf{R}.$$

The Jacobian matrix corresponding to (9.43) is

$$DH(Z,\lambda) = \begin{pmatrix} d_1 A - h^2\ell^2 D_U F & -h^2\ell^2 D_V F & -h^2\ell^2 D_\lambda F \\ -h^2\ell^2 D_U G & d_2 A - h^2\ell^2 D_V G & -h^2\ell^2 D_\lambda G \end{pmatrix},$$

where $D_U F$, $D_V F, \ldots$ are diagonal matrices

$$D_U F(Z,\lambda) = \operatorname{diag}(f_u(U_1,V_1,\lambda),\cdots,f_u(U_N,V_N,\lambda)), \quad D_U F(0,\lambda) = f_u^0 I_{N\times N},$$
$$D_V F(Z,\lambda) = \operatorname{diag}(f_v(U_1,V_1,\lambda),\cdots,f_v(U_N,V_N,\lambda)), \quad D_V F(0,\lambda) = f_v^0 I_{N\times N},$$

and $D_U G$, $D_V G$ are defined similarly. Moreover, we have

$$D_\lambda F(Z,\lambda) = (f_\lambda(U_1,V_1,\lambda),\cdots,f_\lambda(U_N,V_N,\lambda))^T, \qquad D_\lambda F(0,\lambda) \equiv 0.$$

Along the trivial solution curve $Z \equiv 0$, $\lambda \in \mathbf{R}$, we have

$$DH(0,\lambda) = \begin{pmatrix} d_1 A - h^2\ell^2 f_u^0 I & -h^2\ell^2 f_v^0 I & 0 \\ -h^2\ell^2 g_u^0 I & d_2 A - h^2\ell^2 g_v^0 I & 0 \end{pmatrix}.$$

A stationary bifurcation occurs whenever the matrix

$$L_h := \begin{pmatrix} d_1 A - h^2\ell^2 f_u^0 I & -h^2\ell^2 f_v^0 I \\ -h^2\ell^2 g_u^0 I & d_2 A - h^2\ell^2 g_v^0 I \end{pmatrix} \tag{9.44}$$

becomes singular. To determine the eigenvalues of L_h, we note that eigenpairs of A are

$$\mu_k = 2(1 - \cos\frac{k\pi}{N+1}),$$
$$U_k = \left(\sin\frac{k\pi}{N+1},\ \sin\frac{2k\pi}{N+1},\cdots,\ \sin\frac{Nk\pi}{N+1}\right), \qquad k = 1,\ldots,N,$$

see e.g. Isaacson/Keller [168, pp.456]. Define

$$X_k^h := \mathbf{R}^2 U_k = \left\{ \begin{pmatrix} c_1 U_k \\ c_2 U_k \end{pmatrix}; \quad c_1, c_2 \in \mathbf{R} \right\}.$$

We have

$$\mathbf{R}^{2N} = \sum_{k=1}^{N} \oplus X_k^h. \tag{9.45}$$

Moreover, the operator L_h maps X_k^h into itself and for all $c_1,\ c_2 \in \mathbf{R},\ k = 1, \ldots, N$,

$$L_h \begin{pmatrix} c_1 \\ c_2 \end{pmatrix} U_k = \begin{pmatrix} d_1\mu_k - h^2\ell^2 f_u^0 & -h^2\ell^2 f_v^0 \\ -h^2\ell^2 g_u^0 & d_2\mu_k - h^2\ell^2 g_v^0 \end{pmatrix} \begin{pmatrix} c_1 \\ c_2 \end{pmatrix} U_k. \tag{9.46}$$

Consequently, under the decomposition (9.45) L_h is a block diagonal matrix

$$L_h = -h^2\ell^2 \mathrm{diag}(M_1^h, \cdots, M_N^h) \tag{9.47}$$

with

$$M_k^h(\lambda, \ell) := \begin{pmatrix} -\frac{d_1\mu_k}{h^2\ell^2} + f_u^0 & f_v^0 \\ g_u^0 & -\frac{d_2\mu_k}{h^2\ell^2} + g_v^0 \end{pmatrix}, \quad k = 1, \ldots, N. \tag{9.48}$$

Thus spectrum of L_h is the union of eigenvalues of $M_k^h,\ k = 1, \ldots, N$. These properties are analogue of the continuous problem and valid for the special discretization. Suppose that $\lambda = \lambda_0$ is a simple bifurcation point of the continuous problem (9.6). From Section 9.2 we know that there is a $k \in \mathbf{N}$ such that $\det(M_k(\lambda_0, \ell)) = 0$. For $N > k$ we derive from (9.48) and (9.13) that

$$\begin{aligned} \det(M_k^h(\lambda_0, \ell)) &= \det(M_k^h(\lambda_0, \ell)) - \det(M_k(\lambda_0, \ell)) \\ &= O(\left| \frac{\mu_k}{h^2} - k^2\pi^2 \right|) \\ &= O(h^2). \end{aligned} \tag{9.49}$$

As expected, λ_0 is no more a bifurcation point of the discrete problem. Nevertheless, the generalized implicit function theorem in Brezzi *et al.* [48] implies that for $h > 0$ sufficiently small, there is a $\lambda_0^h \in \mathbf{R}$ such that

$$\det(M_k^h(\lambda_0^h, \ell)) = 0 \qquad \text{and} \quad |\lambda_0^h - \lambda_0| = O(h^2).$$

In other words, there is a simple bifurcation point $(0, \lambda_h)$ of the discrete problem in the $O(h^2)$-neighborhood of the bifurcation point $(0, \lambda_0)$ of the continuous problem.

It is well-known that a double bifurcation point of a one-parameter problem may be split into two simple ones after discretizations (cf. [216]). With the length ℓ as the second bifurcation parameter, we want to preserve the double bifurcation in the discrete problem (9.43). This will allow the discrete problem to have qualitatively the same bifurcation scenario as the continuous problem. It is also essential for the accurate numerical approximations of solution branches.

For $\ell = \ell_0$ let $\mathbf{u}_0 = 0$, $\lambda = \lambda_0$ be a double bifurcation point such that zero is an eigenvalue of the matrices M_k and M_m. It is easy to verify that

$$\det(M_k^h(\lambda_0, \ell_0)) = O(h^2), \qquad \det(M_m^h(\lambda_0, \ell_0)) = O(h^2).$$

We show that there exists $(\lambda_0^h,\ \ell_0^h)$ in the neighborhood of $(\lambda_0,\ \ell_0)$ such that

$$\det(M_k^h(\lambda_0^h, \ell_0^h)) = \det(M_m^h(\lambda_0^h, \ell_0^h)) = 0.$$

This corresponds to a double bifurcation of the discrete problem.

Define a mapping $d:\ \mathbf{R}^3 \to \mathbf{R}^2$ by

$$\begin{aligned} d(\lambda, \ell, h) &:= \begin{pmatrix} \det(M_k^h(\lambda, \ell) \\ \det(M_m^h(\lambda, \ell) \end{pmatrix} \\ &= \begin{pmatrix} \left(-\frac{d_1\mu_k}{h^2\ell^2} + f_u^0\right)\left(-\frac{d_2\mu_k}{h^2\ell^2} + g_v^0\right) - f_v^0 g_u^0 \\ \left(-\frac{d_1\mu_m}{h^2\ell^2} + f_u^0\right)\left(-\frac{d_2\mu_m}{h^2\ell^2} + g_v^0\right) - f_v^0 g_u^0 \end{pmatrix}, \end{aligned}$$

and for $h = 0$,

$$d(\lambda, \ell, 0) := \begin{pmatrix} \det(M_k(\lambda, \ell)) \\ \det(M_m(\lambda, \ell)) \end{pmatrix}.$$

If $f,\ g$ are smooth functions, so is $d(\lambda, \ell, h)$ for $\ell \neq 0$. Note that

$$d(\lambda_0, \ell_0, 0) = 0.$$

If the Jacobian

$$\frac{\partial d(\lambda, \ell, 0)}{\partial(\lambda, \ell)} =: \begin{pmatrix} \tilde{d}_{11} & \tilde{d}_{12} \\ \tilde{d}21 & \tilde{d}_{22} \end{pmatrix}$$

with the elements

$$\begin{aligned} \tilde{d}_{11} &= -\frac{k^2\pi^2}{\ell^2}(f_{u\lambda}d_2 + g_{v\lambda}d_1) + f_{u\lambda}g_v + f_u g_{v\lambda} - f_{v\lambda}g_u - f_v g_{u\lambda} \\ \tilde{d}_{12} &= -\frac{4d_1d_2(k\pi)^4}{\ell^5} + \frac{2k^2\pi^2}{\ell^3}(d_1 g_v + d_2 f_u) \\ \tilde{d}_{21} &= -\frac{m^2\pi^2}{\ell^2}(f_{u\lambda}d_2 + g_{v\lambda}d_1) + f_{u\lambda}g_v + f_u g_{v\lambda} - f_{v\lambda}g_u - f_v g_{u\lambda} \\ \tilde{d}_{22} &= -\frac{4d_1d_2(m\pi)^4}{\ell^5} + \frac{2m^2\pi^2}{\ell^3}(d_1 g_v + d_2 f_u) \end{aligned}$$

is nonsingular at $(\lambda_0, \ell_0, 0)$, then by the implicit function theorem for $h > 0$ small enough there is a unique solution curve $(\lambda(h), \ell(h))$ such that

$$d(\lambda(h), \ell(h), h) = 0$$

and

$$|\lambda(h) - \lambda_0| + |\ell(h)) - \ell_0| \leq \text{const} \cdot \|d(\lambda_0, \ell_0, 0)\| = O(h^2).$$

Thus we have proven the following conclusion.

Theorem 9.4.1. *For $\ell = \ell_0$, let $(0, \lambda_0)$ be a double stationary bifurcation point of (9.8) and let $k,\ m \in \mathbf{N}$ with $k \neq m$ be the corresponding wave numbers such that* $\det(M_k(\lambda_0, \ell_0)) = \det(M_m(\lambda_0, \ell_0)) = 0$. *If the Jacobian*

$$\frac{\partial d(\lambda, \ell, 0)}{\partial(\lambda, \ell)}$$

is nonsingular at $(\lambda_0, \ell_0, 0)$, then for sufficiently small $h > 0$ there exists $\ell = \ell(h)$ such that discrete problem (9.43) has a double bifurcation point $(0, \lambda(h))$ in the $O(h^2)$-neighborhood of $(0, \lambda_0)$.

Applying the regularization method in Section 9.3, we obtain the bifurcating solution branches of the discrete problem at $(0, \lambda(h))$, and in turn the approximations of the bifurcation scenario of the original problem.

Remark 9.4.1. Note that it is the finite difference discretization with the uniform meshsize and the property (9.46) of L_h that makes the mapping $d(\lambda, \ell, h)$ well defined and related directly to the zero eigenvalue of the original problem at $h = 0$. If another method for discretization is used, a double bifurcation point may be split in the discrete problem.

9.4.2 The Brusselator equations

For the Brusselator equations the functions $f(u, v, \lambda)$, $g(u, v, \lambda)$ are given in (9.20) and

$$M_k^h := \begin{pmatrix} -\mu_k d_1/h^2\ell^2 + \lambda - 1 & \alpha^2 \\ -\lambda & -\mu_k d_2/h^2\ell^2 - \alpha^2 \end{pmatrix}. \tag{9.50}$$

Theorem 9.4.2. *The bifurcation points of the discrete Brusselator equations are*

$$(0,\ \lambda_p^h) = (0,\ 1 + \frac{\alpha^2 d_1}{d_2} + \frac{d_1 \mu_p}{h^2\ell^2} + \frac{\alpha^2 h^2 \ell^2}{d_2 \mu_p}), \quad p = 1, \ldots, N. \tag{9.51}$$

The null vector $(\tilde{U}_p, \tilde{V}_p)$ of L_h in (9.44) can be chosen as

$$\tilde{V}_p = -(\frac{d_1}{d_2} + \frac{h^2\ell^2}{d_2\mu_p})\tilde{U}_p, \qquad \tilde{U}_p = U_p, \quad p = 1, \ldots, N. \tag{9.52}$$

Furthermore, a double bifurcation, i.e., $\lambda_p^h = \lambda_q^h$ for $p \neq q$, occurs if and only if the equality

$$\mu_p \mu_q = \frac{\alpha^2 h^4 \ell^4}{d_1 d_2} \tag{9.53}$$

is satisfied.

Proof: As indicated above, the operator L_h is singular if and only if there is an integer $p \in \{1, \ldots, N\}$, such that

$$\det(M_p^h) = \frac{\mu_p^2}{h^4\ell^4} d_1 d_2 + \frac{\mu_p}{h^2\ell^2}[d_1\alpha^2 + d_2(1-\lambda)] + \alpha^2 = 0.$$

Solving it for λ, we obtain

$$\lambda_p^h := 1 + \frac{\alpha^2 d_1}{d_2} + \frac{d_1\mu_p}{h^2\ell^2} + \frac{\alpha^2 h^2 \ell^2}{d_2 \mu_p}.$$

Recall the statements (9.46) and (9.47). The null vectors of L_h are of the form $(c_1, c_2)^T U_p$ with $(c_1, c_2)^T$ as a null vector of $M_p^h(\lambda_p^h, \ell)$, which can be chosen as

$$c_1 = 1, \qquad c_2 = -\left(\frac{d_1}{d_2} + \frac{h^2\ell^2}{d_2\mu_p}\right).$$

The statement (9.52) follows immediately.

A double bifurcation occurs if and only if there are $p,\ q \in \{1, \ldots, N\}$, $p \neq q$, such that $\lambda_p^h = \lambda_q^h$, which yields the statement (9.53). ∎

A Hopf bifurcation of the discrete Brusselator equations occurs at $(0, \lambda_0^h)$ if and only if there is a $p \in \{1, \ldots, N\}$ such that both eigenvalues of M_p^h are pure imaginary or equivalently $\mathrm{trace}(M_p^h) = 0$ and $\det(M_p^h) > 0$. These lead directly to the conditions

$$\begin{aligned} \lambda_0^h &= 1 + \alpha^2 + \frac{d_1 + d_2}{h^2\ell^2}\mu_p, \\ \ell^2 &> \frac{2d_2^2\mu_p}{h^2[\alpha^2(d_1 - d_2) + \sqrt{\alpha^4(d_1-d_2)^2 + 4\alpha^2 d_2^2}]} > 0. \end{aligned} \tag{9.54}$$

For example, if $d_1 = 1$, $d_2 = 2$, $\alpha = 4$, the necessary condition (9.25) for a Hopf bifurcation of the continuous problem is

$$\ell^2 > \frac{\pi^2}{2(\sqrt{2}-1)} \approx 11.9137.$$

For the discrete problems, this condition becomes

$$\ell^2 > \begin{cases} 11.8160 & \text{for } h = 0.1, \\ 11.9075 & \text{for } h = 0.025. \end{cases}$$

If we choose $\ell = 11.91$, then there is a Hopf bifurcation point for the discrete problem, but none for the continuous one. Thus a coarse discretization may induce spurious bifurcations which will disappear with the refinement of discretizations.

9.5 Numerical Results for the Brusselator Equations

We discretize the Brusselator equations with the homogeneous boundary conditions

$$u(0) = u(1) = v(0) = v(1) = 0$$

by the central difference scheme with uniform meshsize and apply the continuation-Arnoldi Algorithm 2.3.10 in Chapter 2 to trace nontrivial solution curves. Two different lengths ℓ of the domain are considered.
The following notations are used in Table 9.1 and Figures 9.1-9.5.

NCS	:	number of the continuation steps.
MAXNORM	:	maximum norm of the approximate solution U_1.
ε	:	tolerance in Newton corrector.
θ_1	:	minimum eigenvalue of H_j.
κ_2	:	the two-norm condition number of H_j.
tol	:	stopping criterion for the Arnoldi method.
η	:	number of negative eigenvalues of H_j.
U_i^+, V_i^+	:	discrete solution curve of the Brusselator equations, corresponding to u_i and v_i with positive values.
U_i^-, V_i^-	:	discrete solution curve of the Brusselator equations corresponding to u_i and v_i with negative values.

NCS describes how far the solution is from the bifurcation point. MAXNORM shows how the solution differs from the trivial state. With ε we control how close the approximate solution should go toward the exact solution curve. We monitor bifurcations via θ_1.

9.5.1 The length $\ell = 1$, diffusion rates $d_1 = 1,\ d_2 = 2$

Example 9.5.1. Simple bifurcations: Choosing $\alpha = 4,\ k = 1$ and $h = 0.025$, we find that $(0, \lambda_1^h) = (0, 19.67547)$ is a simple bifurcation point of the discrete problem.

Table 9.1 shows how the continuation-Arnoldi Algorithm performs in tracing the solution curve of the Brusselator equations, where a bifurcation point is detected at $\lambda_1^h \approx 19.676547$.

Figure 9.1 shows that the first solution curve U_1 branching from $(0, \lambda_1^h)$ has a "barely transcritical" form (cf. [129, pp.233]). On the other hand, the second solution curve U_2 branching from $(0, \lambda_2^h) \approx (0, 48.6004)$ is a pitchfork bifurcation. The solution on U_1 and V_1 at $\lambda = 19.6755$, and on U_2 and V_2 at $\lambda = 48.5907$, are shown in Figure 9.1b.

λ	MAXNORM	NCS	η	dim(H_j)	θ_1	κ_2
18.89999832	$.6661E-03$	16	2	26	$.3617E-03$	$.3601E+05$
19.29997347	$.1410E-02$	36	2	45	$.1810E-03$	$.7266E+05$
19.59739626	$.7087E-02$	51	2	42	$.4664E-04$	$.2838E+06$
19.66883761	$.2943E-01$	58	2	45	$.1565E-04$	$.8756E+06$
19.67654448	$.3712E-01$	60	2	46	$.1486E-04$	$.8903E+06$
19.69130612	$.5649E-01$	65	2	45	$.1603E-04$	$.8228E+06$
19.73422780	$.1265E+00$	83	2	44	$.2904E-04$	$.4500E+06$
19.78089004	$.2003E+00$	102	2	45	$.4689E-04$	$.2759E+06$

Table 9.1. Sample result for Example 1, h=0.025, $\varepsilon = 5 \cdot 10^{-5}$, $tol = 5 \cdot 10^{-9}$, $\lambda_1^h = 19.67547$, using the Arnoldi method.

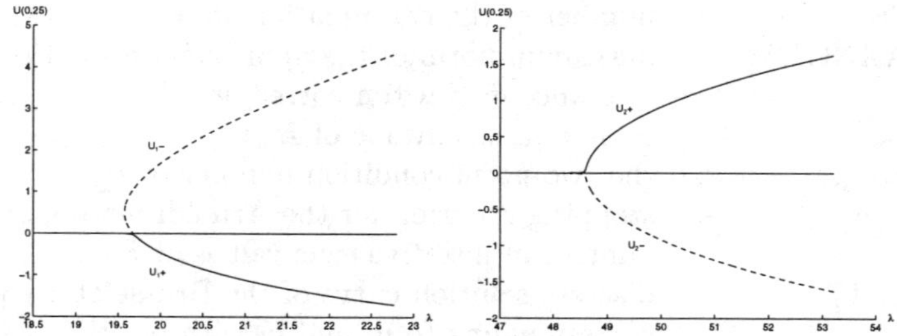

a) Bifurcation diagrams.

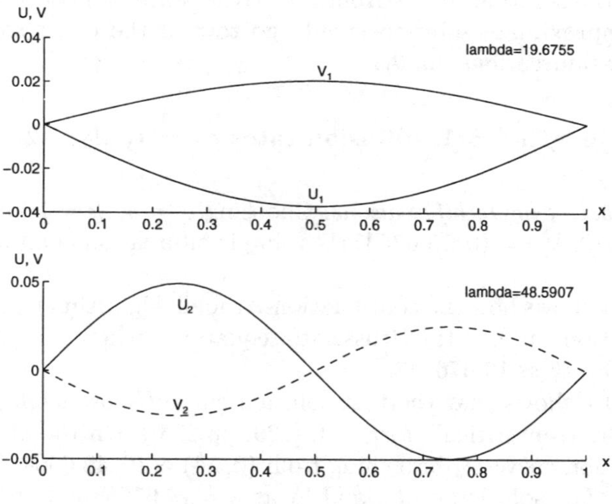

b) Solutions as functions of x.

Fig. 9.1. Transcritical and pitchfork bifurcations of the Brusselator equations at simple bifurcation points $(0, \lambda_1) \approx (0, 19.6755)$ and $(0, \lambda_2) \approx (0, 48.6004)$, respectively.

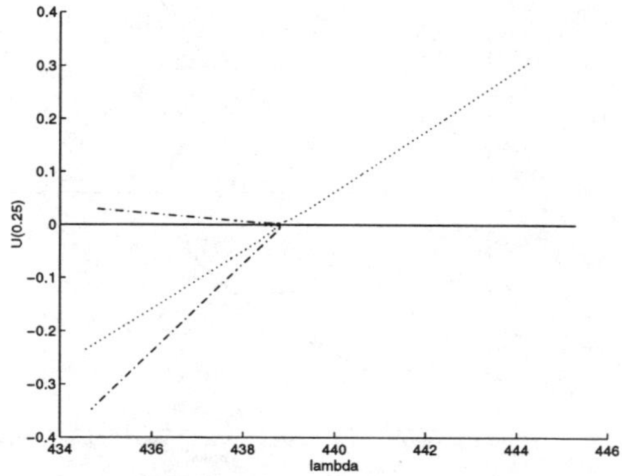

a) Bifurcation diagram at $\lambda_1 = \lambda_2 = 438.8979$.

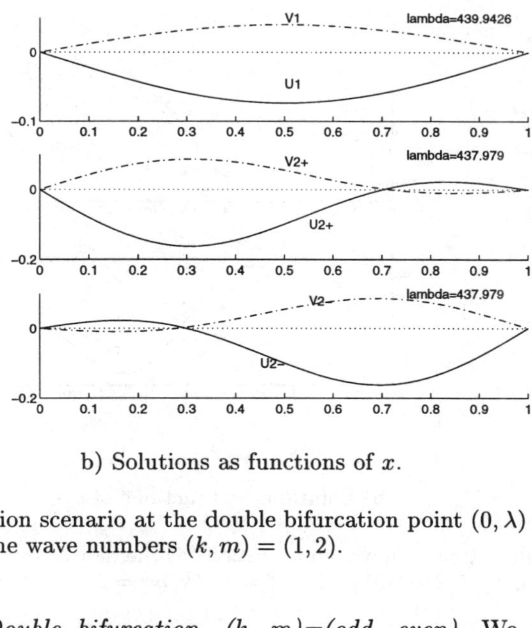

b) Solutions as functions of x.

Fig. 9.2. Bifurcation scenario at the double bifurcation point $(0, \lambda) = (438.8979)$, corresponding to the wave numbers $(k, m) = (1, 2)$.

Example 9.5.2. Double bifurcation, (k, m)=(odd, even). We choose $h = 0.025$, $k = 1$, $m = 2$ and derive from (9.53) that when $\alpha = 27.879599$, the first bifurcation point $(0, \lambda_1^h) = (0, \lambda_2^h) \approx (0, 438.897870)$ is double.

Figure 9.2 shows the bifurcation diagram and the solution curves of the Brusselator equations at $(0, \lambda_1^h) = (0, \lambda_2^h)$. Note that the solution curve of the U_1-mode is transcritical, while the solution curve of the U_2-mode is a nearly pitchfork bifurcation.

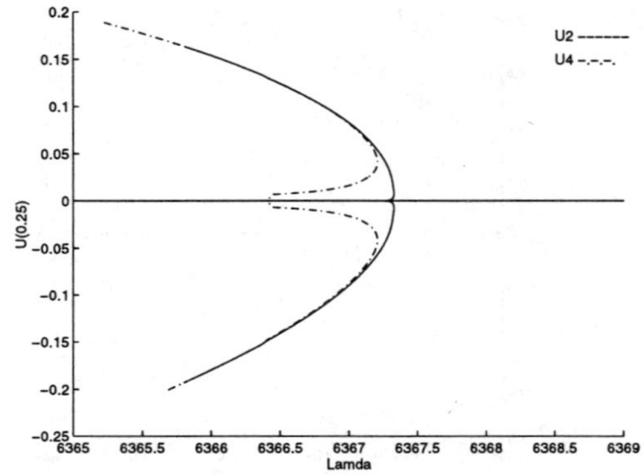

a) Bifurcation diagram at $\lambda_2 = \lambda_4 = 6367.2462431$.

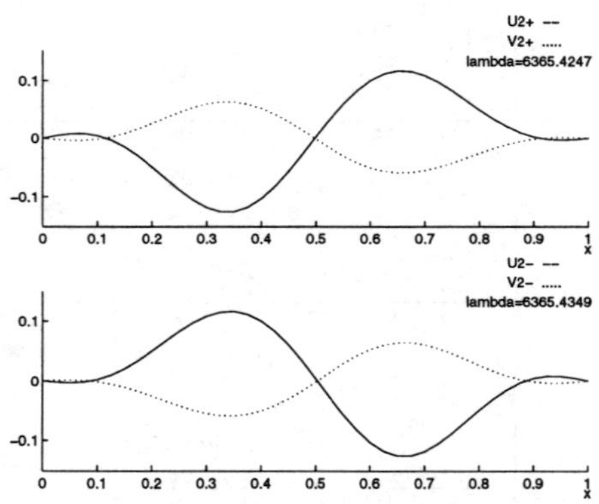

b) Solutions as functions of x.

Fig. 9.3. Bifurcation scenario and solutions at the double bifurcation point $(0, \lambda_2) = (0, \lambda_4) = (0,\ 6367.2462431)$. Here $\ell = 1,\ (k, m) = (2, 4),\ d_1 = 1,\ d_2 = 2$.

Example 9.5.3. Double bifurcation, (k, m)=(even, even). We choose $h = 0.025$, $k = 2$, $m = 4$. Then from (9.53) we have $\alpha = 111.088912$, and the first bifurcation point $(0, \lambda_2^h) = (0, \lambda_4^h) \approx (0, 6367.3896)$ is double.

Figure 9.3a shows the bifurcation diagram and the solution curves of the Brusselator equations at $(0, \lambda_2^h) = (0, \lambda_4^h)$ are pitchfork and turn to the left. Figure 9.3b shows these two solution curves.

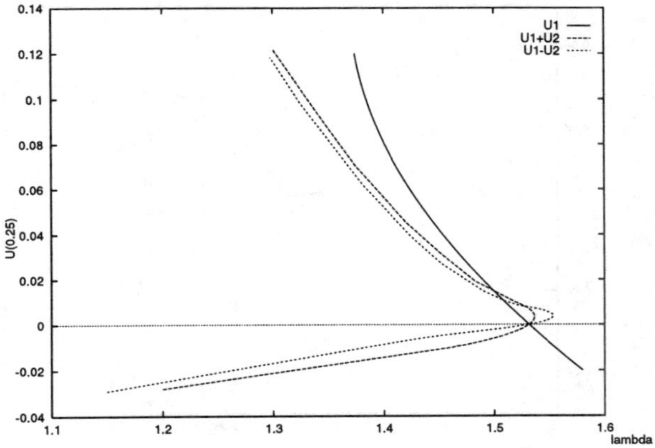

a) Bifurcation diagram at $\lambda_1 = \lambda_2 = 1.5315$.

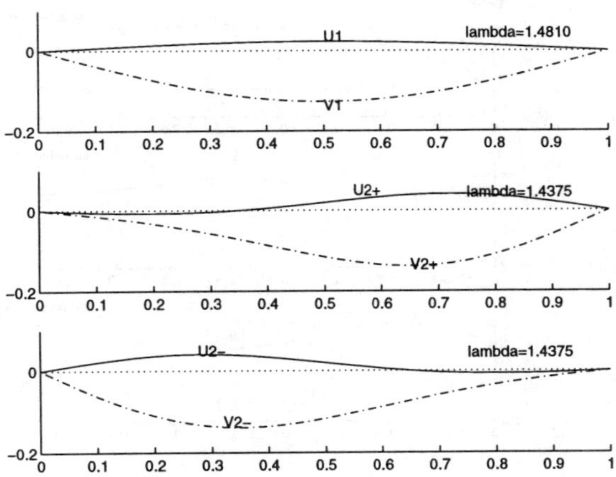

b) Solutions as functions of x.

Fig. 9.4. Bifurcation scenario at the double bifurcation point $(0, \lambda_1) = (0, \lambda_2) = (0,\ 1.5315)$. Here $(k, m) = (1, 2)$, $d_1 = 1$, $d_2 = 2$ and $\alpha = 0.278796$.

9.5.2 The length $\ell = 10$, diffusion rates $d_1 = 1$, $d_2 = 2$

Example 9.5.4. Double bifurcation, (k, m)=(odd, even). For $h = 0.025$ and $k = 1$, $m = 2$ we obtain $\alpha = 0.278796$ from the equation (9.53). Hence, the first double bifurcation occurs at $(0, \lambda_1^h) = (0, \lambda_2^h) \approx (0, 1.531482)$.

Figure 9.4a shows that there are three solution curves branching there from the trivial solution. These solution curves are depicted in Figure 9.4b.

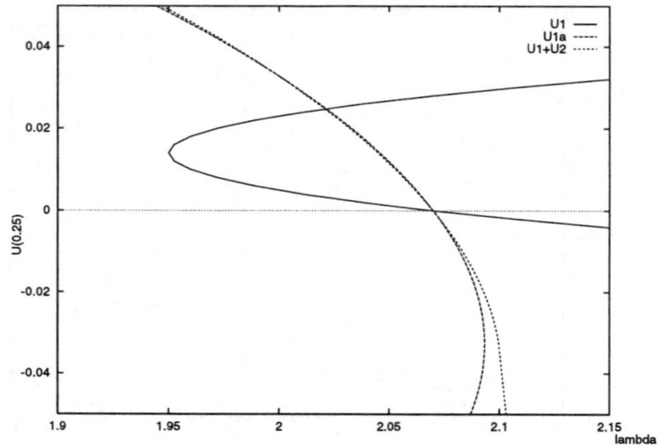

a) Bifurcation diagrams.

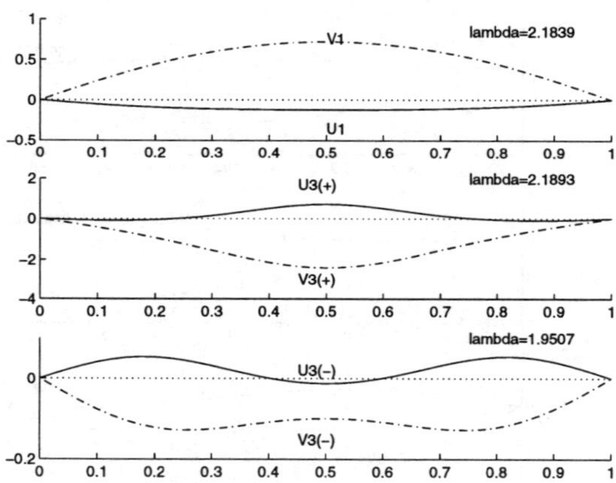

b) Solutions as functions of x.

Fig. 9.5. Bifurcation scenario at the double bifurcation point $(0, \lambda_1) = (0, \lambda_3) = (0,\ 2.07)$. Here $(k, m) = (1, 3)$, $d_1 = 1$, $d_2 = 2$ and $\alpha = 0.41765$.

Example 9.5.5. Double bifurcation, (k, m)=(odd, odd). We choose $h = 0.025$ and $k = 1$, $m = 3$. Then the equation (9.53) implies $\alpha = 0.41765655$, and we obtain a double bifurcation at $(0, \lambda_1^h) = (0, \lambda_3^h) \approx (0, 2.07002636)$. Figure 9.5 shows the bifurcation diagrams and solutions on three bifurcating branches. Note that there is a turning point on one of the solution.

10. Reaction-Diffusion Equations on a Square

We study in this chapter bifurcations of the reaction-diffusion equation

$$\frac{\partial u}{\partial t} = G(u, \lambda, d) \tag{10.1a}$$

with

$$G(u, \lambda, d) := \begin{pmatrix} \Delta u_1 + f_1(u_1, u_2, \lambda) \\ d\Delta u_2 + f_2(u_1, u_2, \lambda) \end{pmatrix}$$

on the unit square $\Omega := [0, 1] \times [0, 1]$ with the homogeneous Dirichlet boundary conditions

$$u_1 = 0, \qquad u_2 = 0 \quad \text{on } \partial\Omega \tag{10.1b}$$

Here $u := (u_1, u_2)^T$ are state variables representing concentrations of immediate products; $\lambda \in \mathbf{R}^p$ is a vector of control parameters and $d \in \mathbf{R}$ is the diffusion rate of the second substance. The functions $f_i : \mathbf{R}^{2+p} \mapsto \mathbf{R}, i = 1, 2,$ describe reactions among the substances. They are supposed to be sufficiently smooth and have a polynomial growth

$$|\partial_x^i f_j(x, \lambda)| \leq c_1 + c_2 ||x||^r \quad \text{for } ||x|| \to \infty, \quad i = 1, 2, 3, \ j = 1, 2$$

for some constants $c_1,\ c_2,\ r \geq 0$. Furthermore, we assume

$$f_i(0, 0, \lambda) = 0 \quad \text{for all } \lambda \in \mathbf{R},\ i = 1, 2. \tag{10.2}$$

Thus the system (10.1) has a quiescent 'trivial' solution with u independent of space and time. In order to investigate the stability of this uniform state, we need to know the location of the eigenvalues of the linearized operator $D_u G(0, \lambda, d)$ in the complex plane. If they all have negative real part, the trivial state is stable to small perturbations. If a parameter is varied continuously then the trivial solution may lose stability at some critical point where one or more eigenvalues of $D_u G(0, \lambda, d)$ cross the imaginary axis. This is a typical local bifurcation of the trivial state. Symmetries and nonlinearities of the system (10.1) may result in new steady or time periodic solutions with small amplitude at nearby values of the critical parameters.

We study at first in Section 10.1 symmetries of reaction-diffusion equations on square domains and then properties of eigenpairs of the Laplacian in Section 10.2. This leads to a direct analysis of linear stabilities of the trivial solution and bifurcation points in the parameter space in Sections 10.3 and 10.4, respectively. Part of the discussion is similarly to those for systems of one-dimensional equations in Chapter 9. We determine in Section 10.5 points where different types of mode interactions take place. We derive kernels of $D_uG(0, \lambda, d)$ and its adjoint at bifurcation points in Section 10.6 and consider briefly in Section 10.7 application of the equivariant Liapunov-Schmidt reduction. By examining the reduced bifurcation equations we obtain the complete bifurcation scenario at simple and double steady state bifurcation points.

10.1 D_4-Symmetry

Let

$$D_4 := \{S_1,\ S_2;\ S_1',\ S_2',\ I,\ R,\ R^2,\ R^3\} \subset O(2) \tag{10.3}$$

be the symmetry group of the unit square $[0,1] \times [0,1]$, consisting of the reflections S_i, S_i', $i = 1,2$ and the rotations R^i, $i = 1,2,3$, see Fig. 5.3 in Chapter 5.

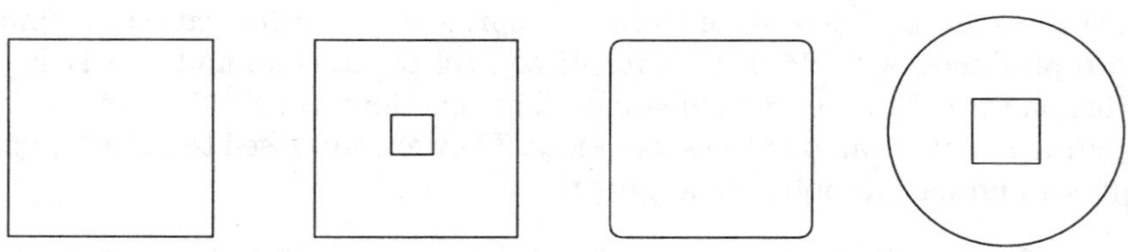

Fig. 10.1. Examples of domains with D_4-symmetry.

For the unit square $\Omega := [0,1] \times [0,1]$ the action of D_4 is generated by

$$S_1(x,y) = (1-x, y), \qquad R(x,y) = (1-y, x) \quad \text{for all } (x,y) \in \Omega. \tag{10.4}$$

For other domains in (10.1) it is more appropriate to shift first the origin of coordinates to the center of the domain and use then the reflection S_1 and the rotation R

$$S_1(x,y) = (-x, y), \qquad R(x,y) = (-y, x) \quad \text{for all } (x,y) \in \Omega \tag{10.5}$$

to generate the action of D_4 in Ω.

For the homogeneous Dirichlet boundary conditions we define the spaces

$$\begin{aligned} X &:= \{u = (u_1, u_2)^T \;\big|\; u_i \in C^{2,s}(\Omega),\ u_i\big|_{\partial\Omega} = 0,\ i = 1,2\}, \\ Y &:= \{u = (u_1, u_2)^T \;\big|\; u_i \in C^{0,s}(\Omega),\ u_i\big|_{\partial\Omega} = 0,\ i = 1,2\}. \end{aligned} \tag{10.6}$$

Here $C^{k,s}$ denote the space of all k-times differentiable functions u such that u and its derivatives are locally Hölder continuous with the exponent $s \in (0,1)$. The operator $G(u, \lambda, d)$ maps X into Y, moreover, it is generic that $D_uG(0, \lambda, d)$ is a Fredholm operator of index zero. We impose the following $L^2(\Omega)$-product in Y

$$\langle u,\ v\rangle = \int_\Omega (u_1v_1 + u_2v_2)dxdy \tag{10.7}$$

for all $u = (u_1, u_2)^T,\ v = (v_1, v_2)^T \in Y$.

Remark 10.1.1. For the homogeneous Neumann boundary conditions we choose the spaces

$$\begin{aligned} \tilde{X} &:= \{u = (u_1, u_2)^T \quad | \quad u_i \in C^{2,s}(\Omega),\ \tfrac{\partial u_i}{\partial n}|_{\partial\Omega} = 0,\ i = 1, 2\}, \\ \tilde{Y} &:= \{u = (u_1, u_2)^T \quad | \quad u_i \in C^{0,s}(\Omega),\ i = 1, 2\}. \end{aligned} \tag{10.8}$$

If Robin boundary conditions are imposed, then we consider weak form of (10.1) in the Sobolev spaces, see Chapter 14.

Functions in $\tilde{X}$ can be extended C^2-continuously to the whole space $\mathbf{R}^2$ by reflections along the sides of Ω and periodic extensions afterwards. This property and the equalities

$$\frac{\partial^2 u}{\partial x \partial y} = 0 \quad \text{at all corners of the square (rectangle) } \Omega$$

ensure regularity of the solutions of $\Delta u = f$ for $f \in C^{0,s}(\Omega)$ and make $D_uG(0, \lambda, d)$ to be a Fredholm operator of index zero.

Consider the reflection group $Z_2 := \{1, -1\}$ and

$$Z_2 \times D_4 = \{\pm S_1,\ \pm S_2;\ \ \pm I,\ \pm S_1',\ \pm S_2',\ \pm R,\ \pm R^2,\ \pm R^3\}.$$

We call a function $f(x, y, \lambda)$ is *odd* in (x, y) if one of the following statements hold for all $x, y \in \mathbf{R}$ and $\lambda \in \mathbf{R}^p$

$$\begin{array}{ll} \text{i)} & f(-x, y, \lambda) = -f(x, y, \lambda), \\ \text{ii)} & f(x, -y, \lambda) = -f(x, y, \lambda), \\ \text{iii)} & f(-x, -y, \lambda) = -f(x, y, \lambda). \end{array} \tag{10.9}$$

Denote

$$\Gamma := \begin{cases} Z_2 \times D_4 & \text{if the function } f_i(u_1, u_2, \lambda) \text{ is odd in } u_1, u_2,\ i = 1, 2; \\ D_4 & \text{otherwise.} \end{cases}$$

Action of Γ in the function space $\left(L^2(\Omega)\right)^2$ is defined as

$$\gamma u = (\gamma u_1, \gamma u_2) \qquad \text{for all } \ \gamma \in \Gamma,\ u = (u_1, u_2) \in \left(L^2(\Omega)\right)^2 \tag{10.10a}$$

with

$$\gamma u = \pm u(\delta^{-1}(x,y)) \text{ for all } \gamma = \pm\delta \in \Gamma,\ \delta \in D_4 \text{ and } u \in L^2(\Omega). \quad (10.10b)$$

The spaces X, Y are obviously Γ-invariant and the L^2-product is also Γ-invariant. The *Γ-equivariance* of the mapping G

$$G(\gamma u, \lambda, d) = \gamma G(u, \lambda, d) \qquad \text{for all} \quad \gamma \in \Gamma,\ u \in X,\ \lambda \in \mathbf{R} \qquad (10.11)$$

can be verified directly on the generator $\pm S_1,\ \pm S_2$ of the group Γ.

If the functions $f_i(u_1, u_2, \lambda),\ i = 1, 2$ are odd in $(u_1,\ u_2)$, or the homogeneous Neumann boundary condition is imposed, then the problem (10.1) can be embedded into periodic boundary value problems. Correspondingly we will have a larger symmetry group of periodic boundary conditions, which is partially responsible for occurrence of high degeneracy in those problems, and results in solutions which are not related directly with the D_4-symmetry of the domain. For further details we refer to Crawford *et al* [73, 76], Gomes/Stewart [133, 132], Healey/Kielhöfer [155] and Mei [217], in particular, on deriving appropriate equivariants and invariants with knowledge of the extension of symmetries. Other nonlinear and twisted symmetries with parameters may also occur in reaction-diffusion problems (cf. Cicogna [66]).

10.2 Eigenpairs of the Laplacian on a Square

Let $\Omega \subset \mathbf{R}^2$ be a bounded D_4-invariant domain with $C^{0,1}$-boundary, e.g. rectangular domains (cf. Hackbusch [146, pp.118]. We consider the eigenvalue problem

$$\begin{aligned} -\Delta u &= \sigma u \qquad \text{in } \Omega \\ u &= 0 \qquad \text{on } \partial\Omega \end{aligned} \qquad (10.12)$$

It is obvious that the Laplacian Δ is D_4-equivariant under the actions (10.4) and (10.5), respectively. If σ_0 is an eigenvalue of $-\Delta$ with E_0 as the corresponding eigenspace, then the space E_0 is finite and D_4-invariant, i.e.,

$$\delta\phi \in E_0 \qquad \text{for all} \quad \phi \in E_0,\ \delta \in D_4.$$

The action of a compact symmetry group on the eigenspace is generally irreducible (cf. Golubitsky/Stewart/Schaeffer [131, pp.82]). Furthermore, an irreducible representation of the D_4-group is either one or two dimensional.

We consider in the sequel the unit square

$$\Omega := [0, 1] \times [0, 1].$$

Since the problem (10.12) on a square can be embedded into a problem with periodic boundary condition, arbitrarily high, but finite, dimensional eigenspace E_0 is possible, (cf. Ashwin [20]). Eigenpairs of the Laplacian with the homogeneous Dirichlet boundary conditions are of the form

$$\sigma = (m_i^2 + n_i^2)\pi^2, \qquad i = 1, 2, \ldots, l$$
$$\phi_i(x, y) = 2 \sin m_i \pi x \ \sin n_i \pi y.$$

A few examples of these eigenvalues are listed in Table 10.1.

eigenvalues	multiplicity
$(1^2 + 1^2)\pi^2$	simple
$(1^2 + 2^2)\pi^2 = (2^2 + 1^2)\pi^2$	double
$(5^2 + 5^2)\pi^2 = (1^2 + 7^2)\pi^2 = (7^2 + 1^2)\pi^2$	3-fold
$(4^2 + 7^2)\pi^2 = (7^2 + 4^2)\pi^2 = (1^2 + 8^2)\pi^2 = (8^2 + 1^2)\pi^2$	4-fold

Table 10.1. Examples of eigenvalues of the Laplacian on the unit square with the homogeneous (Neumann) Dirichlet boundary conditions.

Definition 10.2.1. *Let*

$$\sigma = (m_i^2 + n_i^2)\pi^2, \qquad i = 1, \ldots, k$$

be an eigenvalue of $-\Delta$. *The corresponding eigenfunctions* $\phi_i = 2 \sin m_i \pi x \sin n_i \pi y$, $i = 1, 2$ *are called modes and* m_i, n_i *are called wave numbers. The parity of* ϕ_i *is defined as*

$$(\textit{parity of } (m_i), \textit{parity of } (n_i)) = \begin{cases} (\textit{even, even}) & \textit{if both } m_i,\ n_i \textit{ are even;} \\ (\textit{even, odd}) & \textit{if } m_i \textit{ is even, } n_i \textit{ is odd;} \\ (\textit{odd, even}) & \textit{if } m_i \textit{ is odd, } n_i \textit{ is even;} \\ (\textit{odd, odd}) & \textit{if both } m_i,\ n_i \textit{ are odd.} \end{cases}$$

Remark 10.2.1. The parities of all modes ϕ_i associated to the same eigenvalue $\sigma = (m_i^2 + n_i^2)\pi^2, i = 1, \ldots, k$, are identical. This can be seen from the equality

$$m_1^2 + n_1^2 = m_2^2 + n_2^2.$$

If $(m_1, n_1) =$ (even, even), (even, odd) or (odd, odd), so is (m_2, n_2), respectively.

If $\sigma = (m^2 + m^2)\pi^2$ is a simple eigenvalue and the associated eigenspace is

$$E_0 = \text{span}[\phi], \quad \text{with} \quad \phi = 2 \sin m\pi x \ \sin n\pi y,$$

then the induced action of D_4 in $E_0 \sim \mathbf{R}$ is

$$S_1 = (-1)^{m-1}, \qquad R = (-1)^{m-1}. \tag{10.13}$$

Since eigenspace of the Laplacian is D_4-invariant, the one dimensional eigenspace E_0 consists of D_4-symmetric functions for m =odd and D_4-antisymmetric functions for m =even, respectively.

In the following we consider a two-dimensional eigenspace

$$\begin{aligned} E_0 &= \text{span}[\phi_1,\ \phi_2] \qquad \text{with} \\ \phi_1 &= 2\sin m\pi x \sin n\pi y, \qquad \phi_2 \;=\; 2\sin n\pi x \sin m\pi y. \end{aligned} \tag{10.14}$$

Representations of D_4 in E_0 depend on the parities of m, n. In fact,

$$\begin{aligned} S_1(\alpha\phi_1 + \beta\phi_2) &= (-1)^{m-1}\alpha\phi_1 + (-1)^{n-1}\beta\phi_2, \\ R(\alpha\phi_1 + \beta\phi_2) &= (-1)^{m-1}\alpha\phi_2 + (-1)^{n-1}\beta\phi_1. \end{aligned}$$

Thus the induced representation of D_4 in $\mathbf{R}^2 \sim E_0$ is generated by

$$S_1 = \begin{pmatrix} (-1)^{m-1} & 0 \\ 0 & (-1)^{n-1} \end{pmatrix}, \quad R = \begin{pmatrix} 0 & (-1)^{m-1} \\ (-1)^{n-1} & 0 \end{pmatrix}. \tag{10.15}$$

Here we have explicitly used properties of the trigonometric functions. Similar conclusions hold also for mixed type of boundary conditions, see Chapter 15.

10.3 Linear Stability Analysis

Owing to (10.2), the reaction-diffusion problem has a uniform stationary solution manifold $\{(0, \lambda, d) \mid \lambda, d \in \mathbf{R}\}$. A bifurcation occurs on the manifold of trivial solutions when an eigenvalue of the operator $D_uG(0, \lambda, d)$ crosses the imaginary axis. For any $(\lambda, d) \in \mathbf{R}^2$, simple calculation yields

$$D_uG(0, \lambda, d) = \begin{pmatrix} \Delta + f^0_{1u_1} & f^0_{1u_2} \\ f^0_{2u_1} & d\Delta + f^0_{2u_2} \end{pmatrix}. \tag{10.16}$$

Here and below, if no confusion arises, $D_uG_0, DG_0, f^0_{1u_1}, \ldots$ denote evaluations of the derivatives $D_uG, DG, \frac{\partial f_1}{\partial u_1}, \ldots$ at the point $(0, \lambda, d)$. Similarly as one-dimensional problems in Chapter 9, properties of eigenpairs of the Laplacian allow a classification of curves of bifurcation curves in the parameter space λ, d. First of all, we note that eigenspace

$$\{\sin m\pi x \sin n\pi y \quad | \quad m, n \in \mathbf{N}\}$$

of the Laplacian makes up an orthogonal basis for $L^2(\Omega)$. Thereafter, we can decompose the space X as

$$X = \sum_{m,n\in N} \oplus X_{m,n} \tag{10.17}$$

with

$$X_{m,n} := \left\{ \begin{pmatrix} a \\ b \end{pmatrix} \sin m\pi x \sin n\pi y \;\middle|\; \begin{pmatrix} a \\ b \end{pmatrix} = \begin{pmatrix} 1 \\ 0 \end{pmatrix} \text{ or } \begin{pmatrix} 0 \\ 1 \end{pmatrix} \right\}. \tag{10.18}$$

The subspace $X_{m,n}$, for all $m,\ n \in \mathbf{N}$, is invariant under D_uG_0. More precisely, from the equation

$$D_uG_0 \begin{pmatrix} a \\ b \end{pmatrix} \sin m\pi x \sin n\pi y = \begin{pmatrix} f^0_{1u_1} - c & f^0_{1u_2} \\ f^0_{2u_1} & f^0_{2u_2} - dc \end{pmatrix} \begin{pmatrix} a \\ b \end{pmatrix} \sin m\pi x \sin n\pi y$$

we derive the restriction of D_uG_0 in the subspace $X_{m,n}$ as a 2×2 matrix

$$M(\lambda, d, c) := \begin{pmatrix} f^0_{1u_1} - c & f^0_{1u_2} \\ f^0_{2u_1} & f^0_{2u_2} - dc \end{pmatrix}. \tag{10.19}$$

Thus spectrum of the operator D_uG_0 consists of eigenvalues of the matrix $M(\lambda, d, c)$ for all $c = (m^2 + n^2)\pi^2$, $m,\ n \in \mathbf{N}$. Therefore, $(0, \lambda, d)$ is a bifurcation point if and only if there is an eigenvalue c of $-\Delta$, such that the matrix $M(\lambda, d, c)$ has an eigenvalue on the imaginary axis. If the equality

$$\det(M(\lambda, d, c)) = (c - D_{u_1} f_1^0)(dc - D_{u_2} f_2^0) - D_{u_2} f_1^0 D_{u_1} f_2^0 = 0 \tag{10.20}$$

holds, i.e., zero is an eigenvalue of $D_uG(0, \lambda, d)$, a steady state bifurcation occurs. If there exists a constant $\omega > 0$, such that

$$M(\lambda, d, c)^2 = -\omega I_{2\times 2} \iff M(\lambda, d, c) \sim \begin{pmatrix} 0 & \omega \\ -\omega & 0 \end{pmatrix}, \tag{10.21}$$

then generically the operator $D_uG(0, \lambda, d)$ has a complex pair of eigenvalues crossing the imaginary axis at $\pm i\omega$, i.e., a *Hopf bifurcation* occurs. There are typically branches of time-periodic spatially inhomogeneous solutions near the Hopf bifurcation points. The statement (10.21) is equivalent to

$$\begin{aligned} \operatorname{trace}(M(\lambda, d, c)) &= f^0_{1u_1} + f^0_{2u_2} - c(1 + d) = 0, \\ \det(M(\lambda, d, c)) &= \delta - c(f^0_{1u_1} d + f^0_{2u_2}) + c^2 d > 0, \end{aligned} \tag{10.22}$$

where

$$\delta := \det \begin{pmatrix} f^0_{1u_1} & f^0_{1u_2} \\ f^0_{2u_1} & f^0_{2u_2} \end{pmatrix} = f^0_{1u_1} f^0_{2u_2} - f^0_{2u_1} f^0_{1u_2}. \tag{10.23}$$

Consider the mapping $S:\ \mathbf{R} \to \mathbf{R}^2$ with

$$S(c) := \{(\lambda,\ d) \in \mathbf{R}^2 \mid \lambda,\ d \text{ satisfy } (10.20)\} \subset \mathbf{R}^2 \tag{10.24}$$

for steady state bifurcations and the mapping $H:\ \mathbf{R} \to \mathbf{R}^2$ with

$$H(c) := \{(\lambda,\ d) \in \mathbf{R}^2 \mid \lambda,\ d \text{ satisfy } (10.22)\} \subset \mathbf{R}^2 \tag{10.25}$$

for Hopf bifurcations. Concluding the above discussions we have

Lemma 10.3.1. $(0, \lambda, d) \in X \times \mathbf{R}^2$ *is a steady state bifurcation point of (10.1) if and only if there exist* $m, n \in \mathbf{N}$ *such that* $(\lambda, d) \in S(c)$ *for* $c = (m^2 + n^2)\pi^2$. *Similarly, a Hopf bifurcation occurs at* $(0, \lambda, d) \in X \times \mathbf{R}^2$ *on the trivial solution manifold if and only if there exist* $m, n \in \mathbf{N}$ *such that* $(\lambda, d) \in H(c)$ *for some* $c = (m^2 + n^2)\pi^2$.

The wave numbers m, n are chosen coprime in some studies (cf. Aston[28], Crawford [73] and Gomes/Stewart [133]). Furthermore, if the reaction terms are odd functions, by embedding (10.1) into periodic problem we can establish scaling laws between bifurcating solution branches, see e.g. Mei [220]. As we shall see later on, if both m, n are even, the reduced equations involve additional symmetries which simplify the bifurcation analysis.

For one-parameter problems with D_4-symmetry generic bifurcations are either simple or double. However, for the square domain and the homogeneous (Neumann) Dirichlet boundary conditions high degenerate bifurcations do occur as consequences of the number theoretic degeneracies (cf. Ashwin [20]) and the symmetries hidden in periodic boundary conditions (cf. Crawford *et al* [77], Mei [217]).

Example 10.3.1. **Simple and double steady state bifurcations.** Suppose the equality (10.20) holds for exactly one simple eigenvalue $c = 2m^2\pi^2$ of the Laplacian. Then $(0, \lambda, d)$ is a *simple* steady state bifurcation point of (10.1) and

$$\mathrm{Ker}(D_u G_0) = \mathrm{span}[\psi], \qquad \psi = \begin{pmatrix} D_{u_2} f_1^0 \\ c - D_{u_1} f_1^0 \end{pmatrix} \phi.$$

Here $\phi = 2 \sin m\pi x \ \sin m\pi y$ is the eigenfunction of of the Laplacian associated with the eigenvalue c.

If (10.20) holds for exactly one double eigenvalue $c = (m^2+n^2)\pi^2$, $m \neq n$ of the Laplacian with the corresponding eigenfunctions

$$\phi_1 = 2 \sin m\pi x \sin n\pi y, \qquad \phi_2 = 2 \sin n\pi x \sin m\pi y,$$

then $(0, \lambda, d)$ is a *double* steady state bifurcation point of of (10.1) and

$$\mathrm{Ker}(D_u G_0) = \mathrm{span}[\psi_1, \ \psi_2], \qquad \psi_i = \begin{pmatrix} -D_{u_2} f_1^0 \\ c - D_{u_1} f_1^0 \end{pmatrix} \phi_i, \quad i = 1, 2. \quad (10.26)$$

Here ϕ_1 and ϕ_2 (ψ_1 and ψ_2) are on the same D_4-orbit.

Example 10.3.2. **Simple and double Hopf bifurcations.** Let the statements (10.22) be satisfied for a single eigenvalue $c = (m^2 + n^2)\pi^2$ of the Laplacian. Define

$$\omega_0 := \Big(\det(M(\lambda, d, c))\Big)^{1/2}.$$

We see that $\pm\omega_0 i$ are all the eigenvalues of $D_u G_0$ on the imaginary axis. Scaling the time variable $t \to (1 + \tau)\omega_0 t$, we write the problem (10.1) as

$$\Phi(u, \lambda, d, \tau) := (1+\tau)\omega_0 \frac{\partial u}{\partial t} - G(u, \lambda, d) = 0.$$

If $m = n$, then $(0, \lambda, d)$ is a simple Hopf bifurcation point and the kernel

$$\operatorname{Ker}(\partial_u \Phi(0, \lambda, d, 0)) = \operatorname{span}[\psi_1,\ \psi_2]$$

is two dimensional with

$$\psi_1 := \Big(e_1 \cos t + \frac{1}{\omega_0} M(\lambda, d, c) e_1 \sin t\Big)\phi,$$
$$\psi_2 := \Big(-\frac{1}{\omega_0} M(\lambda, d, c) e_1 \cos t + e_1 \sin t\Big)\phi$$

and $e_1 := (1,0)^T,\ e_2 := (0,1)^T \in \mathbf{R}^2$.

If $m \neq n$, i.e., the multiplicity of c is two, then $(0, \lambda, d)$ is a double Hopf bifurcation point of (10.1) in the sense that the kernel

$$\operatorname{Ker}(\partial_u \Phi(0, \lambda, d, 0)) = \operatorname{span}[\psi_1,\ \psi_2,\ \psi_3,\ \psi_4]$$

is four dimensional with

$$\psi_1 := \Big(e_1 \cos t + \frac{1}{\omega_0} M(\lambda, d, c) e_1 \sin t\Big)\phi_1,$$
$$\psi_2 := \Big(-\frac{1}{\omega_0} M(\lambda, d, c) e_1 \cos t + e_1 \sin t\Big)\phi_1,$$
$$\psi_3 := \Big(e_1 \cos t + \frac{1}{\omega_0} M(\lambda, d, c) e_1 \sin t\Big)\phi_2,$$
$$\psi_4 := \Big(-\frac{1}{\omega_0} M(\lambda, d, c) e_1 \cos t + e_1 \sin t\Big)\phi_2.$$

Remark 10.3.1. Similar discussions hold for other domains and boundary conditions with the D_4-symmetry if we replace the eigenvalue $c = (m^2 + n^2)\pi^2$ and the associated eigenfunctions $\phi_1 = 2\sin m\pi x \sin n\pi y$, $\phi_2 = 2\sin n\pi x \sin m\pi y$ of the Laplacian with the abstract notations c and ϕ_1, ϕ_2, respectively.

10.4 Bifurcation Points in Parameter Space

We have seen from the discussion in Section 10.3 that eigenvalues of the matrix $M(\lambda, d, c)$ in (10.19) reveal stationary and Hopf bifurcations of the reaction-diffusion equation (10.1). On the other hand, if $f_i(u_1, u_2, \lambda)$ are smooth functions of λ, d, so are simple eigenvalues of $M(\lambda, d, c)$ for any given eigenvalue c of the Laplacian. As an example for determining bifurcation points in parameter space we consider in this section reaction terms of the following form

$$f_i(u_1, u_2, \lambda) = \lambda \tilde{f}_i(u_1, u_2), \quad i = 1, 2. \tag{10.27}$$

We neglect the sign ˜ in the sequel if no confusion arises. This kind of reaction terms can be considered as varying one specific parameter and fixing the others, see e.g. the model

$$\begin{aligned} f_1(u_1, u_2, \lambda) &= \lambda(\alpha u_1 + u_2 + \mu u_1^3), \\ f_2(u_1, u_2, \lambda) &= \lambda(u_1 + \gamma u_2 + \nu u_1^2 u_2) \end{aligned}$$

in Cohen/Hoppensteadt/Mirura [68] and Kirchgässner [190], and the Fitzhugh-Nagumo equations for pulse propagations along the nerve axon

$$\begin{aligned} f_1(u_1, u_2, \lambda) &= \lambda[\mu u_2 - \alpha u_1(u_1 + a)(a - u_1)], \\ f_2(u_1, u_2, \lambda) &= \lambda(u_1 - u_2) \end{aligned}$$

in Price/Wambacq/Oosterlinck [245].

10.4.1 Steady state bifurcation points

Let c be an arbitrarily fixed eigenvalue of the Laplacian. We study the set $S(c)$ of steady state bifurcation points. For the reaction terms in (10.27) the equation

$$\delta\lambda^2 - c(f_{1u_1}^0 d + f_{2u_2}^0)\lambda + c^2 d = 0.$$

describes points in $S(c)$. This is a quadratic curve in the (λ, d)-plane. Namely, $S(c)$ consists of either straight lines or nondegenerate cone sections.

1) Straight lines.

If $S(c)$ consists of straight lines, then $\det(M(0, \lambda, d))$ has a factorization of the form

$$\begin{aligned} &\delta\lambda^2 - cf_{1u_1}^0 d\lambda - cf_{2u_2}^0 \lambda + c^2 d \\ &= (\alpha_1\lambda + \beta_1 d + \gamma_1)(\alpha_2\lambda + \beta_2 d + \gamma_2) \qquad (10.28) \\ &= \alpha_1\alpha_2\lambda^2 + (\alpha_1\beta_2 + \alpha_2\beta_1)\lambda d + \beta_1\beta_2 d^2 + (\gamma_1\alpha_2 + \gamma_2\alpha_1)\lambda \\ &\quad + (\gamma_1\beta_2 + \gamma_2\beta_1)d + \gamma_1\gamma_2. \end{aligned}$$

By comparing coefficients we find $\gamma_1\beta_2 + \gamma_2\beta_1 = c^2 > 0$ and

$$\beta_1\beta_2 = \gamma_1\gamma_2 = 0.$$

Hence, we have either

$$\begin{array}{ll} \text{i)} & \gamma_1 = 0 \Rightarrow \gamma_2\beta_1 = c^2 \neq 0 \Rightarrow \beta_2 = 0, \quad \text{or} \\ \text{ii)} & \gamma_2 = 0 \Rightarrow \gamma_1\beta_2 = c^2 \neq 0 \Rightarrow \beta_1 = 0. \end{array} \tag{10.29}$$

Confining the discussion to the first case in (10.29), we obtain

$$\begin{aligned} \gamma_2\beta_1 &= c^2 &\Longrightarrow\quad \beta_1 &= c^2/\gamma_2, \\ \gamma_2\alpha_1 &= -cf^0_{2u_2} &\Longrightarrow\quad \alpha_1 &= -cf^0_{2u_2}/\gamma_2, \\ \alpha_2\beta_1 &= -c\,f^0_{1u_1} &\Longrightarrow\quad \alpha_2 &= -\gamma_2 f^0_{1u_1}/c. \end{aligned}$$

Thereafter, the factorization in (10.28) holds if and only if

$$\delta = \alpha_1\alpha_2 = f^0_{1u_1}\, f^0_{2u_2},$$

which, compared with the definition (10.23) of δ, is equivalent to

$$f^0_{2u_1} \cdot f^0_{1u_2} = 0. \tag{10.30}$$

In this case we have, either from (10.29), (10.30), or by the equivalence, directly from (10.28), that

$$S(c) = \{(\lambda, d) \quad | \quad \det M(\lambda, d, c) = (f^0_{2u_2}\lambda - cd)(f^0_{1u_1}\lambda - c) = 0\}. \tag{10.31}$$

Hence, $S(c)$ is represented as the union of two straight lines if and only if (10.30) is satisfied, see Figure 10.2.

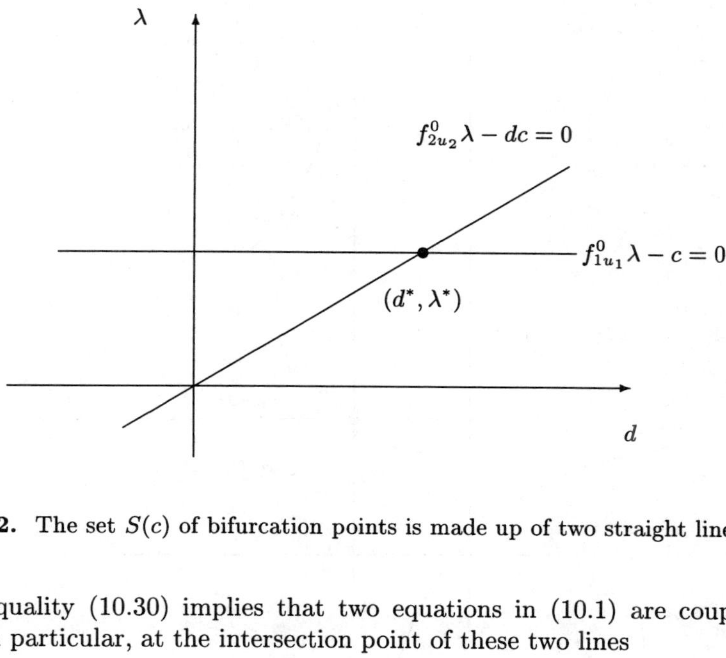

Fig. 10.2. The set $S(c)$ of bifurcation points is made up of two straight lines.

The equality (10.30) implies that two equations in (10.1) are coupled weakly. In particular, at the intersection point of these two lines

$$(\lambda^*, d^*) = (c/f^0_{1u_1},\ f^0_{2u_2}/f^0_{1u_1})$$

a higher singularity occurs, i.e.,

$$M(\lambda^*, d^*, c) = \begin{pmatrix} 0 & 0 \\ \lambda f^0_{2u_1} & 0 \end{pmatrix}.$$

We have either

$$M(\lambda^*, d^*, c) \sim \begin{pmatrix} 0 & 0 \\ 1 & 0 \end{pmatrix} \quad \text{for } |f^0_{1u_2}| + |f^0_{2u_1}| > 0,$$

which leads to a TB-bifurcation at (λ^*, d^*), or

$$M(\lambda^*, d^*, c) = 0 \quad \text{for } f^0_{1u_2} = f^0_{2u_1} = 0,$$

a bifurcation with higher multiplicity at (λ^*, d^*). In the latter case, two equations in (10.1) are coupled merely with higher order terms.

Bifurcation scenario and dynamics at a TB-point are sophisticated. We refer to Guckenheimer/Holmes [144] for further discussions. The D_4-symmetry makes occurence of a TB-bifurcation to be generically double. Thus analysis of bifurcation scenario becomes even more complicated. Center manifold reduction is necessary and Birkhoff normal forms has to be calulated and analyzed.

2) Nondegenerate cone sections.

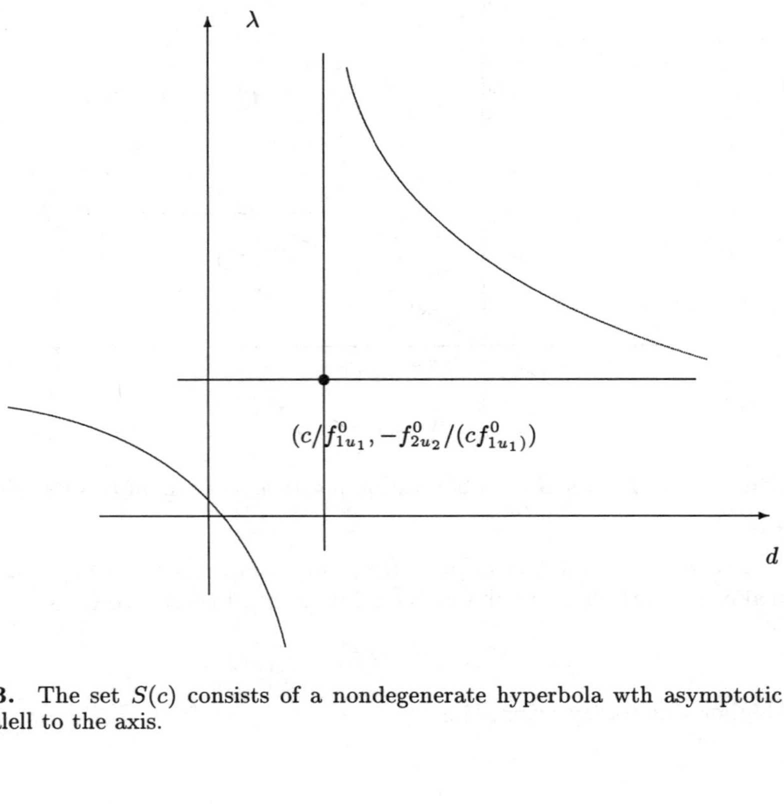

Fig. 10.3. The set $S(c)$ consists of a nondegenerate hyperbola wth asymptotic lines paralell to the axis.

If $S(c)$ consists of nondegenerate cone sections, the coefficients in the defining equation

$$\delta\lambda^2 - c(f^0_{1u_1}d + f^0_{2u_2})\lambda + c^2 d = 0$$

of $S(c)$ are confined to the following three situations

a) $\delta = 0,\ f^0_{1u_1} \neq 0$; b) $\delta \neq 0,\ f^0_{1u_1} = 0$; c) $\delta f^0_{1u_1} \neq 0$.

If the statements

$$\delta = 0, \qquad f^0_{1u_1} \neq 0$$

hold, the defining equation of $S(c)$ reduces to

$$-f^0_{1u_1}\lambda d - f^0_{2u_2}\lambda + cd = 0,$$

which is equivalent to, beside $(\lambda, d) = (0, 0)$, the equation

$$\Big(-\lambda + \frac{c}{f^0_{1u_1}}\Big)\Big(d + \frac{f^0_{2u_2}}{cf^0_{1u_1}}\Big) = \frac{f^0_{2u_2}}{(f^0_{1u_1})^2}.$$

Thus $S(c)$ consists of a nondegenerate hyperbola, see Figure 10.3.

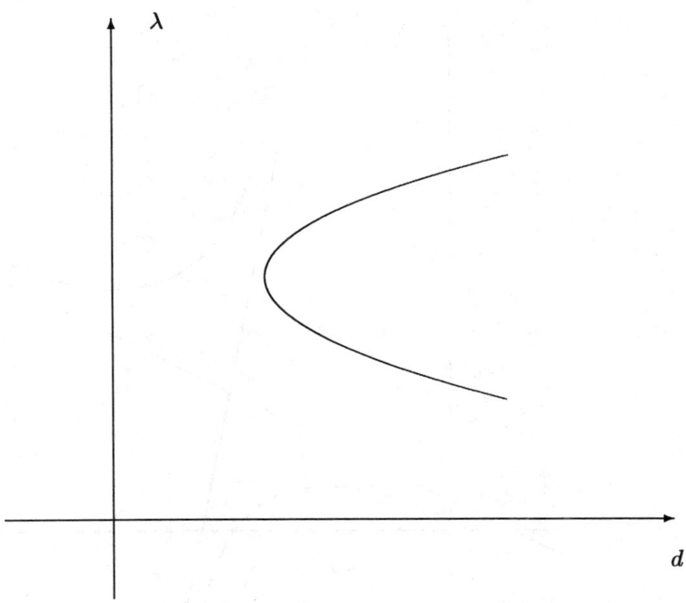

Fig. 10.4. A nondegenerate parabola as the set $S(c)$ of bifurcation points.

If

$$\delta \neq 0, \qquad f^0_{1u_1} = 0,$$

we obtain from the defining equation of $S(c)$ that

$$\lambda^2 - \delta^{-1} c f^0_{2u_2} \lambda + \delta^{-1} c^2 d = 0,$$

i.e.,

$$(\lambda - \frac{c f^0_{2u_2}}{2\delta})^2 = -c^2 \delta^{-2} [\delta d - (f^0_{2u_2})^2/4].$$

This shows that $S(c)$ is a parabola in the (λ, d)-plane, see Figure 10.4.

Finally, for

$$f^0_{1u_1} \cdot \delta \neq 0$$

we rewrite the defining equation of $S(c)$ as

$$\lambda^2 - \delta^{-1} c (f^0_{1u_1} d + f^0_{2u_2}) \lambda + \delta^{-1} c^2 d = 0.$$

One may easily reformulate it into

$$\left[\lambda - \frac{f^0_{1u_1} d + f^0_{2u_2}}{2\delta}\right]^2 - \frac{c^2 (f^0_{1u_1})^2}{4\delta^2} \left[d + \frac{f^0_{2u_2} - 2\delta}{f^0_{1u_1}}\right]^2 = \frac{c^2}{4(\delta\, f^0_{1u_1})^2} f^0_{1u_2} f^0_{2u_1}.$$

Hence, in this case $S(c)$ is again a nondegenerate hyperbola, see Figure 10.5.

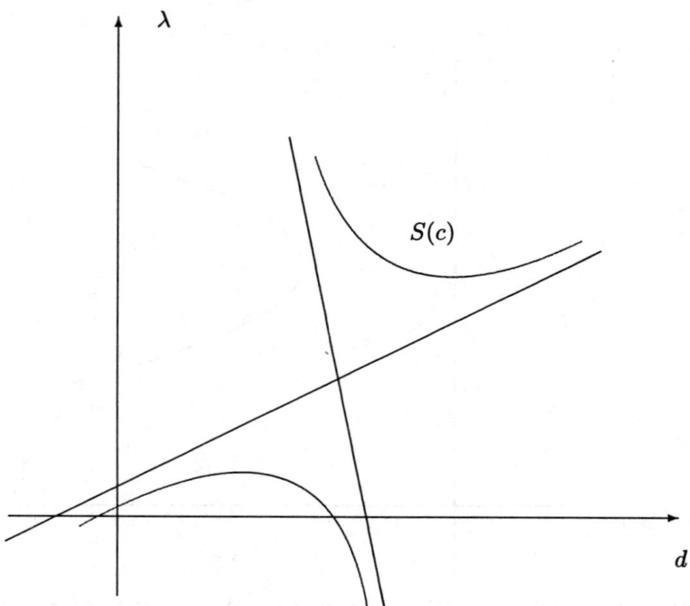

Fig. 10.5. The set $S(c)$ of bifurcation points is a nondegenerate hyperbola.

10.4.2 Hopf bifurcation points

As we have seen in (10.22) and (10.25), Hopf bifurcation points are located on the lines

$$\mathrm{trace}(M(\lambda, d, c)) = \lambda(f^0_{1u_1} + f^0_{2u_2}) - c(1+d) = 0, \;\; c := (m^2+n^2)\pi^2, \;\; m, n \in \mathbf{N},$$

which pass through the point $(\lambda, d) = (0, -1)$ with various slopes. This gives a necessary condition for a Hopf bifurcation

$$f^0_{1u_1} + f^0_{2u_2} \neq 0. \tag{10.32}$$

Furthermore,

$$\det(M(\lambda, d, c)) = \delta\lambda^2 - c(f^0_{1u_1} d + f^0_{2u_2})\lambda + c^2 d > 0.$$

10.5 Mode Interactions

Mode interactions occur at intersection points of bifurcation curves.

10.5.1 Steady/steady state mode interactions

At first we claim $S(c_1) \cap S(c_2) \cap S(c_3) = \emptyset$ for any three different eigenvalues c_i, $i = 1, 2, 3$ of the Laplacian. We recall the characteristic equation of $S(c_i)$

$$\delta\lambda^2 - c_i(f^0_{1u_1} d + f^0_{2u_2})\lambda + c_i^2 d = 0, \;\; i = 1, 2, 3. \tag{10.33}$$

If $(\lambda, d) \in S(c_1) \cap S(c_2) \cap S(c_3)$, we derive from the differences of two equations in (10.33) that

$$-(f^0_{1u_1} d + f^0_{2u_2})\lambda + (c_i + c_j)d = 0, \quad i, \; j = 1, 2, 3, \; i \neq j.$$

It follows from these equations that

$$(c_2 - c_3)d = 0.$$

Since $d > 0$, we have $c_2 = c_3$, contradicting to the assumption. Thus we restrict the discussion to the intersection $S(c_1) \cap S(c_2)$ of two different bifurcation curves.

Recall the two cases for $S(c)$ in Section 10.4.1. We see that the sets $S(c_i)$ of bifurcation points for all eigenvalues c_i of the Laplacian are given by either straight lines for which $f^0_{2u_1} \cdot f^0_{1u_2} = 0$; or nondegenerate cone sections (quadratic curves) for which $f^0_{2u_1} \cdot f^0_{1u_2} \neq 0$.

1) The case $f^0_{2u_1} \cdot f^0_{1u_2} = 0$. Now both $S(c_1)$ and $S(c_2)$ consist of straight lines respectively. Here and below we indicate the curves (and asymptotes) for c_1 by thin lines, for c_2 by thick lines, respectively. In addition to the

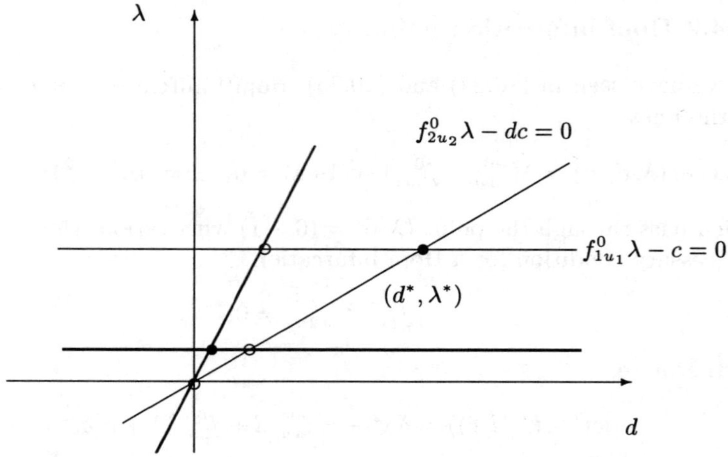

Fig. 10.6. The sets $S(c_1)$ and $S(c_2)$ of bifurcation points.

original intersection points in Figure 10.2, indicated by •, we obtain one or three additional intersection points, indicated by ∘, one of them is the trivial $(\lambda, d) = (0,0)$, see Figure 10.6

For $f^0_{1u_1} \neq 0$, the two nontrivial intersection points are

$$(\lambda_1^+, d_1^+) := \Big(\frac{c_1}{f^0_{1u_1}}, \frac{c_1\, f^0_{2u_2}}{c_2\, f^0_{1u_1}}\Big) \quad \text{and} \quad (\lambda_2^+, d_2^+) := \Big(\frac{c_2}{f^0_{1u_1}}, \frac{c_2\, f^0_{2u_2}}{c_1\, f^0_{1u_1}}\Big) \qquad (10.34)$$

and

$$S(c_1) \cap S(c_2) = \begin{cases} \{(0,0)\} & \text{for } f^0_{1u_1} = 0, \\ \{(0,0),\ (\lambda_1^+, d_1^+),\ (\lambda_2^+, d_2^+)\} & \text{for } f^0_{1u_1} \neq 0. \end{cases}$$

2) The case $f^0_{2u_1} \cdot f^0_{1u_2} \neq 0$. Under this condition $S(c_i), i = 1,2$ consist of nondegenerate cone sections. This case is covered by the following three situations.

a) If $f^0_{1u_1} \neq 0$ and $\delta = 0$, we derive from the defining equations of $S(c_i), i = 1,2$ $(c_1 \neq c_2)$

$$(c_1 - c_2)d = 0 \Longrightarrow d = 0.$$

Hence,

$$S(c_1) \cap S(c_2) = \{(0,0)\}.$$

b) If $\delta \neq 0$ and $f^0_{1u_1} = 0$, the difference of the defining equations (10.33) for $i = 1,2$ yields

$$d = (c_1 + c_2)^{-1}\, f^0_{2u_2}\, \lambda.$$

Substituting this into the equation (10.33) for $i = 1$ we obtain an equation for λ

$$[\delta\lambda - c_1\, f^0_{2u_2} + c_1^2(c_1 + c_2)^{-1}\, f^0_{2u_2}]\lambda = 0.$$

This has exactly one nontrivial solution and

$$S(c_1) \cap S(c_2) = \{(0,0),\ (\lambda_1, d_1)\}$$

with

$$(\lambda_1, d_1) = \frac{c_1 c_2\, f^0_{2u_2}}{\delta(c_1 + c_2)} \left(1,\ \frac{f^0_{2u_2}}{c_1 + c_2}\right), \tag{10.35}$$

see Figure 10.7.

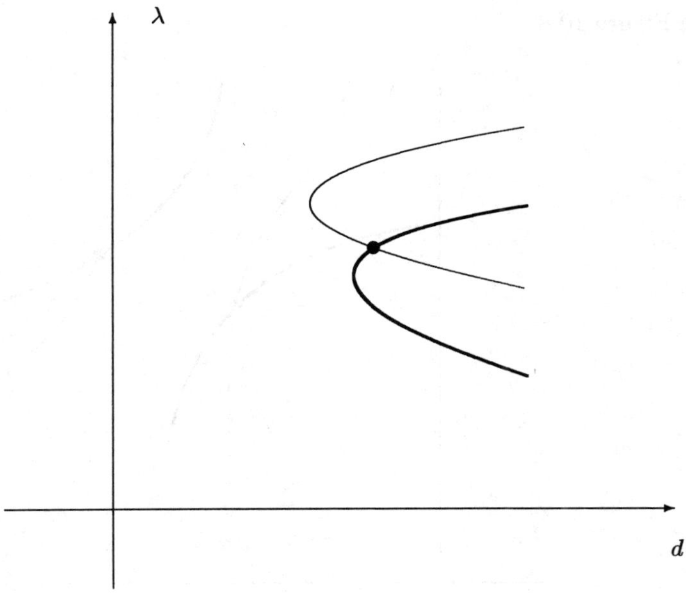

Fig. 10.7. Intersection of the sets $S(c_1)$ and $S(c_2)$ as two nondegenerate parabolas.

c) Finally, if $\delta\, f^0_{1u_1} \neq 0$, the equations (10.33) for $i = 1, 2$ are equivalent to

$$\frac{\delta}{c_1}\lambda^2 - (f^0_{1u_1} d + f^0_{2u_2})\lambda + c_1 d = 0, \tag{10.36a}$$

$$(f^0_{1u_1} d + f^0_{2u_2})\lambda - (c_1 + c_2) d = 0. \tag{10.36b}$$

Taking the sum of (10.36a) and (10.36b) we find

$$d = \frac{\delta\lambda^2}{c_1 c_2}.$$

Substituting this into (10.36b) yields an equation for λ

$$[\delta\, f^0_{1u_1}\,\lambda^2 - c_1c_2(c_1+c_2)\delta\lambda + c_1c_2\, f^0_{2u_2}]\lambda = 0.$$

Except the trivial solution it has two real solutions under proper conditions

$$\lambda_\pm := \frac{c_1c_2(c_1+c_2)\delta \pm \sqrt{[c_1c_2(c_1+c_2)]^2 - 4c_1c_2\delta\, f^0_{2u_2}}}{2\delta\, f^0_{1u_1}}.$$

Thus

$$S(c_1)\cap S(c_2) = \{(0,0),\ (\lambda_+,d_+),\ \lambda_-,d_-)\}, \tag{10.37}$$

see Figure 10.8.

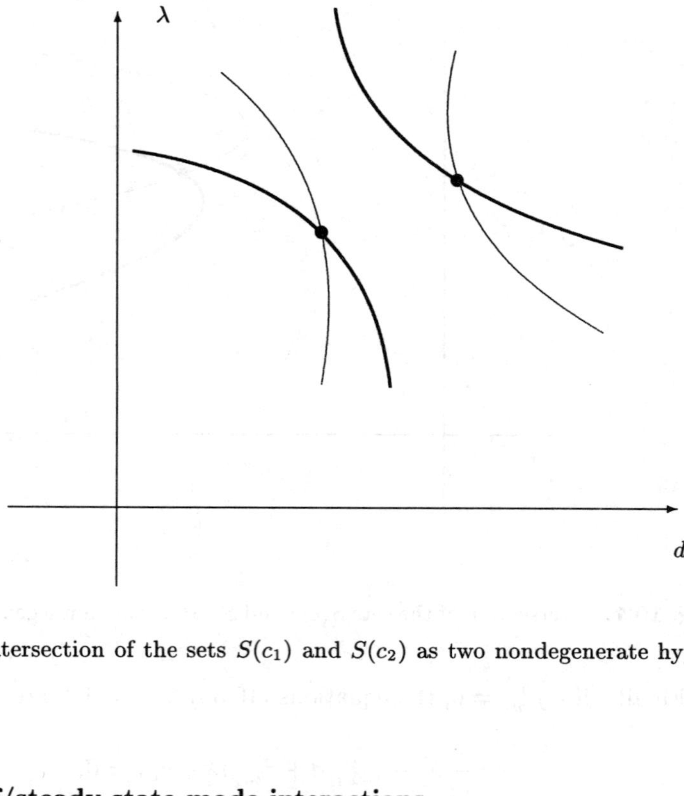

Fig. 10.8. Intersection of the sets $S(c_1)$ and $S(c_2)$ as two nondegenerate hyperbolas.

10.5.2 Hopf/steady state mode interactions

A Hopf/state steady mode inteaction occurs if λ and d satisfy following conditions for some $c_i = (m_i^2+n_i^2)\pi^2,\ m_i,\ n_i \in \mathbf{N},\ i=1,2$

$$\begin{aligned}
\mathrm{trace}(M(\lambda,d,c_1)) &= \lambda(f^0_{1u_1}+f^0_{2u_2}) - c_1(1+d) = 0, && (10.38a)\\
\det(M(\lambda,d,c_1)) &= \delta\lambda^2 - c_1(f^0_{1u_1}d + f^0_{2u_2})\lambda + c_1^2 d > 0, && (10.38b)\\
\det(M(\lambda,d,c_2)) &= \delta\lambda^2 - c_2(f^0_{1u_1}d + f^0_{2u_2})\lambda + c_2^2 d = 0. && (10.38c)
\end{aligned}$$

Since the Hopf bifurcation requires that $f^0_{1u_1} + f^0_{2u_2} \neq 0$, with c_1 and c_2 as parameters the system (10.38) is generally solvable for $d > 0, \lambda \in \mathbf{R}$, in other words, Hopf/steady state mode interaction is generic for the problem (10.1) with the reaction terms (10.27), see e.g., Figure 10.9.

10.5.3 Hopf/Hopf mode interactions

Hopf/Hopf mode interaction can not occur in the system (10.1) and (10.27), since the equations

$$\begin{aligned}\operatorname{trace}(M(\lambda, d, c_1)) &= \lambda(f^0_{1u_1} + f^0_{2u_2}) - c_1(1+d) = 0,\\ \operatorname{trace}(M(\lambda, d, c_2)) &= \lambda(f^0_{1u_1} + f^0_{2u_2}) - c_2(1+d) = 0\end{aligned}$$

have no solution λ, d for $c_1 \neq c_2$.

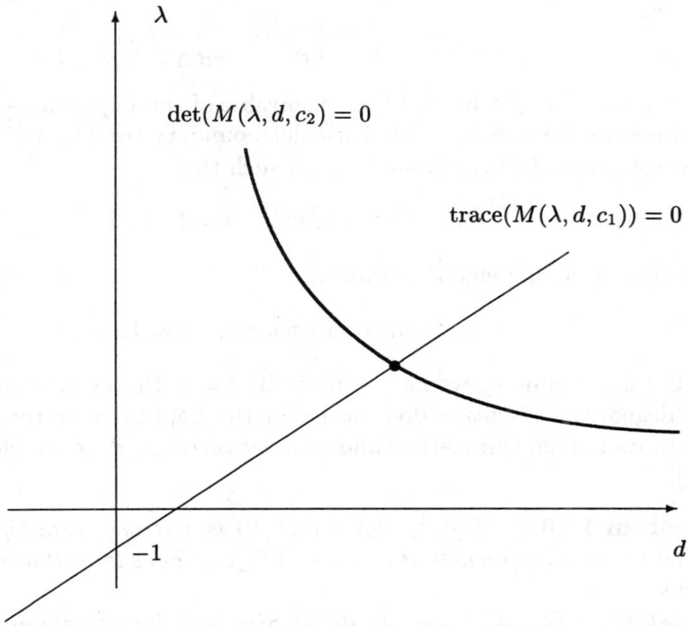

Fig. 10.9. Hopf/steady state mode interaction as an intersection of sets $H(c_1)$ and $S(c_2)$.

10.6 Kernels of D_uG_0 and $(D_uG_0)^*$

We derive the adjoint operator $(D_uG_0)^*$ and determine kernels of the operators D_uG_0 and $(D_uG_0)^*$, which are needed in descriptions of solution manifolds bifurcating from the $S(c)$, see (10.34).

Under the L^2-product in X we derive the adjoint operator for D_uG_0 by partial integrations

$$\begin{aligned}
&\langle v, D_uG_0u\rangle \\
&= \left\langle \begin{pmatrix} v_1 \\ v_2 \end{pmatrix}, \begin{pmatrix} \Delta u_1 + \lambda f^0_{1u_1} u_1 + \lambda f^0_{1u_2} u_2 \\ d\Delta u_2 + \lambda f^0_{2u_1} u_1 + \lambda f^0_{2u_2} u_2 \end{pmatrix} \right\rangle \\
&= \int_\Omega \Big[v_1(\Delta u_1 + \lambda f^0_{1u_1} u_1 + \lambda f^0_{1u_2} u_2) + v_2(d\Delta u_2 + \lambda f^0_{2u_1} u_1 + \lambda f^0_{2u_2} u_2) \Big]\, dxdy \\
&= \int_\Omega \Big[u_1(\Delta v_1 + \lambda f^0_{1u_1} v_1 + \lambda f^0_{2u_1} v_2) + u_2(d\Delta v_2 + \lambda f^0_{1u_1} v_1 + \lambda f^0_{2u_2} v_2) \Big]\, dxdy \\
&= \left\langle \begin{pmatrix} u_1 \\ u_2 \end{pmatrix}, \begin{pmatrix} \Delta v_1 + \lambda f^0_{1u_1} v_1 + \lambda f^0_{2u_1} v_2 \\ d\Delta v_2 + \lambda f^0_{1u_2} v_1 + \lambda f^0_{2u_2} v_2 \end{pmatrix} \right\rangle \qquad \text{for all } u, v \in X.
\end{aligned}$$

Hence

$$(D_uG_0)^* = \begin{pmatrix} \Delta + \lambda f^0_{1u_1} & \lambda f^0_{2u_1} \\ \lambda f^0_{1u_2} & d\Delta + \lambda f^0_{2u_2} \end{pmatrix}. \tag{10.39}$$

Let $c = (m^2 + n^2)\pi^2$ be an l-fold eigenvalue of the Laplacian $-\Delta$ on the unit square with the homogeneous Dirichlet boundary conditions, i.e., there are l different pairs of $(m_i, n_i), i = 1, \ldots, l$, such that

$$c = (m_i^2 + n_i^2)\pi^2, \quad i = 1, \ldots, l.$$

The corresponding eigenfunctions are

$$\phi_i := 2 \sin m_i \pi x \sin n_i \pi y, \quad i = 1, \ldots, l.$$

By the D_4-symmetry we have generically $l = 2$. However, arbitrarily large l, i.e., degenerate situation does occur for the Laplace operator on the square due to its hidden symmetries and number theoretic degeneracies (cf. Ashwin [19]).

Theorem 10.6.1. *If $(0, \lambda_0, d_0) \neq (0, 0, 0)$ is a steady state bifurcation point of (10.1), then the kernels of D_uG_0 and $D_uG_0{}^*$ have the following five possible forms.*
(i) Let $f^0_{1u_2} \cdot f^0_{2u_1} \neq 0$ and $(\lambda_0, d_0) \in S(c)$ hold for exactly one eigenvalue c of the Laplacian. Then

$$\psi_i := \begin{pmatrix} -\lambda_0 f^0_{1u_2} \\ \lambda_0 f^0_{1u_1} - c \end{pmatrix} \phi_i, \qquad \psi_i^* := \begin{pmatrix} -\lambda_0 f^0_{2u_1} \\ \lambda_0 f^0_{1u_1} - c \end{pmatrix} \phi_i, \ i = 1, \ldots, l \tag{10.40}$$

and

$$\begin{aligned} \mathrm{Ker}(D_uG_0) &= \mathrm{span}[\psi_1, \ldots, \psi_l], \\ \mathrm{Ker}(D_uG_0{}^*) &= \mathrm{span}[\psi_1^*, \ldots, \psi_l^*]. \end{aligned} \tag{10.41}$$

Furthermore,

$$X = \mathrm{Ker}(D_uG_0) \oplus M, \quad Y = \mathrm{Ker}(D_uG_0{}^*) \oplus \mathrm{Im}(D_uG_0). \tag{10.42}$$

(ii) Let $f^0_{1u_2} \cdot f^0_{2u_1} = 0$ *and* $(\lambda_0, d_0) \in S(c)$ *hold for exactly one eigenvalue c of the Laplacian and* $(\lambda_0, d_0) \neq (\lambda^*, d^*)$*, see (10.31), i.e., the equation*

$$(\lambda_0\, f^0_{1u_1} - c)(\lambda_0\, f^0_{2u_2} - cd_0) = \det(M(\lambda_0, d_0, c) = 0$$

has at least one nontrivial factor. If $(\lambda_0\, f^0_{1u_1} - c) \neq 0$*, we choose* ψ_i*,* ψ_i^* *as in (10.40). Otherwise, for* $\lambda_0\, f^0_{2u_2} - cd_0 \neq 0$ *we define for* $i = 1, \ldots, l$

$$\psi_i := \begin{pmatrix} \lambda_0 f^0_{2u_2} - cd_0 \\ -\lambda_0 f^0_{2u_1} \end{pmatrix} \phi_i, \qquad \psi_i^* := \begin{pmatrix} \lambda_0 f^0_{2u_2} - cd_0 \\ -\lambda_0 f^0_{1u_2} \end{pmatrix} \phi_i. \tag{10.43}$$

Then (10.41) is again valid with ψ_i*,* ψ_i^* *in (10.40) for* $\lambda_0\, f^0_{1u_1} - c \neq 0$ *or in (10.43) for* $\lambda_0\, f^0_{2u_2} - cd_0 \neq 0$*, respectively.*
(iii) Let $f^0_{1u_2} \cdot f^0_{2u_1} = 0$ *and* $(\lambda_0, d_0) = (\lambda^*, d^*) \in S(c)$ *hold for exactly one eigenvalue c of the Laplacian. Then* $M(\lambda^*, d^*, c) = 0$ *and for* $f^0_{1u_2} = f^0_{2u_1} = 0$ *we have*

$$\mathrm{Ker}(D_uG_0) = \mathrm{Ker}(D_uG_0{}^*) = \mathrm{span}\left[\begin{pmatrix}1\\0\end{pmatrix}\phi_i,\ \begin{pmatrix}0\\1\end{pmatrix}\phi_i,\ i = 1, \ldots, l\right].$$

In addition, if $|f^0_{1u_2}| + |f^0_{2u_1}| > 0$*, then the eigenvalue zero of* D_uG_0 *has different geometric and algebraic multiplicities, and*

$$\left.\begin{aligned} \mathrm{Ker}(D_uG_0) &= \mathrm{span}\left[\begin{pmatrix}1\\0\end{pmatrix}\phi_i,\ i = 1, \ldots, l\right], \\ \mathrm{Ker}(D_uG_0{}^*) &= \mathrm{span}\left[\begin{pmatrix}0\\1\end{pmatrix}\phi_i,\ i = 1, \ldots, l\right], \end{aligned}\right\} \quad \textit{for}\ \ f^0_{1u_2} = 0,\ \ f^0_{2u_1} \neq 0,$$

and

$$\left.\begin{aligned} \mathrm{Ker}(D_uG_0) &= \mathrm{span}\left[\begin{pmatrix}0\\1\end{pmatrix}\phi_i,\ i = 1, \ldots, l\right], \\ \mathrm{Ker}(D_uG_0{}^*) &= \mathrm{span}\left[\begin{pmatrix}1\\0\end{pmatrix}\phi_i,\ i = 1, \ldots, l\right], \end{aligned}\right\} \quad \textit{for}\ \ f^0_{1u_2} \neq 0,\ \ f^0_{2u_1} = 0.$$

(iv) Let $f^0_{1u_2} \cdot f^0_{2u_1} \cdot \delta \neq 0$ *and* $(\lambda_0, d_0) \in S(c_1) \cap S(c_2)$ *for different* $c_1 \neq c_2$*, where* (λ_0, d_0) *is chosen to be one among the following points:* (λ_+, d_+) *or* (λ_-, d_-) *for* $f^0_{1u_1} \neq 0$ *or* (λ_1, d_1) *for* $f^0_{1u_1} = 0$*, see (10.37) and (10.35), respectively. Then*

$$\begin{aligned} \mathrm{Ker}(D_uG_0) &= \mathrm{span}[\psi_i(c_j),\ i = 1, \ldots, l,\ j = 1, 2], \\ \mathrm{Ker}(D_uG_0{}^*) &= \mathrm{span}[\psi_i^*(c_j),\ i = 1, \ldots, l,\ j = 1, 2], \end{aligned} \tag{10.44}$$

where the $\psi_i(c_j)$*,* $\psi_i^*(c_j)$ *are defined as (10.40) with c replaced by* c_j*,* $j = 1, 2$ *respectively.*
(v) Let $f^0_{1u_1} \neq 0$*,* $f^0_{1u_2} \cdot f^0_{2u_1} = 0$ *and* $(\lambda_0, d_0) = (\lambda_i^+, d_i^+) \in S(c_1) \cap S(c_2)$ *for different* $c_1 \neq c_2$ *as defined in (10.34). Then* $(\lambda_i^+\, f^0_{2u_2} - c_i d_i^+) \neq 0$ *for* $i = 1, 2$*. Define the* $\psi_i(c_j)$*,* $\psi_i^*(c_j)$*,* $i = 1, \ldots, l$ *by (10.43) for* $c = c_j$*,* $j = 1, 2$ *and* $(\lambda_0, d_0) = (\lambda_i^+, d_i^+)$ *for* $i = 1$ *or* $i = 2$*. We have again (10.44) (with the changed basis vectors* $\psi_i(c_j)$*,* $\psi_i^*(c_j)$*).*

Proof: We start by showing that (i)–(v) are the only possibilities for $(\lambda_0, d_0) \neq (0,0)$.

Let $(\lambda_0, d_0) \neq (0,0)$ be a point in $S(c_1) \cap S(c_2)$ and $c_1 \neq c_2$. Recall the discussions in Section 10.5. If $f^0_{1u_2} \cdot f^0_{2u_1} = 0$, we have necessarily $f^0_{1u_1} \neq 0$ and $\lambda^+_i f^0_{1u_1} - c_i = 0, i = 1, 2$. Since $(\lambda^+_i, d^+_i) = (\lambda^*(c_i), d^*(c_i))$, we obtain $\lambda^+_i f^0_{2u_2} - c_i d^+_i \neq 0$, i.e., the case (v). If $f^0_{1u_2} \cdot f^0_{2u_1} \neq 0$, then only the cases iv) and v) are relevant. Possible choices for (λ_0, d_0) are the combinations $f^0_{1u_1} = 0$, $\delta \neq 0$ and $f^0_{1u_1} \cdot \delta \neq 0$, respectively, which corresponds the case (iv).

The homogeneous Dirichlet boundary conditions in a rectangular domain allows a C^2-continuous extension of the function in X beyond the boundary of Ω by reflections of anti-symmetries along the sides, see e.g. Crawford *et al* [77] and Mei [217]. Thereafter a periodic extension yields the $C^{2,s}$-continuity in $\mathbf{R}^2$. After extension the function u has the property

$$\Delta u = 0 \quad \text{on } \partial\Omega \quad \text{for all } u \in X.$$

On the other hand, solution u of the equation

$$\Delta u = f \quad \text{in } \Omega, \qquad u|_{\partial\Omega} = 0$$

is in the space $C^{2,s}(\Omega)$ for all $f \in C^{0,s}(\Omega)$. Applying recursively this regularity property yields smoothness of its solutions. Similar conclusion holds for the system (10.1). Furthermore, $D_u G_0$ is a Fredholm operator of index zero and we have the decomposition (10.42) of X (cf. Berger [35, pp.101]).

Now we want to discuss the five cases (i)—(v).

(i): The inequality $f^0_{1u_2} \cdot f^0_{2u_1} \neq 0$ and the defining equation (10.20) imply that $\lambda_0 f^0_{1u_1} - c \neq 0$. Then the two rows of $M(\lambda_0, d_0, c)$ and $M^*(\lambda_0, d_0, c)$ are linearly dependent. Therefore, we consider the first row of $M(\lambda_0, d_0, c)$ and $M(\lambda_0, d_0, c)^T$ and obtain the ψ_i and ψ^*_i in (10.40), respectively. This yields in turn the null spaces $\mathrm{Ker}(D_u G_0)$ and $\mathrm{Ker}(D_u G_0{}^*)$ in (10.41).

(ii): If $f^0_{1u_2} \cdot f^0_{2u_1} = 0$ and $(\lambda_0, d_0) \neq (\lambda^*, d^*)$, by (10.31) exactly one of the factors is zero for (λ_0, d_0). Choosing the nonzero factor, we may argue as above to yield (10.40) for $f^0_{1u_2} \cdot f^0_{2u_1} \neq 0$ and (10.43) for $\lambda_0 f^0_{1u_1} - c = 0$, respectively.

(iii): In this case $M(\lambda^*, d^*, c) = M^*(\lambda^*, d^*, c) = 0$ or similar to

$$\begin{pmatrix} 0 & 0 \\ 1 & 0 \end{pmatrix}.$$

Therefore, its null vectors can be chosen directly.

(iv): By $f^0_{1u_2} \cdot f^0_{2u_1} \neq 0$ we have necessarily $\lambda_0 f^0_{1u_1} - c_i \neq 0$, and hence we may use the argumentation as in (i) to obtain (10.44) for each $c = c_1$ and c_2.

(v): Since $\lambda^+_i f^0_{2u_2} - c_i d^+_i \neq 0$ for $i = 1$ or $i = 2$, the arguments in (ii) are applicable to this case. ∎

10.7 Equivariant Liapunov-Schmidt Reduction

Let $(0, \lambda_0, d_0)$ be a bifurcation point of (10.1). We exclude TB-points for the following discussion. Note that $D_uG(0, \lambda_0, d_0) : X \to Y$ is a Fredholm operator of index zero. We have the decompositions

$$X = \mathrm{Ker}(D_uG_0) \oplus M, \qquad Y = \mathrm{Ker}(D_uG_0^*) \oplus \mathrm{Im}(D_uG_0) \tag{10.45}$$

with respect to the $L^2(\Omega)$-product in Y

$$\langle u,\ v\rangle = \int_\Omega (u_1v_1 + u_2v_2)\ dx\ dy \quad \text{for all } u = \begin{pmatrix} u_1 \\ u_2 \end{pmatrix},\ v = \begin{pmatrix} v_1, \\ v_2 \end{pmatrix} \in Y.$$

Moreover, the subspaces $\mathrm{Ker}(D_uG_0)$, $\mathrm{Im}(D_uG_0)$ and their complementaries M and $\mathrm{Ker}(D_uG_0^*)$ are Γ-invariant, respectively. Denote

$$\mathrm{Ker}(D_uG_0) = \mathrm{span}[\psi_1, \ldots, \psi_k],\ \mathrm{Ker}(D_uG_0^*) = \mathrm{span}[\psi_1^*, \ldots, \psi_k^*]. \tag{10.46}$$

Here we have generically either $k \in \{1, 2\}$ for $(0, \lambda_0, d_0)$ on a single curve $S(c)$ of bifurcation points or $k \in \{3, 4\}$ for $(0, \lambda_0, d_0)$ at an intersection of two curves $S(c_1)$, $S(c_2)$ of bifurcation points. Moreover, the functions ψ_i, $\psi_i^*, i = 1, \ldots, k$ are of the form

$$\psi_i = \begin{pmatrix} f^0_{1u_2} \\ c/\lambda_0 - f^0_{1u_1} \end{pmatrix} \phi_i, \qquad \psi_i^* = \begin{pmatrix} f^0_{2u_1} \\ c/\lambda_0 - f^0_{1u_1} \end{pmatrix} \phi_i \tag{10.47}$$

for some eigenvalue c of the Laplacian. We normalize these basis vectors such that

$$\langle \psi_i,\ \psi_j\rangle = \langle \psi_i^*,\ \psi_j^*\rangle = \delta_{ij},\ \ i,\ j = 1, \ldots, k. \tag{10.48}$$

Define

$$Qu := u - \sum_{i=1}^k \langle \psi_i,\ u\rangle \psi_i \quad \text{for all} \ \ u \in X$$

$$\tilde{Q}u := u - \sum_{i=1}^k \langle \psi_i^*,\ u\rangle \psi_i^* \quad \text{for all } u \in Y.$$

These mappings are projections

$$Q:\ X \to M \qquad \text{and} \qquad \tilde{Q}:\ Y \to \mathrm{Im}(D_uG_0)$$

and commute with actions of the group Γ. After writing an elements (u, λ, d) in the space $X \times \mathbf{R} \times \mathbf{R}$ into the form

$$\begin{aligned} u &= \sum_{i=1}^k \alpha_i \psi_i + w, \quad w \in M, \\ \lambda &= \lambda_0 + \beta, \\ d &= d_0 + e \end{aligned}$$

we know from the equivariant Liapunov-Schmidt method that stationary bifurcation scenario of (10.1) at $(0, \lambda_0, d_0)$ is described by the reduced bifurcation equations

$$\begin{aligned} &B(\alpha, \beta, e) \qquad (10.49) \\ &= \left(\left\langle \psi_i^*, G\Big(\sum_{i=1}^{k} \alpha\psi_i + w(\alpha, \beta, e), \ \lambda_0 + \beta, \ d_0 + e\Big)\right\rangle, \ i = 1, \ldots, k\right) \\ &= 0, \end{aligned}$$

where $\alpha := (\alpha_1, \ldots, \alpha_k) \in \mathbf{R}^k$, and $w(\alpha, \beta, e)$ is the unique solution of

$$\tilde{Q}G(\sum_{i=1}^{k} \alpha_i\psi_i + w, \ \lambda_0 + \beta, \ d_0 + e) = 0. \qquad (10.50)$$

Under the induced action of Γ in $\mathbf{R}^k$, the system (10.49) is Γ-equivariant. We use these symmetry properties to derive necessary structure of (10.49). The induced symmetries often reveal directly the bifurcation scenario. Moreover, some bifurcation is a direct consequence of symmetry of the governing equations. In other cases, as mode interactions, symmetry and nonlinearity interact and lead to more sophisticated bifurcation behavior.

10.8 Simple and Double Bifurcations

In this section we study solutions of the reduced bifurcation equations at simple and double bifurcation points. These are generic bifurcations for one-parameter problems with D_4-symmetries. We fix the diffusion rate as $d = d_0$ and study bifurcation of (10.1) with respect to the parameter λ.

10.8.1 Simple bifurcations

Let $k = 1$ in (10.46), i.e., $(0, \lambda_0, d_0)$ is a simple bifurcation point corresponding to a simple eigenvalue $c = 2m^2\pi^2$ of the Laplacian $-\Delta$. By (10.49) we obtain a scalar reduced bifurcation equation

$$B(\alpha, \beta) = \langle \psi_1^*, \ G(\alpha_1\psi_1 + w(\alpha_1, \beta), \ \lambda_0 + \beta, \ d_0) \rangle = 0. \qquad (10.51)$$

This equation is Γ-equivariant under the following induced actions of Γ in the kernel of $D_uG_0 = \text{span}[\psi_1]$

$$\Gamma = \begin{cases} I & \text{for } m = \text{odd}; \\ Z_2 & \text{for } m = \text{even}; \\ Z_2 & \text{if } G(u, \lambda, d) \text{ is odd in } u. \end{cases} \qquad (10.52)$$

Hence the mapping $B(\alpha_1, \beta)$ can be rewritten as

$$B(\alpha_1, \beta) := \begin{cases} \alpha_1 \tilde{B}(\alpha_1, \beta) & \text{for } \Gamma = I; \\ \alpha_1 \tilde{B}(\alpha_1^2, \beta) & \text{for } \Gamma = Z_2. \end{cases}$$

Taylor expansion of $B(\alpha_1, \beta)$ in the neighborhood of $(\alpha_1, \beta) = (0, 0)$ is of the form

$$B(\alpha_1, \beta) = \begin{cases} a_1\alpha_1^2 + a_2\alpha_1\beta + h.o.t. & \text{for } \Gamma = I; \\ a_2\alpha_1\beta + a_3\alpha_1^3 + a_4\alpha_1\beta^2 + h.o.t. & \text{for } \Gamma = Z_2. \end{cases} \tag{10.53}$$

Here we have used the fact $B(0, \beta) = 0$ for all $\beta \in \mathbf{R}$, implied by the statement $G(0, \lambda, d) = 0$ for all $\lambda,\ d \in \mathbf{R}$. Furthermore,

$$\begin{aligned} a_1 &= \frac{1}{2}\langle \psi_1^*,\ D_{uu}G_0\psi_1\psi_1 \rangle, \\ a_2 &= \langle \psi_1^*,\ D_{u\lambda}G_0\psi_1\psi_1 \rangle \\ a_3 &= \frac{1}{6}\langle \psi_1^*,\ D_{uuu}G_0\psi_1^3 + 3D_{uu}G_0\psi w_{20} \rangle, \\ a_4 &= \frac{1}{2}\langle \psi_1^*, D_{u\lambda\lambda}G_0\psi_1 + 2D_{uu}G_0\psi w_{11} \rangle \end{aligned}$$

with $w_{20},\ w_{11}$ as coefficients of the Taylor expansion of the solution $w(\alpha_1, \beta)$ of (10.50)

$$w(\alpha_1, \beta) = w_{20}\alpha_1^2 + w_{11}\alpha_1\beta + h.o.t.$$

Generically the coefficients $a_1,\ a_2,\ a_3,\ a_4$ do not vanish. By (10.51) the term $\alpha_1\beta^2$ is of higher order than α^3 and $\alpha_1\beta$ along its solution curves. Therefore, solution of (10.51) can be determined qualitatively by the truncated equations

$$\begin{aligned} a_1\alpha_1^2 + a_2\alpha_1\beta &= 0 \qquad \text{for } \Gamma = I, \\ a_2\alpha_1\beta + a_3\alpha_1^3 &= 0 \qquad \text{for } \Gamma = Z_2, \end{aligned} \tag{10.54}$$

respectively. Both equations show that there is exactly one nontrivial solution branch of (10.1) at the simple bifurcation point $(0, \lambda_0, d_0)$.

The bifurcation is transcritical if parity of the corresponding mode $\phi_1 = 2\sin m\pi x \sin m\pi y$ is even and the reaction terms $f_i(u, \lambda)$ are not odd in u. Otherwise, the bifurcation is pitchfork and (sub-) supercritical for $a_2 \cdot a_3 < 0\ (> 0)$.

10.8.2 Double bifurcations induced by the D_4-symmetries

Let $(0, \lambda_0, d_0)$ be a double bifurcation point of (10.1), corresponding to a double eigenvalue of the Laplacian $-\Delta$

$$c = (m^2 + n^2)\pi^2, \quad m \neq n.$$

Namely,

$$(\lambda_0, d_0) \in S(c) \quad \text{and} \quad \operatorname{Ker}(D_u G_0) = \operatorname{span}[\psi_1, \ \psi_2].$$

We fix $d = d_0$ and consider λ as the bifurcation parameter. The reduced bifurcation equations (10.49) are

$$B(\alpha_1, \alpha_2, \beta) = \Big(\langle \psi_i^*, \ G(\alpha_1 \psi_1 + \alpha_2 \psi_2 + w(\alpha_1, \alpha_2, \beta), \ \lambda_0 + \beta, \ d_0) \rangle, \ i = 1, 2 \Big). \tag{10.55}$$

Since the mapping

$$B(\alpha_1, \alpha_2, \beta) = \begin{pmatrix} b_1(\alpha_1, \alpha_2, \beta) \\ b_2(\alpha_1, \alpha_2, \beta) \end{pmatrix} \tag{10.56}$$

is Γ-equivariant under induced D_4-actions (10.15) in $\mathbf{R}^2$, we utilize this property to specify its structure.

1) $(m, n) =$**(even, even).** By (10.15) the induced actions of Γ are actions

$$S = \begin{pmatrix} -1 & 0 \\ 0 & -1 \end{pmatrix}, \qquad R = \begin{pmatrix} 0 & -1 \\ -1 & 0 \end{pmatrix}. \tag{10.57}$$

It follows from the Γ-equivariance of $B(\alpha_1, \alpha_2, \beta)$ that

$$b_2(\alpha_1, \alpha_2, \beta) = b_1(\alpha_2, \alpha_1, \beta), \qquad b_1(-\alpha_1, -\alpha_2, \beta) = -b_1(\alpha_1, \alpha_2, \beta).$$

Taking into account scales of the solutions and the low order terms in the Taylor expansion

$$b_1(\alpha_1, \alpha_2, \beta) = (a_1 \alpha_1 + a_2 \alpha_2)\beta + a_3 \alpha_1^3 + a_4 \alpha_1^2 \alpha_2 + a_5 \alpha_1 \alpha_2^2 + a_6 \alpha_2^3 + h.o.t.,$$

we determine bifurcation scenario of (10.1) at the double bifurcation point $(0, \lambda_0, d_0)$ by the truncated system

$$\begin{aligned} &(a_1 \alpha_1 + a_2 \alpha_2)\beta + a_3 \alpha_1^3 + a_4 \alpha_1^2 \alpha_2 + a_5 \alpha_1 \alpha_2^2 + a_6 \alpha_2^3 = 0, \\ &(a_2 \alpha_1 + a_1 \alpha_2)\beta + a_6 \alpha_2^3 + a_5 \alpha_1^2 \alpha_2 + a_4 \alpha_1 \alpha_2^2 + a_3 \alpha_2^3 = 0. \end{aligned} \tag{10.58}$$

Note that the function $w(\alpha_1, \alpha_2, \beta)$ in (10.55) is of the form

$$w(\alpha_1, \alpha_2, \beta) = w_{110} \alpha_1^2 + 2 w_{120} \alpha_1 \alpha_2 + w_{220} \alpha_2^2 + 2 w_{101} \alpha_1 \beta + 2 w_{011} \alpha_2 \beta + h.o.t.$$

We derive

$$\begin{aligned} &G(\alpha_1 \psi_1 + \alpha_2 \psi_2 + w(\alpha_1, \alpha_2, \beta), \lambda_0 + \beta, d_0) \\ &= \frac{1}{2} D_{uu} G_0 (\alpha_1 \psi_1 + \alpha_2 \psi_2)^2 + D_{u\lambda} G_0 (\alpha_1 \psi_1 + \alpha_2 \psi_2)\beta \\ &\quad + \Big(D_{uu} G_0 \psi_1 w_{110} + \frac{1}{6} D_{uuu} G_0 \psi_1^3 \Big) \alpha_1^3 \\ &\quad + \Big[D_{uu} G_0 (2\psi_1 w_{120} + \psi_2 w_{110}) + \frac{1}{2} D_{uuu} G_0 \psi_1^2 \psi_2 \Big] \alpha_1^2 \alpha_2 \\ &\quad + \Big[D_{uu} G_0 (\psi_1 w_{220} + 2\psi_2 w_{120}) + \frac{1}{2} D_{uuu} G_0 \psi_1 \psi_2^2 \Big] \alpha_1 \alpha_2^2 \\ &\quad + \Big(D_{uu} G_0 \psi_2 w_{220} + \frac{1}{6} D_{uuu} G_0 \psi_2^3 \Big) \alpha_2^3 + \cdots . \end{aligned} \tag{10.59}$$

Thereafter, we obtain the coefficients a_i, $i = 1, \dots, 6$ from the Taylor expansion of (10.55) at $(\alpha_1, \alpha_2, \beta) = (0, 0, 0)$

$$\begin{aligned}
a_i &= \langle \psi_1^*,\ D_{u\lambda} G_0 \psi_i \rangle, \qquad\qquad i = 1, 2,\\
a_3 &= \langle \psi_1^*,\ D_{uu} G_0 \psi_1 w_{110} + \frac{1}{6} D_{uuu} G_0 \psi_1^3 \rangle,\\
a_4 &= \langle \psi_1^*,\ D_{uu} G_0 (2\psi_1 w_{120} + \psi_2 w_{110}) + \frac{1}{2} D_{uuu} G_0 \psi_1^2 \psi_2 \rangle,\\
a_5 &= \langle \psi_1^*,\ D_{uu} G_0 (\psi_1 w_{220} + 2\psi_2 w_{120}) + \frac{1}{2} D_{uuu} G_0 \psi_1 \psi_2^2 \rangle,\\
a_6 &= \langle \psi_1^*,\ D_{uu} G_0 \psi_2 w_{220} + \frac{1}{6} D_{uuu} G_0 \psi_2^3 \rangle.
\end{aligned} \tag{10.60}$$

For the eigenfunctions

$$\phi_1 = 2 \sin m\pi x \sin n\pi y, \qquad \phi_2 = 2 \sin n\pi x \sin m\pi y$$

of the the Laplacian $-\Delta$, elementary calculations yield the properties

$$\begin{aligned}
&a)\ \ (\phi_i^4,\ 1) = 9/4,\ \ (\phi_i^2,\ \phi_j^2) = 1,\ \ (\phi_i,\ \phi_j^3) = 0, \ \ \text{for all } i \neq j,\ i, j = 1, 2;\\
&b)\ \ (\phi_i,\ \phi_j, \phi_k) = 0, \ \ \text{for all } i, j, k = 1, 2 \ \text{ and one of } m, n \text{ is even;}\\
&c)\ \ (\phi_i^3,\ 1) = \frac{128}{9mn\pi},\quad (\phi_i^2,\ \phi_j) = \frac{8(m^2 - 4n^2 + 4mn)(n^2 - 4m^2 + 4mn)}{\pi^2 mn (m^2 - 4n^2)(n^2 - 4m^2)}\\
&\quad \text{for all } i \neq j,\ i, j = 1, 2 \ \text{ if both } m, n \text{ are odd.}
\end{aligned} \tag{10.61}$$

Here $(\cdot,\ \cdot)$ is the inner product in $L^2(\Omega)$. Thereafter, the statements in (10.60) can be simplified as

$$\begin{aligned}
a_2 &= 0\\
a_4 &= \langle \psi_1^*,\ D_{uu} G_0 (2\psi_1 w_{120} + \psi_2 w_{110}) \rangle,\\
a_6 &= \langle \psi_1^*,\ D_{uu} G_0 \psi_2 w_{220} \rangle.
\end{aligned}$$

Furthermore, if the reaction terms $f_i(u, \lambda)$, $i = 1, 2$ are odd in u, then $D_{uu} G_0 = 0$ and the coefficients a_i, $i = 3, 4, 5, 6$ can be calculated directly via

$$\begin{aligned}
a_3 &= \frac{1}{6} \langle \psi_1^*,\ D_{uuu} G_0 \psi_1^3 \rangle,\\
a_4 &= 0,\\
a_5 &= \frac{1}{2} \langle \psi_1^*,\ D_{uuu} G_0 \psi_1 \psi_2^2 \rangle,\\
a_6 &= 0.
\end{aligned}$$

The truncated bifurcation equations (10.58) reduce to

$$\begin{aligned}
a_1 \alpha_1 \beta + a_3 \alpha_1^3 + a_5 \alpha_1 \alpha_2^2 &= 0,\\
a_1 \alpha_2 \beta + a_5 \alpha_1^2 \alpha_2 + a_3 \alpha_2^3 &= 0.
\end{aligned} \tag{10.62}$$

These equations are consistent with those in (6.43) derived by scaling techniques in Chapter 6. We solve the system (10.62) directly and obtain eight nontrivial isolated solutions

$$\begin{aligned}&\pm\left(\left(-\tfrac{a_1}{a_3}\beta\right)^{1/2}, 0\right), \quad \pm\left(0, \left(-\tfrac{a_1}{a_3}\beta\right)^{1/2}\right),\\ &\pm\left(\left(-\frac{a_1}{a_3+a_5}\beta\right)^{1/2}, \quad \pm\left(-\frac{a_1}{a_3+a_5}\beta\right)^{1/2}\right).\end{aligned} \tag{10.63}$$

Depending on the signs of a_1/a_3 and $a_1/(a_3+a_5)$, these solutions are real only for $\beta \geq 0$ or $\beta \leq 0$, respectively. Thus bifurcations of (10.1) at $(0, \lambda_0, d_0)$ are pitchfork.

For the general system (10.58), we consider the sum and the difference of its two equations and derive

$$[(a_1+a_2)\beta + (a_3+a_6)(\alpha_1^2 - \alpha_1\alpha_2 + \alpha_2^2) + (a_4+a_5)\alpha_1\alpha_2](\alpha_1+\alpha_2) = 0,$$

$$[(a_1-a_2)\beta + (a_3-a_6)(\alpha_1^2 + \alpha_1\alpha_2 + \alpha_2^2) + (a_4-a_5)\alpha_1\alpha_2](\alpha_1-\alpha_2) = 0.$$

These give us immediately four nontrivial solutions

$$\begin{aligned}&a) \quad \alpha_1 = \alpha_2 = \pm\left(-\frac{a_1+a_2}{a_3+a_4+a_5+a_6}\beta\right)^{1/2},\\ &b) \quad \alpha_1 = -\alpha_2 = \pm\left(-\frac{a_1-a_2}{a_3+a_5-a_4-a_6}\beta\right)^{1/2}.\end{aligned} \tag{10.64}$$

The other four nontrivial solutions are described by the system

$$\begin{aligned}(a_1-a_2)\beta + (a_3-a_6)(\alpha_1^2 + \alpha_1\alpha_2 + \alpha_2^2) + (a_4-a_5)\alpha_1\alpha_2 &= 0\\ (a_1+a_2)\beta + (a_3+a_6)(\alpha_1^2 - \alpha_1\alpha_2 + \alpha_2^2) + (a_4+a_5)\alpha_1\alpha_2 &= 0.\end{aligned} \tag{10.65}$$

Denote

$$\gamma_1 := \alpha_1^2 + \alpha_2^2, \qquad \gamma_2 := \alpha_1\alpha_2.$$

We obtain from (10.65) a linear system for γ_1, γ_2 and

$$\gamma_1 = -\frac{a_1(a_3-a_5) + a_2(a_4-a_6)}{a_3(a_3-a_5) + a_6(a_4-a_6)}\beta,$$

$$\gamma_2 = \frac{a_1a_3 - a_1a_6}{a_3(a_3-a_5) + a_6(a_4-a_6)}\beta.$$

Thus the four solutions of (10.65) are

$$\begin{aligned}&\pm\frac{1}{2}\left(\sqrt{\gamma_1+2\gamma_2} + \sqrt{\gamma_1-2\gamma_2}, \quad \sqrt{\gamma_1+2\gamma_2} - \sqrt{\gamma_1-2\gamma_2}\right),\\ &\pm\frac{1}{2}\left(\sqrt{\gamma_1+2\gamma_2} - \sqrt{\gamma_1-2\gamma_2}, \quad \sqrt{\gamma_1+2\gamma_2} + \sqrt{\gamma_1-2\gamma_2}\right).\end{aligned} \tag{10.66}$$

Note the factor $\sqrt{\beta}$ in these solutions. We see that they are all pitchfork branches.

2) $(m,\ n)$ =**(odd, odd).** In this case the D_4-actions in $\mathbf{R}^2$ reduce to

$$S = I, \qquad R = \begin{pmatrix} 0 & 1 \\ 1 & 0 \end{pmatrix}. \tag{10.67}$$

If the reaction terms $f_i(u, \lambda)$, $i = 1, 2$ are odd in u, combination of Z_2-symmetry and the D_4-action (10.67) gives again the symmetry (10.57). Thereafter, bifurcation scenario of (10.1) is the same as the case that both m and n are even.

For general reaction terms, we derive from the rotation symmetry in (10.67)

$$b_1(\alpha_1, \alpha_2, \beta) = b_2(\alpha_1, \alpha_2, \beta).$$

Consider lower order terms in the Taylor expansion of $b_1(\alpha_1, \alpha_2, \beta)$

$$b_1(\alpha_1, \alpha_2, \beta) = a_1\alpha_1\beta + a_2\alpha_2\beta + a_3\alpha_1^2 + a_4\alpha_1\alpha_2 + a_5\alpha_2^2 + h.o.t.$$

and the truncated equations

$$\begin{aligned} a_1\alpha_1\beta + a_2\alpha_2\beta + a_3\alpha_1^2 + a_4\alpha_1\alpha_2 + a_5\alpha_2^2 = 0, \\ a_2\alpha_1\beta + a_1\alpha_2\beta + a_5\alpha_1^2 + a_4\alpha_1\alpha_2 + a_3\alpha_2^2 = 0. \end{aligned} \tag{10.68}$$

The coefficients a_i, $i = 1, \ldots, 5$ are calculated via

$$\begin{aligned} a_i &= \langle \psi_1^*,\ D_{u\lambda}G_0\psi_i \rangle, \quad i = 1, 2 \\ a_3 &= \langle \psi_1^*,\ \frac{1}{2}D_{uu}G_0\psi_1\psi_1 \rangle, \\ a_4 &= \langle \psi_1^*,\ D_{uu}G_0\psi_1\psi_2 \rangle, \\ a_5 &= \langle \psi_1^*,\ \frac{1}{2}D_{uu}G_0\psi_2\psi_2 \rangle. \end{aligned}$$

With the ansatz $\alpha_1 = \alpha_2$ we derive a nontrivial solution of (10.68) immediately

$$\alpha_1 = \alpha_2 = -\frac{a_1 + a_2}{a_3 + a_4 + a_5}\beta. \tag{10.69}$$

The other two nontrivial solutions are

$$\begin{aligned} \alpha_1 &= -\alpha_2 - \frac{a_1 - a_2}{a_3 - a_5}\beta, \\ \alpha_2 &= \frac{1}{2(a_3 - a_5)(a_3 - a_4 + a_5)}\Big[\beta(a_1 - a_2)(a_4 - a_3 - a_5) \\ &\pm \Big(\beta^2(a_1 - a_2)^2(a_3 + a_5 - a_4)^2 \\ &+ (a_3 - a_4 + a_5)(a_1 - a_2)\beta[a_1(a_3 - a_5) - a_3(a_1 - a_2)\beta]\Big)^{1/2}\Big] \end{aligned} \tag{10.70}$$

On the other hand, by the definition (10.47) of the null elements ψ_i, ψ_i^*, $i = 1, 2$ and by the definition (10.7) of the $L^2(\Omega)$-product in Y and the properties (10.61) of eigenfunctions of the Laplacian, we have

$$a_2 = 0, \qquad a_4 = 2a_5. \tag{10.71}$$

Hence, the four nontrivial solutions in (10.69) and (10.70) of the truncated bifurcation equations (10.68) can be written as

$$\begin{aligned} &-\frac{a_1\beta}{a_3+3a_5}(1,\ 1),\\ &\frac{a_1\beta}{2(a_3-a_5)}\left(-1-\sqrt{\frac{a_3-5a_5}{a_3-a_5}},\ -1+\sqrt{\frac{a_3-5a_5}{a_3-a_5}}\right),\\ &\frac{a_1\beta}{2(a_3-a_5)}\left(-1+\sqrt{\frac{a_3-5a_5}{a_3-a_5}},\ -1-\sqrt{\frac{a_3-5a_5}{a_3-a_5}}\right). \end{aligned} \tag{10.72}$$

These solutions are all transcritical branches.

3) $(m,n) =$ **(even, odd) or (odd, even).** For these parities of the modes ψ_1, ψ_2, the induced representation of D_4 in $\mathrm{Ker}(D_uG_0) \sim \mathbf{R}^2$ is reducible and generated by

$$S_1 = \begin{pmatrix} -1 & 0 \\ 0 & 1 \end{pmatrix}, \qquad R = \begin{pmatrix} 0 & -1 \\ 1 & 0 \end{pmatrix}.$$

Furthermore,

$$S_1R = \begin{pmatrix} 0 & 1 \\ 1 & 0 \end{pmatrix}, \quad RS_1 = \begin{pmatrix} 0 & -1 \\ -1 & 0 \end{pmatrix}$$

and

$$S_1R^2S_1 = -I.$$

By S_1R-symmetry we have

$$b_2(\alpha_1,\alpha_2,\beta) = b_1(\alpha_2,\alpha_1,\beta), \tag{10.73}$$

while the S_1- and R-symmetries imply

$$\begin{aligned} b_1(-\alpha_1,\alpha_2,\beta) &= -b_1(\alpha_1,\alpha_2,\beta),\\ b_1(-\alpha_2,\alpha_1,\beta) &= -b_2(\alpha_1,\alpha_2,\beta). \end{aligned}$$

Hence we have the Taylor expansion

$$b_1(\alpha_1,\alpha_2,\beta) = \alpha_1\tilde{b}_1(\alpha_1^2,\alpha_2^2,\beta) = a_1\alpha_1\beta + \alpha_1(a_2\alpha_1^2 + a_3\alpha_2^2) + h.o.t.$$

Consider the truncated bifurcation equations

$$\begin{aligned} a_1\alpha_1\beta + \alpha_1(a_2\alpha_1^2 + a_3\alpha_1^2) &= 0,\\ a_1\alpha_2\beta + \alpha_2(a_3\alpha_1^2 + a_2\alpha_1^2) &= 0. \end{aligned} \tag{10.74}$$

We obtain eight nontrivial solutions

$$\begin{aligned}&\left(\pm\left(-\tfrac{a_1}{a_2}\beta\right)^{1/2}, 0\right), \quad \left(0, \pm\left(-\tfrac{a_1}{a_2}\beta\right)^{1/2}\right),\\&\left(\pm\left(-\frac{a_1}{a_2+a_3}\beta\right)^{1/2}, \pm\left(-\frac{a_1}{a_2+a_3}\beta\right)^{1/2}\right).\end{aligned} \tag{10.75}$$

These solutions are real only at one side of $\lambda = \lambda_0$, i.e., either for $\beta \geq 0$ or for $\beta \leq 0$. Correspondingly the original problem (10.1) has pitchfork bifurcations at $(0, \lambda_0, d_0)$.

Remark 10.8.1. It is known that secondary bifurcations in the neighborhood of $(0, \lambda_0, d_0)$ can be observed by examining the Jacobian

$$D_{(\alpha_1,\alpha_2)}B(\alpha_1,\alpha_2,\beta) = \begin{pmatrix} \frac{\partial b_1}{\partial \alpha_1}(\alpha_1,\alpha_2,\beta) & \frac{\partial b_1}{\partial \alpha_2}(\alpha_1,\alpha_2,\beta) \\ \frac{\partial b_2}{\partial \alpha_1}(\alpha_1,\alpha_2,\beta) & \frac{\partial b_2}{\partial \alpha_2}(\alpha_1,\alpha_2,\beta) \end{pmatrix}$$

of the system (10.58), (10.68) with (10.71) and (10.74) along their solution branches (10.64), (10.66), (10.69), (10.70) or (10.72), and (10.75) respectively. By elementary calculations we find that there is no secondary bifurcation on these solution branches near $(0, \lambda_0, d_0)$.

11. Normal Forms for Hopf Bifurcations

In this chapter we study Hopf bifurcations of the reaction-diffusion equations

$$\frac{\partial u}{\partial t} = G(u, \lambda, d) \tag{11.1a}$$

with

$$G(u, \lambda, d) := \begin{pmatrix} \Delta u_1 + f_1(u_1, u_2, \lambda) \\ d\Delta u_2 + f_2(u_1, u_2, \lambda) \end{pmatrix} \tag{11.1b}$$

on the square $\Omega := [0, \pi] \times [0, \pi]$ and appropriate boundary conditions. Assume the functions $f_i : \mathbf{R}^{2+p} \mapsto \mathbf{R}$, $i = 1, 2$ to be sufficiently smooth and

$$f_i(0, 0, \lambda) = 0 \quad \text{for all } \lambda \in \mathbf{R},\ i = 1, 2. \tag{11.2}$$

We are interested in branches of timely periodic and spatially inhomogeneous solutions near Hopf bifurcations on the trivial solution curve $u \equiv 0$.

11.1 Introduction

As we have seen in Chapter 10, the problem (11.1) is D_4-equivariant. By the theory of Golubitsky *et al* [129, 131] if a system has a group of symmetries Γ, then the dimension of the center eigenspace at a Hopf bifurcation must be either twice the dimension of a real irreducible representation of Γ or it is the dimension of a complex irreducible representation of Γ. As a consequence for systems with D_4-symmetry the admissible dimensions of a center eigenspace are two and four. The former corresponds to the classical Hopf bifurcation without symmetry and yields generically a unique periodic solution branch. The later has a larger dimensional space. There can be many different types of periodic solutions; even quasiperiodic solutions and chaos bifurcating from the trivial solutions at Hopf bifurcation, see e.g. Swift [288] for further details.

However, as has been found in many different contexts, notably in steady-state bifurcations, often partial differential equations have center eigenspaces which are *too big* from the "generic-with-symmetry" point of view. For example, Crawford [73] finds that firstly, center eigenspaces at bifurcation with dimensions that are simply too big and secondly, they can have an admissible dimension, but which do not carry an irreducible representation of the symmetry group as we have seen in Chapter 10. To solve this problem we may look at genericity within a suitable class of problems and then get reducible representations on the center eigenspace (cf. Crawford/Gollub/Lane [76]). This was noticed by Shaw [278] in the context of solutions of Schroedinger's equation for a square-well potential. The center eigenspace may even be of arbitrarily large dimension due to the effects of rotation symmetry in the extension of the problem to $\mathbf{R}^2$, see e.g. [20, 73, 196].

The system (11.1) with the homogeneous Dirichlet (Neumann) boundary is an example for such problems which are periodically extendible, or even extendible problems on the plane with Euclidean symmetry. Crawford *et al* [73, 76] and Gomes *et al* [77, 132] investigate the normal forms of problems where the representation of the group on the center eigenspace takes into account the presence of hidden symmetries of the problem. They find that at low order, the bifurcation equations have extra symmetry which is broken to the generic symmetry only at higher orders, depending on wave numbers at the bifurcation.

In this chapter we consider Robin (or Cauchy) boundary conditions

$$au_i(x) + b\frac{\partial u_i}{\partial n}(x) = 0 \quad \text{for all} \quad x \in \partial\Omega, \quad i = 1, 2. \tag{11.3}$$

Here $\frac{\partial}{\partial n}$ is the outward normal derivative on the boundary. The constants $a \geq 0$ and $b \in \mathbf{R}$ satisfy $a + |b| > 0$. The cases $a = 0$ and $b = 0$ correspond to Neumann and Dirichlet boundary conditions respectively. It is possible to generalize the boundary conditions such that a and b do depend on i, but this requires restriction of the bifurcating modes and introduces more complexity. These boundary conditions exclude the hidden translation symmetry. Nevertheless, there are reducible center eigenspaces for the D_4-symmetry and therefore not generic in the usual sense of equivariant bifurcation theory. The reason for this lies in the special structure of the linear part of the problem, which has similar properties as the Laplacian. At bifurcation we take explicit account of this structure and derive a normal form on an eigenspace carrying a reducible representation of the symmetry group.

By generalizing the normal form of Swift [288] for Hopf bifurcation with the generic action of $D_4 \times \mathbf{S}^1$ on $\mathbf{C}^2$ which is in turn working with a normal form derived by Golubitsky and Stewart [130], we derive a Birkhoff normal form and analyze its generic (codimension one) bifurcation behavior as a function of the third order coefficients.

We discuss conditions on coefficients of the normal form to ensure generic branching behavior at Hopf bifurcations. Furthermore, we give an algorithm to find the number of branches and their symmetries. As an example, we use the normal form analysis to interpret the bifurcation diagram of the Brusselator equations with Robin boundary conditions, reduced by the numerical/computer algebraic Liapunov-Schmidt method in Chapter 6.

11.2 Domain Symmetries and Their Extensions

The reaction-diffusion problem (11.1) (or other domains as well, e.g. hexagon and rhombus) can be extended from Ω to the whole plane and becomes equivariant under the Euclidean group on the plane, i.e., the group $\mathcal{E}(2)$ of transformations on $\mathbf{R}^2$ generated by

$$R_\psi \begin{pmatrix} x_1 \\ x_2 \end{pmatrix} := \begin{pmatrix} \cos\psi & -\sin\psi \\ \sin\psi & \cos\psi \end{pmatrix} \begin{pmatrix} x_1 \\ x_2 \end{pmatrix},$$
$$S_2 \begin{pmatrix} x_1 \\ x_2 \end{pmatrix} := \begin{pmatrix} x_2 \\ x_1 \end{pmatrix},$$
$$\tau_{(y_1,y_2)} \begin{pmatrix} x_1 \\ x_2 \end{pmatrix} := \begin{pmatrix} x_1 + y_1 \\ x_2 + y_2 \end{pmatrix}.$$

Here R_ψ is a rotation about the origin, S_2 is reflection in the line $x_1 = x_2$ and $\tau_{(y_1,y_2)}$ is a translation. As noted by Crawford [74, 75] we can write any element in $\gamma \in \mathcal{E}(2)$ as

$$\gamma = \tau_a S_2^k R_\psi \quad \text{for some} \;\; a \in \mathbf{R}^2, \; \psi \in [0, 2\pi) \;\; \text{and} \;\; k = 1 \text{ or } 2.$$

These actions of $\mathcal{E}(2)$ on $\mathbf{R}^2$ induce a natural action on function $u \in L^2(\mathbf{R}^2)$

$$\gamma \cdot u(x) := u(\gamma^{-1}x),$$

and on functions $u = (u_1, u_2) \in \left(L^2(\mathbf{R}^2)\right)^2$ as

$$\gamma \cdot u(x) := (\gamma u_1, \; \gamma u_2).$$

Actions of the D_4-group on the domain $\Omega = [0, \pi]^2$ are generated by

$$\begin{aligned} R: \; (x_1, x_2) &\to (x_2, \pi - x_1), \\ S_2: \; (x_1, x_2) &\to (x_2, x_1). \end{aligned}$$

These group elements can be trivially extended to act on the plane $\mathbf{R}^2$ and treated as a subgroup of $\mathcal{E}(2)$.

Crawford [74, 75] shows that the center eigenspace for linear problems with the $\mathcal{E}(2)$-symmetry is spanned by exponentials of the form

$$u(x) = Ue^{k \cdot x} + c.c.$$

for $|k| = \kappa$, the critical wave number at bifurcation and U a constant vector. Here *c.c.* represents the complex conjugate of the first term. This follows from the fact that the function space is a direct sum of the subspaces spanned by these functions and each subspace is invariant for $\mathcal{E}(2)$-equivariant operators, see Chapters 9 and 10 for similar discussion on the homogeneous Dirichlet boundary conditions. Generically there is a unique critical wave number for one-parameter problems. Imposing Robin boundary conditions (11.3) leads to the center eigenspace consisting of eigenfunctions

$$\sum_k a_k U_k \cos(l_{1,k} x_1 + \theta_{1,k}) \cos(l_{2,k} x_2 + \theta_{2,k}) \tag{11.4}$$

with a constant vector $U_k \in \mathbf{R}^2$ and $\|U_k\| = 1$, $a_k \in \mathbf{R}$ and $l_{1,k}^2 + l_{2,k}^2 = \kappa^2$. The constants $l_{i,k}$ link the wave numbers of the Neumann and Dirichlet problems whereas $\theta_{i,k}$ are phase shifts to ensure the Robin boundary conditions (11.3). More precisely, $l_{i,k}$ satisfy

$$2a\, b\, l_{i,k} \cos(l_{i,k}\pi) + (a^2 - b^2 l_{i,k}^2) \sin(l_{i,k}\pi) = 0$$

and $\theta_{i,k}$ are solution of

$$\cos(\theta_{i,k}) = \frac{b l_{i,k}}{a^2 + b^2 l_{i,k}^2}, \qquad \sin(\theta_{i,k}) = -\frac{a}{a^2 + b^2 l_{i,k}^2}, \quad i = 1, 2$$

in $[0, 2\pi)$, where a and b are given constants in the boundary conditions (11.3). These sets of constants are calculated numerically, see e.g. Chapter 14 and Ashwin/Mei [25] for details.

Definition 11.2.1. *A continuous function* $u : \Omega = [0, \pi]^2 \to \mathbf{R}^2$ *is* separable *if it can be written in the form*

$$u(x_1, x_2) = U f(x_1) g(x_2) \quad \textit{for all} \quad x := (x_1, x_2) \in \Omega$$

for a constant vector $U \in \mathbf{R}^2$ *and two continuous function* f, $g : [0, \pi] \to \mathbf{R}$ *with* $f(\pi - x) = \pm f(x)$ *and* $g(\pi - x) = \pm g(x)$.

We note that the square domain does not force symmetries of the form $f(\pi - x) = -f(x)$ for the full nonlinear equations. However, this is the case for solutions of the linearized equations at a trivial solution.

Definition 11.2.2. *Let* $f : [0, \pi] \to \mathbf{R}$ *be a continuous function and* $p \in \{0, 1\}$ *such that*

$$f(\pi - x_1) = (-1)^p f(x_1).$$

We say that f *has parity* p*. For a separable function* $u(x) = f(x_1)g(x_2)$ *we define its parity to be the sum of the parities of* f *and* g*. Note that the parity is defined modulo two and so we refer to parity as being* even *or* odd.

An important property is that parity of eigenfunctions is well defined for Robin boundary conditions and stays constant on varying a and b.

11.3 Actions of D_4 on the Center Eigenspace

For corank-2 steady state bifurcations, the center eigenspace can be identified with $(a_1, a_2) \in \mathbf{R}^2$, and the action of D_4 on the basis (11.4) is given by

$$\begin{aligned}
&R(\cos(l_{1,k}x_1 + \theta_{1,k})\cos(l_{2,k}x_2 + \theta_{2,k})) \\
&= (-1)^{p_1}\cos(l_{2,k}x_1 + \theta_{2,k})\cos(l_{1,k}x_2 + \theta_{1,k}), \\
&R(\cos(l_{2,k}x_1 + \theta_{1,k})\cos(l_{1,k}x_2 + \theta_{2,k})) \\
&= (-1)^{p_2}\cos(l_{1,k}x_1 + \theta_{1,k})\cos(l_{2,k}x_2 + \theta_{2,k}), \\
&S_2 f(x_1, x_2) = f(x_2, x_1) \quad \text{for all} \quad f \in C(\Omega),
\end{aligned}$$

where $p_i \in \{0, 1\}, i = 1, 2$. If $p_1 + p_2$ is even then the bifurcating mode has even parity, otherwise it has odd parity. Moreover, these modes have a separable form.

At a Hopf bifurcation point $(0, \lambda_0, d_0)$ the linearization $D_u G_0$ has an even dimensional center eigenspace corresponding to eigenvalues $\pm i\omega$, and generically these eigenspaces are spanned by functions of the form

$$\{U_k \cos(l_{1,k}x_1 + \theta_{1,k})\cos(l_{2,k}x_2 + \theta_{2,k})e^{i\omega t} + c.c. \quad | \quad l_{1,k}^2 + l_{2,k}^2 = \kappa^2\}$$

with $U_k \in \mathbf{C}^2$, $k = 1, 2, \ldots$ as appropriate constant vectors. We assume that for $\lambda = \lambda_0 + \beta$ with $|\beta| \ll 1$, the operator $D_u G(0, \lambda, d_0)$ has a complex pair of eigenvalues goes through the imaginary axis with a non-zero rate. After appropriately scaling of the bifurcation parameter these eigenvalues can be written as

$$\beta \pm i\omega.$$

By Ashwin/Mei [25] we can generically exclude "number-theoretic" degeneracies and assume there is a Hopf bifurcation at $\lambda = 0$ with critical wave number κ and center eigenspace is of dimension 2 or 4. For this case we denote $p = l_{1,1}$, $q = l_{2,1}$, $\theta_1 = \theta_{1,1}$ and $\theta_2 = \theta_{2,1}$. We note that the case $p = q$ gives a normal form on $\mathbf{C}$ which is that for standard Hopf bifurcation without symmetry. Thus we concentrate on the case $p \neq q$. We use the notations p, q instead of m, n in the earlier chapters because they are not integers for Robin boundary conditions.

Define the spatial dependent eigenvectors

$$\begin{aligned}
\phi_1(x) &:= \cos(px_1 + \theta_1)\cos(qx_2 + \theta_2), \\
\phi_2(x) &:= \cos(qx_1 + \theta_2)\cos(px_2 + \theta_1).
\end{aligned}$$

The exponential of the adjoint $D_u G_0^*$ induces an action of $\mathbf{S}^1$ on the center eigenspace given by $e^{sD_u G_0^*}$ for $s \in [0, 2\pi\omega)$, see Theorem 7.4.1 and Elphick *et al* [99]. Thus there is a natural complex structure on the eigenspace by $\mathcal{I} = e^{\omega D_u G_0^*/2}$. For the generic case, choosing $U \in \mathbf{R}^2$ such that

$$U\phi_k(x)e^{i\omega t} + c.c., \quad k = 1, 2$$

is in the center eigenspace, we define $V = \mathcal{I}U$ and coordinates $(z_+,\ z_-) \in \mathbf{C}^2$ for the center eigenspace by

$$\begin{aligned} u(x,t) := \ & z_+(\phi_1(x)U + i\phi_2(x)V)e^{i\omega t} \\ & + z_-(\phi_2(x)U + i\phi_1(x)V)e^{i\omega t} + c.c. \end{aligned} \tag{11.5}$$

The induced action of D_4 on $\mathbf{C}^2$ is given by

$$\begin{aligned} R:\ & (z_+,\ z_-) \to (\pm iz_+,\ \mp iz_-), \\ S_2:\ & (z_+,\ z_-) \to (z_-,\ z_+), \end{aligned}$$

for modes with odd parity and

$$\begin{aligned} R:\ & (z_+,\ z_-) \to \pm(z_-,\ z_+), \\ S_2:\ & (z_+,\ z_-) \to (z_-,\ z_+). \end{aligned}$$

for modes with even parity. We discuss on the case where $R(z_+, z_-) = (iz_+, -iz_-)$ for odd parity and $R(z_+, z_-) = (z_-, z_+)$ for even parity. The case $R(z_+, z_-) = -(z_-, z_+)$ leads to the same normal form, just with a slightly different interpretation of the solution branches.

The natural $\mathbf{S}^1$ action on the center eigenspace becomes a symmetry of the Birkhoff normal form to all orders, although there will typically be error terms not in Birkhoff normal form which break this symmetry. As in Golubitsky *et al* [131] we shall take the approach of examining the bifurcation by using the Birkhoff normal form of the vector field on the center manifold and so including this extra $\mathbf{S}^1$ symmetry. The bifurcation behavior in the full system is then the normal form bifurcation behavior perturbed by a generic $\mathbf{S}^1$ symmetry breaking perturbation. Such perturbations will cause hyperbolic structures (for example, hyperbolic periodic orbits) in the dynamics to persist. As before, the action of $D_4 \times \mathbf{S}^1$ is reducible for even parity and irreducible for odd parity.

This $\mathbf{S}^1$ symmetry acts via temporal phase shift as indicated before. The induced action on $\mathbf{C}^2$ is

$$T_\psi:\ (z_+,\ z_-) \to (e^{i\psi}z_+,\ e^{i\psi}z_-)$$

for Hopf bifurcations with both odd and even parity mode. The action of $D_4 \times \mathbf{S}^1$ on $\mathbf{C}^2$ is the same as Swift's [288] for the case of odd parity and gives an action of $\mathbf{Z}_2 \times \mathbf{S}^1$ for even parity. Upon performing a center manifold reduction we derive that the dynamics on the center manifold is determined by an ordinary differential equation on $\mathbf{C}^2$:

$$\begin{aligned} \dot{z}_+ &= (\beta + i\omega)z_+ + f_+(z_+, z_-), \\ \dot{z}_- &= (\beta + i\omega)z_- + f_-(z_+, z_-), \end{aligned}$$

where $f_\pm(z_+, z_-)$ are k times differentiable functions with zero linear part, and equivariant under the appropriate action of $D_4 \times \mathbf{S}^1$ (cf. Vanderbauwhede/Iooss [298]).

11.4 The Normal Form

The ring of invariants for both actions of $D_4 \times \mathbf{S}^1$ on $\mathbf{C}^2$ are generated by

$$|z_+|^2 + |z_-|^2, \qquad |z_+|^2|z_-|^2$$

for odd parity, and the extra generators

$$z_+\overline{z}_- + z_-\overline{z}_+, \qquad |z_+|^2 z_+\overline{z}_- + |z_-|^2 z_-\overline{z}_+$$

for even parity. The equivariants are the module over this ring. For odd parity the equivariants are generated by

$$\begin{pmatrix} z_+ \\ z_- \end{pmatrix}, \quad \begin{pmatrix} |z_+|^2 z_+ \\ |z_-|^2 z_- \end{pmatrix}, \quad \begin{pmatrix} \overline{z}_+ z_-^2 \\ \overline{z}_- z_+^2 \end{pmatrix},$$

while for even parity there are three additional generators

$$\begin{pmatrix} z_- \\ z_+ \end{pmatrix}, \quad \begin{pmatrix} |z_-|^2 z_- \\ |z_+|^2 z_+ \end{pmatrix}, \quad \begin{pmatrix} \overline{z}_- z_+^2 \\ \overline{z}_+ z_-^2 \end{pmatrix}.$$

This is proven by defining

$$z^t = \begin{cases} z^t & \text{for } t \geq 0; \\ \overline{z}^{-t} & \text{for } t < 0 \end{cases}$$

and considering the action of the groups on sums of homogeneous monomials of the form $z_+^l z_-^m |z_+|^{2p}|z_-|^{2q}$. For the invariants, the $\mathbf{S}^1$-action means that p, q are arbitrary and $l + m = 0$. A minimal set of generators is then given as above. For the equivariants, the S_2-action swaps the components, so considering monomials in the first component gives again p, q arbitrary and $l + m = 1$. It is a routine calculation to show the equivariants are as above, see Chapters 12 and 13 for more details.

Thus, C^k-continuous ($k > 4$) vector fields commuting with the action of $D_4 \times \mathbf{S}^1$ can be written in the form

$$\begin{pmatrix} \dot{z}_+ \\ \dot{z}_- \end{pmatrix} = A_1 \begin{pmatrix} z_+ \\ z_- \end{pmatrix} + A_2 \begin{pmatrix} |z_+|^2 z_+ \\ |z_-|^2 z_- \end{pmatrix} + A_3 \begin{pmatrix} \overline{z}_+ z_-^2 \\ \overline{z}_- z_+^2 \end{pmatrix}$$
$$+ A_4 \begin{pmatrix} z_- \\ z_+ \end{pmatrix} + A_5 \begin{pmatrix} |z_-|^2 z_- \\ |z_+|^2 z_+ \end{pmatrix} + A_6 \begin{pmatrix} \overline{z}_- z_+^2 \\ \overline{z}_+ z_-^2 \end{pmatrix}.$$

with $A_1 \in C^{k-1}$ and $A_i \in C^{k-3}$ ($i = 2, \ldots, 6$) as complex valued functions of the invariants ($A_4 = A_5 = A_6 = 0$ for odd parity), in particular, $A_4(0) = 0$. Truncating at cubic order gives

$$\begin{aligned} \dot{z}_+ &= (\beta + i\omega)z_+ + [A(|z_+|^2 + |z_-|^2) + B|z_+|^2]z_+ + C\overline{z}_+ z_-^2 \\ &\quad + [X_1(|z_+|^2 + |z_-|^2) + X_2|z_-|^2]z_- + X_3\overline{z}_- z_+^2, \\ \dot{z}_- &= (\beta + i\omega)z_- + [A(|z_+|^2 + |z_-|^2) + B|z_-|^2]z_- + C\overline{z}_- z_+^2 \\ &\quad + [X_1(|z_+|^2 + |z_-|^2) + X_2|z_+|^2]z_+ + X_3\overline{z}_+ z_-^2 \end{aligned} \tag{11.6}$$

with six complex coefficients A, B, C, X_1, X_2 and X_3. For odd parity, $X_1 = X_2 = X_3 = 0$ and the equations reduce to Swift's [288, equation (25)].

11.5 Analysis of the Normal Form

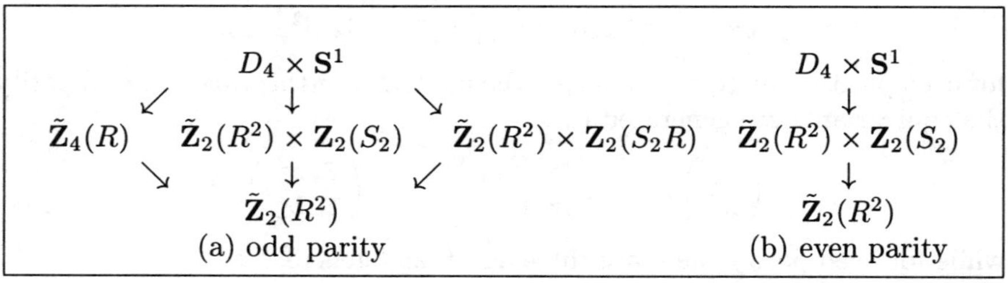

Table 11.1. Isotropy lattices for the actions of $D_4 \times \mathbf{S}^1$ on $\mathbf{C}^2$.

Table 11.1 shows the isotropy lattices for the two different actions of $D_4 \times \mathbf{S}^1$ on $\mathbf{C}^2$; as usual, the isotropy subgroups are classified up to conjugacy. Note that there are three maximal isotropy subgroups for the odd parity (irreducible), but only one maximal isotropy subgroup for the even parity. In both cases, the kernel of the action is $\tilde{\mathbf{Z}}_2(R^2)$ corresponding to equivariance under temporal phase shift of half a period and spatial rotation by half a period. This equivariance will be lost in the original equations by the introduction of terms that are not in the center eigenspace at bifurcation.

Isotropy	Fix	$\dim_{\mathbf{C}}$ Fix
$D_4 \times \mathbf{S}^1$	$(0,0)$	0
$\tilde{\mathbf{Z}}_2(R^2) \times \mathbf{Z}_2(S_2)$	(z,z)	1
$\tilde{\mathbf{Z}}_2(R^2)$	(z,w)	2

(a) even parity

Isotropy	Fix	$\dim_{\mathbf{C}}$ Fix	Name of solutions
$D_4 \times \mathbf{S}^1$	$(0,0)$	0	homogeneous equilibrium
$\tilde{\mathbf{Z}}_4(R)$	$(z,0)$	1	rotating wave
$\tilde{\mathbf{Z}}_2(R^2) \times \mathbf{Z}_2(S_2)$	(z,z)	1	edge oscillation
$\tilde{\mathbf{Z}}_2(R^2) \times \mathbf{Z}_2(S_2R)$	(z,iz)	1	vertex oscillation
$\tilde{\mathbf{Z}}_2(R^2)$	(z,w)	2	submaximal solution

(b) odd parity,

Table 11.2. Representative fixed point spaces for the actions of $D_4 \times \mathbf{S}^1$ on $\mathbf{C}^2$.

Table 11.2 gives the size and some representative elements of the fixed point spaces of the isotropy subgroups displayed in Table 11.1, as well as the names assigned by Swift [288]. By the Equivariant Hopf Lemma [130], there are generically branches at bifurcation with isotropy Σ if $\dim_{\mathbf{R}} \mathrm{Fix}(\Sigma) = 2$ ($\dim_{\mathbf{C}} \mathrm{Fix}(\Sigma) = 2$). This implies that Σ is maximal for the actions discussed. For odd parity, we obtain the three maximal isotropy types of Golubitsky and Stewart while for even parity there is only one. This is a noticeable difference in the branching behavior of bifurcation; we only obtain branches with $\tilde{\mathbf{Z}}_2(R^2) \times \mathbf{Z}_2(S_2)$ symmetry for even parity. However, for even parity there may exist up to four distinct group orbits of periodic solutions with *submaximal* symmetry, as opposed to only one group orbit of submaximal solutions for odd parity.

11.5.1 Odd parity

We summarize here briefly the results in Swift [288]. Since for odd parity the coefficients X_1, X_2 and X_3 vanish, we obtain from (11.6)

$$\begin{aligned} \dot{z}_+ &= (\beta + i\omega)z_+ + [A(|z_+|^2 + |z_-|^2) + B|z_+|^2]z_+ + C\bar{z}_+ z_-^2, \\ \dot{z}_- &= (\beta + i\omega)z_- + [A(|z_+|^2 + |z_-|^2) + B|z_-|^2]z_- + C\bar{z}_- z_+^2. \end{aligned} \tag{11.7}$$

The nondegeneracy conditions of the normal form (11.7) are

$$\begin{aligned} &R_a := Re(A + B/2 + C/2) \neq 0, \quad &&R_b := Re(A + B/2 - C/2) \neq 0, \\ &R_c := Re(A + B) \neq 0, &&|B|^2 \neq |C|^2 \neq |\mathrm{Re}(B\bar{C})|, \end{aligned} \tag{11.8}$$

and the direction of branching of the three maximal isotropy types are given by the signs of the real quantities R_a, R_b and R_c. For

$$|B|^2 > |C|^2 > |\mathrm{Re}(B\bar{C})|, \tag{11.9}$$

Swift [288] shows the existence of branches of solutions with submaximal isotropy. Table 11.3 summarizes characteristics of bifurcating solution branches.

Nondegeneracy	Fix	Bif. direction	Name of solutions
$R_a \neq 0$	(z, z)	Sign(R_a)	Edge waves
$R_b \neq 0$	(z, iz)	Sign(R_b)	Vertex waves
$R_c \neq 0$	$(z, 0)$	Sign(R_c)	Rotating waves
$\|B\|^2 > \|C\|^2 > \|\mathrm{Re}(B\bar{C})\|$	(z_1, z_2)		Submaximal oscillations

Table 11.3. Solution branches at a Hopf bifurcation with irreducible $\mathbf{D}_4 \times \mathbf{S}^1$-symmetry.

11.5.2 Even parity

Using the following coordinate transformation from (z_+, z_-) to (u, v, w, ψ)

$$\begin{aligned} u &= |z_+|^2 - |z_-|^2, \\ v &= 2\text{Im}(z_+\overline{z}_-) = \frac{z_+\overline{z}_- - z_-\overline{z}_+}{i}, \\ w &= 2\text{Re}(z_+\overline{z}_-) = z_+\overline{z}_- + z_-\overline{z}_+, \\ e^{i\psi} &= \frac{z_+\overline{z}_-}{|z_+\overline{z}_-|}, \end{aligned}$$

we write (11.6) as

$$\begin{aligned} \dot{z}_+ &= (\beta + i\omega)z_+ + \alpha_u u z_+ + i\alpha_v v z_- + \alpha_w w z_- + \beta_u u z_- + i\beta_v v z_+ + \beta_w w z_+, \\ \dot{z}_- &= (\beta + i\omega)z_- - \alpha_u u z_- - i\alpha_v v z_+ + \alpha_w w z_+ - \beta_u u z_+ - i\beta_v v z_- + \beta_w w z_-. \end{aligned} \tag{11.10}$$

Note that the coordinate (u, v, w, ψ) is different from Swift [288] in which u and w are interchanged. The reason for this is to ensure that the coordinate singularity is at a point of maximal isotropy. The parameters α_u, α_v, α_w, β_u, β_v, β_w are complex numbers; $A_r, A_i \ldots, X_{3r}, X_{3i}$ are real and imaginary parts of A, B, C and X_1, X_2, X_3, respectively. These parameters are related by

$$\begin{aligned} A = A_r + iA_i &= -\alpha_u + \alpha_v + \alpha_w, \\ B = B_r + iB_i &= 2\alpha_u - \alpha_v - \alpha_w, \\ C = C_r + iC_i &= -\alpha_v + \alpha_w, \\ X_1 = X_{1r} + iX_{1i} &= \beta_u - \beta_v + \beta_w, \\ X_2 = X_{2r} + iX_{2i} &= -2\beta_u + \beta_v - \beta_w, \\ X_3 = X_{3r} + iX_{3i} &= \beta_v + \beta_w. \end{aligned}$$

We denote the real and imaginary parts of the α's and β's by

$$\begin{aligned} \alpha_u &= R_u + iI_u, \quad \alpha_v = R_v + iI_v, \quad \alpha_w = R_w + iI_w, \\ \beta_u &= S_u + iJ_u, \quad \beta_v = S_v + iJ_v, \quad \beta_w = S_w + iJ_w, \end{aligned}$$

and define r by

$$r^2 = u^2 + v^2 + w^2.$$

We rewrite (11.10) in the form of a modified "Euler equation"

$$\begin{aligned} \frac{1}{2}\dot{u} &= u(R_u r + \lambda) + (I_w - I_v)vw - J_v vr + (S_u + S_w)uw, \\ \frac{1}{2}\dot{v} &= v(R_v r + \lambda) + (I_u - I_w)uw + J_u ur + (S_v + S_w)vw, \\ \frac{1}{2}\dot{w} &= w(R_w r + \lambda) + (I_v - I_u)uv + S_w w^2 - S_u u^2 - S_v v^2 \\ \dot{\psi} &= \omega + O(r). \end{aligned} \tag{11.11}$$

When all parameters are zero except for I_u, I_v and I_w, this is Euler's equation for the motion of a rigid body. As a consequence of the $\mathbf{S}^1$ symmetry of the normal form, the ψ equation uncouples from those for u, v and w and so *periodic solutions* in (z_+, z_-)–space correspond to *fixed points* in (u, v, w)–space.

Because (11.11) is homogeneous except for the λ term, we follow Swift [288] further and reduce it to an *associated spherical system* by changing the coordinate from (u, v, w) into (r, θ, ϕ) via the transformation

$$u + iv = r \sin\theta e^{i\phi}, \qquad w = r\cos\theta.$$

Note that for $\sin\theta = 0$, ϕ is not defined; this coordinate singularity is the "pole" of the sphere. We derive the following system for r, θ and ϕ

$$\begin{aligned}
\dot r = \frac{r}{2}\Big[4\lambda + r\Big(&4A_r + 3B_r - C_r + (B_r + C_r)\sin^2\theta\cos 2\phi \\
&- (X_{2i} + X_{3i})\sin^2\theta\sin 2\phi + (3C_r - B_r)\cos^2\theta \\
&+ (4X_{1r} + 2X_{2r} + 2X_{3r})\cos\theta\Big)\Big]
\end{aligned} \tag{11.12a}$$

$$\begin{aligned}
\dot\theta = \frac{r\sin\theta}{2}\Big\{&\cos\theta\Big[(-X_{2i} - X_{3i})\sin\phi\cos\phi + (B_r + C_r)\cos^2\phi - 2C_r\Big] \\
&+ (B_i + C_i)\sin\phi\cos\phi - (X_{2r} + X_{3r})\cos^2\phi \\
&- 2X_{1r} - X_{2r} + X_{3r}\Big\}
\end{aligned} \tag{11.12b}$$

$$\begin{aligned}
\dot\phi = \frac{r}{2}\Big\{&\Big[(X_{2r} + X_{3r})\cos\theta - (B_r + C_r)\Big]\sin 2\phi \\
&+\Big[(B_i + C_i)\cos\theta - (X_{2i} + X_{3i})\Big]\cos 2\phi \\
&+ (B_i - 3C_i)\cos\theta - 4X_{1i} - 3X_{2i} + X_{3i}\Big\}.
\end{aligned} \tag{11.12c}$$

This system is defined for $r \geq 0$, $\theta \in [0, 2\pi)$ and $\phi \in [0, \pi)$. The action of the reflection and rotation is given by

$$\begin{aligned}
S_2 &: (r,\ \theta,\ \phi) \to (r,\ 2\pi - \theta,\ \phi) \\
R &: (r,\ \theta,\ \phi) \to (r,\ \pi - \theta,\ \pi - \phi) \quad \text{(odd parity only).}
\end{aligned}$$

As noted by Swift [288], the equations for θ and ϕ can be made independent of $r > 0$ by scaling time, and then we can find periodic solutions of the full problem (steady state solutions for (r, θ, ϕ)) in two stages:

1. Solve the *associated steady state spherical system* $(\dot\theta, \dot\phi) = (0, 0)$ for $r > 0$.
2. Solve the radial equation $\dot r = 0$ at zeroes of the associated spherical system.

For A_r, B_r and C_r asymptotically close to zero compared to the other coefficients and $X_k \equiv 0$, this system describes the motion of a rigid body. Van Gils and Silber [294] have also investigated the dynamics of this system with

$X_k \equiv 0$ using other techniques. In addition to those found by Swift, we believe there may be new types of heteroclinic orbit or quasiperiodic behavior near bifurcation. The equation $\dot{\theta} = 0$ is always satisfied if

$$\sin\theta = 0. \tag{11.13}$$

This corresponds to the existence of branch with maximal symmetry $\mathbf{Z}_2$ that bifurcates. For this solution branch we have $u = v = 0$ and $w = r$. Namely, our reason for choosing u, v, w differently from Swift is to ensure that this maximal solution is at the pole of the coordinates. Alternatively, the equation (11.12b) has another equilibrium at

$$\cos\theta = -\frac{(B_i + C_i)\sin\phi\cos\phi - (X_{2r} + X_{3r})\cos^2\phi - 2X_{1r} - X_{2r} + X_{3r}}{(-X_{2i} - X_{3i})\sin\phi\cos\phi + (B_r + C_r)\cos^2\phi - 2C_r} \tag{11.14}$$

corresponding to periodic solutions with trivial symmetry; i.e., submaximal symmetry solutions. This is different from Swift's equation for $\cos\theta$ due to our interchanging the definitions of u and w. If we were to keep to Swift's coordinates, problems begin at this point as there is no easy solution giving $\cos\theta$ when terms $X_k \neq 0$ are included. Substituting (11.14) into $\dot{\phi} = 0$ we get the following equation for ϕ

$$\alpha\cos 4\phi + \beta\sin 4\phi + \gamma\cos 2\phi + \delta\sin 2\phi + \epsilon = 0 \tag{11.15}$$

where the constants $\alpha, \ldots, \epsilon$ are given by

$$\alpha = \frac{1}{4}\mathrm{Im}[(\bar{B} + \bar{C})(X_2 + X_3)], \tag{11.16a}$$

$$\beta = \frac{1}{8}[|B|^2 + |C|^2 - |X_2|^2 - |X_3|^2 + 2\mathrm{Re}(\bar{B}C - \bar{X}_2X_3)], \tag{11.16b}$$

$$\gamma = \mathrm{Im}[(\bar{B} + \bar{C})X_1 + C\bar{X}_3 + \bar{B}X_2], \tag{11.16c}$$

$$\delta = \frac{1}{4}[|B|^2 + |X_3|^2 - 3(|C|^2 + |X_2|^2) - 2\mathrm{Re}(B\bar{C} + 2X_1(\bar{X}_2 + \bar{X}_3) + X_2\bar{X}_3)], \tag{11.16d}$$

$$\epsilon = \frac{1}{4}\mathrm{Im}[(\bar{B} - 3\bar{C})(4X_1 + 3X_2 - X_3)]. \tag{11.16e}$$

For odd parity ($X_k = 0$), we have $\alpha = \gamma = \epsilon = 0$ and so either $\sin 2\phi = 0$ (corresponding to maximal branches) or

$$\cos 2\phi = -\frac{\delta}{2\beta}.$$

Equation (11.15) for ϕ can be written as a second order polynomial of $\exp(i2\phi)$ and as such there exists a closed form solution to this equation. There may be zero, two or four nondegenerate (isolated) solutions ϕ in the interval $[0, \pi)$ which we write as $\phi_1 < \phi_2$ (and $\phi_1 < \phi_2 < \phi_3 < \phi_4$ if they exist).

After solving this equation, we solve (11.14) with $\phi = \phi_i$ which will have zero or two (symmetrically placed) solutions according to the right hand side having modulus greater than or less than unity.

Given a pair (θ_{ij}, ϕ_i) that solves these equations, the equation $\dot{r} = 0$ (11.12a) implies there will be a branch of solutions in the direction sign(l_i)

$$\begin{aligned} l_i := -4\Big[&4A_r + 3B_r - C_r + \Big((B_r + C_r)\cos 2\phi_i - (X_{2i} + X_{3i})\sin 2\phi_i\Big)\sin^2\theta_{ij} \\ &+ (3C_r - B_r)\cos^2\theta_{ij} + (4X_{1r} + 2X_{2r} + 2X_{3r})\cos\theta_{ij}\Big]^{-1} \end{aligned} \tag{11.17}$$

and correspondingly

$$r = l_i \lambda.$$

Since the solutions indexed by j are symmetrically related, l_i is independent of j.

We summarize the calculations for finding the number and direction of branching of periodic solutions at a Hopf bifurcation point with the following algorithm.

Algorithm 11.5.1. Procedure to find periodic solution branches

1. *Find the complex coefficients* (A, B, C, X_1, X_2, X_3) *in the normal form using center manifold or Liapunov-Schmidt reduction.*
2. *Calculate* $\alpha, \ldots, \epsilon$ *via equations (11.16a)—(11.16e).*
3. *Determine the solutions (including multiplicities)* ϕ_i *of the equation (11.15). We exclude the case that*
   (I) *the solutions* ϕ_i *are not isolated.*
4. *For a given solution* ϕ_i *of (11.15), calculate the right hand side of the equation (11.14) and determine its solutions* θ_{ij} *if the equation is well defined, i.e., its right hand side is in the interval* $[-1,\ 1]$. *We exclude the case*
   (II) $|\cos\theta_{ij}| = 1$.
5. *Compute* l_i *using equation (11.17) and observe that the bifurcating branch corresponding to submaximal solutions with* (θ_{ij}, ϕ_i) *is subcritical for* $l_i > 0$ *and supercritical for* $l_i < 0$. *The value of* r *is then given to lowest order by* $r = l_i\lambda$. *We exclude the case*
   (III) $l_i = 0$.

This procedure will work for normal form coefficients which do not satisfy any of the degeneracies marked (I)-(III). It is clear that there is an open everywhere dense set of parameters for which none of (I)-(III) satisfied, and so we have a procedure that will work for a generic set of third order coefficients.

11.6 Brusselator Equations

As an example of a normal form derived from reaction-diffusion equations on the square, we briefly present the reduced bifurcation equations for the Brusselator equation

$$\begin{aligned}\frac{\partial u_1}{\partial t} &= \Delta u_1 + a - (b+\lambda+1)u_1 + u_1^2 u_2 \\ \frac{\partial u_2}{\partial t} &= d\Delta u_2 + (b+\lambda)u_1 - u_1^2 u_2\end{aligned} \tag{11.18}$$

on the square domain $\Omega := [0,\pi]^2$ with Robin boundary conditions

$$\mu(u_i - u_i^s) + (1-\mu)\frac{\partial u_i}{\partial n} = 0 \ \text{ on } \partial\Omega. \tag{11.19}$$

Here u^s is the homogeneous state $(u_1, u_2) = (a, (b+\lambda)/a)$ and n is the outward normal derivative.

Because an analytical reduction is not possible for the boundary conditions assumed, a hybrid numerical-analytical Liapunov-Schmidt reduction is used. Furthermore, the normal form is determined by third order terms. Liapunov-Schmidt method yields the same reduced equations as the center manifold reduction for this example, see Chapters 6, 7, Ashwin/Böhmer/Mei [22, 25] and Chossat/Golubitsky [63].

Setting $(u_1, u_2) = (a, (b+\lambda)/a) + (\tilde{u}_1, \tilde{u}_2)$ and scaling time $t \leftarrow (\omega_0 + \tau)t$ we obtain the following operator equation for periodic solutions near the trivial state after dropping the tilde $\tilde{}$:

$$\Phi(u,\lambda,\tau) := -\omega_0 \frac{\partial u}{\partial t} + Lu + R(u,\lambda,\tau) = 0, \tag{11.20}$$

where $u := (u_1, u_2)^T$,

$$L := \begin{pmatrix} \Delta + b - 1 & a^2 \\ -b & d\Delta - a^2 \end{pmatrix}, \tag{11.21}$$

and

$$R(u,\lambda,\tau) := \begin{pmatrix} 1 \\ -1 \end{pmatrix} \left(\frac{b+\lambda}{a} u_1^2 + 2a u_1 u_2 + u_1^2 u_2 + \lambda u_1 \right) - \tau \frac{\partial u}{\partial t}. \tag{11.22}$$

The operator L maps $X := (C_0^{2,\alpha}(\Omega))^2$ into $Y := (C^{0,\alpha}(\Omega))^2$. Here the subscript 0 indicates that functions in X satisfy the boundary condition (11.19). Alternatively, we can consider a weak form of the equations (11.18) in Sobolev spaces. Let

$$C_{2\pi} := \{u(\cdot) \in C(\mathbf{R}, Y) \ \mid \ u(s+2\pi) = u(s)\} \ \text{ and}$$

$$C_{2\pi}^1 := \{u(\cdot),\ \tfrac{\partial u}{\partial s}(\cdot) \in C(\mathbf{R}, X) \ \mid \ u(s+2\pi) = u(s)\}.$$

The operator Φ maps $C^1_{2\pi}$ into $C_{2\pi}$ and

$$\Phi(0,\lambda,\tau)=0 \quad \text{for all} \quad \lambda,\ \tau\in[0,\infty).$$

Fixing the parameters a, b and d at a Hopf bifurcation point, we study periodic solution branches of (11.18) bifurcating from the trivial solution curve upon varying λ, i.e., steady solutions of the operator equation (11.20).

11.6.1 Linear stability analysis

Consider the eigenvalue problem

$$-\Delta u=\kappa u \tag{11.23}$$

with the Robin boundary conditions (11.19). To concentrate on the discussion of Hopf bifurcations of Brusselator equations, we state briefly that this eigenvalue problem is solvable and postpone the analysis to Chapter 14, see also Ashwin/Mei [25]. An eigenvalue of (11.23) is of the form $\kappa(\mu)=\kappa_x(\mu)^2+\kappa_y(\mu)^2$ and each $\kappa_x,\ \kappa_y$ satisfies the following equation for κ

$$2\mu(1-\mu)\kappa\cos(\kappa\pi)+[\mu^2-(1-\mu)^2\kappa^2]\sin(\kappa\pi)=0. \tag{11.24}$$

Generic multiplicity of this eigenvalue is two. The corresponding eigenfunctions are

$$\begin{aligned}\phi_1 &:= [\mu\sin(\kappa_x(\mu)x)+(1-\mu)\kappa_x(\mu)\cos(\kappa_x(\mu)x)]\\ &\quad\cdot[\mu\sin(\kappa_y(\mu)y)+(1-\mu)\kappa_y(\mu)\cos(\kappa_y(\mu)y)],\\ \phi_2 &:= [\mu\sin(\kappa_y(\mu)x)+(1-\mu)\kappa_y(\mu)\cos(\kappa_y(\mu)x)]\\ &\quad\cdot[\mu\sin(\kappa_x(\mu)y)+(1-\mu)\kappa_x(\mu)\cos(\kappa_x(\mu)y)].\end{aligned}$$

The statement $\phi_1=\phi_2$ holds if and only if $\kappa_x=\kappa_y$, implying the multiplicity one.

The Laplacian operator in $C_0^{2,\alpha}(\Omega)$ is self-adjoint and elliptic, its eigenfunctions make up a basis for $C_0^{2,\alpha}(\Omega)$, see Chapter 14 By similar discussion as in Chapter 10, in particular, the statement (10.22), we see that there is Hopf bifurcation at $\lambda=0$ if for an eigenvalue $\kappa(\mu)$ of $-\Delta$ the parameters a_0, b_0, d_0 satisfy

$$b_0-1-a_0^2-(1+d_0)\kappa(\mu)=0, \tag{11.25a}$$
$$\omega_0^2:=(1+\kappa(\mu))(a_0^2+d_0\kappa(\mu))-b_0d_0\kappa(\mu)>0. \tag{11.25b}$$

If the eigenvalue $\kappa(\mu)$ has the multiplicity two, the kernel $\text{Ker}(D_u\Phi_0)$ is 4 dimensional. Let L^* be the adjoint operator of L in (11.21) with respect to the inner product

$$\langle u,\ v\rangle=\frac{2}{\pi^3}\int_0^{2\pi}\int_\Omega(u,v)\,dx\,dy\,dt \quad \text{for all} \quad u,v\in C_{2\pi}. \tag{11.26}$$

We have

$$\mathrm{Ker}(D_u\Phi_0) = \mathrm{span}[\psi_1, \ldots, \psi_4], \qquad \mathrm{Ker}(D_u\Phi_0^*) = \mathrm{span}[\psi_1^*, \ldots, \psi_4^*]$$

with

$$\begin{aligned}
\psi_1 &= (e_1 \cos t + Pe_1 \sin t)\phi_1, \qquad \psi_2 = (-Pe_1 \cos t + e_1 \sin t)\phi_1,\\
\psi_3 &= (e_1 \cos t + Pe_1 \sin t)\phi_2, \qquad \psi_4 = (-Pe_1 \cos t + e_1 \sin t)\phi_2,\\
\psi_1^* &= \frac{1}{2e_1^T P^T e_2}(P^T e_2 \cos t + e_2 \sin t)\phi_1,\\
\psi_2^* &= \frac{1}{2e_1^T P^T e_2}(-e_2 \cos t + P^T e_2 \sin t)\phi_1,\\
\psi_3^* &= \frac{1}{2e_1^T P^T e_2}(P^T e_2 \cos t - e_2 \sin t)\phi_2,\\
\psi_4^* &= \frac{1}{2e_1^T P^T e_2}(e_2 \cos t + P^T e_2 \sin t)\phi_2,
\end{aligned}$$

and $e_1 := (1,0)^T$, $e_2 := (0,1)^T \in \mathbf{R}^2$ and

$$P = (P_{ij})_{i,j=1,2} := \frac{1}{\omega_0}\begin{pmatrix} b_0 - 1 - \kappa(\mu) & a_0^2 \\ -b_0 & -d_0\kappa(\mu) - a_0^2 \end{pmatrix}.$$

Define the projection from $C_{2\pi}^1$ into $\mathrm{Im}(D_u\Phi_0) \cap C_{2\pi}^1$ by

$$Qw := w - \sum_{i=1}^{4} \langle \psi_i^*, w\rangle \psi_i, \quad w \in C_{2\pi}.$$

We write element u in X as

$$u = \psi + w$$

with $\psi \in \mathrm{Ker}(D_u\Phi_0)$ and $w \in M$ the complementary of $\mathrm{Ker}(D_u\Phi_0)$). Thereafter, we apply Algorithm 6.6.1 to the equation

$$\Phi(u, \lambda, \tau) = D_u\Phi_0 w + R(\eta + w, \lambda, \tau) = 0$$

and calculate the reduced bifurcation equations up to the third order of (ψ, λ, τ). We remind that what we aim is to compute the complex quantities A, B, C, X_1, X_2 and X_3 at a Hopf bifurcation point, though the Liapunov-Schmidt reduction process itself can be tedious. Detailed techniques are discussed in Ashwin/Böhmer/Mei [22]. After obtaining the normal form (11.6) we and determine branching behavior with Algorithm 11.5.1.

11.6.2 Bifurcation scenario

Odd parity

Table 11.4 shows the branching information as the homotopy parameter μ in the boundary conditions varies from 0 to 1. By choosing $d_0 = 0.05$, $a_0 = 1$ we consider the curve of Hopf points from the $(1,2)$-mode of Neumann problem to the $(2,3)$-mode of Dirichlet problem.

μ	R_a	R_b	R_c	$\lvert B\rvert^2$	$\lvert C\rvert^2$	$\mathrm{Re}(B\bar{C})$
0	−0.872348	−0.927833	−0.495963	0.693339	0.00891380	0.0295561
1/9	−0.875804	−0.937010	−0.493289	0.739228	0.00800229	0.0350549
2/9	−0.877861	−0.943525	−0.487108	0.796978	0.00688819	0.0413364
3/9	−0.878745	−0.947059	−0.476676	0.871397	0.00571162	0.0488723
4/9	−0.879868	−0.948792	−0.461760	0.970963	0.00484463	0.0586257
5/9	−0.885721	−0.953975	−0.443769	1.11287	0.00493039	0.0724611
6/9	−0.908012	−0.976895	−0.427953	1.33706	0.00662581	0.0937503
7/9	−0.973323	−1.04780	−0.426371	1.74616	0.0100083	0.128246
8/9	−1.12849	−1.21264	−0.456123	2.60500	0.0144037	0.184467
1	−1.39711	−1.48415	−0.511478	4.38337	0.0189617	0.264659

Table 11.4. Variation of branching coefficients along a homotopy from the (1,2) mode of Neumann problem at $\mu = 0$ to the (2,3) mode of Dirichlet problem at $\mu = 1$. We choose $d_0 = 0.05$ and $a_0 = 1$ in the Brusselator equations.

In Figure 11.1 the coefficients which determines branching behavior in the λ direction are depicted along the homotopy path from the (1,2)-mode Neumann to the (2,3)-mode Dirichlet problem with the constants $d_0 = 0.1$ and $a_0 = 1$.

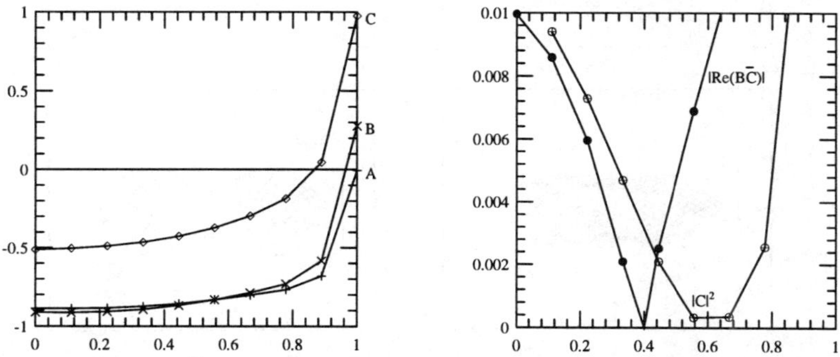

Fig. 11.1. The changing signs of R_c at $\mu \sim 0.85$ and R_b at $\mu \sim 0.94$ indicate that two maximal isotropy types of solutions change direction of bifurcation. The submaximal solution branch existing for $\mu < 0.44$ no longer exists for $\mu > 0.44$, when $|B|^2 > |C|^2 > \mathrm{Re}(B\bar{C})$ is no longer satisfied.

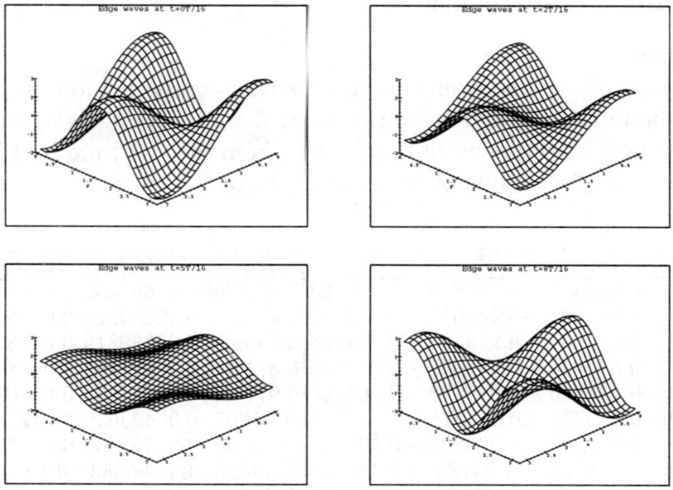

Fig. 11.2. The edge oscillations with (anti-) reflection symmetries parallel to the edges.

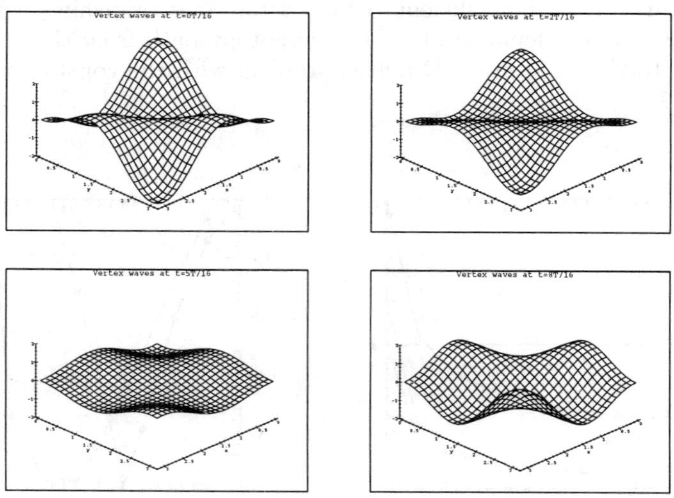

Fig. 11.3. The vertex oscillations with (anti-) reflection symmetries parallel to the diagonals of the square.

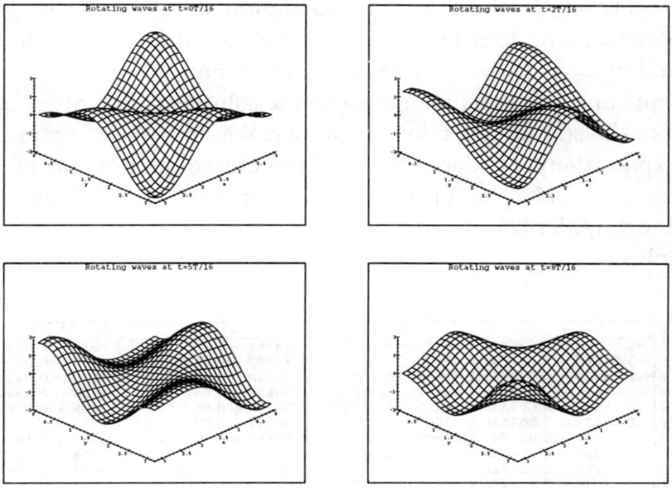

Fig. 11.4. The rotating oscillations with the symmetry generated by a spatial quarter-turn coupled with phase shift of a quarter of period.

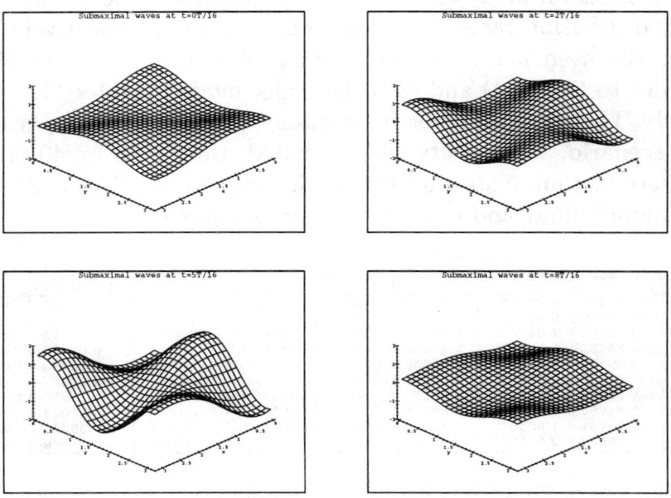

Fig. 11.5. The submaximal oscillations. These solutions have no symmetries except the coupling of a half spatial turn with a phase shift of half a period.

Figures 11.2–11.5 show the spatiotemporal development of patterns of the component u_2 on four branches at the Hopf bifurcation with $d_0 = 0.05$, $a_0 = 1$ and $\mu = 1/3$. We plot four stages in half a period of three maximal isotropic and one submaximal isotropic solutions. The other half period follows from the spatial reflection symmetry. We draw these picture by examining the spatiotemporal dependence of the critical modes and by analyzing the normal form (11.7) as in Swift [288]. For more detailed discussions we refer to Ashwin/Mei [25].

Even parity

μ	A	B	C	$X_1 = -2X_2 = 2X_3$
0	$-2.28506 - 0.02634i$	$1.40773 - 0.06248i$	$0.14184 + 0.00928i$	$0.00000 + 0.00000i$
1/9	$-2.28660 - 0.02773i$	$1.43026 - 0.08191i$	$0.12962 + 0.01898i$	$-0.00920 - 0.00981i$
2/9	$-2.28642 - 0.02545i$	$1.45898 - 0.10496i$	$0.11620 + 0.03116i$	$-0.02063 - 0.02246i$
3/9	$-2.28496 - 0.01901i$	$1.49673 - 0.13175i$	$0.10126 + 0.04623i$	$-0.03452 - 0.03861i$
4/9	$-2.28403 - 0.00853i$	$1.54805 - 0.16133i$	$0.08396 + 0.06409i$	$-0.05059 - 0.05872i$
5/9	$-2.28877 + 0.00382i$	$1.62067 - 0.19015i$	$0.06244 + 0.08307i$	$-0.06727 - 0.08265i$
6/9	$-2.31316 + 0.01066i$	$1.72865 - 0.20966i$	$0.03353 + 0.09821i$	$-0.08063 - 0.10914i$
7/9	$-2.39389 - 0.00803i$	$1.90197 - 0.20602i$	$-0.00358 + 0.10190i$	$-0.08415 - 0.13767i$
8/9	$-2.61188 - 0.10182i$	$2.21470 - 0.17007i$	$-0.03309 + 0.10064i$	$-0.06981 - 0.17877i$
1	$-2.97648 - 0.38048i$	$2.75807 - 0.10249i$	$-0.02528 + 0.13237i$	$-0.01403 - 0.25490i$

Table 11.5. Calculated values of the complex coefficients of the normal form (11.6) that determine the branching behavior in the λ direction of periodic solutions near Hopf bifurcation. Here $d_0 = 0.05$, $a_0 = 1$.

Tables 11.5 and 11.6 show the coefficients of the normal form (11.6) obtained as a function of μ, the homotopy parameter in boundary condition, for the fixed diffusion rates $d_0 = 0.05$ and $d_0 = 0.01$, respectively. We show results for the even parity branches going from the $(1,3)$ and $(2,4)$ Neumann modes to the $(2,4)$ and $(3,5)$ Dirichlet modes (Tables 11.5 and 11.6), respectively. There are no mode interactions which might complicate the bifurcation scenario. The results are calculated using a 40 by 40 grid for the spatial discretization. Note that the coefficients $X_1 = -2X_2 = 2X_3$ are zero at the Neumann limit and become non-zero for $\mu \neq 0$.

μ	A	B	C	$X_1 = -2X_2 = 2X_3$
0	-4.69443+ 0.96305i	2.95487-0.74399i	-0.13504-0.33405i	0.00000+ 0.00000i
1/9	-4.73780+ 0.99776i	2.99837-0.79350i	-0.14463-0.31677i	-0.00100-0.00081i
2/9	-4.79001+ 1.04133i	3.05536-0.85072i	-0.15651-0.28904i	-0.00511-0.00390i
3/9	-4.85233+ 1.09567i	3.13034-0.91711i	-0.17197-0.24639i	-0.01487-0.01064i
4/9	-4.92590+ 1.16299i	3.22953-0.99450i	-0.19315-0.18334i	-0.03446-0.02279i
5/9	-5.01349+ 1.24552i	3.36253-1.08484i	-0.22375-0.09518i	-0.07005-0.04195i
6/9	-5.12985+ 1.34472i	3.54877-1.18923i	-0.27006+ 0.01636i	-0.12838-0.06760i
7/9	-5.34685+ 1.46328i	3.84318-1.30675i	-0.34310+ 0.13068i	-0.21002-0.09442i
8/9	-5.94761+ 1.63667i	4.42698-1.45787i	-0.45809+ 0.20979i	-0.30365-0.12271i
1	-7.49637+ 2.00104i	5.66954-1.75650i	-0.61536+ 0.28048i	-0.41589-0.19121i

Table 11.6. As in Table 11.5 except these results are calculated for the branch connecting the $(2,4)$ Neumann mode to the $(3,5)$ Dirichlet mode and $a_0 = 1$, $d_0 = 0.01$.

For the Neumann boundary conditions, there is a hidden symmetry arising from the fact that solutions can be smoothly extended to solutions on the torus obtained by identifying opposite edges of the square. By looking at the Neumann boundary conditions as a symmetry restriction of the problem with periodic boundary conditions, it becomes generic by symmetry arguments that the center eigenspace of even parity Hopf bifurcation can be two (complex) dimensional. This explains why $X_1 = X_2 = X_3 = 0$ in the results for $\mu = 0$.

It can be seen from the results that very surprisingly, $X_1 = -2X_2 = 2X_3$ in all of these cases. Numerical experiments on other mode bifurcations imply that this is always the case for these equations. This means that instead of the three additional third order equivariants, there is a degeneracy that forces them to appear in the combination

$$\begin{pmatrix} |z_-|^2 z_- + 2|z_+|^2 z_- + \bar{z}_- z_+^2 \\ |z_+|^2 z_+ + 2|z_-|^2 z_+ + \bar{z}_+ z_-^2 \end{pmatrix}.$$

This is the derivative $(D_{\bar{z}_+} g(z_+, \bar{z}_+, z_-, \bar{z}_-),\ D_{z_+} g(z_+, \bar{z}_+, z_-, \bar{z}_-))$ of the invariant function

$$g(z_+, \bar{z}_+, z_-, \bar{z}_-) := (|z_+|^2 + |z_-|^2)(\bar{z}_+ z_- + \bar{z}_- z_+),$$

with z_+, $\bar{z}_+$, z_-, $\bar{z}_-$ as independent variables. We discuss some reasons for this later in this section.

Tables 11.7 and 11.8 show possible bifurcating periodic solutions on homotopy of boundary conditions. Some submaximal solutions are annihilated and emanate, while solutions with maximal isotropy group do persist along the homotopy. The signs of l determine the bifurcation direction on changing λ. We denote no solution by "n.s.". In Table 11.7, there are two group orbits of submaximal solutions up until $\mu = 5/9$; the solutions ϕ_3 and ϕ_4 of the ϕ-equation are annihilated. From $\mu = 7/9$ there are again two group orbits of solutions.

In Table 11.8 there are up to 4 solutions for ϕ-equation, and for each of these there are up to two solutions of θ along the homotopy path starting from (2,4) mode of Neumann problem to (3,5) mode of Dirichlet problem. The value of l determines the direction of branching. The solutions with maximal isotropy type for $\mu = 0$ all have unique continuations to $\mu = 1$; there are no secondary bifurcations on this branch.

11.6.3 Nonlinear degeneracy

Numerical results on the Brusselator equations with Robin boundary conditions show an extra degeneracy of the new third order terms, i.e., $X_1 = -2X_2 = 2X_3$. We now show that this arises because the linearized problem has a basis of eigenfunctions that are separable.

It turns out that a different coordinate system enables us to investigate this degeneracy more easily. We define

μ	ϕ_1	θ_{11}	θ_{12}	l_1	ϕ_2	θ_{21}	θ_{22}	l_2
	ϕ_3	θ_{31}	θ_{32}	l_3	ϕ_4	θ_{41}	θ_{42}	l_4
0	0	1.570796	4.712389	1.139826	1.129473	n.s.	n.s.	n.s.
	1.570796	1.570796	4.712389	1.139826	2.0121197	n.s.	n.s.	n.s.
1/9	0.006723964	1.577545	4.705641	1.167746	1.151941	n.s.	n.s.	n.s.
	1.565340	1.536620	4.746565	0.6110883	1.988384	n.s.	n.s.	n.s.
2/9	0.01551817	1.585313	4.697872	1.208514	1.177226	n.s.	n.s.	n.s.
	1.559913	1.485305	4.797880	0.6188707	1.959732	n.s.	n.s.	n.s.
3/9	0.02674702	1.593917	4.689269	1.268420	1.205644	n.s.	n.s.	n.s.
	1.556158	1.405486	4.877699	0.6289309	1.923841	n.s.	n.s.	n.s.
4/9	0.04042879	1.602729	4.680456	1.357882	1.237870	n.s.	n.s.	n.s.
	1.557934	1.272077	5.011109	0.6412422	1.876156	n.s.	n.s.	n.s.
5/9	0.05562084	1.610297	4.672888	1.494196	1.276148	n.s.	n.s.	n.s.
	1.575267	0.9971433	5.286042	0.6545510	1.805353	n.s.	n.s.	n.s.
6/9	0.06969786	1.614028	4.669158	1.703950	1.326229	n.s.	n.s.	n.s.
	n.s.	n.s.	n.s.	n.s.	n.s.	n.s.	n.s.	n.s.
7/9	0.07934294	1.610901	4.672285	2.015995	1.389855	2.849157	3.434028	0.7346098
	n.s.	n.s.	n.s.	n.s.	n.s.	n.s.	n.s.	n.s.
8/9	0.08591220	1.599423	4.683762	2.474485	1.434446	2.174610	4.108575	0.6918723
	n.s.	n.s.	n.s.	n.s.	n.s.	n.s.	n.s.	n.s.
1	0.09436033	1.576897	4.706289	4.338267	1.456928	1.774929	4.508256	0.6385345
	n.s.	n.s.	n.s.	n.s.	n.s.	n.s.	n.s.	n.s.

Table 11.7. This table shows the possible solutions ϕ, θ and l along the homotopy path corresponding to the values in Table 11.5.

μ	ϕ_1	θ_{11}	θ_{12}	l_1	ϕ_2	θ_{21}	θ_{22}	l_2
	ϕ_3	θ_{31}	θ_{32}	l_3	ϕ_4	θ_{41}	θ_{42}	l_4
0	0	1.570796	4.712389	0.5748576	1.570796	1.570796	4.712389	0.5748576
	n.s.	n.s.	n.s.	n.s.	n.s.	n.s.	n.s.	n.s.
1/9	0.0003176902	1.571003	4.712183	0.5748991	1.574077	1.586851	4.696334	0.3158247
	n.s.	n.s.	n.s.	n.s.	n.s.	n.s.	n.s.	n.s.
2/9	0.001548596	1.571839	4.711346	0.5764834	1.577760	1.612483	4.670702	0.3140639
	n.s.	n.s.	n.s.	n.s.	n.s.	n.s.	n.s.	n.s.
3/9	0.004275930	1.573794	4.709391	0.5807234	1.579042	1.641955	4.641230	0.3124132
	n.s.	n.s.	n.s.	n.s.	n.s.	n.s.	n.s.	n.s.
4/9	0.009336105	1.577652	4.705533	0.5895024	1.575964	1.675929	4.607257	0.3112329
	n.s.	n.s.	n.s.	n.s.	n.s.	n.s.	n.s.	n.s.
5/9	0.01769188	1.584513	4.698672	0.6057593	1.566185	1.715654	4.567532	0.3110555
	n.s.	n.s.	n.s.	n.s.	n.s.	n.s.	n.s.	n.s.
6/9	0.02964825	1.595335	4.687851	0.6327034	1.548260	1.760115	4.523070	0.3120088
	n.s.	n.s.	n.s.	n.s.	n.s.	n.s.	n.s.	n.s.
7/9	0.04271166	1.609049	4.674136	0.6658259	1.526578	1.800929	4.482256	0.3105186
	n.s.	n.s.	n.s.	n.s.	n.s.	n.s.	n.s.	n.s.
8/9	0.05160103	1.619921	4.663264	0.6592501	1.513119	1.822905	4.460281	0.2897166
	n.s.	n.s.	n.s.	n.s.	n.s.	n.s.	n.s.	n.s.
1	0.05906714	1.623301	4.659884	0.5487133	1.504274	1.827378	4.455808	0.2337781
	n.s.	n.s.	n.s.	n.s.	n.s.	n.s.	n.s.	n.s.

Table 11.8. This table is generated with the values in Table 11.6.

$$z_x = \frac{z_+ + z_-}{2},$$
$$z_y = \frac{z_+ - z_-}{2},$$

and note that elements in the center eigenspace (11.5) can be written

$$u = z_x \Psi_x + z_y \Psi_y + \text{c.c.},$$

where

$$\Psi_x = (\phi_1 + \phi_2)(U + iV)e^{i\omega t}/2,$$
$$\Psi_y = (\phi_1 - \phi_2)(U - iV)e^{i\omega t}/2.$$

with $\phi_1 = f_k(x_1)f_l(x_2)$ and $\phi_2 = f_l(x_1)f_k(x_2)$ as before. In this case, equations (11.6) become

$$\begin{aligned} \dot{z}_x &= (\beta + i\omega) z_x + f_x(z_x, z_y) \\ \dot{z}_y &= (\beta + i\omega) z_y + f_y(z_x, z_y) \end{aligned} \tag{11.27}$$

with

$$\begin{aligned} f_x(z_x, z_y) := {} & (2A + B + C)|z_x|^2 z_x + 2(A + B - C)|z_y|^2 z_x + (B + C)\bar{z}_x z_y^2 \\ & + (2X_1 + X_2 + X_3)|z_x|^2 z_x + 2(X_1 + X_2 - X_3)|z_y|^2 z_x \\ & + (X_2 + X_3)\bar{z}_x z_y^2, \\ f_y(z_x, z_y) := {} & (2A + B + C)|z_y|^2 z_y + 2(A + B - C)|z_x|^2 z_y + (B + C)\bar{z}_y z_x^2 \\ & -(2X_1 + X_2 + X_3)|z_y|^2 z_y - 2(X_1 + X_2 - X_3)|z_x|^2 z_y \\ & -(X_2 + X_3)\bar{z}_y z_x^2. \end{aligned}$$

In these coordinates, the degeneracy $X_1 = -2X_2 = 2X_3$ means that the only term that breaks the symmetry of the odd parity representation is of the form

$$\begin{pmatrix} |z_x|^2 z_x \\ -|z_y|^2 z_y \end{pmatrix}.$$

We can see that this is the case by considering the spatial dependence in the inner product giving these coefficients. Noting that complex conjugation does not affect the spatial dependence of the eigenfunctions, we use $(|z_x|^2 z_x)_y$ to signify the coefficient of $|z_x|^2 z_x$ in f_y. The Liapunov-Schmidt reduction gives rise to the following formulae

$$\begin{aligned} C_1 &:= (|z_x|^2 z_x)_x - (|z_y|^2 z_y)_y \\ &= K_1 \int_\Omega 4\Big((\phi_1^2 + \phi_2^2) P_1 \phi_1 \phi_2 + \phi_1 \phi_2 P_1 (\phi_1^2 + \phi_2^2)\Big) dx dy, \\ C_2 &:= (|z_y|^2 z_x)_x - (|z_x|^2 z_y)_y \\ &= K_2 \int_\Omega \Big(4\phi_1 \phi_2 P_2 (\phi_1^2 + \phi_2^2) - (\phi_1^2 + \phi_2^2) P_2 \phi_1 \phi_2\Big)\, dx dy, \\ C_3 &:= (z_y^2 \bar{z}_x)_x - (z_x^2 \bar{z}_y)_y \\ &= K_3 \int_\Omega \Big(4\phi_1 \phi_2 P_3 (\phi_1^2 + \phi_2^2) - (\phi_1^2 + \phi_2^2) P_3 \phi_1 \phi_2\Big)\, dx dy. \end{aligned}$$

Here $P_i,\ i = 1, 2, 3$ are linear operators that are of the form $c_1 I + c_2 D_u G_0^{-1}$ with $c_1,\ c_2 \in \mathbf{R}$. The constants K_i come from the reaction terms and the temporal parts of the eigenfunctions.

We write the linear operator into $P_2 = L_d + L_r$ with $L_d := L_1 + L_2 (= \Delta)$ representing the self-adjoint spatial operator and L_r the reaction operator, in particular, L_1 and L_2 are self-adjoint, and L_r behaves like a matrix operator, moreover

$$\langle u,\ L_r v\rangle_2 = U M_r V \int_\Omega f(x)g(x)\ dx_1 dx_2$$

for all $u(x) = Uf(x),\ v(x) = Vg(x) \in C(\Omega, \mathbf{R}^2),\ U,\ V \in \mathbf{R}^2$ and $f,\ g \in C(\Omega, \mathbf{R})$. Here $\langle \cdot,\ \cdot\rangle_k$ denotes the L^2-product in $C([0,\pi]^k, \mathbf{R}^2)$. Based on this and the self-adjointness of L_d and the separable property

$$\phi_1 = U f_k(x_1) f_l(x_2), \qquad \phi_2 = U f_k(x_2) f_l(x_1),$$

we derive

$$\begin{aligned}
C_2 &= K_2[\langle 4\phi_1\phi_2,\ P_2(\phi_1^2+\phi_2^2)\rangle_2 - \langle(\phi_1^2+\phi_2^2),\ P_2\phi_1\phi_2\rangle_2] \\
&= 2K_2[\langle \phi_1\phi_2,\ L_d\phi_1^2\rangle_2 - \langle\phi_1^2,\ L_d\phi_1\phi_2\rangle_2] \\
&= K_2\Big\{\Big[\langle f_k(x_1)f_l(x_1), L_1 f_k^2(x_1)\rangle_1 - \langle f_k(x_1)^2, L_1 f_k(x_1)f_l(x_1)\rangle_1\Big] \\
&\quad \cdot\langle f_k(x_2)f_l(x_2), f_l^2(x_2)\rangle_1 + \langle f_k(x_1)f_l(x_1), f_k^2(x_1)\rangle_1 \cdot \\
&\quad \Big[\langle f_k(x_2)f_l(x_2),\ L_2 f_l^2(x_2)\rangle_1 - \langle f_l(x_2)^2, L_2 f_k(x_2)f_l(x_2)\rangle_1\Big]\Big\} \\
&= 0 \quad (\text{ due to the self-adjointness of } L_i,\ i=1,2).
\end{aligned}$$

Similarly C_3 can be shown to be zero and these imply that $X_1+X_2-X_3=0$ and $X_2+X_3=0$, and consequently,

$$X_1 = -2X_2 = 2X_3.$$

We note that the above argument cannot be adapted to get $C_1 = 4X_1 = 0$ and this is supported by the numerical results we have presented.

12. Steady/Steady State Mode Interactions

In this chapter we study mode interactions in the stationary reaction-diffusion equation

$$G(u, \lambda, d) := \begin{pmatrix} \Delta u_1 + f_1(u, \lambda) \\ d\Delta u_2 + f_2(u, \lambda) \end{pmatrix} = 0, \quad u := (u_1, u_2) \tag{12.1}$$

in D_4-symmetric domains, in particular, the unit square. We impose the homogeneous Dirichlet boundary conditions to these equations. We distinguish mode interactions resulted from multiple parameters from multiple bifurcations induced by symmetries in the problem. More precisely, we treat multiple bifurcations as a special case of mode interactions since this kind of linear degeneracy occurs as a consequence of the geometric property of the problem.

Assume

$$f_i(0, \lambda) = 0, \ \ i = 1, 2, \ \lambda \in \mathbf{R}.$$

Bifurcation points of (12.1) on the trivial solution $\{(0, \lambda, d); \ \lambda, d \in \mathbf{R}\}$ are described by the curves $S(c)$ in Chapter 10. Simple and double bifurcations along a single curve $S(c)$ have been discussed in Section 10.8. To study mode interactions at intersections of two different curves $S(c_1)$, $S(c_2)$ of bifurcation points and general multiple bifurcations on a single curve $S(c)$, it is crucial to exploit the group theoretic concepts in Chapter 5 in order to simplify the bifurcation analysis. We shall use the diffusion rate d as the second parameter. Alternatively in Chapter 9 length of the interval was used as the second parameter for analyzing double bifurcations.

12.1 Induced Actions

As we have seen in Section 10.3, a bifurcation point $(0, \lambda_0, d_0)$ of the stationary problem (12.1) is described by the equations

$$\det(M(\lambda, d, c)) = (c - D_{u_1} f_1^0)(dc - D_{u_2} f_2^0) - D_{u_2} f_1^0 D_{u_1} f_2^0 = 0 \tag{12.2}$$

for some eigenvalue c of the Laplacian.

Example 12.1.1. Consider the Brusselator equations with $(B \longleftrightarrow \lambda)$

$$f_1(u_1, u_2, B) = (B-1)u_1 + A^2 u_2 + \frac{B}{A}u_1^2 + 2Au_1u_2 + u_1^2u_2,$$
$$f_2(u_1, u_2, B) = -Bu_1 - A^2 u_2 - \left(\frac{B}{A}u_1^2 + 2Au_1u_2 + u_1^2u_2\right)$$

on the square domain $\Omega := [0, \pi] \times [0, \pi]$ with the homogeneous Dirichlet boundary conditions. Steady state bifurcation points in the parameter space $B,\ d$ with fixed value $A = 2$ are depicted in Figure 12.1 for the first few modes.

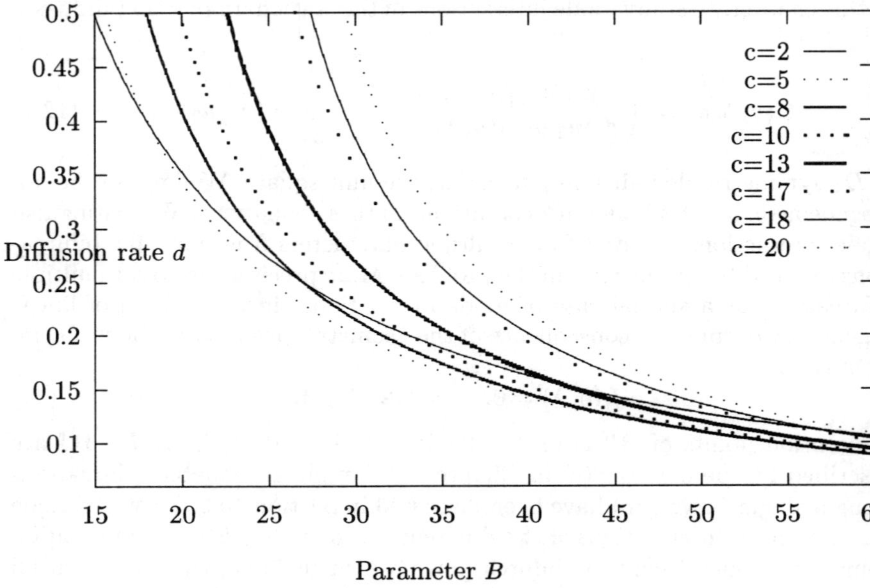

Fig. 12.1. Curves of bifurcation points of the Brusselator equations in the parameter plane (B, d) with $A = 2$.

If we vary the parameters λ and d in (12.1), mode interaction occurs when the equality $\det(M(c, \lambda_0, d_0)) = 0$ holds for two different eigenvalues $c_1,\ c_2$ of the Laplacian $-\Delta$. There are three generic situations for steady state mode interactions with two, three, and four modes respectively.

1. Both c_1 and c_2 are simple eigenvalues of $-\Delta$, e.g. for $c_1 = 2m^2\pi^2$, $c_2 = 2n^2\pi^2$ and $i = 1, 2$ we have

$$\begin{aligned} &\mathrm{Ker}(D_uG_0) = \mathrm{span}[\psi_1,\ \psi_2], && \psi_i = \begin{pmatrix} D_{u_2}f_1^0 \\ c_i - D_{u_1}f_1^0 \end{pmatrix}\phi_i, \\ &\phi_1 = 2\sin m\pi x \sin m\pi y, && \phi_2 = 2\sin n\pi x \sin n\pi y. \end{aligned} \tag{12.3}$$

Representations of D_4 in $\mathrm{Ker}(D_uG_0) \sim \mathbf{R}^2$ are given via

$$S_1 = \begin{pmatrix} (-1)^{m-1} & \\ & (-1)^{n-1} \end{pmatrix}, \qquad R = \begin{pmatrix} (-1)^{m-1} & \\ & (-1)^{n-1} \end{pmatrix} \quad (12.4)$$

2. The eigenvalue c_1 is of the multiplicity two, while c_2 is simple eigenvalue. Denote $c_1 = (m^2+n^2)\pi^2$, $m \neq n$ and $c_1 = 2k^2\pi^2$. We have

$$\mathrm{Ker}(D_u G_0) = \mathrm{span}[\psi_1,\ \psi_2,\ \psi_3] \quad (12.5)$$

with

$$\psi_i = \begin{pmatrix} D_{u_2} f_1^0 \\ c_2 - D_{u_1} f_1^0 \end{pmatrix} \phi_i,\ i = 1,2, \qquad \psi_3 = \begin{pmatrix} D_{u_2} f_1^0 \\ c_1 - D_{u_1} f_1^0 \end{pmatrix} \phi_3$$

and

$$\begin{aligned} \phi_1 &= 2\sin m\pi x \sin n\pi y, \\ \phi_2 &= 2\sin n\pi x \sin m\pi y, \\ \phi_3 &= 2\sin k\pi x \sin k\pi y. \end{aligned}$$

The action of D_4 in $\mathrm{Ker}(D_u G_0) \sim \mathbf{R}^3$ is reducible and generated by

$$S_1 = \begin{pmatrix} (-1)^{m-1} & 0 & 0 \\ 0 & (-1)^{n-1} & 0 \\ 0 & 0 & (-1)^{k-1} \end{pmatrix},$$

and

$$R = \begin{pmatrix} 0 & (-1)^{m-1} & 0 \\ (-1)^{n-1} & 0 & 0 \\ 0 & 0 & (-1)^{k-1} \end{pmatrix}$$

3. Both c_1 and c_2 are double eigenvalues of $-\Delta$, e.g. $c_i = (m_i^2 + n_i^2)$, $m_i \neq n_i$, $i = 1,2$. We have

$$\mathrm{Ker}(D_u G_0) = \mathrm{span}[\psi_1,\ \psi_2,\ \psi_3,\ \psi_4]$$

with

$$\psi_i := \begin{pmatrix} D_{u_2} f_1^0 \\ c_1 - D_{u_1} f_1^0 \end{pmatrix} \phi_i,\ i = 1,2, \qquad \psi_i := \begin{pmatrix} D_{u_2} f_1^0 \\ c_2 - D_{u_1} f_1^0 \end{pmatrix} \phi_i,\ i = 3,4$$

and

$$\phi_{2(i-1)+1} = 2\sin m_i\pi x\ \sin n_i\pi y, \phi_{2(i-1)+2} = 2\sin n_i\pi x\ \sin m_i\pi y, i = 1,2.$$

The representation of D_4 in $\mathrm{Ker}(D_u G_0) \sim \mathbf{R}^4$ is reducible and given by

$$S_1 = \begin{pmatrix} (-1)^{m_1-1} & 0 & 0 & 0 \\ 0 & (-1)^{n_1-1} & 0 & 0 \\ 0 & 0 & (-1)^{m_2-1} & 0 \\ 0 & 0 & 0 & (-1)^{n_2-1} \end{pmatrix}$$

and

$$R = \begin{pmatrix} 0 & (-1)^{m_1-1} & 0 & 0 \\ (-1)^{n_1-1} & 0 & 0 & 0 \\ 0 & 0 & 0 & (-1)^{m_2-1} \\ 0 & 0 & (-1)^{n_2-1} & 0 \end{pmatrix}.$$

Evidently, representations of D_4 in $\mathrm{Ker}(D_uG_0)$ depend on parities of the wave numbers $m,\ n$. So do the bifurcation scenarios, as we will see later,

Remark 12.1.1. If the reaction terms $f_i(u,\lambda),\ i=1,2$ are odd in u, the additional Z_2-symmetry should be included. Its actions in $\mathrm{Ker}(D_uG_0)\sim \mathbf{R}^k$ is equal to $\pm I$.

12.2 Interaction of Two D_4-Modes

Let (λ_0, d_0) be an intersection point of two different bifurcation curves $S(c_1),\ S(c_2)$ with $c_1=2m^2\pi^2,\ c_2=2n^2\pi^2$. Namely, $(0,\lambda_0,d_0)$ is a double bifurcation point of (12.1) and the statement in (12.3) holds. Assume the Liapunov-Schmidt reduction has been performed. We consider the necessary structure of the reduced bifurcation equations

$$B(\alpha_1,\alpha_2,\beta,e) \tag{12.6}$$
$$:= \begin{pmatrix} \langle\psi_1^*,\ G(\alpha_1\psi_1+\alpha_2\psi_2+w(\alpha_1,\alpha_2,\beta,e),\ \lambda_0+\beta,\ d_0+e)\rangle \\ \langle\psi_2^*,\ G(\alpha_1\psi_1+\alpha_2\psi_2+w(\alpha_1,\alpha_2,\beta,e),\ \lambda_0+\beta,\ d_0+e)\rangle \end{pmatrix} = 0.$$

Here $\psi_i^*,\ i=1,2$ are null vectors of $D_uG_0^*$. They are given similarly as ψ_i in (12.3). The element $w(\alpha_1,\alpha_2,\beta,e)$ in the complement space M of $\mathrm{Ker}(D_uG_0)$ is the unique solution of the equation

$$QG(\alpha_1\psi_1+\alpha_2\psi_2+w,\ \lambda_0+\beta,\ d_0+e)=0. \tag{12.7}$$

12.2.1 Interaction of two even modes

If the modes ψ_1 and ψ_2 have the parities $(m,n)=$ (even, even), actions of D_4 in $\mathrm{Ker}(D_uG_0)\sim\mathbf{R}^2$ reduce to

$$S_1=R=\begin{pmatrix} -1 & 0 \\ 0 & -1 \end{pmatrix}.$$

The reduced bifurcation equations

$$B(\alpha_1,\alpha_2,\beta,e)=\begin{pmatrix} b_1(\alpha_1,\alpha_2,\beta,e) \\ b_2(\alpha_1,\alpha_2,\beta,e) \end{pmatrix}=0 \tag{12.8}$$

have the property

$$b_i(-\alpha_1,-\alpha_2,\beta,e)=-b_i(\alpha_1,\alpha_2,\beta,e). \tag{12.9}$$

The statement $b_i(0,0,\beta,e)=0$ holds for all $\beta,\ e\in\mathbf{R}$. In addition, we have $D_{\alpha_j}b_i(0,0,0,0)=0,\ i,j=1,2$. Therefore,

$$\begin{aligned} b_i(\alpha_1,\alpha_2,\beta,e) = & (a_{i1}\alpha_1 + a_{i2}\alpha_2)\beta + (\tilde{a}_{i1}\alpha_1 + \tilde{a}_{i2}\alpha_2)e \\ & + a_{i3}\alpha_1^3 + a_{i4}\alpha_1^2\alpha_2 + a_{i5}\alpha_1\alpha_2^2 + a_{i6}\alpha_2^3 \\ & + h.o.t, \qquad i = 1,2. \end{aligned}$$

Here we have taken into account the different scales of α, β, e in solutions of the reduced bifurcation equations and collected terms with the factors β^2, βe and e^2 into the higher order terms. The coefficients a_{ij}, $i = 1,2$, $j = 1,\ldots,6$ and $\tilde{a}_{ij}$, i, $j = 1,2$ are calculated by

$$\begin{aligned} a_{ij} &= \langle\psi_i^*,\ D_{u\lambda}G_0\psi_j\rangle, \qquad \tilde{a}_{ij} = \langle\psi_i^*,\ D_{ud}G_0\psi_j\rangle, \quad i,j = 1,2, \\ a_{i3} &= \langle\psi_i^*,\ D_{uu}G_0\psi_1 w_{110} + \frac{1}{6}D_{uuu}G_0\psi_1^3\rangle, \\ a_{i4} &= \langle\psi_i^*,\ D_{uu}G_0(2\psi_1 w_{120} + \psi_2 w_{110}) + \frac{1}{2}D_{uvu}G_0\psi_1^2\psi_2\rangle, \\ a_{i5} &= \langle\psi_i^*,\ D_{uu}G_0(\psi_1 w_{220} + 2\psi_2 w_{120}) + \frac{1}{2}D_{uuu}G_0\psi_1\psi_2^2\rangle, \\ a_{i6} &= \langle\psi_i^*,\ D_{uu}G_0\psi_2 w_{220} + \frac{1}{6}D_{uuu}G_0\psi_2^3\rangle, \end{aligned} \tag{12.10}$$

see also (10.60). For the integers $m \neq n$ with $(m,n) =$ (even, even), the functions $\phi_1 = 2\sin m\pi x \sin m\pi y$, $\phi_2 = 2\sin n\pi x \sin n\pi y$ have the following properties

$$\begin{gathered} \langle\phi_1,\ \phi_1\rangle = 1, \\ \langle\phi_1,\ \phi_2\rangle = \langle\phi_1^3,\ \phi_2\rangle = \langle\phi_2^3,\ \phi_1\rangle = \langle\phi_i,\ \phi_j,\ \phi_j\rangle = 0, \quad i,\ j,\ k = 1,2. \end{gathered} \tag{12.11}$$

Hence,

$$\begin{aligned} a_{i2} &= \tilde{a}_{i2} = 0, \\ a_{14} &= \langle\psi_1^*,\ D_{uu}G_0(2\psi_1 w_{120} + \psi_2 w_{110})\rangle, \\ a_{25} &= \langle\psi_2^*,\ D_{uu}G_0(\psi_1 w_{220} + 2\psi_2 w_{120})\rangle, \\ a_{23} &= \langle\psi_2^*,\ D_{uu}G_0\psi_1 w_{110}\rangle, \\ a_{16} &= \langle\psi_1^*,\ D_{uu}G_0\psi_2 w_{220}\rangle. \end{aligned} \tag{12.12}$$

We consider the truncated bifurcation equations

$$\begin{aligned} (a_{11}\beta + \tilde{a}_{11}e)\alpha_1 + a_{13}\alpha_1^3 + a_{14}\alpha_1^2\alpha_2 + a_{15}\alpha_1\alpha_2^2 + a_{16}\alpha_2^3 = 0, \\ (a_{21}\beta + \tilde{a}_{21}e)\alpha_2 + a_{23}\alpha_1^3 + a_{24}\alpha_1^2\alpha_2 + a_{25}\alpha_1\alpha_2^2 + a_{26}\alpha_2^3 = 0. \end{aligned} \tag{12.13}$$

Isolated nontrivial solutions of this system lead to solution branches of (12.1) bifurcating at $(0,\lambda_0,d_0)$. Recall that in Subsection 10.8.2 the modes ψ_1, ψ_2 are on the same group orbit and a double bifurcation is caused by the D_4-symmetry. But here the modes ψ_1, ψ_2 are not related by any symmetries. Structure of the truncated bifurcation equations (12.13) is more complicated than those in Subsection 10.8.2. Solutions of the system (12.13) are not directly available.

If $D_{uu}G_0 = 0$, e.g. when the reaction terms are odd in u, the coefficients a_{ij} in (12.10) can be calculated directly with the knowledge of ψ_1, ψ_2. The equations (12.13) reduce to

$$\begin{aligned}(a_{11}\beta + \tilde{a}_{11}e)\alpha_1 + a_{13}\alpha_1^3 + a_{15}\alpha_1\alpha_2^2 = 0,\\ (a_{21}\beta + \tilde{a}_{21}e)\alpha_1 + a_{24}\alpha_1^2\alpha_2 + a_{26}\alpha_2^3 = 0.\end{aligned} \tag{12.14}$$

Under the nondegeneracy conditions

$$a_{13} = 0, \qquad a_{26} \neq 0, \qquad a_{15}a_{24} - a_{13}a_{26} \neq 0$$

we obtain firstly four nontrivial solutions of (12.14)

$$\left(\pm\left(\frac{a_{11}\beta + \tilde{a}_{11}e}{a_{13}}\right)^{1/2}, 0\right), \qquad \left(0, \ \pm\left(\frac{a_{21}\beta + \tilde{a}_{21}e}{a_{26}}\right)^{1/2}\right). \tag{12.15}$$

These lead to solutions corresponding to the pure modes ψ_1, ψ_2, respectively. The other four nontrivial solutions are $\pm(\alpha_1, \ \mp\alpha_2)$ with

$$\begin{aligned}\alpha_1 &= \left(\frac{(a_{11}a_{26} - a_{15}a_{21})\beta + (\tilde{a}_{11}a_{26} - a_{15}\tilde{a}_{21})e}{a_{15}a_{24} - a_{13}a_{26}}\right)^{1/2},\\ \alpha_2 &= \left(-\frac{(a_{11}a_{24} - a_{13}a_{21})\beta + (\tilde{a}_{11}a_{24} - a_{13}\tilde{a}_{21})e}{a_{15}a_{24} - a_{13}a_{26}}\right)^{1/2}.\end{aligned} \tag{12.16}$$

They are solved from the equations

$$\begin{aligned}a_{11}\beta + \tilde{a}_{11}e + a_{13}\alpha_1^2 + a_{15}\alpha_2^2 = 0,\\ a_{21}\beta + \tilde{a}_{21}e + a_{24}\alpha_1^2 + a_{26}\alpha_2^2 = 0.\end{aligned}$$

This kind of mode interaction has been investigated by Aston in [28]. There are mixed modes solutions at $(0, \lambda_0, d_0)$, i.e., the modes ψ_1, ψ_2 interact and yield these new solution branches. Moreover, secondary branches occur as mixed and pure modes intersect at $\beta = -(\tilde{a}_{11}a_{26} - a_{15}\tilde{a}_{21})/(a_{11}a_{26} - a_{15}a_{21})$, and $\beta = -\tilde{a}_{11}a_{24} - \tilde{a}_{21}a_{13}$. For further discussions on this system we refer to Aston [28].

12.2.2 Interaction of an even mode with an odd mode

Let us consider modes with parities $(m, n) =$ (odd, even) (similar discussion holds for $(m, n) =$ (even, odd)). The induced actions of D_4 in $\mathrm{Ker}(D_uG_0) \sim \mathbf{R}^2$ are

$$S_1 = R = \begin{pmatrix} 1 & 0 \\ 0 & -1 \end{pmatrix}. \tag{12.17}$$

This implies the following form of the reduced bifurcation equations (12.8)

$$B(\alpha_1, \alpha_2, \beta, e) = \begin{pmatrix} b_1(\alpha_1, \alpha_2^2, \beta, e) \\ \alpha_2 b_2(\alpha_1, \alpha_2^2, \beta, e) \end{pmatrix} = 0. \tag{12.18}$$

Taking the Taylor expansion of $B(\alpha_1, \alpha_2, \beta, e)$ at the origin and truncating at the second order terms

$$\begin{aligned}(a_{11}\beta + \tilde{a}_{11}e)\alpha_1 + a_{12}\alpha_1^2 + a_{13}\alpha_2^2 = 0,\\ (a_{21}\beta + \tilde{a}_{21}e)\alpha_2 + a_{22}\alpha_1\alpha_2 = 0.\end{aligned} \tag{12.19}$$

Here $a_{i1},\ \tilde{a}_{i1},\ i = 1, 2$ are same as (12.10) and $a_{12},\ a_{13},\ a_{22}$ are defined by

$$a_{12} = \frac{1}{2}\langle \psi_1^*,\ D_{uu}G_0\psi_1\psi_1\rangle, \qquad a_{13} = \frac{1}{2}\langle \psi_1^*,\ D_{uu}G_0\psi_2\psi_2\rangle,$$
$$a_{22} = \langle \psi_2^*,\ D_{uu}G_0\psi_1\psi_2\rangle.$$

These coefficients do not vanish generically. For (m, n) =(odd, even) the eigenfunctions $\phi_1 = 2\sin m\pi x \sin m\pi y$, $\phi_2 = 2\sin n\pi x \sin n\pi y$ of the Laplacian have the properties

$$\langle \phi_1^3,\ 1\rangle = \frac{128}{9m^2\pi^2}, \qquad \langle \phi_1,\ \phi_2^2\rangle = \frac{8n^4}{m^2(n^2 - m^2)^2\pi^2}. \tag{12.20}$$

Furthermore, by

$$\psi_i = \begin{pmatrix} f^0_{1u_2} \\ c_i - f^0_{1u_2}\end{pmatrix}\phi_i, \qquad \psi_i^* = \begin{pmatrix} f^0_{2u_1} \\ c_i - f^0_{1u_1}\end{pmatrix}\phi_i, \quad i = 1, 2$$

we have for $i, j, k = 1, 2$

$$\langle \psi_k^*,\ D_{uu}G_0\psi_i\psi_j\rangle = (f^0_{2u_1},\ c_k - f^0_{1u_1})\begin{pmatrix} g_1 \\ g_2\end{pmatrix}\langle \phi_i,\ \phi_j\phi_k\rangle, \tag{12.21}$$

$$\begin{aligned} g_k := {} & D_{u_1u_1}f_k^0(f^0_{1u_2})^2 + D_{u_1u_2}f_k^0 f^0_{1u_2}(c_1 + c_2 - 2f^0_{1u_2}) \\ & + D_{u_2u_2}f_k^0(c_1 - f^0_{1u_2})(c_2 - f^0_{1u_2}).\end{aligned}$$

The system (12.19) has the following nontrivial solutions

$$\left(-\frac{a_{11}\beta + \tilde{a}_{11}e}{a_{12}},\ 0\right), \tag{12.22}$$

$$\left(-\frac{a_{21}\beta + \tilde{a}_{21}e}{a_{12}}, \pm\left(\frac{a_{21}\beta + \tilde{a}_{21}e}{a_{13}a_{22}}\Big[(a_{11} - a_{12}a_{21})\beta + (\tilde{a}_{11} - a_{12}\tilde{a}_{21})e\Big]\right)^{1/2}\right).$$

The solutions (12.22) show that the pure ψ_1 mode solution, which has the D_4-symmetry, persists at $(0, \lambda_0, d_0)$, while in the ranges where the parameters $\beta,\ e$ satisfy

$$\frac{1}{a_{13}a_{22}}(a_{21}\beta + \tilde{a}_{21}e)\Big[(a_{11} - a_{12}a_{21})\beta + (\tilde{a}_{11} - a_{12}\tilde{a}_{21})e\Big] \geq 0, \tag{12.23}$$

interaction of ψ_1 and ψ_2 takes place and perturbs the pure ψ_2 mode solution to a mixed mode solution. The ψ_2-wave is annihilated when the condition (12.23) is violated. Secondary bifurcation occurs at

$$(a_{11} - a_{12}a_{21})\beta + (\tilde{a}_{11} - a_{12}\tilde{a}_{21})e = 0,$$

where the mixed mode merges into the pure ψ_1 mode solution.

Remark 12.2.1. If the function f_i in (12.1) is odd in u, the second derivative $D_{uu}f_i$ vanishes at $(0, \lambda_0, d_0)$, $i = 1, 2$, and the system (12.19) becomes undetermined. We need to include cubic orders in the Taylor expansion of the reduced bifurcation equations

$$\begin{aligned}(a_{11}\beta + \tilde{a}_{11}e)\alpha_1 + \bar{a}_{12}\alpha_2^3 + \bar{a}_{13}\alpha_1\alpha_2^2 = 0,\\ (a_{21}\beta + \tilde{a}_{21}e)\alpha_2 + \bar{a}_{22}\alpha_1\alpha_2 + \bar{a}_{23}\alpha_2^3 = 0,\end{aligned} \tag{12.24}$$

where

$$\begin{aligned}\bar{a}_{12} &= \frac{1}{6}\langle\psi_1^*,\ D_{uuu}G_0\psi_1^3\rangle, & \bar{a}_{13} &= \frac{1}{2}\langle\psi_1^*,\ D_{uuu}G_0\psi_1\psi_2^2\rangle,\\ \bar{a}_{22} &= \frac{1}{6}\langle\psi_2^*,\ D_{uuu}G_0\psi_1^2\psi_2\rangle, & \bar{a}_{23} &= \frac{1}{6}\langle\psi_2^*,\ D_{uuu}G_0\psi_2^3\rangle.\end{aligned} \tag{12.25}$$

The system (12.24) has similar structure as the reduced bifurcation equations at a double bifurcation with irreducible D_4-actions, see (10.75). However, the reflection symmetry $(\alpha_1,\ \alpha_2) \to (\alpha_2,\ \alpha_1)$ is broken in (12.24). This is the difference between multiple bifurcations resulted from symmetries and from multiple parameters. Nontrivial solutions of (12.24) are

$$\left(\pm\left(\frac{a_{11}\beta + \tilde{a}_{11}e}{\bar{a}_{12}}\right)^{1/2},\ 0\right), \qquad \left(0,\ \pm\left(\frac{a_{21}\beta + \tilde{a}_{21}e}{\bar{a}_{23}}\right)^{1/2}\right) \tag{12.26}$$

and

$$(\pm\alpha_1^0,\ \pm\alpha_2^0) \tag{12.27}$$

with

$$\begin{aligned}\alpha_1^0 &:= \left(-\frac{(a_{11}\bar{a}_{23} - a_{21}\bar{a}_{13})\beta(\bar{a}_{11}\bar{a}_{23} - \bar{a}_{21}\bar{a}_{13})\beta}{\bar{a}_{12}\bar{a}_{23} - \bar{a}_{22}\bar{a}_{13}}\right)^{1/2},\\ \alpha_2^0 &:= \left(-\frac{(a_{21}\bar{a}_{12} - a_{11}\bar{a}_{22})\beta(\bar{a}_{12}\bar{a}_{21} - \bar{a}_{11}\bar{a}_{22})\beta}{\bar{a}_{12}\bar{a}_{23} - \bar{a}_{22}\bar{a}_{13}}\right)^{1/2}.\end{aligned}$$

12.2.3 Interaction of two odd modes

For the parity (m, n) =(odd odd), the D_4-group acts in $\mathrm{Ker}(D_uG_0) \sim \mathbf{R}^2$ trivially,

$$S_1 = R = I. \tag{12.28}$$

The reduced bifurcation equations

$$B(\alpha_1, \alpha_2, \beta, e) = \begin{pmatrix} b_1(\alpha_1, \alpha_2, \beta, e)\\ b_2(\alpha_1, \alpha_2, \beta, e)\end{pmatrix} = 0$$

do not have any symmetry and are independent from each other. Thus generically we can truncate their Taylor expansions at the second (lowest) order

$$\begin{aligned}(a_{11}\beta + \tilde{a}_{11}e)\alpha_1 + a_{12}\alpha_1^2 + a_{13}\alpha_1\alpha_2 + a_{14}\alpha_2^2 &= 0,\\ (a_{21}\beta + \tilde{a}_{21}e)\alpha_2 + a_{22}\alpha_1^2 + a_{23}\alpha_1\alpha_2 + a_{24}\alpha_2^2 &= 0.\end{aligned} \tag{12.29}$$

The coefficients $a_{i,j}$ and $\tilde{a}_{ij}$ are given via

$$\begin{aligned} a_{i1} &= \langle \psi_i^*,\ D_{u\lambda}G_0\psi_i\rangle,\\ \tilde{a}_{i1} &= \langle \psi_i^*,\ D_{ud}G_0\psi_i\rangle,\\ a_{i2} &= \langle \psi_i^*,\ D_{uu}G_0\psi_1\psi_1\rangle, \qquad i = 1,2,\\ a_{i3} &= 2\langle \psi_i^*,\ D_{uu}G_0\psi_1\psi_2\rangle,\\ a_{i4} &= \langle \psi_i^*,\ D_{uu}G_0\psi_2\psi_2\rangle. \end{aligned}$$

Denote $\mathbf{f}(u,\lambda) = (f_1(u,\lambda),\ f_2(u,\lambda))^T$. We have for $i,j,k = 1,2$

$$\begin{aligned} \langle \psi_i^*,\ D_{u\lambda}G_0\psi_i\rangle &= \begin{pmatrix} f^0_{2u_1} \\ c_i - f^0_{1u_1}\end{pmatrix}^T D_{u\lambda}\mathbf{f}_0 \begin{pmatrix} f^0_{1u_2} \\ c_i - f^0_{1u_1}\end{pmatrix},\\ \langle \psi_i^*,\ D_{ud}G_0\psi_i\rangle &= \begin{pmatrix} f^0_{2u_1} \\ c_i - f^0_{1u_1}\end{pmatrix}^T D_{ud}G_0 \begin{pmatrix} f^0_{1u_2} \\ c_i - f^0_{1u_1}\end{pmatrix} = -c_i(c_i - f^0_{1u_1})^2,\\ \langle \psi_i^*, D_{uu}G_0\psi_j\psi_k\rangle &= \begin{pmatrix} f^0_{2u_1} \\ c_i - f^0_{1u_1}\end{pmatrix}^T D_{uu}\mathbf{f}_0 \begin{pmatrix} f^0_{1u_2} \\ c_j - f^0_{1u_1}\end{pmatrix}\begin{pmatrix} f^0_{2u_2} \\ c_k - f^0_{1u_1}\end{pmatrix}\langle \phi_i, \phi_j\phi_k\rangle. \end{aligned}$$

Simple calculations show that for (m,n) =(odd odd)

$$\langle \phi_1^3,\ 1\rangle = \frac{128}{9m^2\pi^2}, \qquad \langle \phi_1^2,\ \phi_2\rangle = \frac{128m^4}{n^2\pi^2(n^2-4m^2)}.$$

Example 12.2.1. Consider the Brusselator equations. An interaction of two D_4-modes with the odd parity occurs at $A = 6$, $(B,d) = (57,\ 1)$ with the wave numbers $m = 1$ and $n = 3$. The truncated bifurcation equations (12.29) are

$$\begin{aligned} -0.0156214\alpha_2^2 - 0.000312\alpha_1\alpha_2 - 0.021194\beta\alpha_1 - 0.045836\alpha_1^2 - 1.144448e\alpha_1 &= 0,\\ -0.0211722\alpha_2^2 - 0.336953\alpha_1\alpha_2 - 0.190746\beta\alpha_2 - 0.013309\alpha_1^2 - 7.248352e\alpha_2 &= 0. \end{aligned}$$

12.3 Mode Interactions of Three Modes

If a curve $S(c_1)$ of double bifurcation points intersects a curve $S(c_2)$ of simple bifurcation point at $(0,\lambda_0,d_0)$, interactions of the two modes ψ_1, ψ_2 on $S(c_1)$ with the simple mode ψ_3 on $S(c_2)$ generate more sophisticated scenario than the interactions of two single D_4-modes. First of all, the kernel of D_uG_0 becomes three dimensional

$$\mathrm{Ker}(D_uG_0) = \mathrm{span}[\psi_1,\ \psi_2,\ \psi_3]. \tag{12.30}$$

Let $(\alpha_1, \alpha_2, \alpha_3) \in \mathbf{R}^3$ be the coordinates of an element v in $\mathrm{Ker}(D_u G_0)$ and let β, e be the variations of the parameters λ, d from λ_0 and d_0, respectively. We consider the reduced bifurcation equations

$$B(\alpha, \beta, e) = \begin{pmatrix} b_1(\alpha, \beta, e) \\ b_2(\alpha, \beta, e) \\ b_3(\alpha, \beta, e) \end{pmatrix} = 0, \qquad \alpha := (\alpha_1, \alpha_2, \alpha_3) \in \mathbf{R}^3. \tag{12.31}$$

Lowest orders of $b_i(\alpha, \beta, e)$ in α_i, $i = 1, 2, 3$ are

$$\begin{aligned} b_i(\alpha, \beta, e) &= (a_1\beta + a_2 e)\alpha_i + h.o.t., \quad i = 1, 2 \\ b_3(\alpha, \beta, e) &= (b_1\beta + b_2 e)\alpha_3 + h.o.t. \end{aligned} \tag{12.32}$$

To describe actions of D_4 in $\mathrm{Ker}(D_u G_0) \sim \mathbf{R}^3$ we recall

$$\begin{aligned} \psi_i &= \begin{pmatrix} f^0_{1u_2} \\ c_1 - f^0_{1u_1} \end{pmatrix} \phi_i, \qquad c_1 = (m^2 + n^2)\pi^2, \quad i = 1, 2, \\ \psi_3 &= \begin{pmatrix} f^0_{1u_2} \\ c_2 - f^0_{1u_1} \end{pmatrix} \phi_3, \qquad c_2 = 2k^2\pi^2 \end{aligned} \tag{12.33}$$

and

$$\phi_1 = 2\sin m\pi x \sin n\pi y, \quad \phi_2 = 2\sin n\pi x \sin m\pi y, \quad \phi_3 = 2\sin k\pi x \sin k\pi y,$$

where ϕ_1, ϕ_2 are eigenfunctions of the Laplacian $-\Delta$ associated with the double eigenvalue c_1, while ϕ_3 is associated with the simple eigenvalue c_2.

12.3.1 Induced actions

Similarly as in the previous sections, parities of the wave numbers m, n and k determine actions of D_4 in $\mathrm{Ker}(D_u G_0) \sim \mathbf{R}^3$. We have

$$S_1 = \begin{pmatrix} (-1)^{m-1} & & \\ & (-1)^{n-1} & \\ & & (-1)^{k-1} \end{pmatrix}$$

and

$$R = \begin{pmatrix} 0 & (-1)^{m-1} & 0 \\ (-1)^{n-1} & 0 & 0 \\ 0 & 0 & (-1)^{k-1} \end{pmatrix}.$$

Since the subspaces $\mathrm{span}[\psi_1, \psi_2]$ and $\mathrm{span}[\psi_3]$ are D_4-invariant, the representation of D_4 in $\mathrm{Ker}(D_u G_0) \sim \mathbf{R}^3$ is reducible. It is straightforward to verify the various types of actions of D_4 in $\mathbf{R}^3$ in Table 12.1, where the parities even and odd are abbreviated as "e" and "o", respectively.

There are eight possibilities for interactions of three modes. Exploiting the D_4-equivariance and the induced actions in $\mathbf{R}^3$ reveals systematically the structure of the reduced bifurcation equations. Correspondingly, the reduced bifurcation equations of mode interactions with three modes are divided into six different classes in Table 12.1.We mention that computer algebra can be utilized to generate effectively invariants and equivariants for finite groups (cf. Gatermann/Guyard [120]). As examples we consider two cases in the following subsections and another one in Chapter 16 in conjunction with variation of boundary conditions. The other cases can be studied similarly. We note that $S(c_1)$, $S(c_2)$ are curves of double and simple bifurcation points, respectively. If we fix the diffusion rate d to d_0 and consider bifurcations with respect to the parameter λ, the discussion in Section 10.8 shows that there are, e.g. four for (m,n) =(even,odd), and one nontrivial solution branches at all points on $S(c_i)$, $i=1,2$, respectively, except at $\lambda=\lambda_0$, where mode interaction occurs. Furthermore, locally there is no secondary bifurcation with respect to the bifurcation parameter λ for any fixed diffusion rate d_0. However, taking d as the second bifurcation parameter and varying it around d_0, secondary bifurcations may occur on $S(c_1)$, in particular, at intersections of $S(c_1)$ and $S(c_2)$.

12.3.2 Interactions of the modes (m,n,k) =(even, odd, odd)

First of all, the induced actions of D_4 in $\mathbf{R}^3$ are generated by

$$S_1=\begin{pmatrix}-1&0&0\\0&1&0\\0&0&1\end{pmatrix},\quad R=\begin{pmatrix}0&-1&0\\1&0&0\\0&0&1\end{pmatrix}.$$

Thereafter, the equivariance $S_1B(\alpha,\beta,e)=B(\alpha,\beta,e)$ implies that $b_1(\alpha,\beta,e)$ is an odd function in α_1, and $b_i(\alpha,\beta,e)$, $i=2,3$ are even in α_1. Furthermore, it follows from

$$\gamma B(\alpha,\beta,e)=B(\gamma\alpha,\beta,e)\quad\text{with}\quad \gamma:=S_1R^2=\begin{pmatrix}1&0&0\\0&-1&0\\0&0&1\end{pmatrix},$$

that both $b_i(\alpha,\beta,e)$, $i=1,3$ are even in α_2. We rewrite $b_i(\alpha,\beta,e)$, $i=1,3$ as

$$\begin{aligned}b_1(\alpha_1,\alpha_2,\alpha_3,\beta,e)&=\alpha_1\tilde{b}_1(\alpha_1^2,\alpha_2^2,\alpha_3,\beta,e),\\ b_3(\alpha_1,\alpha_2,\alpha_3,\beta,e)&=\tilde{b}_3(\alpha_1^2,\alpha_2^2,\alpha_3,\beta,e).\end{aligned}$$

Similarly, from the equivariance

$$\gamma B(\alpha,\beta,e)=B(\gamma\alpha,\beta,e)\quad\text{with}\quad \gamma:=S_1R=\begin{pmatrix}0&1&0\\1&0&0\\0&0&1\end{pmatrix},$$

we derive

$$\begin{aligned}b_2(\alpha_1,\alpha_2,\alpha_3,\beta,e)&=b_1(\alpha_2,\alpha_1,\alpha_3,\beta,e),\\ b_3(\alpha_1,\alpha_2,\alpha_3,\beta,e)&=b_3(\alpha_2,\alpha_1,\alpha_3,\beta,e).\end{aligned}$$

Cases	Parity of (m, n, k)	Generators of D_4 actions in $\mathbf{R}^3$	Group	Reduced bifurcation
1)	(e, o, o)	$S_1 = \begin{pmatrix} -1 & 0 & 0 \\ 0 & 1 & 0 \\ 0 & 0 & 1 \end{pmatrix}, \quad R = \begin{pmatrix} 0 & -1 & 0 \\ 1 & 0 & 0 \\ 0 & 0 & 1 \end{pmatrix}$	D_4	$\begin{pmatrix} \alpha_1 b_1(\alpha_1^2, \alpha_2^2, \alpha_3, \beta, e) \\ \alpha_2 b_2(\alpha_2^2, \alpha_1^2, \alpha_3, \beta, e) \\ b_3((\alpha_1\alpha_2)^2, \alpha_1^2 + \alpha_2^2, \alpha_3, \beta, e) \end{pmatrix}$
2)	(e, o, e)	$S_1 = \begin{pmatrix} -1 & 0 & 0 \\ 0 & 1 & 0 \\ 0 & 0 & -1 \end{pmatrix}, \quad R = \begin{pmatrix} 0 & -1 & 0 \\ 1 & 0 & 0 \\ 0 & 0 & -1 \end{pmatrix}$	D_4	$\begin{pmatrix} \alpha_1 b_1(\alpha_1^2, \alpha_2^2, \alpha_3^2, \beta, e) \\ \alpha_2 b_2(\alpha_2^2, \alpha_1^2, \alpha_3^2, \beta, e) \\ \alpha_3 b_3((\alpha_1\alpha_2)^2, \alpha_1^2 + \alpha_2^2, \alpha_3^2, \beta, e) \end{pmatrix}$
1)'	(o, e, o)	$S_1 = \begin{pmatrix} 1 & 0 & 0 \\ 0 & -1 & 0 \\ 0 & 0 & 1 \end{pmatrix}, \quad R = \begin{pmatrix} 0 & 1 & 0 \\ -1 & 0 & 0 \\ 0 & 0 & 1 \end{pmatrix}$	D_4	same as 1)
2)'	(o, e, e)	$S_1 = \begin{pmatrix} 1 & 0 & 0 \\ 0 & -1 & 0 \\ 0 & 0 & -1 \end{pmatrix}, \quad R = \begin{pmatrix} 0 & 1 & 0 \\ -1 & 0 & 0 \\ 0 & 0 & -1 \end{pmatrix}$	D_4	same as 2)
3)	(e, e, e)	$S_1 = -I, \quad R = -\begin{pmatrix} 0 & 1 & 0 \\ 1 & 0 & 0 \\ 0 & 0 & 1 \end{pmatrix}$	$Z_2 \times Z_2$	$\begin{pmatrix} \sum_{i=1}^3 \alpha_i b_{1i}(\alpha_1^2, \alpha_2^2, \alpha_3^2) + \alpha_1\alpha_2\alpha_3 \mathbf{b}_{14}(\alpha_1^2, \alpha_2^2, \alpha_3^2) \\ b_1(\alpha_1, \alpha_2, \alpha_3) \\ \alpha_3 b_3((\alpha_1\alpha_2)^2, \alpha_1^2 + \alpha_2^2, \alpha_3^2) \end{pmatrix}$
4)	(e, e, o)	$S_1 = \begin{pmatrix} -1 & 0 & 0 \\ 0 & -1 & 0 \\ 0 & 0 & 1 \end{pmatrix}, \quad R = \begin{pmatrix} 0 & -1 & 0 \\ -1 & 0 & 0 \\ 0 & 0 & 1 \end{pmatrix}$	$Z_2 \times Z_2$	$\begin{pmatrix} \alpha_1 b_{11}(\alpha_1^2, \alpha_2^2, \alpha_3) + \alpha_2 b_{12}(\alpha_1^2, \alpha_2^2, \alpha_3) \\ b_1(\alpha_1, \alpha_2, \alpha_3) \\ \alpha_3 b_3((\alpha_1\alpha_2)^2, \alpha_1^2 + \alpha_2^2, \alpha_3^2) \end{pmatrix}$
5)	(o, o, e)	$S_1 = \begin{pmatrix} 1 & 0 & 0 \\ 0 & 1 & 0 \\ 0 & 0 & -1 \end{pmatrix}, \quad R = \begin{pmatrix} 0 & 1 & 0 \\ 1 & 0 & 0 \\ 0 & 0 & -1 \end{pmatrix}$	$Z_2 \times Z_2$	$\begin{pmatrix} b_1(\alpha_1, \alpha_2, \alpha_3) \\ b_2(\alpha_2, \alpha_1, \alpha_3) \\ \alpha_3 b_3((\alpha_1\alpha_2)^2, \alpha_1^2 + \alpha_2^2, \alpha_3^2) \end{pmatrix}$
6)	(o, o, o)	$S_1 = I, \quad R = \begin{pmatrix} 0 & 1 & 0 \\ 1 & 0 & 0 \\ 0 & 0 & 1 \end{pmatrix}$	Z_2	$\begin{pmatrix} b_1(\alpha_1, \alpha_2, \alpha_3) \\ b_1(\alpha_2, \alpha_1, \alpha_3) \\ \alpha_3 b_3((\alpha_1\alpha_2)^2, \alpha_1^2 + \alpha_2^2, \alpha_3) \end{pmatrix}$

Table 12.1: Induced actions of D_4 in $\mathbf{R}^3 \sim \mathrm{Ker}(D_u G_0)$.

Lemma 12.3.1. *If a smooth function $g(\alpha_1, \alpha_2)$ has the property*

$$g(\alpha_1, \alpha_2) = g(\alpha_2, \alpha_1),$$

then it can be written as functions of the variables $\alpha_1\alpha_2$ and $\alpha_1 + \alpha_2$, i.e.,

$$g(\alpha_1, \alpha_2) = \tilde{g}(\alpha_1\alpha_2,\ \alpha_1 + \alpha_2).$$

Proof: Consider the Taylor expansion $g(\alpha_1, \alpha_2) = \sum_{i,j=0}^{\infty} c_{ij}\alpha_1^i\alpha_2^j$ of g at the origin. We see immediately from the symmetry $g(\alpha_1, \alpha_2) = g(\alpha_2, \alpha_1)$ that

$$c_{ij} = c_{ji} \qquad \text{for all } i, j \in \mathbf{N}.$$

Hence

$$g(\alpha_1, \alpha_2) = \sum_{i\le j} \tilde{c}_{ij}(\alpha_1^i\alpha_2^j + \alpha_1^j\alpha_2^i) = \sum_{i\le j} \tilde{c}_{ij}(\alpha_1\alpha_2)^i(\alpha_1^{j-i} + \alpha_2^{j-i}),$$

where

$$\tilde{c}_{ij} = \begin{cases} c_{ij} & \text{for } i \neq j; \\ \dfrac{1}{2}c_{ij} & \text{for } i = j. \end{cases}$$

For all $n \geq 2$ we have

$$\alpha_1^n + \alpha_2^n = (\alpha_1 + \alpha_2)(\alpha_1^{n-1} + \alpha_2^{n-1}) - \alpha_1\alpha_2(\alpha_1^{n-2} + \alpha_2^{n-2}).$$

It follows from induction that

$$g(\alpha_1, \alpha_2) = \sum \hat{c}_{ij}(\alpha_1\alpha_2)^i(\alpha_1 + \alpha_2)^j =: \tilde{g}(\alpha_1\alpha_2,\ \alpha_1 + \alpha_2). \qquad \blacksquare$$

Writing $b_3(\alpha, \beta)$ as

$$b_3(\alpha, \beta) = \tilde{b}_3((\alpha_1\alpha_2)^2,\ \alpha_1^2 + \alpha_2^2,\ \alpha_3,\ \beta,\ e),$$

we obtain the following form of the reduced bifurcation equations

$$B(\alpha, \beta, e) = \begin{pmatrix} \alpha_1 b_1(\alpha_1^2,\ \alpha_2^2,\ \alpha_3,\ \beta,\ e) \\ \alpha_2 b_1(\alpha_2^2,\ \alpha_1^2,\ \alpha_3,\ \beta,\ e) \\ b_3((\alpha_1\alpha_2)^2,\ \alpha_1^2 + \alpha_2^2,\ \alpha_3,\ \beta,\ e) \end{pmatrix} = 0. \tag{12.34}$$

Bifurcation scenario of (12.34) is described generically by the lower order terms of the Taylor expansions at the origin

$$\begin{aligned} \alpha_1(a_1\beta + a_2 e + a_3\alpha_1^2 + a_4\alpha_2^2 + a_5\alpha_3) &= 0, \\ \alpha_2(a_1\beta + a_2 e + a_4\alpha_1^2 + a_3\alpha_1^2 + a_5\alpha_3) &= 0, \\ (b_1\beta + b_2 e)\alpha_3 + b_3(\alpha_1^2 + \alpha_2^2) + b_4\alpha_3^2 &= 0. \end{aligned} \tag{12.35}$$

12.3.3 Interactions of the modes (m, n, k) =(even, odd, even)

The induced D_4-action is generated by

$$S_1 = \begin{pmatrix} -1 & 0 & 0 \\ 0 & 1 & 0 \\ 0 & 0 & -1 \end{pmatrix}, \qquad R = \begin{pmatrix} 0 & -1 & 0 \\ 1 & 0 & 0 \\ 0 & 0 & -1 \end{pmatrix}. \tag{12.36}$$

Exploiting equivariance of the reduced bifurcation equations with respect to the group elements S_1 and

$$R^2 = \begin{pmatrix} -1 & & \\ & -1 & \\ & & 1 \end{pmatrix}, \quad R^2S_1 = \begin{pmatrix} 1 & & \\ & -1 & \\ & & -1 \end{pmatrix}, \quad S_1R = \begin{pmatrix} 0 & 1 & 0 \\ 1 & 0 & 0 \\ 0 & 0 & 1 \end{pmatrix},$$

respectively, we obtain the statements

$$\begin{aligned} b_1(\alpha_1, \alpha_2, \alpha_3) &= -b_1(-\alpha_1, \alpha_2, -\alpha_3) \\ &= -b_1(-\alpha_1, -\alpha_2, \alpha_3) \\ &= b_1(\alpha_1, -\alpha_2, -\alpha_3), \end{aligned} \tag{12.37a}$$

$$\begin{aligned} b_3(\alpha_1, \alpha_2, \alpha_3) &= -b_3(-\alpha_1, \alpha_2, -\alpha_3) \\ &= b_3(-\alpha_1, -\alpha_2, \alpha_3) \\ &= -b_3(\alpha_1, -\alpha_2, -\alpha_3) \end{aligned} \tag{12.37b}$$

and

$$b_2(\alpha_1, \alpha_2, \alpha_3) = b_1(\alpha_2, \alpha_1, \alpha_3), \qquad b_3(\alpha_1, \alpha_2, \alpha_3) = b_3(\alpha_2, \alpha_1, \alpha_3). \tag{12.37c}$$

Here and in the sequel we suppress the dependence of b_i on the parameters β and e when no confusion arises. Consider the Taylor expansion

$$b_1(\alpha_1, \alpha_2, \alpha_3) = \sum_{i,j,k \in \mathbf{N}} c_{ijk} \alpha_1^i \alpha_2^j \alpha_3^k.$$

We derive from the statements in (12.37a) the possibilities

$$i + j = \text{odd}, \qquad i + k = \text{odd}, \qquad j + k = \text{even}.$$

Correspondingly we have either

$$i = \text{even}, \qquad j = \text{odd}, \qquad k = \text{odd}$$

or

$$i = \text{odd}, \qquad j = \text{even}, \qquad k = \text{even}.$$

The former is excluded by the presence of a linear term in α_1 in $b_1(\alpha, \beta, e)$, see (12.32). The latter implies that the function $b_1(\alpha_1, \alpha_2, \alpha_3)$ is of the form

$$b_1(\alpha_1, \alpha_2, \alpha_3) = \alpha_1 \tilde{b}_1(\alpha_1^2, \alpha_2^2, \alpha_3^2).$$

Similarly, we have

$$b_3(\alpha_1, \alpha_2, \alpha_3) = \alpha_3 \tilde{b}_3(\alpha_1^2, \alpha_2^2, \alpha_3^2).$$

Moreover, it follows from the statements in (12.37c) and Lemma 12.3.1 that

$$\begin{aligned} b_2(\alpha_1, \alpha_2, \alpha_3) &= \alpha_2 \tilde{b}_2(\alpha_2^2, \alpha_1^2, \alpha_3^2), \\ b_3(\alpha_1, \alpha_2, \alpha_3) &= \alpha_3 \tilde{b}_3(\alpha_1^2 \alpha_2^2, \alpha_1^2 + \alpha_2^2, \alpha_3^2). \end{aligned}$$

In other words, the reduced bifurcation equations are of the form

$$B(\alpha, \beta, e) = \begin{pmatrix} \alpha_1 b_1(\alpha_1^2, \alpha_2^2, \alpha_3^2, \beta, e) \\ \alpha_2 b_2(\alpha_2^2, \alpha_1^2, \alpha_3^2, \beta, e) \\ \alpha_3 b_3((\alpha_1 \alpha_2)^2, \alpha_1^2 + \alpha_2^2, \alpha_3^2, \beta, e) \end{pmatrix} = 0.$$

We are interested in the lower order terms in the Taylor expansion of $B(\alpha, \beta, e)$,

$$\begin{aligned} \alpha_1(a_1\beta + a_2 e + a_3\alpha_1^2 + a_4\alpha_2^2 + a_5\alpha_3^2) &= 0, \\ \alpha_2(a_1\beta + a_2 e + a_4\alpha_1^2 + a_3\alpha_1^2 + a_5\alpha_3^2) &= 0, \\ \alpha_3(b_1\beta + b_2 e + b_3(\alpha_1^2 + \alpha_2^2) + b_4\alpha_2^2) &= 0. \end{aligned}$$

Structure of the reduced bifurcation equations for other cases in Table 12.3.2 can be derived similarly. Note that the function $b_i(\alpha, \beta, e)$ for reducible actions of D_4 has less symmetry. But terms of the quadratic order in the Taylor expansion of $b_i(\alpha, \beta, e)$ become important for the determination of bifurcation scenario. This reduction of orders in truncating Taylor expansions simplifies the analysis of bifurcation too.

Remark 12.3.1. We postpone analysis of bifurcating solution curves at a three-mode interaction to Chapter 16, where a homotopy of boundary conditions is considered.

12.4 Interactions of Four Modes

Let $S(c_1)$, $S(c_2)$ be two curves of double bifurcation points and

$$c_1 = (m^2 + n^2)^2 \pi^2, \; c_2 = (k^2 + l^2)\pi^2, \; c_1 \neq c_2, \; m \neq n \; \text{ and } \; k \neq l. \tag{12.38}$$

If $(0, \lambda_0, d_0)$ is an intersection of these two curves, then it is a bifurcation point of the multiplicity four and

$$\text{Ker}(D_u G_0) = \text{span}[\psi_1, \; \psi_2, \; \psi_3, \; \psi_4] \tag{12.39}$$

with

$$\begin{aligned}
\psi_i &= \begin{pmatrix} f^0_{1u_2} \\ c_1 - f^0_{1u_1} \end{pmatrix} \phi_i, & \psi_{i+2} &= \begin{pmatrix} f^0_{1u_2} \\ c_2 - f^0_{1u_1} \end{pmatrix} \phi_{i+2}, \quad i = 1,2, \\
\phi_1 &= 2 \sin mx \sin ny, & \phi_2 &= 2 \sin nx \sin my, \\
\phi_3 &= 2 \sin kx \sin ly, & \phi_4 &= 2 \sin lx \sin ky.
\end{aligned} \tag{12.40}$$

Action of D_4 in $\mathrm{Ker}(D_u G_0) \sim \mathbf{R}^4$ is reducible and generated by

$$S_1 = \begin{pmatrix} (-1)^{m-1} & & & \\ & (-1)^{n-1} & & \\ & & (-1)^{l-1} & \\ & & & (-1)^{k-1} \end{pmatrix},$$

$$R = \begin{pmatrix} 0 & (-1)^{m-1} & 0 & 0 \\ (-1)^{n-1} & 0 & 0 & 0 \\ 0 & 0 & 0 & (-1)^{k-1} \\ 0 & 0 & (-1)^{l-1} & 0 \end{pmatrix}.$$

Similarly to the discussions in Section 12.3, we choose the coordinates $(\alpha_1, \alpha_2, \alpha_3, \alpha_4)$ of v in $\mathrm{Ker}(D_u G_0)$ via

$$v = \sum_{i=1}^{4} \alpha_i \phi_i \in \mathrm{Ker}(D_u G_0) \longleftrightarrow \alpha := (\alpha_1, \alpha_2, \alpha_3, \alpha_4)^T \in \mathbf{R}^4. \tag{12.41}$$

We derive structure of the reduced bifurcation equations

$$B(\alpha, \beta, e) = \begin{pmatrix} b_1(\alpha_1, \alpha_2, \alpha_3, \alpha_4, \beta, e) \\ b_2(\alpha_1, \alpha_2, \alpha_3, \alpha_4, \beta, e) \\ b_3(\alpha_1, \alpha_2, \alpha_3, \alpha_4, \beta, e) \\ b_4(\alpha_1, \alpha_2, \alpha_3, \alpha_4, \beta, e) \end{pmatrix} = 0 \tag{12.42}$$

with the induced actions of D_4 in $\mathbf{R}^4$ for different parities of m, n, k, l. Table 12.2 shows the isotropy lattice of D_4 in $\mathbf{R}^4$.

$$\begin{array}{c} D_4 \\ \downarrow \\ \mathbf{Z}_2(S_1) \times \mathbf{Z}_2(R) \\ \downarrow \\ \mathbf{Z}_2(R) \end{array}$$

Table 12.2. Isotropy lattices for the actions of D_4 on $\mathbf{R}^4$.

12.4.1 Interactions of the modes $(m,\ n,\ k,\ l)$ = (even, odd, even, odd)

Suppose the parity (m, n, k, l) =(even, odd, even, odd) for the four modes $\psi_i,\ i = 1, \ldots, 4$ at $(0, \lambda_0, d_0)$. Action of D_4 in $\mathbf{R}^4$ is generated by

$$S_1 = \begin{pmatrix} -1 & & & \\ & 1 & & \\ & & -1 & \\ & & & 1 \end{pmatrix}, \qquad R = \begin{pmatrix} & -1 & & \\ 1 & & & \\ & & & -1 \\ & & 1 & \end{pmatrix}. \tag{12.43}$$

The D_4-equivariance of $B(\alpha, \beta, e)$ and the action of

$$S_1 R = \begin{pmatrix} & 1 & & \\ 1 & & & \\ & & & 1 \\ & & 1 & \end{pmatrix}$$

imply

$$b_{2+i}(\alpha_1, \alpha_2, \alpha_3, \alpha_4) = b_{1+i}(\alpha_2, \alpha_1, \alpha_4, \alpha_3), \quad i = 0, 2.$$

We examine the Taylor expansion of $b_1(\alpha_1, \alpha_2, \alpha_3, \alpha_4)$ at $\alpha = 0$

$$b_1(\alpha) = \sum_{i,j,p,q=0}^{\infty} b_{1ijpq} \alpha_1^i \alpha_2^j \alpha_3^p \alpha_4^q.$$

By the actions of S_1 and $R^2 S_1$ we obtain

$$i + p = \text{odd}, \qquad j + q = \text{even}.$$

Thus the integers $i,\ j,\ p,\ q$ have following possible parities

$$(i,\ j,\ p,\ q) \in \left\{ \begin{array}{c} \text{(even, even, odd, even)}, \\ \text{(even, odd, odd, odd)}, \\ \text{(odd, even, even, even)}, \\ \text{(odd, odd, even, odd)} \end{array} \right\}.$$

Writing the functions $b_i(\alpha),\ i = 1, 3$ as

$$\begin{aligned} b_i(\alpha) &= \alpha_1 b_{i1}(\alpha_1^2, \alpha_2^2, \alpha_3^2, \alpha_4^2) + \alpha_3 b_{i2}(\alpha_1^2, \alpha_2^2, \alpha_3^2, \alpha_4^2) \\ &\quad + \alpha_1 \alpha_2 \alpha_4 b_{i3}(\alpha_1^2, \alpha_2^2, \alpha_3^2, \alpha_4^2) + \alpha_3 \alpha_2 \alpha_4 b_{i4}(\alpha_1^2, \alpha_2^2, \alpha_3^2, \alpha_4^2) \\ &= \alpha_1 \tilde{b}_i(\alpha_1^2, \alpha_2^2, \alpha_3^2, \alpha_4^2, \alpha_2 \alpha_4) + \alpha_3 \tilde{b}_{i+1}(\alpha_1^2, \alpha_2^2, \alpha_3^2, \alpha_4^2, \alpha_2 \alpha_4) \end{aligned}$$

and omitting the tilde $\tilde{}$, we obtain the following form of the reduced bifurcation equations

$$B(\alpha) = \begin{pmatrix} \alpha_1 b_1(\alpha_1^2, \alpha_2^2, \alpha_3^2, \alpha_4^2, \alpha_2\alpha_4) + \alpha_3 b_2(\alpha_1^2, \alpha_2^2, \alpha_3^2, \alpha_4^2, \alpha_2\alpha_4) \\ \alpha_2 b_1(\alpha_2^2, \alpha_1^2, \alpha_4^2, \alpha_3^2, \alpha_1\alpha_3) + \alpha_4 b_2(\alpha_2^2, \alpha_1^2, \alpha_4^2, \alpha_3^2, \alpha_1\alpha_3) \\ \alpha_1 b_3(\alpha_1^2, \alpha_2^2, \alpha_3^2, \alpha_4^2, \alpha_2\alpha_4) + \alpha_3 b_4(\alpha_1^2, \alpha_2^2, \alpha_3^2, \alpha_4^2, \alpha_2\alpha_4) \\ \alpha_2 b_3(\alpha_2^2, \alpha_1^2, \alpha_4^2, \alpha_3^2, \alpha_1\alpha_3) + \alpha_4 b_4(\alpha_2^2, \alpha_1^2, \alpha_4^2, \alpha_3^2, \alpha_1\alpha_3) \end{pmatrix} = 0. \tag{12.44}$$

Taylor expansion of $B(\alpha)$ follows from those of its first and third components. Moreover, their truncations at the cubic order are determined if coefficients in the following equalities are available

$$b_i(\alpha_1^2, \alpha_2^2, \alpha_3^2, \alpha_4^2, \alpha_2\alpha_4) = b_{i0}(0) + \sum_{j=1}^{4} b_{ij}\alpha_j^2 + b_{i5}\alpha_2\alpha_4, \quad i = 1, \ldots, 4. \tag{12.45}$$

We consider now (12.44) as a system of cubic polynomials with the components b_i, $i = 1, 2, 3, 4$ in (12.45). Since the subspaces span$[e_1, e_3]$, span$[e_2, e_4]$, span$[e_1 - e_2, e_3 - e_4]$ and span$[e_1 + e_2, e_3 + e_4] \subset \mathbf{R}^4$ are invariant for the vector fields $B(\alpha)$, we obtain some special solutions of (12.44).

1. Let $\alpha_2 = \alpha_4 = 0$. The system (12.44) reduces to

$$\alpha_1 b_1(\alpha_1^2, 0, \alpha_3^2, 0, 0) + \alpha_3 b_2(\alpha_1^2, 0, \alpha_3^2, 0, 0) = 0,$$
$$\alpha_1 b_3(\alpha_1^2, 0, \alpha_3^2, 0, 0) + \alpha_3 b_4(\alpha_1^2, 0, \alpha_3^2, 0, 0) = 0.$$

This yields eight nontrivial solutions $(\alpha_1, 0, \alpha_3, 0)$ for (12.44). Applying the symmetry R to those solutions, we obtain eight other solutions of the form $(0, \alpha_1, 0, \alpha_3)$.

2. Let $\alpha_1 = \alpha_2 \neq 0$, $\alpha_3 = \alpha_4 \neq 0$. We obtain from (12.44) another system for α_1, α_3:

$$\alpha_1 b_1(\alpha_1^2, \alpha_1^2, \alpha_3^2, \alpha_3^2, \alpha_1\alpha_3) + \alpha_3 b_2(\alpha_1^2, \alpha_1^2, \alpha_3^2, \alpha_3^2, \alpha_1\alpha_3) = 0,$$
$$\alpha_1 b_3(\alpha_1^2, \alpha_1^2, \alpha_3^2, \alpha_3^2, \alpha_1\alpha_3) + \alpha_3 b_4(\alpha_1^2, \alpha_1^2, \alpha_3^2, \alpha_3^2, \alpha_1\alpha_3) = 0.$$

With b_i in (12.45) this system of two cubic polynomials has eight nontrivial solutions of (12.44). The action of S yields eight other solutions on the group orbits.

Remark 12.4.1. Since the system (12.44) generically does not have solutions of the form $(\alpha_1, \alpha_2, 0, 0)$ or $(0, 0, \alpha_3, \alpha_4)$, the four modes ψ_1, ψ_2 and ψ_3, ψ_4 interact well at $(0, \lambda_0, d_0)$ such that all bifurcating solution branches involve modes on both curves $S(c_1)$ and $S(c_2)$. In other words, we have merely mixed modes solutions at the interaction point.

12.4.2 Interactions of the modes (m, n, k, l) = (even, even, even, odd)

If parity of the four modes at $(0, \lambda_0, d_0)$ is (m, n, k, l) =(even, even, even, odd), action of D_4 in $\mathbf{R}^4$ is generated by

$$S_1 = \begin{pmatrix} -1 & & & \\ & -1 & & \\ & & -1 & \\ & & & 1 \end{pmatrix}, \quad R = \begin{pmatrix} & -1 & & \\ -1 & & & \\ & & & -1 \\ & & 1 & \end{pmatrix},$$

We exploit the D_4-equivariance of the reduced bifurcation equations to determine structure of $b_i(\alpha)$, $i = 1,2,3,4$. First of all by S_1R-symmetry we obtain

$$b_{2+i}(\alpha_1,\alpha_2,\alpha_3,\alpha_4) = b_{1+i}(\alpha_2,\alpha_1,\alpha_4,\alpha_3), \quad i = 0,2.$$

Let

$$b_1(\alpha) = \sum b_{1ijpq}\alpha_1^i\alpha_2^j\alpha_3^p\alpha_4^q.$$

We derive from actions of the group elements S_1,

$$R^2 = \begin{pmatrix} 1 & & & \\ & 1 & & \\ & & -1 & \\ & & & -1 \end{pmatrix} \quad \text{and} \quad S_1R = \begin{pmatrix} -1 & & & \\ & -1 & & \\ & & 1 & \\ & & & -1 \end{pmatrix}$$

that the power of $\alpha_1^i\alpha_2^j\alpha_3^p\alpha_4^q$ is of the pattern

$$(i,\ j,\ p,\ q) \in \left\{ \begin{array}{c} \text{(even, odd, even, even)}, \\ \text{(odd, even, even, even)}, \\ \text{(even, even, odd, odd)}, \\ \text{(odd, odd, odd, odd)} \end{array} \right\}.$$

Hence we write the function $b_1(\alpha)$ as

$$\begin{aligned} b_1(\alpha) &= \alpha_1 b_{11}(\alpha_1^2,\alpha_2^2,\alpha_3^2,\alpha_4^2) + \alpha_2 b_{12}(\alpha_1^2,\alpha_2^2,\alpha_3^2,\alpha_4^2) \\ &\quad + \alpha_3\alpha_4 b_{13}(\alpha_1^2,\alpha_2^2,\alpha_3^2,\alpha_4^2,\alpha_1\alpha_2). \end{aligned}$$

Similarly, the function $b_3(\alpha) = \sum b_{3ijpq}\alpha_1^i\alpha_2^j\alpha_3^p\alpha_4^q$ has the property

$$(i,\ j,\ p,\ q) \in \left\{ \begin{array}{c} \text{(even, odd, even, odd)}, \\ \text{(odd, even, even, odd)}, \\ \text{(odd, odd, odd, even)}, \\ \text{(even, even, odd, even)} \end{array} \right\}$$

and can be written as

$$\begin{aligned} b_3(\alpha) &= \alpha_2\alpha_4\tilde{b}_{31}(\alpha_1^2,\alpha_2^2,\alpha_3^2,\alpha_4^2) + \alpha_1\alpha_4\tilde{b}_{32}(\alpha_1^2,\alpha_2^2,\alpha_3^2,\alpha_4^2) \\ &\quad + \alpha_1\alpha_2\alpha_3\tilde{b}_{33}(\alpha_1^2,\alpha_2^2,\alpha_3^2,\alpha_4^2) + \alpha_3\tilde{b}_{34}(\alpha_1^2,\alpha_2^2,\alpha_3^2,\alpha_4^2) \\ &= \alpha_3 b_{31}(\alpha_1^2,\alpha_2^2,\alpha_3^2,\alpha_4^2,\alpha_1\alpha_2) + \alpha_4\alpha_1 b_{32}(\alpha_1^2,\alpha_2^2,\alpha_3^2,\alpha_4^2) \\ &\quad + \alpha_4\alpha_2 b_{33}(\alpha_1^2,\alpha_2^2,\alpha_3^2,\alpha_4^2). \end{aligned}$$

So we consider the truncated Taylor expansion of the reduced bifurcation equations

$$\begin{aligned} &\alpha_1 b_{11}(\alpha_1^2,\alpha_2^2,\alpha_3^2,\alpha_4^2) + \alpha_2 b_{12}(\alpha_1^2,\alpha_2^2,\alpha_3^2,\alpha_4^2) \\ &\qquad + \alpha_3\alpha_4 b_{13}(\alpha_1^2,\alpha_2^2,\alpha_3^2,\alpha_4^2,\alpha_1\alpha_2) = 0, \\ &\alpha_2 b_{11}(\alpha_2^2,\alpha_1^2,\alpha_4^2,\alpha_3^2) + \alpha_1 b_{12}(\alpha_2^2,\alpha_1^2,\alpha_4^2,\alpha_3^2) \\ &\qquad + \alpha_3\alpha_4 b_{13}(\alpha_2^2,\alpha_1^2,\alpha_4^2,\alpha_3^2,\alpha_1\alpha_2) = 0, \\ &\alpha_3 b_{31}(\alpha_1^2,\alpha_2^2,\alpha_3^2,\alpha_4^2,\alpha_1\alpha_2) + \alpha_4\alpha_1 b_{32}(\alpha_1^2,\alpha_2^2,\alpha_3^2,\alpha_4^2) \\ &\qquad + \alpha_4\alpha_2 b_{33}(\alpha_1^2,\alpha_2^2,\alpha_3^2,\alpha_4^2) = 0, \\ &\alpha_4 b_{31}(\alpha_2^2,\alpha_1^2,\alpha_4^2,\alpha_3^2,\alpha_1\alpha_2) + \alpha_3\alpha_2 b_{32}(\alpha_2^2,\alpha_1^2,\alpha_4^2,\alpha_3^2) \\ &\qquad + \alpha_3\alpha_1 b_{33}(\alpha_2^2,\alpha_1^2,\alpha_4^2,\alpha_3^2) = 0. \end{aligned} \tag{12.46}$$

The structure of these equations makes some solutions directly available. For examples,

1. Let $\alpha_3 = \alpha_4 = 0$. The system (12.46) reduces to
$$\begin{aligned} \alpha_1 b_{11}(\alpha_1^2, \alpha_2^2, 0, 0) + \alpha_2 b_{12}(\alpha_1^2, \alpha_2^2, 0, 0) = 0, \\ \alpha_2 b_{11}(\alpha_2^2, \alpha_1^2, 0, 0) + \alpha_1 b_{12}(\alpha_2^2, \alpha_1^2, 0, 0) = 0. \end{aligned} \tag{12.47}$$
These are exactly the reduced bifurcation equations along the curve $S(c_1)$ of double bifurcation points, and can be solved as (10.58). Therefore, the bifurcation solution branches along $S(c_1)$ persist at the intersection $S(c_1) \cap S(c_2)$. Moreover, if the reaction terms $f_i(u, \lambda)$ are odd in u, then
$$b_{12}(\alpha_1^2, \alpha_2^2, 0, 0) = 0.$$
We obtain from (12.47)
$$\begin{aligned} \alpha_1 b_{11}(\alpha_1^2, \alpha_2^2, 0, 0) = 0, \\ \alpha_2 b_{11}(\alpha_2^2, \alpha_1^2, 0, 0) = 0. \end{aligned} \tag{12.48}$$
Solutions of this system can be given directly as those in (10.63).
2. For $\alpha_1 = \alpha_2 = 0$ the system (12.46) is equivalent to
$$\begin{aligned} \alpha_3 \alpha_4 b_{13}(0, 0, \alpha_3^2, \alpha_4^2, 0) = 0, \\ \alpha_3 b_{31}(0, 0, \alpha_3^2, \alpha_4^2, 0) = 0, \\ \alpha_4 b_{31}(0, 0, \alpha_4^2, \alpha_3^2, 0) = 0. \end{aligned} \tag{12.49}$$
Since the inequality $b_{13}(0) \neq 0$ holds generically, the system (12.49) is overdetermined and has no nontrivial solutions. Consequently, bifurcating solution branches along $S(c_2)$ are destroyed in the sense there are no solutions with merely the modes ψ_3, ψ_4.
3. Consider the case $\alpha_1 = \alpha_2$, $\alpha_3 = \alpha_4$, we derive from (12.46)
$$\begin{aligned} &\alpha_1 \Big(b_{11}(\alpha_1^2, \alpha_1^2, \alpha_3^2, \alpha_3^2) + b_{12}(\alpha_1^2, \alpha_1^2, \alpha_3^2, \alpha_3^2) \Big) \\ &\qquad + \alpha_3^2 b_{13}(\alpha_1^2, \alpha_1^2, \alpha_3^2, \alpha_3^2, \alpha_1^2) = 0, \\ &\alpha_3 \Big[b_{31}(\alpha_1^2, \alpha_1^2, \alpha_3^2, \alpha_3^2, \alpha_1^2) + \alpha_1 \Big(b_{32}(\alpha_1^2, \alpha_1^2, \alpha_3^2, \alpha_3^2) \\ &\qquad + b_{33}(\alpha_1^2, \alpha_1^2, \alpha_3^2, \alpha_3^2) \Big) \Big] = 0. \end{aligned}$$
This system gives six additional nontrivial solutions of (12.46). Two of them are of the form $\alpha_1 \neq 0$, $\alpha_3 = 0$, which are described also by (12.47).
4. Similarly, by setting $\alpha_1 = \alpha_2$, $\alpha_3 = -\alpha_4$ we reduce the system (12.46) to
$$\begin{aligned} &\alpha_1 \Big(b_{11}(\alpha_1^2, \alpha_1^2, \alpha_3^2, \alpha_3^2) + b_{12}(\alpha_1^2, \alpha_1^2, \alpha_3^2, \alpha_3^2) \Big) \\ &\qquad - \alpha_3 b_{13}(\alpha_1^2, \alpha_1^2, \alpha_3^2, \alpha_3^2, \alpha_1^2) = 0, \\ &\alpha_3 \Big[b_{31}(\alpha_1^2, \alpha_1^2, \alpha_3^2, \alpha_3^2, \alpha_1^2) - \alpha_1 \Big(b_{32}(\alpha_1^2, \alpha_1^2, \alpha_3^2, \alpha_3^2) \\ &\qquad + b_{33}(\alpha_1^2, \alpha_1^2, \alpha_3^2, \alpha_3^2) \Big) \Big] = 0. \end{aligned}$$

This leads to other six nontrivial solutions of (12.46).

12.5 Reactions with Z_2-Symmetry

We consider in this section a special type of reactions

$$f_i(u,\lambda) = \lambda \tilde{f}_i(u) \quad \text{and} \quad \tilde{f}_i(-u) = -\tilde{f}_i(u), \qquad i = 1,2, \tag{12.50}$$

see also Section 10.4. We will neglect the sign $\tilde{}$ in the sequel. With this kind of reaction terms the system (12.1) with (12.50) is $D_4 \times Z_2$-equivariant. We use the scaling techniques, discussed in Section 6.7, to analyze the bifurcation scenario of (12.1) at a multiple bifurcation point $(0, \lambda(d), d)$. Assume

$$(\lambda(d), d) \in S(c), \quad (\lambda, d) \notin \bigcup_{\tilde{c} \neq c} S(\tilde{c}) \quad \text{for all } \tilde{c} = (\tilde{m}^2 + \tilde{n}^2)\pi^2, \;\; \tilde{m}, \tilde{n} \in \mathbf{N}. \tag{12.51}$$

In other words, $(0, 0, \lambda(d), d)$ is on a single curve $S(c)$ of bifurcation points. The multiple bifurcations on a single curve $S(c)$ result from (geometric) symmetries and can be tackled more easily than mode interactions at $S(c_1) \cap S(c_2)$ ($c_1 \neq c_2$) which are induced by (algebraic degeneracies) the multiple parameters in the problem.

Let $\{(\lambda_0(d), d) \mid d \in [-d_0, d_0]\}$ be a piece of a smooth curve in $S(c)$ and the statement (12.51) is satisfied with an l_c-fold eigenvalue c of the Laplacian. We want to determine smooth solution manifold $\{(u(t,d), \lambda(t,d), d) \mid d \in [-d_0, d_0],\ t \in [-t_0, t_0]\}$ of (12.1) such that

$$\begin{aligned} &u(0,d) = 0, \quad \lambda(0,d) = \lambda_0(d), \quad d \in [-d_0, d_0], \\ &\frac{\partial}{\partial t}(u(t,d), \lambda(t,d))\Big|_{t=0} \neq (0,0). \end{aligned} \tag{12.52}$$

By Theorem 10.6.1 there are smooth functions $\alpha_i(t,d)$, $\beta(t,d) \in \mathbf{R}, i = 1, \ldots, l_c$ and $w(t,d) \in M$, such that

$$\begin{aligned} u(t,d) &= t \sum_{i=1}^{l_c} \alpha_i(t,d)\psi_i + t w(t,d), \\ \lambda(t,d) &= \lambda_0(d) + t\beta(t,d), \qquad w(t,d) \in M. \end{aligned} \tag{12.53}$$

Here and in the sequel, we use ψ_i to denote the basis element of $\mathrm{Ker}(D_u G(0, \lambda, d))$ for $(\lambda, d) \in S(c)$, see (10.40) and Section 10.6 for more details. Let

$$\delta := \det \begin{pmatrix} f^0_{1u_1} & f^0_{1u_2} \\ f^0_{2u_1} & f^0_{2u_2} \end{pmatrix} = f^0_{1u_1} f^0_{2u_2} - f^0_{2u_1} f^0_{1u_2}.$$

Theorem 12.5.1. *If the conditions (12.51), (12.52) and the nondegeneracy condition*

$$a_0 := \delta\, f^0_{1u_1}\, \lambda^2 - 2c\delta\lambda + c^2\, f^0_{2u_2} \neq 0 \tag{12.54}$$

are satisfied, solution manifolds of (12.1) are of either the form

$$\begin{aligned} u(t,d) &= t^2 \sum_{i=1}^{l_c} \alpha_i(t,d)\psi_i + t^3 w(t,d), \\ \lambda(t,d) &= \lambda_0(d) + t\beta(t,d) \quad \text{with } \beta(0,d) \neq 0,\ w(t,d) \in M \end{aligned} \tag{12.55}$$

or

$$\begin{aligned} u(t,d) &= t \sum_{i=1}^{l_c} \alpha_i(t,d)\psi_i + t^3 w(t,d), \\ \lambda(t,d) &= \lambda_0(d) + t^2\beta(t,d), \qquad \sum_{i=1}^{l_c} |\alpha_i(0,d)| > 0,\ w(t,d) \in M. \end{aligned} \tag{12.56}$$

Proof: Differentiating the equation

$$G(u(t,d), \lambda(t,d), d) = 0$$

with respect to t yields

$$D_u G \dot{u} + D_\lambda G \dot{\lambda} = 0, \tag{12.57}$$

where the dot "$\cdot$" represents the differentiation with respect to t. For $t = 0$ we obtain

$$D_u G_0 w(0,d) = 0.$$

Since $w(0,d) \in M = \mathrm{Ker}(D_u G_0)^\perp$, we have

$$w(0,d) = 0. \tag{12.58}$$

We differentiate (12.57) with respect to t

$$D_{uu} G \dot{u}^2 + 2D_{u\lambda} G \dot{u}\dot{\lambda} + D_{\lambda\lambda} G \dot{\lambda}^2 + D_u G \ddot{u} + D_\lambda G \ddot{\lambda} = 0. \tag{12.59}$$

Note that the property (12.50) implies $D_{uu} G_0 = 0$, $D_{\lambda\lambda} G_0 = 0$ and $D_\lambda G_0 = 0$. Denote

$$\alpha_i = \alpha_i(0,d), \qquad \beta = \beta(0,d), \qquad \dot{w} = \dot{w}(0,d). \tag{12.60}$$

At $t = 0$ the equation (12.59) reduces to

$$D_{u\lambda} G_0 \left(\sum_{i=1}^{l_c} \alpha_i \psi_i \right) \beta + D_u G_0 \dot{w} = 0. \tag{12.61}$$

We take at both sides of the equation the L^2-product with ψ_j^*, $j = 1, \ldots, l_c$, see (10.40),

$$(-\lambda f^0_{2u_1},\ \lambda f^0_{1u_1} - c)\begin{pmatrix} f^0_{1u_1} & f^0_{1u_2} \\ f^0_{2u_1} & f^0_{2u_2} \end{pmatrix}\begin{pmatrix} -\lambda f^0_{1u_2} \\ \lambda f^0_{1u_1} - c \end{pmatrix}\left\langle \phi_j,\ \sum_{i=1}^{l_c} \alpha_i \phi_i \right\rangle \beta = 0.$$

It follows from the orthogonality of the eigenfunctions ϕ_j, $j = 1, \ldots, l_c$ that

$$(\delta f^0_{1u_1} \lambda^2 - 2c\delta\lambda + c^2 f^0_{2u_2})\alpha_j \beta = 0, \qquad j = 1, \ldots, l_c.$$

In view of the nondegeneracy condition (12.54), we see

$$\alpha_j \beta = 0, \qquad j = 1, \ldots, l_c.$$

Thus we have either

$$\text{(i)} \qquad \beta(0,d) \neq 0, \qquad \alpha_j = 0, \qquad j = 1, \ldots, l_c \tag{12.62}$$

or

$$\text{(ii)} \qquad \beta = 0 \ \text{ with } \ \sum_{i=1}^{l_c} |\alpha_j| > 0. \tag{12.63}$$

In both cases we can simplify (12.61) into

$$D_u G_0 \dot{w} = 0 \implies \dot{w} = 0.$$

Thereafter, we scale the parameterization of $\alpha_i(t,d)$, $\beta(t,d)$ and $w(t,d)$ in (12.53) and obtain the conclusions (12.55), (12.56). ■

Remark 12.5.1. The solution manifold in (12.55) corresponds to the trivial solution branch $\{(0, \lambda_0(d), d);\ d \in [-d_0, d_0]\}$ (cf. Mei [213, 216]). We restrict the following discussion to the situation (12.56).

Differentiating (12.59) with respect to t at $t = 0$,

$$D_{uuu}G_0\Big(\sum_{i=1}^{l_c} \alpha_i \psi_i\Big)^3 + 6\beta D_{u\lambda}G_0\Big(\sum_{i=1}^{l_c} \alpha_i \psi_i\Big) + 6 D_u G_0 w = 0. \tag{12.64}$$

Note that

$$D_{uuu}G_0 v^3 = \lambda D^3 \begin{pmatrix} f_1 \\ f_2 \end{pmatrix}\Bigg|_{(0,\lambda(d),d)} v^3 \quad \text{for all} \quad v \in \mathrm{Ker}(D_u G(0, \lambda(d), d)).$$

Define

$$b_0 := \begin{pmatrix} -\lambda f^0_{2u_2} \\ \lambda f^0_{1u_1} - c \end{pmatrix}^T D^3 \begin{pmatrix} f_1 \\ f_2 \end{pmatrix}\Bigg|_{(0,\lambda(d),d)} \begin{pmatrix} -\lambda f^0_{1u_2} \\ \lambda f^0_{1u_1} - c \end{pmatrix}^3. \tag{12.65}$$

We consider the L^2-product of $\psi^*_j, j = 1, \ldots, l_c$ with (12.64), i.e.,

$$b_0 \left\langle \phi_j,\ \Big(\sum_{i=1}^{l_c} \alpha_i \phi_i\Big)^3 \right\rangle + 6 a_0 \beta \alpha_j = 0, \quad j = 1, \ldots, l_c. \tag{12.66}$$

Thanks to the following properties of the eigenfunctions ϕ_i, $i = 1, \ldots, l_c$

$$\begin{aligned}&\langle \phi_i^2,\ \phi_i^2\rangle = \frac{9}{4},\quad \langle \phi_i^2,\ \phi_{i_1}^2\rangle = 1,\ \text{for all } i \neq i_1 \text{ and } i,\ i_1 = 1,\ldots,l_c,\\ &\langle \phi_i,\ \phi_{i_1}\phi_{i_2}\phi_{i_3}\rangle = 0,\ \text{ for all } i_1, i_2, i_3 \in \{1,\ldots,l_c\}\backslash\{i\},\ i = 1,\ldots,l_c,\end{aligned} \tag{12.67}$$

we obtain the *algebraic bifurcation equations*

$$\Big[6\beta a_0 + \frac{3b_0}{4}\Big(3\alpha_j^2 + 4\sum_{\substack{i=1\\ i\neq j}}^{l_c}\alpha_i^2\Big)\Big]\alpha_j = 0, \qquad j = 1,\ldots,l_c. \tag{12.68}$$

Recall that the statement $\alpha_i \neq 0$ holds at least for one i in (12.56). For $a_0 \cdot b_0 \neq 0$ we have $\beta(0,d) \neq 0$. Moreover, by choosing

$$\beta_0 := \operatorname{sign}(\beta(0,d)) = \operatorname{sign}(-a_0 b_0)$$

and taking a local normalization

$$\beta(t,d) \equiv \beta_0,$$

we can write the solution manifolds (12.56) as

$$\begin{aligned} u(t,d) &= t\sum_{i=1}^{l_c}\alpha_i(t,d)\psi_i + t^3 w(t,d),\\ \lambda(t,d) &= \lambda_0(d) + \beta_0 t^2, \qquad w(t,d) \in M. \end{aligned} \tag{12.69}$$

To determine this solution branch, we consider an enlarged system as in Sections 4.4 and 6.7, i.e.,

$$F(w,\alpha,t,d) := \begin{pmatrix} G(t\sum_{i=1}^{l_c}\alpha_i\psi_i + t^3 w,\ \lambda_0(d) + \beta_0 t^2,\ d)/t^3 \\ \langle \psi_1, w\rangle \\ \vdots \\ \langle \psi_{l_c}, w\rangle \end{pmatrix} = 0. \tag{12.70}$$

Here $\alpha := (\alpha_1, \ldots, \alpha_{l_c})$ and F maps $X \times \mathbf{R}^{l_c}$ into $Y \times \mathbf{R}^{l_c}$. At $t = 0$, the mapping F, correspondingly, the enlarged system (12.70) is defined by its limit

$$F(w,\alpha,0,d) := \begin{pmatrix} D_u G_0 w + \beta_0 D_{u\lambda} G_0\Big(\sum_{i=1}^{l_c}\alpha_i\psi_i\Big) + \frac{1}{6}D_{uuu}G_0\Big(\sum_{i=1}^{l_c}\alpha_i\psi_i\Big)^3 \\ \langle \psi_1,\ w\rangle \\ \vdots \\ \langle \psi_{l_c},\ w\rangle \end{pmatrix} = 0. \tag{12.71}$$

The solvability condition of (12.71) is that the L^2-product of ψ_j^*, $j = 1, \ldots, l_c$ with the first equation in (12.71) should vanish, namely

$$\frac{b_0}{6}\left\langle \phi_j, \left(\sum_{i=1}^{l_c} \alpha_i \phi_i\right)^3 \right\rangle + a_0 \beta_0 \alpha_j = 0, \quad j = 1, \ldots, l_c.$$

This gives us the system (12.66) again. Using the statements (12.67), we derive the following equivalent form

$$g_j(\alpha_1, \ldots, \alpha_{l_c}) := \left[1 + \frac{b_0}{8 a_0 \beta_0}\left(3\alpha_j^2 + 4\sum_{\substack{i=1 \\ i \neq j}}^{l_c} \alpha_i^2\right)\right]\alpha_j = 0, \; j = 1, \ldots, l_c. \quad (12.72)$$

We remind that the coefficients a_0, b_0, β_0 depend on the parameters $\lambda(d), d$ and we aim to solve the system (12.72) along the curve $(\lambda_0(d), d)$.

Lemma 12.5.1. *Under the nondegenerate condition $a_0 \cdot b_0 \neq 0$ the system (12.72) has exactly $3^{l_c} - 1$ different nonvanishing and isolated solutions. More precisely, for $k = 1, 2, \ldots, l_c$, the system (12.72) has $2^{k-1}\binom{l_c}{k}$ different solutions of the form*

$$\alpha_{i_1} = \pm\alpha_{i_2} = \cdots = \pm\alpha_{i_k} = \pm\left(\frac{-8 a_0 \beta_0}{(4k-1) b_0}\right)^{1/2}, \; \alpha_j = 0, \; j \notin \{i_1, \ldots, i_k\}. \quad (12.73)$$

Here the indices $i_1, \ldots, i_k \in \{1, \ldots, l_c\}$ are different from each other, representing a permutation of k integers in $\{1, \ldots, l_c\}$.

Proof: For arbitrary k, and $i_1, \ldots, i_k \in \{1, \ldots, l_c\}$, it is easy to verify that the $\alpha_1, \ldots, \alpha_{l_c}$ in (12.73) satisfy (12.72). Permutations of $i_1, \ldots, i_k$ in $\{1, \ldots, l_c\}$ and combinations of $\pm\alpha_{i_1}, \ldots, \pm\alpha_{i_k}$ yield exactly $2^{k-1}\binom{l_c}{k}$ different solutions. In this way we obtain for $k = 1, \ldots, l_c$ altogether

$$\sum_{k=1}^{l_c} 2^{k-1}\binom{l_c}{k} = 3^{l_c} - 1$$

different nonvanishing solution of (12.72). Bezout's Theorem (cf. [47]) shows that these are also the all nonvanishing solutions of (12.72). To show that these solutions are isolated, we note in (12.73)

$$\alpha_{i_1}^2 = \cdots = \alpha_{i_k}^2 = -8 a_0 \beta_0 / [(4k-1) b_0], \qquad \alpha_j = 0, \; j \notin \{i_1, \ldots, i_k\}.$$

Define a mapping $\mathbf{g} : \mathbf{R}^{l_c} \mapsto \mathbf{R}^{l_c}$ as

$$\mathbf{g}(\alpha_1, \ldots, \alpha_{l_c}) := (g_1(\alpha_1, \ldots, \alpha_{l_c}), \; \ldots, \; g_{l_c}(\alpha_1, \ldots, \alpha_{l_c}))^T.$$

At a solution point, the derivative of $\mathbf{g}$ is equal to

$$Dg = A + \frac{b_0}{a_0\beta_0}\alpha\alpha^T, \tag{12.74}$$

where $\alpha := (\alpha_1, \ldots, \alpha_{l_c})^T$ and $A :=$diag$(a_1, \ldots, a_{l_c})$ with

$$a_j := 1 + \frac{b_0}{8a_0\beta_0}\Big[-3\alpha_j^2 + 4\sum_{i=1}^{l_c}\alpha_i^2\Big], \quad j = 1, \ldots, l_c.$$

Since

$$\sum_{i=1}^{l_c}\alpha_i^2 = k\alpha_{i_1}^2 = -\frac{8a_0\beta_0 k}{(4k-1)b_0},$$

we have

$$a_j = \begin{cases} 2/(4k-1), & \text{for } j \in \{i_1, \ldots, l_k\}; \\ -1/(4k-1), & \text{for } j \notin \{i_1, \ldots, i_k\}. \end{cases}$$

Hence, for $k \geq 1$,

$$\begin{aligned} \frac{b_0}{a_0\beta_0}\alpha^T A^{-1}\alpha &= \frac{b_0}{a_0\beta_0}\sum_{j=1}^{l_c} a_j^{-1}\alpha_j^2 \\ &= -\frac{b_0}{a_0\beta_0}k a_{i_1}^{-1}\alpha_{i_1}^2 \\ &= -\frac{b_0}{a_0\beta_0}\cdot k \cdot \frac{4k-1}{2}\cdot\frac{8ka_0\beta_0}{(4k-1)b_0} \\ &= -4k \\ &\neq -1. \end{aligned}$$

It follows from the Sherman/Morrison formula (cf. Golub/Van Loan [128, pp.51]) that Dg is nonsingular. Correspondingly, the solution of (12.72) is isolated. ∎

Remark 12.5.2. The reduced stability of a bifurcating solution branch may be studied with the properties of eigenvalues of the matrix Dg, see e.g. Allgower/Böhmer/Mei [9], McLeod/Sattinger [209], Mei [223], etc.

Theorem 12.5.2. *Let $(0, \lambda(d), d)$ be on the curve $S(c)$ of bifurcation points and the statement (12.51) hold. If the nondegenerate condition $a_0 \cdot b_0 \neq 0$ is satisfied, see also (12.54) and (12.65), then the system (12.1) has exactly $(3^{l_c} - 1)/2$ different nontrivial solution manifolds $(u(t,d), \lambda(t,d), d)$ bifurcating from $\{(0, \lambda(d), d);\ d \in [-d_0, d_0]\}$. These solutions are of the form*

$$\begin{aligned} u(t,d) &= t\sum_{i=1}^{l_c}\alpha_i(t,d)\psi_i + t^3 w(t,d), \\ \lambda(t,d) &= \lambda_0(d) + t^2\beta_0, \quad t \in [-t_0, t_0],\ w(t,d) \in M, \end{aligned} \tag{12.75}$$

where $(\alpha_1(0,d), \ldots, \alpha_{l_c}(0,d))$ satisfies (12.72).

Proof: Let $(\alpha_1^0, \ldots, \alpha_{l_c}^0)$ be a solution of (12.72). According to the formulation of (12.72) and Theorem 10.6.1, there is a unique $w^0 \in M$, such that

$$D_u G_0 w^0 + D_{u\lambda} G_0 \Big(\sum_{i=1}^{l_c} \alpha_i^0 \psi_i \Big) + \frac{1}{6} D_{uuu} G_0 \Big(\sum_{i=1}^{l_c} \alpha_i^0 \psi_i \Big)^3 = 0.$$

Thus $(w^0, \alpha_1^0, \ldots, \alpha_{l_c}^0, 0, d)$ satisfies the equation (12.71). Let us consider the derivative $D_{(v,\alpha_1,\ldots,\alpha_{l_c})} F(w^0, \alpha_1^0, \ldots, \alpha_{l_c}^0, 0, d)$ and the equation

$$D_{(v,\alpha_1,\ldots,\alpha_{l_c})} F(w^0, \alpha_1^0, \ldots, \alpha_{l_c}^0, 0, d)(v, \beta_1, \ldots, \beta_{l_c}) = (h, s_1, \ldots, s_{l_c})$$

for $(v, \beta_1, \ldots, \beta_{l_c})$ in $Y \times \mathbf{R}^{l_c}$, more precisely,

$$\begin{aligned} & D_u G_0 v + D_{u\lambda} G_0 \Big(\sum_{i=1}^{l_c} \beta_i \psi_i \Big) + \frac{1}{2} D_{uuu} G_0 (\sum_{i=1}^{l_c} \alpha_i^0 \psi_i)^2 \sum_{i=1}^{l_c} \beta_i \psi_i = h, \\ & \langle \psi_k ,\ v \rangle = s_k, \qquad k = 1, \ldots, l_c. \end{aligned} \tag{12.76}$$

Thanks to the decomposition (10.41) of the space X, the unknowns $\beta_1, \ldots, \beta_{l_c}$ can be computed from

$$\Big\langle \psi_j^*,\ D_{u\lambda} G_0 \Big(\sum_{i=1}^{l_c} \beta_i \psi_i \Big) + \frac{1}{2} D_{uuu} G_0 \Big(\sum_{i=1}^{l_c} \alpha_i^0 \psi_i \Big)^2 \Big(\sum_{i=1}^{l_c} \beta_i \psi_i \Big) \Big\rangle = \langle \psi_j^* ,\ h \rangle,$$

with $j = 1, \ldots, l_c$ or in the matrix form

$$D\mathbf{g}_0 (\beta_1, \ldots, \beta_{l_c})^T = (\langle \psi_1^*, h \rangle,\ \ldots,\ \langle \psi_{l_c}, h \rangle)^T. \tag{12.77}$$

Since $D\mathbf{g}_0$ is nonsingular, see Lemma 12.5.1, the system (12.77) is uniquely solvable. Substituting its solution into (12.76), we obtain a uniquely solvable system for v. Thus the operator $D_{(v,\alpha_1,\ldots,\alpha_{l_c})} F(w^0, \alpha_1^0, \ldots, \alpha_{l_c}^0, 0, d)$ is nonsingular. We apply the implicit function theorem to the mapping F at the point $(w^0, \alpha_1^0, \ldots, \alpha_{l_c}^0, 0, d)$ and obtain a constant $t_0 > 0$ and a unique mapping $\{(w(t,d), \alpha_1(t,d), \ldots, \alpha_{l_c}(t,d), t, d),\ t \in [-t_0, t_0]\}$, such that

$$\begin{aligned} F(w(t,d),\ \alpha_1(t,d), \ldots, \alpha_{l_c}(t,d),\ t,\ d) = 0, \\ (w(0,d),\ \alpha_1(0,d), \ldots, \alpha_{l_c}(0,d)) = (w^0,\ \alpha_1^0, \ldots, \alpha_{l_c}^0). \end{aligned} \tag{12.78}$$

Substituting $w(t,d), \alpha_1(t,d), \ldots, \alpha_{l_c}(t,d)$ into (12.69) yields a solution branch of (12.1). Owing to the oddness of G (resp. F) in u, two points $\pm(w^0, \alpha_1^0, \ldots, \alpha_{l_c}^0)$ lead to the same solution manifold of (12.70) (resp. (12.1)). Thus, there are exactly $(3^{l_c} - 1)/2$ different solution manifolds of (12.1) bifurcating at any given point on $\{(0, \lambda_0(d), d) \mid d \in [-d_0, d_0]\}$, corresponding to $3^{l_c} - 1$ different isolated solutions of (12.72). ■

13. Hopf/Steady State Mode Interactions

We study in this chapter Hopf/steady state mode interactions of the reaction-diffusion equation

$$\frac{\partial u}{\partial t} = G(u, \lambda, d) \tag{13.1a}$$

with $u := (u_1, u_2)$ and

$$G(u, \lambda, d) := \begin{pmatrix} \Delta u_1 + f_1(u, \lambda) \\ d\Delta u_2 + f_2(u, \lambda) \end{pmatrix} \tag{13.1b}$$

on square domains Ω with the homogeneous Dirichlet boundary conditions

$$u\big|_{\partial\Omega} = 0.$$

We assume again that the functions f_1, f_2 are sufficiently smooth and satisfy

$$f_1(0, \lambda) = 0, \qquad f_2(0, \lambda) = 0 \quad \text{for all} \quad \lambda \in \mathbf{R}^p, \tag{13.1c}$$

see also (1.2) for other conditions. We are particularly interested in realization of the center manifold reduction of (13.1) for studying of dynamics near a Hopf/steady state mode interaction. We consider in Section 13.1 points where a Hopf/steady state mode interaction occurs, and in Section 13.2 the induced action of D_4 in the critical eigenspace. Section 13.3 describes the normal form of the reduced equations on the center manifold. We study the bifurcation scenario of (13.1) for a special Hopf/steady state mode interaction in Section 13.4 and discuss calculations of the normal form in Section 13.5.

13.1 Hopf/Steady State Mode Interactions

Similarly as in Chapter 10, we may consider the problem (13.1) either in Hölder space (10.6) or in the Sobolev space $X := \left(H_0^1(\Omega)\right)^2$ with the definition domain $D(G) := \left(H^2(\Omega)\right)^2 \cap X$ of the mapping G. As we have seen in Section 10.3, linearization of $G(u, \lambda, d)$ on the trivial solution $(0, \lambda, d)$ is

$$D_uG(0, \lambda, d) = \begin{pmatrix} \Delta + f^0_{1u_1} & f^0_{1u_2} \\ f^0_{2u_1} & d\Delta + f^0_{2u_2} \end{pmatrix}. \tag{13.2}$$

Under the $L^2(\Omega)$-product

$$\langle u,\, v\rangle := \int_\Omega (\bar{u}_1 v_1 + \bar{u}_2 v_2)\, dxdy \text{ for all } u = (u_1, u_2),\ v_1 = (v_1, v_2) \in X \quad (13.3)$$

the adjoint operator of $D_uG(0, \lambda, d)$ is

$$D_uG(0, \lambda, d)^* = \begin{pmatrix} \Delta + f^0_{1u_1} & f^0_{2u_1} \\ f^0_{1u_2} & d\Delta + f^0_{2u_2} \end{pmatrix}. \quad (13.4)$$

Moreover, the spectrum of $D_uG(0, \lambda, d)$ is the union of eigenvalues of the matrices

$$M(\lambda, d, c) := \begin{pmatrix} f^0_{1u_1} - c & f^0_{1u_2} \\ f^0_{2u_1} & f^0_{2u_2} - dc \end{pmatrix} \quad (13.5)$$

for all eigenvalues c of the Laplacian. Here $f^0_{iu_j},\ i, j = 1, 2$ represent the partial derivatives of $f_1,\ f_2$ at $(0, \lambda, d)$.

We recall that a stationary bifurcation occurs at a point $(0, \lambda_0, d_0)$ if one of the eigenvalues of $D_uG(0, \lambda_0, d_0)$ vanishes, i.e., the equation

$$\det(M(\lambda_0, d_0, c)) = (f^0_{1u_1} f^0_{2u_2} - f^0_{2u_1} f^0_{1u_2}) - c(f^0_{1u_1} d_0 + f^0_{2u_2}) + c^2 d_0 = 0 \quad (13.6)$$

is satisfied for an eigenvalue c of the Laplacian. Similarly, a Hopf bifurcation point $(0, \lambda_0, d_0)$ is characterized by the conditions

$$\begin{aligned} &\operatorname{trace}(M(\lambda_0, d_0, c)) = f^0_{1u_1} + f^0_{2u_2} - c(1 + d_0) = 0, \\ &\det(M(\lambda_0, d_0, c)) = (f^0_{1u_1} f^0_{2u_2} - f^0_{2u_1} f^0_{1u_2}) - c(f^0_{1u_1} d_0 + f^0_{2u_2}) + c^2 d_0 > 0 \end{aligned} \quad (13.7)$$

for an eigenvalue c of the Laplacian. If at a point $(0, \lambda_0, d_0)$ the statements (13.6) and (13.7) hold simultaneously for two eigenvalues $c_i,\ i = 1, 2$, i.e., $(0, \lambda_0, d_0)$ is an intersection point of curves of stationary and Hopf bifurcation points, then a Hopf/steady-state mode interaction occurs. We consider here the case that on the imaginary axis, in addition to the (semi-) simple eigenvalue zero, the operator $D_uG(0, \lambda_0, d_0)$ has exactly one pair of pure imaginary eigenvalues. Hence, the statements (13.6) and (13.7) hold for a single c or two different eigenvalues $c_i,\ i = 1, 2$ of the Laplacian. Thanks to the D_4-symmetry, these eigenvalues are either simple or double. Therefore, the center eigenspace are generically three, four, five or six dimensional.

Example 13.1.1. Consider again the Brusselator equations

$$\begin{pmatrix} f_1(u_1, u_2, B) \\ f_2(u_1, u_2, B) \end{pmatrix} = \begin{pmatrix} B - 1 & A^2 \\ -B & -A^2 \end{pmatrix} \begin{pmatrix} u_1 \\ u_2 \end{pmatrix} + \begin{pmatrix} 1 \\ -1 \end{pmatrix} \left(\frac{B}{A} u_1^2 + 2Au_1u_2 + u_1^2 u_2\right).$$

on the square domain $\Omega := [0, \pi]^2$ with the homogeneous Dirichlet boundary conditions

$$u_i = 0, \quad i = 1, 2 \text{ on } \partial\Omega.$$

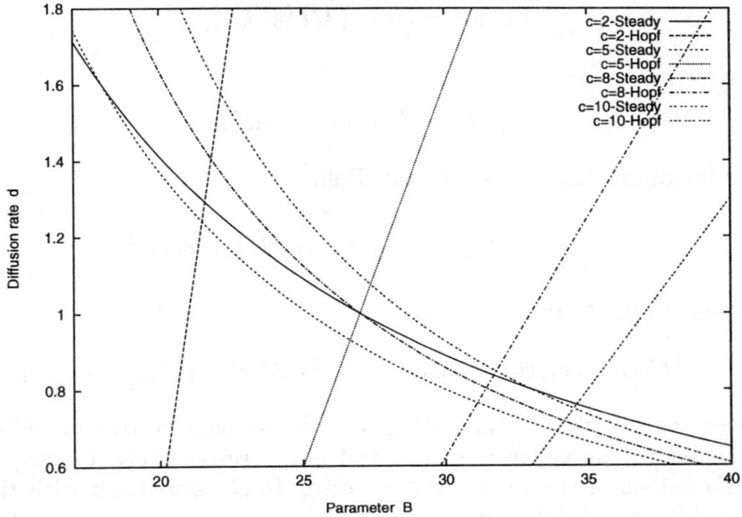

Fig. 13.1. A subset of stationary and Hopf bifurcation points of Brusselator equations with $A = 4$.

Here we have shifted the homogeneous state $(u_1^s, u_2^s) = (A, B/A)$ to the origin. In the parameter space (B, d) the stationary bifurcation points (13.6) lie on the curves

$$d = \frac{(1+c)A^2}{Bc - (1-c)c}, \qquad c = (m^2 + n^2), \quad m,\ n \in \mathbf{N}.$$

If B varies in a given range, the minimum of d_0, as a function of c, can be determined easily. On the other hand, Hopf bifurcation (13.7) occurs on the straight lines

$$d_0 = c^{-1}(B - 1 - A^2 - c), \qquad c = (m^2 + n^2), \quad m,\ n \in \mathbf{N}.$$

Figure 13.1 shows a subset of stationary and Hopf bifurcation points. The envelope of these curves for $B \in [17, 35]$ gives the neutral stability of the system. Intersections of the straight lines with the curves correspond to Hopf/steady state mode interactions.

To study the bifurcation scenario and dynamics at a Hopf/steady state mode interaction, we consider the reduced equations on the center manifolds. In the sequel we choose the domain to be the unit square

$$\Omega := [0, 1] \times [0, 1].$$

Let the statements (13.7) and (13.6) hold at $\lambda = \lambda_0,\ d = d_0$ for eigenvalues $c_1 = (m_1^2 + n_1^2)\pi^2$ and $c_2 = (m_2^2 + n_2^2)\pi^2$, respectively. The purely imaginary eigenvalues of the 2×2 matrix $M(\lambda_0, d_0, c_1)$ are given as

$$\pm\omega_0 i = \sqrt{\det\left(M(\lambda_0, d_0, c_1)\right)}.$$

Moreover,

$$\left(M(\lambda_0, d_0, c_1)\right)^2 = -\omega_0^2 I_{2\times 2}.$$

Let e be an arbitrary vector in $\mathbf{R}^2$. Define

$$\psi_{re} := e, \qquad \psi_{im} := \frac{1}{\omega_0} M(\lambda_0, d_0, c_1)e.$$

It is easy to verify that

$$M(\lambda_0, d_0, c_1)\psi_{re} = \omega_0 \psi_{im}, \qquad M(\lambda_0, d_0, c_1)\psi_{im} = -\omega_0 \psi_{re}.$$

In other words, $\psi_1 := \psi_{re} - i\psi_{im}$ and $\bar{\psi}_1$ are eigenvectors of $M(\lambda_0, d_0, c_1)$ associated to the eigenvalues $i\omega_0$ and $-i\omega_0$, respectively. Correspondingly, eigenfunctions of the differential operator $D_u G_0$ associated with the eigenvalues $\pm i\omega_0$ can be chosen as

$$\phi_1 := 2\psi_1 \sin m_1\pi x \sin n_1 \pi y, \qquad \phi_2 := 2\psi_1 \sin n_1 \pi x \sin m_1 \pi y$$

and their complex conjugate $\bar{\phi}_1$, $\bar{\phi}_2$, respectively. Now we have

$$D_u G_0 \mathrm{Re}(\phi_i) = \omega_0 \mathrm{Im}(\phi_i), \qquad D_u G_0 \mathrm{Im}(\phi_i) = -\omega_0 \mathrm{Re}(\phi_i), \; i = 1, 2.$$

Note that if $m_1 \neq n_1$, the eigenvalues $i\omega_0$ and $-i\omega_0$ are forced to be double by the D_4-symmetry, respectively.

Corresponding to the eigenvalue c_2 of the Laplacian, the matrix $M(\lambda_0, d_0, c_2)$ is singular. Assume that zero is a simple eigenvalue of $M(\lambda_0, d_0, c_2)$ and ψ_2 is the associated eigenvector. We define

$$\phi_3 := 2\psi_2(\sin m_2\pi x \sin n_2 \pi y + i \sin n_2 \pi x \sin m_2 \pi y)$$

and obtain the kernel of $D_u G_0$ as

$$\mathrm{Ker}(D_u G_0) = \mathrm{span}[\mathrm{Re}(\phi_3),\; \mathrm{Im}(\phi_3)].$$

13.2 Induced Actions

We recall

$$S_1(x, y) = (1 - x, y), \qquad R(x, y) = (1 - y, x)$$

as generators of the D_4-symmetry of the unit square. With the above chosen eigenvectors ϕ_i, $i = 1, 2, 3$ the induced actions of the D_4-group in the center eigenspace X_0 of the operator $D_u G_0$ can be derived for different combinations of the wave numbers m_i, n_i, $i = 1, 2$.

1. If $m_1 = n_1$, $m_2 = n_2$, then $\pm\omega_0 i$ and zero are simple eigenvalues of D_uG_0. We have $\phi_1 = \phi_2$, $\mathrm{Re}(\phi_3) = \mathrm{Im}(\phi_3)$ and the center eigenspace of D_uG_0 is three dimensional
$$X_0 = \mathrm{span}[\mathrm{Re}(\phi_1),\ \mathrm{Im}(\phi_1),\ \mathrm{Re}(\phi_3)]. \tag{13.8}$$
The action of the D_4-group in X_0 is generated by
$$S_1 = R = \begin{pmatrix} (-1)^{m_1-1} & & \\ & (-1)^{m_1-1} & \\ & & (-1)^{m_2-1} \end{pmatrix}.$$
2. If $m_1 = n_1$ and $m_2 \neq n_2$, then $\pm\omega_0 i$ are simple eigenvalues of D_uG_0, while the eigenvalue zero is of the multiplicity two. We obtain a four dimensional center eigenspace
$$X_0 = \mathrm{span}[\mathrm{Re}(\phi_1),\ \mathrm{Im}(\phi_1),\ \mathrm{Re}(\phi_3),\ \mathrm{Im}(\phi_3)]. \tag{13.9}$$
Generators of the D_4-group in X_0 are
$$S_1 = \begin{pmatrix} (-1)^{m_1-1} & & & \\ & (-1)^{m_1-1} & & \\ & & (-1)^{m_2-1} & \\ & & & (-1)^{n_2-1} \end{pmatrix},$$
and
$$R = \begin{pmatrix} (-1)^{m_1-1} & & & \\ & (-1)^{m_1-1} & & \\ & & & (-1)^{m_2-1} \\ & & (-1)^{n_2-1} & \end{pmatrix}.$$
3. If $m_1 \neq n_1$ and $m_2 = n_2$, then $\pm\omega_0 i$ are double eigenvalues of D_uG_0, while zero is a simple eigenvalue. The center eigenspace is five dimensional
$$X_0 = \mathrm{span}[\mathrm{Re}(\phi_1),\ \mathrm{Im}(\phi_1),\ \mathrm{Re}(\phi_2),\ \mathrm{Im}(\phi_2),\ \mathrm{Re}(\phi_3)]. \tag{13.10}$$
The induced action of D_4 in X_0 is generated by
$$S_1 = \begin{pmatrix} (-1)^{m_1-1} & & & & \\ & (-1)^{n_1-1} & & & \\ & & (-1)^{m_1-1} & & \\ & & & (-1)^{n_1-1} & \\ & & & & (-1)^{m_2-1} \end{pmatrix},$$
and
$$R = \begin{pmatrix} & (-1)^{m_1-1} & & & \\ (-1)^{n_1-1} & & & & \\ & & & (-1)^{m_1-1} & \\ & & (-1)^{n_1-1} & & \\ & & & & (-1)^{m_2-1} \end{pmatrix}.$$

4. Finally, if $m_1 \neq n_1$, $m_2 \neq n_2$, then eigenvalues $\pm\omega_0 i$ and zero are all of the multiplicity two. The center eigenspace is six dimensional

$$X_0 = \text{span}[\text{Re}(\phi_1),\ \text{Im}(\phi_1),\ \text{Re}(\phi_2),\ \text{Im}(\phi_2),\ \text{Re}(\phi_3),\ \text{Im}(\phi_3)]. \quad (13.11)$$

The induced action of D_4 in X_0 is generated by

$$S_1 = \begin{pmatrix} (-1)^{m_1-1} & & & & & \\ & (-1)^{n_1-1} & & & & \\ & & (-1)^{m_1-1} & & & \\ & & & (-1)^{n_1-1} & & \\ & & & & (-1)^{m_2-1} & \\ & & & & & (-1)^{n_2-1} \end{pmatrix}$$

and

$$R = \begin{pmatrix} & (-1)^{m_1-1} & & & & \\ (-1)^{n_1-1} & & & & & \\ & & & (-1)^{m_1-1} & & \\ & & (-1)^{n_1-1} & & & \\ & & & & & (-1)^{m_2-1} \\ & & & & (-1)^{n_2-1} & \end{pmatrix}.$$

In the case $m_1 = n_1$, $m_2 = n_2$ the D_4-group acts trivially on X_0. This corresponds to the Hopf/steady state mode interaction discussed in Gomes/Stewart [132], Guckenheimer/Holmes [144] and Golubitsky/Stewart/Schaeffer [131].

(m_1, n_1, m_2, n_2)	Generators of D_4 in $\mathbf{R}^6$		Groups
(even,odd,even,odd)	$S_1 = (-e_1, e_2, -e_3, e_4, -e_5, e_6)$,	$R = (e_2, -e_1, e_4, -e_3, e_6, -e_5)$	D_4
(even,odd,odd,even)	$S_1 = (-e_1, e_2, -e_3, e_4, e_5, -e_6)$,	$R = (e_2, -e_1, e_4, -e_3, -e_6, e_5)$	D_4
(odd,even,even,odd)	$S_1 = (e_1, -e_2, e_3, -e_4, -e_5, e_6)$,	$R = (-e_2, e_1, -e_4, e_3, e_6, -e_5)$	D_4
(odd,even,odd,even)	$S_1 = (e_1, -e_2, e_3, -e_4, e_5, -e_6)$,	$R = (-e_2, e_1, -e_4, e_3, -e_6, e_5)$	D_4
(even,even,even,odd)	$S_1 = (-e_1, -e_2, -e_3, -e_4, -e_5, e_6)$,	$R = (-e_2, -e_1, -e_4, -e_3, e_6, -e_5)$	D_4
(even,even,odd,even)	$S_1 = (-e_1, -e_2, -e_3, -e_4, e_5, -e_6)$,	$R = (-e_2, -e_1, -e_4, -e_3, -e_6, e_5)$	D_4
(even,odd,even,even)	$S_1 = (-e_1, e_2, -e_3, e_4, -e_5, -e_6)$,	$R = (e_2, -e_1, e_4, -e_3, -e_6, -e_5)$	D_4
(odd,even,even,even)	$S_1 = (e_1, -e_2, e_3, -e_4, -e_5, -e_6)$,	$R = (-e_2, e_1, -e_4, e_3, -e_6, -e_5)$	D_4
(odd,odd,even,odd)	$S_1 = (e_1, e_2, e_3, e_4, -e_5, e_6)$,	$R = (e_2, e_1, e_4, e_3, e_6, -e_5)$	D_4
(odd,odd,odd,even)	$S_1 = (e_1, e_2, e_3, e_4, e_5, -e_6)$,	$R = (e_2, e_1, e_4, e_3, -e_6, e_5)$	D_4
(even,odd,odd,odd)	$S_1 = (-e_1, e_2, -e_3, e_4, e_5, e_6)$,	$R = (e_2, -e_1, e_4, -e_3, e_6, e_5)$	D_4
(odd,even,odd,odd)	$S_1 = (e_1, -e_2, e_3, -e_4, e_5, e_6)$,	$R = (-e_2, e_1, -e_4, e_3, e_6, e_5)$	D_4
(even,even,even,even)	$S_1 = -I$,	$R = -(e_2, e_1, e_4, e_3, e_6, e_5)$	$Z_2 \times Z_2$
(even,even,odd,odd)	$S_1 = (-e_1, -e_2, -e_3, -e_4, e_5, e_6)$,	$R = (-e_2, -e_i, -e_4, -e_3, e_6, e_5)$	$Z_2 \times Z_2$
(odd,odd,even,even)	$S_1 = (e_1, e_2, e_3, e_4, -e_5, -e_6)$,	$R = (e_2, e_1, e_4, e_3, -e_6, -e_5)$	$Z_2 \times Z_2$
(odd,odd,odd,odd)	$S_1 = I$,	$R = (e_2, e_1, e_4, e_3, e_6, e_5)$	Z_2

Table 13.1. Induced actions of the D_4-group in six dimensional center eigenspaces.

Here and in Section 13.3 we consider the situation $m_1 \neq n_1$, $m_2 \neq n_2$ with six dimensional center eigenspace. Depending on the parities of m_i, n_i, we obtain the induced actions of D_4 in $\mathbf{R}^6 \sim X_0$ as in the Table 13.1.

13.3 Normal Forms

By the normal form theory in Chapter 7, we can reduce directly the problem (13.1) into the normal form on the center manifold. To this end, we study at first the necessary structure of the normal form.

We choose a basis

$$\begin{aligned} \phi_1 &= \psi_1\big(\sin m_1\pi x \sin n_1\pi y + i\ \sin n_1\pi x \sin m_1\pi y\big), \\ \phi_2 &= \psi_1\big(\sin m_1\pi x \sin n_1\pi y - i\ \sin n_1\pi x \sin m_1\pi y\big), \\ \phi_3 &= \psi_2\big(\sin m_2\pi x \sin n_2\pi y + i\ \sin n_2\pi x \sin m_2\pi y\big). \end{aligned} \tag{13.12}$$

Here $\psi_1 \in \mathbf{C}^2$ and $\psi_2 \in \mathbf{R}^2$ represent the eigenvector of $M(\lambda_0, d_0, c_1)$ and $M(\lambda_0, d_0, c_2)$ associated to the eigenvalue $\omega_0 i$ and zero, respectively. The eigenvectors ψ_1, ψ_2 are scaled so that all ϕ_i, $i = 1, 2, 3$ have the norm one.

By the statement (13.4) the center eigenspace of $D_u G_0^*$ is also six dimensional. In fact, let ψ_1^* be the eigenvector of $M(\lambda_0, d_0, c_1)^T$ corresponding to the eigenvalue $-\omega_0 i$ and ψ_2^* be the eigenvector of $M(\lambda_0, d_0, c_2)^T$ associated to the eigenvalue zero. Define

$$\begin{aligned} \phi_1^* &:= \psi_1^*\big(\sin m_1\pi x \sin n_1\pi y + i\ \sin n_1\pi x \sin m_1\pi y\big), \\ \phi_2^* &:= \psi_1^*\big(\sin m_1\pi x \sin n_1\pi y - i\ \sin n_1\pi x \sin m_1\pi y\big), \\ \phi_3^* &:= \psi_2^*\big(\sin m_2\pi x \sin n_2\pi y + i\ \sin n_2\pi x \sin m_2\pi y\big). \end{aligned}$$

We have

$$D_u G_0^* \phi_i^* = -\omega_0 i \phi_i^*,\ i = 1, 2 \qquad D_u G_0^* \phi_3^* = 0.$$

Furthermore, ϕ_i^* can be chosen so that

$$\langle \phi_i^*,\ \phi_i \rangle = 1, \qquad \langle \bar{\phi}_i^*,\ \phi_i \rangle = 0, \tag{13.13}$$

see Golubitsky/Schaeffer [129, pp.346]. We obtain from these properties a projection P from X onto X_0

$$Pu = \sum_{i=1}^{3} \Big[\langle \phi_i^*, u \rangle \phi_i + \overline{\langle \phi_i^*, u \rangle \phi_i} \Big], \quad u \in X.$$

An element $v \in X_0$ will be identified with a vector $(z_1, z_2, z_3) \in \mathbf{C}^2$ via the relation

$$v = \sum_{i=1}^{3} \Big(z_i \phi_i + \overline{z_i \phi_i} \Big) \in X_0 \longleftrightarrow (z_1, z_2, z_3) \in \mathbf{C}^3. \tag{13.14}$$

Induced actions of the D_4-group in $\mathbf{C}^3$ can be derived similarly as before. As examples, we list the following three cases.

1) If $(m_1, n_1, m_2, n_2) =$(odd, even, odd, even), the action of D_4 is irreducible and

$$S_1(z_1, z_2, z_3) = (z_2, z_1, \bar{z}_3), \qquad R(z_1, z_2, z_3) = (-iz_1, iz_2, -iz_3). \tag{13.15}$$

2) If $(m_1, n_1, m_2, n_2) =$(even, odd, even, odd), then D_4 is irreducible and

$$S_1(z_1, z_2, z_3) = -(z_2, z_1, \bar{z}_3), \qquad R(z_1, z_2, z_3) = (iz_1, -iz_2, iz_3). \tag{13.16}$$

3) If $(m_1, n_1, m_2, n_2) =$(even, even, odd, odd), D_4 is reducible in $\mathbf{C}^3$ and as $Z_2 \times Z_2$, which is generated by

$$S_1(z_1, z_2, z_3) = (-z_1, -z_2, \bar{z}_3), \quad R(z_1, z_2, z_3) = (-iz_2, iz_1, i\bar{z}_3). \tag{13.17}$$

In the coordinates z_1, z_2, z_3 the restriction of $D_u G_0$ in X_0 reduces to

$$L_0 = \begin{pmatrix} \omega_0 i & & \\ & \omega_0 i & \\ & & 0 \end{pmatrix}, \qquad L_0^* = \begin{pmatrix} -\omega_0 i & & \\ & -\omega_0 i & \\ & & 0 \end{pmatrix}.$$

The S^1-symmetry generated by $e^{tL_0^*}$ is reduced to

$$\theta(z_1, z_2, z_3) = (e^{i\theta} z_1,\ e^{i\theta} z_2,\ z_3), \quad \text{for all } \theta \in [0, 2\pi),\ (z_1, z_2, z_3) \in \mathbf{C}^3. \tag{13.18}$$

Consider a generic form of the reduced equations

$$\frac{dz_i}{dt} = g_i(z, \beta), \quad i = 1, 2, 3 \tag{13.19}$$

with $z := (z_1, z_2, z_3)$ as flow in the space $\mathbf{C}^3$ and $\beta := (\beta_1, \beta_2) \in \mathbf{R}^2$ as increments of parameters λ, d from λ_0, d_0, respectively.

Wave numbers $(m_1, n_1, m_2, n_2) =$(odd, even, odd, even):

We study the Hopf/steady state mode interaction for the wave numbers $(m_1, n_1, m_2, n_2) =$(odd, even, odd, even). To this end, we note at first the actions (13.15), (13.18) and the $D_4 \times S^1$-equivariance

$$\gamma(g_1(z, \beta),\ g_2(z, \beta),\ g_3(z, \beta)) = (g_1(\gamma z, \beta),\ g_2(\gamma z, \beta),\ g_3(\gamma z, \beta))$$

for all $z \in \mathbf{C}^3$, $\beta \in \mathbf{R}^2$, $\gamma \in D_4 \times S^1$. Thereafter, we derive the following structure of low order terms in the Taylor expansion of $g_i(z, \beta)$, see e.g., Golubitsky/Stewart/Schaeffer [131, Chapter 20] and Steindl [284],

$$\begin{aligned} g_1(z, \beta) &= (\omega_0 i + A_1\beta_1 + A_2\beta_2) z_1 + \left(C_1|z_1|^2 + C_2|z_2|^2 + C_3|z_3|^2\right) z_1 \\ &\quad + C_4 z_2 z_3^2 + C_5 \bar{z}_1 z_2^2 + C_6 z_2 \bar{z}_3^2 + h.o.t., \\ g_2(z, \beta) &= (\omega_0 i + A_1\beta_1 + A_2\beta_2) z_2 + \left(C_2|z_1|^2 + C_1|z_2|^2 + C_3|z_3|^2\right) z_2 \\ &\quad + C_4 z_1 \bar{z}_3^2 + C_5 z_1^2 \bar{z}_2 + C_6 z_1 \bar{z}_3^2 + h.o.t., \\ g_3(z, \beta) &= (B_1\beta_1 + B_2\beta_2) z_2 + \left(D_1|z_1|^2 + D_2|z_2|^2 + D_3|z_3|^2\right) z_3 \\ &\quad + D_4 z_1 \bar{z}_2 \bar{z}_3 + D_5 \bar{z}_1 z_2 z_3 + D_6 \bar{z}_3^3 + h.o.t. \end{aligned} \tag{13.20}$$

Here A_i, B_i, C_j, $D_i \in \mathbf{C}$, $i = 1, 2, j = 1, \ldots, 6$ and $D_3, \ldots, D_6 \in \mathbf{R}$ are constants and $D_1 = \bar{D}_2$. The mapping $g_2(z, \beta)$ is obtained by $g_1(z, \beta)$ via reflection symmetry. For other parities of the wave numbers, additional terms may be needed in (13.20), see e.g. Chapter 11.

We note that for (m_1, n_1, m_2, n_2) =(odd, even, odd, even) the actions of $D_4 \times S^1$ in (13.15), (13.18) and the normal form of the reduced equation coincide with those in Steindl [284]. We adapt the discussion in Steindl [284] and summarize the group data in Table 13.1.

Generally, a combination of spatial rotations and temporal phase shifts leads to rotation waves

$$R\theta z = \left(e^{(-\frac{\pi}{2}+\theta)i} z_1,\ e^{(\frac{\pi}{2}+\theta)i} z_2,\ e^{(\frac{\pi}{2})i} z_3\right).$$

Choosing $\theta = \pi/2$, we obtain a Z_4-group

$$Z_4\{R\frac{\pi}{2}\} = \left\{\left(R\frac{\pi}{2}\right)^i ;\ i = 1, 2, 3, 4\right\}, \qquad \left(R\frac{\pi}{2}\right) z = (z_1,\ -z_2,\ iz_3).$$

Similarly,

$$\begin{aligned}
Z_2\{R^2\pi\} &= \{R^2\pi;\ I\}, & R^2\pi z &= (z_1, z_2, -z_3),\\
Z_2\{\pi S R^2\} &= Z_2\{S R^2 \pi\} = \{\pi S R^2;\ I\}, & \pi S R^2 z &= (z_2, z_1, -\bar{z}_3),\\
Z_4\{SR\} &= \left\{(SR)^i;\ i = 1, 2, 3, 4\right\}, & SRz &= (iz_2,\ -iz_1,\ -i\bar{z}_3),\\
Z_4\{\pi SR\} &= \left\{(\pi SR)^i;\ i = 1, 2, 3, 4\right\}, & \pi SRz &= (-iz_2,\ iz_1,\ i\bar{z}_3).
\end{aligned}$$

Note that symmetry of a solution is described by the conjugacy class of its isotropy subgroup. All conjugate solutions have the same properties and can be represented by one of them on the group orbit.

No.	orbit	isotropy group Σ	$\mathrm{Fix}(\Sigma)$	$\dim(\mathrm{Fix}(\Sigma))$	description
0	$(0,0,0)$	$D_4 \times S^1$	$(0,0,0)$	0	trivial solution
1	$(0,0,a)$	$Z_2\{S_1\} \times S^1$	$(0,0,x)$	1	rectangular solutions
2	$(0,0,(i+1)a)$	$Z_2\{S_1R\} \times S^1$	$(0,0,(i+1)x)$	1	triangular solutions
3	$(a,a,0)$	$Z_2\{S_1\} \times Z_2\{R^2\pi\}$	$(z_1,z_1,0)$	2	standing waves
4	$(a,-ia,0)$	$Z_2\{S_1R\} \times Z_2\{R^2\pi\}$	$(z_1,-iz_1,0)$	2	standing waves
5	$(a,0,0)$	$Z_4\{R\frac{\pi}{2}\}$	$(z_1,0,0)$	2	rotating waves
6	(a,a,b)	$Z_2\{S\}$	(z_1,z_1,x)	3	
7	(a,a,ib)	$Z_2\{\pi SR^2\} = Z_2\{SR^2\pi\}$	(z_1,z_1,ix)	3	
8	$(a,ia,(1-i)b)$	$Z_4\{SR\}$	$(z_1,iz_1,(1-i)x)$	3	
9	$(a,ia,(1+i)b)$	$Z_4\{\pi SR\}$	$(z_1,iz_1,(1+i)x)$	3	
10	$(0,0,b)$	S^1	$(0,0,z)$	2	asymmetric stationary solutions
11	$(a,b,0)$	$Z_2\{R^2\pi\}$	$(z_1,z_2,0)$	4	modulated waves

Table 13.2: Orbit data for $D_4 \times S^1$ acting on $\mathbf{C}^3$.

13.4 Bifurcation Scenario

Since solutions of the reduced equations (13.19) reveal local dynamics of the original problem (13.1) near Hopf/steady state mode interactions, we are interested in the solutions of the reduced equations in neighborhoods of the origin $(0,0)$. In this section we investigate the bifurcation scenario for the wave numbers (m_1, n_1, m_2, n_2) =(odd, even, odd, even), based on the normal form (13.20). To simplify the discussions, we substitute the variables as

$$z_1 = e^{i\omega_0 t}\tilde{z}_1, \quad z_2 = e^{i\omega_0 t}\tilde{z}_2, \quad z_3 = \tilde{z}_3 \tag{13.21}$$

and the parameters

$$\begin{aligned} A_1\beta_1 + A_2\beta_2 &= \mu_1 + i\nu_1, \\ B_1\beta_1 + B_2\beta_2 &= \mu_2 + i\nu_2. \end{aligned} \tag{13.22}$$

Thereafter, we truncate the reduced system (13.19) and write it into a canonical form

$$\frac{d\tilde{z}}{dt} = \tilde{g}(\tilde{z}, \mu, \nu) \tag{13.23}$$

where we have used the notation $\mu := (\mu_1,\ \mu_2)$, $\nu := (\nu_1, \nu_2)$ and omitted the $\tilde{}$ in $\tilde{z}$.

$$\begin{aligned} \tilde{g}_1 &= (\mu_1 + i\nu_1)z_1 + \left(C_1|z_1|^2 + C_2|z_2|^2 + C_3|z_3|^2\right) z_1 \\ &\quad + C_4 z_2 z_3^2 + C_5 \bar{z}_1 z_2^2 + C_6 z_2 \bar{z}_3^2, \\ \tilde{g}_2 &= (\mu_1 + i\nu_1)z_2 + \left(C_2|z_1|^2 + C_1|z_2|^2 + C_3|z_3|^2\right) z_2 \\ &\quad + C_4 z_1 \bar{z}_3^2 + C_5 z_1^2 \bar{z}_2 + C_6 z_1 \bar{z}_3^2, \\ \tilde{g}_3 &= (\mu_2 + i\nu_2)z_2 + \left(D_1|z_1|^2 + D_2|z_2|^2 + D_3|z_3|^2\right) z_3 \\ &\quad + D_4 z_1 \bar{z}_2 \bar{z}_3 + D_5 \bar{z}_1 z_2 z_3 + D_6 \bar{z}_3^3. \end{aligned}$$

Under nondegeneracy conditions (cf. Golubitsky/Stewart/Schaeffer [131, pp. 466]), higher order terms will not alert the existence and the parameterization of the bifurcating solution branches.

Let (z_0, μ_0, ν_0) be a stationary solution, i.e.,

$$\tilde{g}(z, \mu, \nu) = 0 \qquad \text{with } \tilde{g} := (\tilde{g}_1,\ \tilde{g}_2,\ \tilde{g}_3). \tag{13.24}$$

We classify solutions of (13.23) in the neighborhood of the origin as follows.

- If $\nu_1 = 0$, the solution is stationary or periodic with the fixed period $T = 2\pi/\omega_0$.
- If $\nu_1 \neq 0$ and $\nu_2 = 0$, the solution is periodic with the period $2\pi/(\omega_0 + \nu_1)$.
- If $\nu_1\nu_2 \neq 0$, then the solution is quasi-periodic involving the periods $2\pi/(\omega_0 + \nu_1)$ and $2\pi/\nu_2$.

Via the transformation (13.21) we obtain the corresponding solutions of the reduced equations (13.19), and in turn of the original problem (13.1). Existence of such solutions of the original problem is ensured by the fact that for

$\|\mu\| + \|\nu\| \ll 1$, the reduced system (13.19) is merely a high order perturbation of (13.23). Thus a description of bifurcation scenario of the original problem at a Hopf/steady state mode interaction with the wave numbers (m_1, n_1, m_2, n_2) =(odd, even, odd, even) follows from an analysis of stationary solutions of (13.23), its (Hopf) secondary bifurcations and tertiary bifurcations.

Let us consider solutions of (13.24) and their reduced stabilities. For simplicity, we denote $\tilde{g}_i$ again with g_i, $i = 1, 2, 3$. We recall that the stability of a stationary solution (z_0, μ_0, ν_0) is determined by eigenvalues of the Jacobian matrix $Dg(z_0, \mu_0, \nu_0)$. If the real parts of all eigenvalues are negative, the solution is stable. Otherwise if one of the eigenvalues has positive real part, the solution is unstable. Eigenvalues on the imaginary axis lead to further bifurcations.

Restricting the system (13.24) to the fixed point subspace of various isotropy subgroups Σ, one can solve the problem

$$\text{Find } (z, \mu, \nu) \in (\mathbf{C}^3)^{\Sigma} \times \mathbf{R}^2 \times \mathbf{R}^2 \quad \text{such that} \quad g^{\Sigma}(z, \mu, \nu) = 0 \tag{13.25}$$

directly as described in Golubitsky/Stewart/Schaeffer[131, Chapter 20]. Based on the discussions in Steindl [284], we summarize the bifurcation scenario as follows.

(0) $D_4 \times S^1$-symmetry: We obtain the trivial solution

$$z = 0.$$

Eigenvalues of $Dg(0, \mu, \nu)$ are

$$\mu_1 \pm i\nu_1 \text{ (double)}, \ \mu_2 \pm i\nu_2,$$

indicating that this solution is unstable.

(1) $Z_2\{S_1\} \times S^1$-symmetry: Representative for elements in the fixed point subspace of $\Sigma := Z_2\{S\} \times S^1$ is z of the form $z = (0, 0, x)$. Hence, the equation (13.25) is reduced to

$$(\mu_2 + i\nu_2)x + (D_3 + D_6)x^3 = 0.$$

Its nontrivial solutions are $\nu_2 = 0$, $x = \pm\left(-\frac{\mu_2}{D_3 + D_6}\right)^{1/2}$. Correspondingly, solutions of the original problem are called *rectangular* since their nodal lines are parallel to the sides of the square domain. Eigenvalues of the Jacobian Dg along the solution curves are

$$\left\{\begin{array}{c} \mu_1 + \mathrm{Re}(C_3 - C_4 - C_6)x^2 \pm i[\nu_1 + \mathrm{Im}(C_3 - C_4 - C_6)]x^2, \\ \mu_1 + \mathrm{Re}(C_3 + C_4 + C_6)x^2 \pm i[\nu_1 + \mathrm{Im}(C_3 + C_4 + C_6)]x^2, \\ -4D_6 x^2, \qquad 2(D_3 + D_6)x^2. \end{array}\right\}$$

A secondary bifurcation occurs if one of the eigenvalues cross the imaginary axis, e.g. there are secondary Hopf bifurcations at

$$\mu_1 = -\mathrm{Re}[C_3 \pm (C_4 + C_6)]x^2.$$

(2) $Z_2\{SR\} \times S^1$-symmetry: Elements with the $\Sigma := Z_2\{SR\} \times S^1$-symmetry are represented by $z = (0, 0, \frac{\sqrt{2}}{2}(1-i)x)$. Solutions with such symmetries are called *triangular* since they have the reflection symmetry with respect to a diagonal of the square domain.

The equation (13.25) is reduced to

$$(\mu_2 + i\nu_2)x + (D_3 - D_6)x^3 = 0.$$

Its nontrivial solutions are $\nu_2 = 0,\ x = \pm\left(-\frac{\mu_2}{D_3 - D_6}\right)^{1/2}$. Eigenvalues of the Jacobian Dg along these solution branches are

$$\left\{\begin{array}{c} \mu_1 + \mathrm{Re}(C_3 + C_4 - C_6)x^2 \pm i[\nu_1 + \mathrm{Im}(C_3 + C_4 - C_6)]x^2, \\ \mu_1 + \mathrm{Re}(C_3 - C_4 + C_6)x^2 \pm i[\nu_1 + \mathrm{Im}(C_3 - C_4 + C_6)]x^2, \\ 4D_6x^2, \qquad 2(D_3 - D_6)x^2. \end{array}\right\}$$

(3) $Z_2\{S_1\} \times Z_2\{R^2\pi\}$-symmetry: Solutions with such symmetries are of the form $(z_1, z_1, 0)$. The equation (13.25) in the fixed point subspace is

$$(\mu_1 + i\nu_1)z_1 + (C_1 + C_2 + C_5)|z_1|^2 z_1 = 0.$$

Choosing $z = (x, x, 0)$ as representative on the group orbit, we write the equation as

$$\begin{array}{l} \mu_1 + \mathrm{Re}(C_1 + C_2 + C_5)x^2 = 0, \\ \nu_1 + \mathrm{Im}(C_1 + C_2 + C_5)x^2 = 0. \end{array} \tag{13.26}$$

This system is solvable for parameters $(\mu_1, \nu_1) = \Big(\mathrm{Re}(C_1 + C_2 + C_5)s,\ \mathrm{Im}(C_1 + C_2 + C_5)s\Big)$, $s \in \mathbf{R}$. The Jacobian Dg along the solution branches has two real eigenvalues

$$\{0,\ \ 2\mathrm{Re}(C_1 + C_2 + C_5)x^2\}.$$

and four (complex) eigenvalues

$$\left\{\begin{array}{l} \mu_2 + 2\mathrm{Re}(D_1)x^2 \pm \Big((D_4 + D_5)^2x^4 - \nu_2^2\Big)^{1/2}, \\ x^2\Big[\mathrm{Re}(C_1 - C_2 - 3C_5) \\ \quad \pm \Big((\mathrm{Re}(C_1 - C_2 + C_5))^2 + 8\mathrm{Im}(C_5)\mathrm{Im}(C_1 - C_2 - C_5)\Big)^{1/2}\Big]. \end{array}\right\}$$

The eigenvalue zero results from the S^1-orbit of solutions. A secondary bifurcation occurs if one of the eigenvalues, e.g.,

$$\mu_2 + 2\mathrm{Re}(D_1)x^2 \pm \Big((D_4 + D_5)^2x^4 - \nu_2^2\Big)^{1/2}$$

crosses the imaginary axis.

(4) $Z_2\{S_1R\} \times Z_2\{R^2\pi\}$-symmetry: Solutions with these symmetries are of the form $(z_1, -iz_1, 0)$. Correspondingly, the reduced equation (13.25) becomes

$$(\mu_1 + i\nu_1)z_1 + (C_1 + C_2 - C_5)|z_1|^2 z_1 = 0.$$

For the representative $z = (x, x, 0)$, $x \neq 0$ on the group orbit we obtain the equations

$$\begin{aligned} \mu_1 + \mathrm{Re}(C_1 + C_2 - C_5)x^2 &= 0, \\ \nu_1 + \mathrm{Im}(C_1 + C_2 - C_5)x^2 &= 0. \end{aligned} \tag{13.27}$$

Along solution curves of this system eigenvalues of the Jacobian Dg are

$$\left\{ \begin{array}{l} 0, \qquad\qquad 2\mathrm{Re}(C_1 + C_2 - C_5)x^2, \\ \mu_2 + 2\mathrm{Re}(D_1)x^2 \pm \Big((D_4 - D_5)^2 x^4 - \nu_2^2\Big)^{1/2}, \\ x^2 \Big[\mathrm{Re}(C_1 - C_2 + 3C_5) \\ \quad \pm \Big((\mathrm{Re}(-C_1 + C_2 + C_5))^2 - 8\mathrm{Im}(C_5)\mathrm{Im}(C_1 - C_2 + C_5)\Big)^{1/2}\Big]. \end{array} \right\}$$

Secondary bifurcations occur when the following statement holds

$$\mu_2 + 2\mathrm{Re}(D_1)x^2 \pm \Big((D_4 - D_5)^2 x^4 - \nu_2^2\Big)^{1/2} = 0.$$

Here x satisfies (13.27).

(5) $Z_4\{R^2\pi\}$-symmetry: Elements in the fixed point subspace of $\Sigma := Z_4\{R^2\pi\}$ are of the form $(z_1, 0, 0)$ and the equation $g^\Sigma(z, \mu, \nu) = 0$ is

$$(\mu_1 + i\nu_1)z_1 + C_1|z_1|^2 z_1 = 0.$$

Equivalently, for the representative $(x, 0, 0)$ of the group orbit we have

$$\begin{aligned} \mu_1 + \mathrm{Re}(C_1)x^2 &= 0, \\ \nu_1 + \mathrm{Im}(C_1)x^2 &= 0. \end{aligned}$$

Eigenvalues of the Jacobian Dg are

$$\left\{ \begin{array}{l} 0, \qquad\qquad 2\mathrm{Re}(C_1)x^2, \\ \mu_2 + \mathrm{Re}(D_1)x^2 \pm i(\nu_2 + \mathrm{Im}(D_1))x^2, \\ x^2 \Big[\mathrm{Re}(-C_1 + C_2) \pm \Big((|C_5|^2 - (\mathrm{Im}(C_1 + C_2))^2\Big)^{1/2}\Big]. \end{array} \right\}$$

There is a secondary Hopf bifurcation at

$$\mu_2 = -\mathrm{Re}(D_1)x^2, \qquad \nu_2 \neq -\mathrm{Im}(D_1)x^2$$

and a double steady state secondary bifurcation at

$$\mu_2 = -\mathrm{Re}(D_1)x^2, \qquad \nu_2 = -\mathrm{Im}(D_1)x^2.$$

(6) $Z_2\{S_1\}$-symmetry: The action (13.15) of the reflection symmetry S_1 yields immediately

$$(\mathbf{C}^3)^{\Sigma} = \{(z_1,\ z_1, x) \ \mid\ z_1 \in \mathbf{C},\ x \in \mathbf{R}\} \quad \text{for } \Sigma := \{S_1;\ I\}.$$

We consider the representative (x, x, y), $x \cdot y \neq 0$ on the group orbit and write the reduced equations as

$$\begin{aligned}
\mu_1 x + \mathrm{Re}(C_1 + C_2 + C_5)x^3 + \mathrm{Re}(C_3 + C_4 + C_6)xy^2 &= 0,\\
\nu_1 x + \mathrm{Im}(C_1 + C_2 + C_5)x^3 + \mathrm{Im}(C_3 + C_4 + C_6)xy^2 &= 0,\\
\mu_2 y + (D_3 + D_6)y^3 + (2\mathrm{Re}(D_1) + D_4 + D_5)x^2 y &= 0,\\
\nu_2 y &= 0.
\end{aligned}$$

Since $x \cdot y \neq 0$, we have $\nu_2 = 0$ and

$$\begin{aligned}
\mu_1 &= -\mathrm{Re}(C_1 + C_2 + C_5)x^2 - \mathrm{Re}(C_3 + C_4 + C_6)y^2,\\
\nu_1 &= -\mathrm{Im}(C_1 + C_2 + C_5)x^2 - \mathrm{Im}(C_3 + C_4 + C_6)y^2,\\
\mu_2 &= -(2\mathrm{Re}(D_1) + D_4 + D_5)x^2 - (D_3 + D_6)y^2.
\end{aligned}$$

This is a linear system for x^2, y^2 and the parameters μ_1, μ_2, ν_1. Its solution manifold is generic two dimensional. For example, we can solve x^2 and y^2 from the first two equations and then substitute them into the third equation to derive μ_2. This gives us solutions as functions of the parameters μ_1, ν_1. Eigenvalues of the Jacobian Dg on the solution manifolds consists of zero and those of the 2×2 matrix

$$2\begin{pmatrix} x^2\mathrm{Re}(C_1 + C_2 + C_5) & xy\mathrm{Re}(C_3 + C_4 + C_6) \\ xy(2\mathrm{Re}(D_1) + D_4 + D_5) & y^2(D_3 + D_6) \end{pmatrix}$$

and of the 3×3 matrix

$$\begin{pmatrix} -2\mathrm{Re}[2C_5x^2+(C_4+C_6)y^2] & 2\mathrm{Im}[(C_1-C_2-C_5)x^2-(C_4+C_6)y^2] & 4xy\mathrm{Re}(C_4-C_6) \\ 2\mathrm{Im}[2C_5x^2+(C_4+C_6)y^2] & 2\mathrm{Re}[(C_1-C_2-C_5)x^2-(C_4+C_6)y^2] & 4xy\mathrm{Im}(C_4-C_6) \\ xy(D_5-D_4) & -2xy\mathrm{Im}(D_1) & -2(D_4+D_5)x^2-4D_6y^2 \end{pmatrix}.$$

(7) $Z_2\{\pi S_1 R^2\}$-symmetry: Note that $Z_2\{\pi S_1 R^2\} = Z_2\{S_1 R^2 \pi\}$ and

$$(\mathbf{C}^3)^{\Sigma} = \{(z_1, z_1, ix) \ \mid\ z_1 \in \mathbf{C},\ x \in \mathbf{R}\} \quad \text{for } \Sigma := Z_2\{\pi S_1 R^2\}.$$

Taking (x, x, iy), $x \cdot y \neq 0$ as a representative of the group orbit we derive from the reduced equations (13.25)

$$\begin{aligned}
\mu_1 &= -\mathrm{Re}(C_1 + C_2 + C_5)x^2 + \mathrm{Re}(-C_3 + C_4 + C_6)y^2,\\
\nu_1 &= -\mathrm{Im}(C_1 + C_2 + C_5)x^2 + \mathrm{Im}(-C_3 + C_4 + C_6)y^2,\\
\mu_2 &= (-2\mathrm{Re}(D_1) + D_4 + D_5)x^2 - (D_3 + D_6)y^2,\\
\nu_2 &= 0.
\end{aligned}$$

Owing to the S^1-orbit, zero is an eigenvalue of Dg. Stability of the solution manifolds is determined by eigenvalues of the matrix

$$2\begin{pmatrix} x^2\mathrm{Re}(C_1+C_2+C_5) & xy\mathrm{Re}(C_3-C_4-C_6) \\ xy(2\mathrm{Re}(D_1)-D_4-D_5) & y^2(D_3+D_6) \end{pmatrix}$$

and of the 3×3 matrix

$$\begin{pmatrix} 2\mathrm{Re}[-2C_5x^2+(C_4+C_6)y^2] & 2\mathrm{Im}[(C_1-C_2-C_5)x^2+(C_4+C_6)y^2] & 4xy\mathrm{Re}(-C_4+C_6) \\ 2\mathrm{Im}[2C_5x^2-(C_4+C_6)y^2] & 2\mathrm{Re}[(C_1-C_2-C_5)x^2+(C_4+C_6)y^2] & 4xy\mathrm{Im}(C_4-C_6) \\ xy(D_5-D_4) & 2xy\mathrm{Im}(D_1) & 2(D_4+D_5)x^2-4D_6y^2 \end{pmatrix}.$$

(8) $Z_4\{S_1R\}$**-symmetry:** These are rotation symmetries which force the solution to be of the form $(z_1, iz_1, (1-i)y/\sqrt{2})$. For the representative $(x, ix, (1-i)y/\sqrt{2})$, $x \cdot y \neq 0$ of the group orbit we obtain

$$\begin{aligned} \mu_1 &= -\mathrm{Re}(C_1+C_2-C_5)x^2 - \mathrm{Re}(C_3+C_4-C_6)y^2, \\ \nu_1 &= -\mathrm{Im}(C_1+C_2-C_5)x^2 - \mathrm{Im}(C_3+C_4-C_6)y^2, \\ \mu_2 &= -(2\mathrm{Re}(D_1)+D_4-D_5)x^2 - (D_3-D_6)y^2, \\ \nu_2 &= 0. \end{aligned}$$

Eigenvalues of the Jacobian Dg along its solution manifolds consist of zero and those of the matrices

$$2\begin{pmatrix} x^2\mathrm{Re}(C_1+C_2-C_5) & \sqrt{2}xy\mathrm{Re}(-C_3-C_4+C_6) \\ \sqrt{2}xy(-\mathrm{Re}(D_1)-D_4+D_5) & y^2(D_3-D_6) \end{pmatrix}$$

and

$$\begin{pmatrix} 2x^2\mathrm{Re}(C_1-C_2+C_5)+2y^2\mathrm{Re}(-C_4+C_6) & y^2\mathrm{Im}(C_4-C_6)]-2x^2\mathrm{Im}(C_5) & -2\sqrt{2}xy\mathrm{Im}(C_4+C_6) \\ 4x^2\mathrm{Im}(C_1-C_2+C_5)+4y^2\mathrm{Re}(-C_4+C_6) & 4x^2\mathrm{Re}(C_5)+y^2\mathrm{Re}(C_4-C_6)] & 4\sqrt{2}xy\mathrm{Re}(C_4+C_6) \\ 2\sqrt{2}xy\mathrm{Im}(D_1) & \sqrt{2}xy(D_4+D_5) & 2(D_5-D_4)x^2+4D_6y^2 \end{pmatrix}.$$

(9) $Z_4\{\pi S_1R\}$**-symmetry:** These are also rotation symmetries and

$$(\mathbf{C}^3)^\Sigma = \{(z_1, iz_1, \frac{1+i}{\sqrt{2}}y) \mid z_1 \in \mathbf{C},\ y \in \mathbf{R}\} \quad \text{for } \Sigma := Z_4\{\pi S_1 R\}.$$

Choosing $(x, ix, (1+i)y/\sqrt{2})$, $x \cdot y \neq 0$ as a representative we obtain from (13.25)

$$\begin{aligned} \mu_1 &= -\mathrm{Re}(C_1+C_2-C_5)x^2 - \mathrm{Re}(C_3-C_4+C_6)y^2, \\ \nu_1 &= -\mathrm{Im}(C_1+C_2-C_5)x^2 - \mathrm{Im}(C_3-C_4+C_6)y^2, \\ \mu_2 &= -(\mathrm{Re}(D_1)-D_4+D_5)x^2 - (D_3-D_6)y^2, \\ \nu_2 &= 0. \end{aligned}$$

Eigenvalues of the Jacobian Dg along its solution manifolds consist of zero and those of the matrix

$$2\begin{pmatrix} x^2\mathrm{Re}(C_1+C_2-C_5) & \sqrt{2}xy\mathrm{Re}(C_3-C_4-C_6) \\ \sqrt{2}xy(\mathrm{Re}(D_1)-D_4+D_5) & y^2(D_3-D_6) \end{pmatrix}$$

and of the 3×3 matrix

$$\begin{pmatrix} 2x^2\mathrm{Re}(C_1-C_2+C_5)+2y^2\mathrm{Re}(C_4-C_6) & 4x^2\mathrm{Im}(C_5)-2y^2\mathrm{Im}(C_4-C_6)] & -2\sqrt{2}xy\mathrm{Im}(C_4+C_6) \\ 2x^2\mathrm{Im}(-C_1+C_2-C_5)+2y^2\mathrm{Re}(-C_4+C_6) & 4x^2\mathrm{Re}(C_5)+2y^2\mathrm{Re}(C_4-C_6)] & -2\sqrt{2}xy\mathrm{Re}(C_4+C_6) \\ 2\sqrt{2}xy\mathrm{Im}(D_1) & \sqrt{2}xy(D_4+D_5) & 2(D_4-D_5)x^2+4D_6y^2 \end{pmatrix}.$$

13.5 Calculations of the Normal Form

We consider in this section calculations of the truncated normal form

$$\frac{dz_i}{dt} = g_i(z,\beta,e), \quad i=1,2,3 \tag{13.28}$$

with

$$\begin{aligned} g_1(z,\beta,e) &= (\omega_0 i + A_1\beta + A_2 e)z_1 + \left(C_1|z_1|^2 + C_2|z_2|^2 + C_3|z_3|^2\right) z_1 \\ &\quad + C_4 z_2 z_3^2 + C_5\bar{z}_1 z_2^2 + C_6 z_2\bar{z}_3^2, \\ g_2(z,\beta,e) &= (\omega_0 i + A_1\beta + A_2 e)z_2 + \left(C_2|z_1|^2 + C_1|z_2|^2 + C_3|z_3|^2\right) z_2 \\ &\quad + C_4 z_1\bar{z}_3^2 + C_5 z_1^2\bar{z}_2 + C_6 z_1\bar{z}_3^2, \\ g_3(z,\beta,e) &= (B_1\beta + B_2 e)z_2 + \left(D_1|z_1|^2 + D_2|z_2|^2 + D_3|z_3|^2\right) z_3 \\ &\quad + D_4 z_1\bar{z}_2\bar{z}_3 + D_5\bar{z}_1 z_2 z_3 + D_6\bar{z}_3^3. \end{aligned} \tag{13.29}$$

Expand the center manifold $h(z,\beta,e)$ formally into Taylor series at $(z,\beta,e) = (0,0,0)$

$$h(z,\beta,e) = \sum_{|i|=1}^{3} h_i z_1^{i_1}\bar{z}_1^{i_2} z_2^{i_3}\bar{z}_2^{i_4} z_3^{i_5}\bar{z}_3^{i_6}\beta^{i_7}e^{i_8} + h.o.t. \tag{13.30}$$

Here $i = (i_1,\ldots,i_8) \in \mathbf{N}^8$ represents multi-index in the summation and $|i| = \sum_{j=1}^{8} i_j$. It follows from

$$h(0,\beta,e) \equiv 0 \quad \text{for all} \quad \beta,\ e\in\mathbf{R}$$

that $h_i = 0$ for all $i = (0,\ldots,0,i_7,i_8) \in \mathbf{N}^8$. On the other hand, $h_i = 0$ for all i with $|i| = 1$. The actions of S_1 and R in (13.15) imply

$$\begin{aligned} h_i(x,y) &= S_1 h_j(x,y) = h_j(1-x,y), \quad j = (i_3,i_4,i_1,i_2,i_6,i_5,i_7,i_8), \\ h_i(x,y) &= h_i(1-x,1-y), \quad i_1+i_2+i_3+i_4+i_5+i_6 = 1(\mathrm{mod})2. \end{aligned}$$

From the equality $h = \bar{h}$ we derive

$$h_i = \bar{h}_j \quad \text{for all} \quad i\in\mathbf{N}^8 \quad \text{and} \quad j = (i_3,i_4,i_1,i_2,i_6,i_5,i_7,i_8).$$

Substituting

$$u = \sum_{i=1}^{3} (z_i\phi_i + \overline{z_i\phi_i}) + h(z, \beta, e)$$

into the original equation (13.1)

$$\begin{aligned}&\sum_{i=1}^{3} \left[(\phi_i + D_{z_i}h(z,\beta,e))g_i(z,\beta,e) + (\bar{\phi}_i + D_{\bar{z}_i}h(z,\beta,e))\bar{g}_i(z,\beta,e)\right] \\ &= G(\textstyle\sum_{i=1}^{3}(z_i\phi_i + \overline{z_i\phi_i}) + h(z,\beta,e),\ \lambda_0 + \beta,\ d_0 + e)\end{aligned} \tag{13.31}$$

and replacing the terms $g_i(z, \beta, e)$ and $h(z, \beta, e)$ with the statements (13.29) and (13.30), respectively, we expand the both sides of (13.31) into Taylor series in z, β, e at the origin of $\mathbf{C}^3 \times \mathbf{R}^2$. Thereafter, by comparing coefficients of the terms

$$z_1^{i_1} \bar{z}_1^{i_2} z_2^{i_3} \bar{z}_2^{i_4} z_3^{i_5} \bar{z}_3^{i_6} \beta^{i_7} e^{i_8}, \quad i = (i_1, \ldots, i_8) \in \mathbf{N}^8,$$

we obtain equations for h_i, $|i| \leq 3$. Solvabilities of these equations yield the coefficients A_i, B_j, C_k and D_l in the truncated normal form (13.28). Denote

$$L := D_u G_0.$$

We summarize the calculations of these coefficients in the following tables.

terms	indices	equations	solvability	h_i used in
$z_1\beta$	$(1,0,0,0,0,0,1,0)$	$(L-\omega_0 i)h_i = A_1\phi_1 - D_{u\lambda}G_0\phi_1$	$A_1 = \langle \phi_1^*, D_{u\lambda}G_0\phi_1\rangle$	
$z_1 e$	$(1,0,0,0,0,0,0,1)$	$(L-\omega_0 i)h_i = A_2\phi_1 - D_{ud}G_0\phi_1$	$A_2 = \langle \phi_1^*, D_{ud}G_0\phi_1\rangle$	
$z_3\beta$	$(0,0,0,0,1,0,1,0)$	$Lh_i = B_1\phi_3 - D_{u\lambda}G_0\phi_3$	$B_1 = \langle \phi_3^*, D_{u\lambda}G_0\phi_3\rangle$	
$z_3 e$	$(0,0,0,0,1,0,0,1)$	$Lh_i = B_2\phi_3 - D_{ud}G_0\phi_3$	$B_2 = \langle \phi_3^*, D_{ud}G_0\phi_3\rangle$	
$\lvert z_1\rvert^2$	$(1,1,0,0,0,0,0,0)$	$Lh_i = -D_{uu}G_0\phi_1\bar{\phi}_1$	$D_{uu}G_0\phi_1\bar{\phi}_1 \in \mathrm{Im}(L)$ holds	C_1, C_2, D_1
z_1^2	$(2,0,0,0,0,0,0,0)$	$(L-2\omega_0 i)h_i = -\frac{1}{2}D_{uu}G_0\phi_1^2,$	$D_{uu}G_0\phi_1^2 \in \mathrm{Im}(L)$ holds	C_1, C_5
$z_1\bar{z}_2$	$(1,0,0,1,0,0,0,0)$	$Lh_i = -D_{uu}G_0\phi_1\bar{\phi}_2$	$D_{uu}G_0\phi_1\bar{\phi}_2 \in \mathrm{Im}(L)$ holds	C_2, C_5
$z_1 z_2$	$(1,0,1,0,0,0,0,0)$	$(L-2\omega_0 i)h_i = -D_{uu}G_0\phi_1\phi_2$	$(L-2\omega_0 i)$ nonsingular	C_2
$\lvert z_3\rvert^2$	$(0,0,0,0,1,1,0,0)$	$Lh_i = -D_{uu}G_0\phi_3\bar{\phi}_3$	$D_{uu}G_0\phi_3\bar{\phi}_3 \in \mathrm{Im}(L)$ holds	C_3, D_3
$z_1 z_3$	$(1,0,0,0,1,0,0,0)$	$(L-\omega_0 i)h_i = -D_{uu}G_0\phi_1\phi_3$	$D_{uu}G_0\phi_1\phi_3 \in \mathrm{Im}(L)$ holds	C_3, C_6, D_1, D_5
$z_1\bar{z}_3$	$(1,0,0,0,0,1,0,0)$	$(L-\omega_0 i)h_i = -D_{uu}G_0\phi_1\bar{\phi}_3$	$D_{uu}G_0\phi_1\bar{\phi}_3 \in \mathrm{Im}(L)$ holds	C_3, C_4, D_1
z_3^2	$(0,0,0,0,2,0,0,0)$	$Lh_i = -D_{uu}G_0\phi_3^2$	$D_{uu}G_0\phi_3\phi_3 \in \mathrm{Im}(L)$ holds	C_4, C_6, D_3
$\bar{z}_2 z_3$	$(0,0,0,1,1,0,0,0)$	$(L+\omega_0 i)h_i = -D_{uu}G_0\bar{\phi}_2\phi_3$	$D_{uu}G_0\bar{\phi}_2\phi_3 \in \mathrm{Im}(L)$ holds	D_5
$\bar{z}_2\bar{z}_3$	$(0,0,0,1,0,1,0,0)$	$(L+\omega_0 i)h_i = -D_{uu}G_0\bar{\phi}_2\bar{\phi}_3$	$D_{uu}G_0\bar{\phi}_2\bar{\phi}_3 \in \mathrm{Im}(L)$ holds	D_4

Table 13.3: Coefficients of the second order terms in the Taylor expansion of the reduced equations and their properties.

terms	involved indices	equations	solvability
$\|z_1\|^2 z_1$	$i=(2,1,0,0,0,0,0,0)$ $j=(1,1,0,0,0,0,0,0)$ $k=(2,0,0,0,0,0,0,0)$	$(L-\omega_0 i)h_i = C_1\phi_1 - \frac{1}{2}D_{uuu}G_0\phi_1^2\bar{\phi}_1 - D_{uu}G_0(\bar{\phi}_1 h_k + \phi_1 h_j)$	$C_1 = \left\langle \phi_1^*, \frac{1}{2}D_{uuu}G_0\phi_1^2\bar{\phi}_1 + D_{uu}G_0(\bar{\phi}_1 h_k + \phi_1 h_j)\right\rangle$
$\|z_2\|^2 z_1$	$i=(1,0,1,1,0,0,0,0)$ $j=(1,1,0,0,0,0,0,0)$ $k=(1,0,0,1,0,0,0,0)$ $l=(1,0,1,0,0,0,0,0)$	$(L-\omega_0 i)h_i = C_2\phi_1 - D_{uuu}G_0\phi_1\phi_2\bar{\phi}_2 - D_{uu}G_0(\phi_1 S_1 h_j + \phi_2 h_k + \bar{\phi}_2 h_l)$	$C_2 = \left\langle \phi_1^*, D_{uuu}G_0\phi_1\phi_2\bar{\phi}_2 + D_{uu}G_0(\phi_1 S_1 h_j + \phi_2 h_k + \bar{\phi}_2 h_l)\right\rangle$
$\|z_3\|^2 z_1$	$i=(1,0,0,0,1,1,0,0)$ $j=(0,0,0,0,1,1,0,0)$ $k=(1,0,0,0,0,1,0,0)$ $l=(1,0,0,0,1,0,0,0)$	$(L-\omega_0 i)h_i = C_3\phi_1 - D_{uuu}G_0\phi_1\phi_3\bar{\phi}_3 - D_{uu}G_0(\phi_1 h_j + \phi_3 h_k + \phi_3 h_l)$	$C_3 = \left\langle \phi_1^*, D_{uuu}G_0\phi_1\phi_3\bar{\phi}_3 + D_{uu}G_0(\phi_1 h_j + \phi_3 h_k + \phi_3 h_l)\right\rangle$
$z_1\bar{z}_3^2$	$i=(1,0,0,0,0,2,0,0)$ $j=(0,0,0,0,2,0,0,0)$ $k=(1,0,0,0,0,1,0,0)$	$(L-\omega_0 i)h_i = C_4\phi_2 - \frac{1}{2}D_{uuu}G_0\phi_1\bar{\phi}_3^2 - D_{uu}G_0(\phi_1 S_1 h_j + \phi_3 h_k)$	$C_4 = \left\langle \phi_2^*, \frac{1}{2}D_{uuu}G_0\phi_1\bar{\phi}_3^2 + D_{uu}G_0(\phi_1 S_1 h_j + \phi_3 h_k)\right\rangle$
$z_1^2\bar{z}_2$	$i=(2,0,0,1,0,0,0,0)$ $j=(1,0,0,1,0,0,0,0)$ $k=(2,0,0,0,0,0,0,0)$	$(L-\omega_0 i)h_i = C_5\phi_2 - \frac{1}{2}D_{uuu}G_0\phi_1^2\bar{\phi}_2 - D_{uu}G_0(\phi_1 h_j + \bar{\phi}_2 h_k)$	$C_5 = \left\langle \phi_2^*, \frac{1}{2}D_{uuu}G_0\phi_1^2\bar{\phi}_2 + D_{uu}G_0(\phi_1 h_j + \bar{\phi}_2 h_k)\right\rangle$
$z_1 z_3^2$	$i=(1,0,0,0,2,0,0,0)$ $j=(0,0,0,0,2,0,0,0)$ $k=(1,0,0,0,1,0,0,0)$	$(L-\omega_0 i)h_i = C_6\phi_2 - \frac{1}{2}D_{uuu}G_0\phi_1\phi_3^2 - D_{uu}G_0(\phi_1 h_j + \phi_3 h_k)$	$C_6 = \left\langle \phi_2^*, \frac{1}{2}D_{uuu}G_0\phi_1\phi_3^2 + D_{uu}G_0(\phi_1 h_j + \phi_3 h_k)\right\rangle$

Table 13.4: Coefficients of the cubic order terms in the Taylor expansion of the reduced equations.

terms	involved indices	equations	solvability
$\lvert z_1\rvert^2 z_3$	$i=(1,1,0,0,1,0,0,0)$ $j=(1,0,0,0,0,1,0,0)$ $k=(1,0,0,0,1,0,0,0)$ $l=(1,1,0,0,0,0,0,0)$	$Lh_i = D_1\phi_3 - D_{uuu}G_0\phi_1\bar\phi_1\phi_3 - D_{uu}G_0(\phi_1\bar h_j + \bar\phi_1 h_k + \phi_3 h_l)$	$D_1 = \left\langle \phi_3^*, D_{uuu}G_0\phi_1\bar\phi_1\phi_3 + D_{uu}G_0(\phi_1\bar h_j + \bar\phi_1 h_k + \phi_3 h_l)\right\rangle$
$\lvert z_3\rvert^2 z_3$	$i=(0,0,0,0,2,1,0,0)$ $j=(0,0,0,0,1,1,0,0)$ $k=(0,0,0,0,2,0,0,0)$	$Lh_i = D_3\phi_3 - \frac{1}{2}D_{uuu}G_0\phi_3^2\bar\phi_3 - D_{uu}G_0(\phi_3 h_j + \bar\phi_3 h_k)$	$D_3 = \left\langle \phi_3^*, \frac{1}{2}D_{uuu}G_0\phi_3^2\bar\phi_3 + D_{uu}G_0(\phi_3 h_j + \bar\phi_3 h_k)\right\rangle$
$z_1\bar z_2\bar z_3$	$i=(1,0,0,1,0,1,0,0)$ $j=(0,0,0,1,0,1,0,0)$ $k=(1,0,0,0,0,1,0,0)$ $l=(1,0,0,1,0,0,0,0)$	$Lh_i = D_4\phi_3 - D_{uuu}G_0\phi_1\bar\phi_2\bar\phi_3 - D_{uu}G_0(\phi_1 h_j + \bar\phi_2 h_k + \bar\phi_3 h_l)$	$D_4 = \left\langle \phi_3^*, D_{uuu}G_0\phi_1\bar\phi_2\bar\phi_3 + D_{uu}G_0(\phi_1 h_j + \bar\phi_2 h_k + \bar\phi_3 h_l)\right\rangle$
$z_1\bar z_2 z_3$	$i=(1,0,0,1,1,0,0,0)$ $j=(0,0,0,1,1,0,0,0)$ $k=(1,0,0,0,1,0,0,0)$ $l=(1,0,0,1,0,0,0,0)$	$Lh_i = D_5\bar\phi_3 - D_{uuu}G_0\phi_1\bar\phi_2\phi_3 - D_{uu}G_0(\phi_1 h_j + \bar\phi_2 h_k + \phi_3 h_l)$	$D_5 = \left\langle \phi_3^*, D_{uuu}G_0\phi_1\bar\phi_2\phi_3 + D_{uu}G_0(\phi_1 h_j + \bar\phi_2 h_k + \phi_3 h_l)\right\rangle$
$\bar z_3^3$	$i=(0,0,0,0,0,3,0,0)$ $j=(0,0,0,0,2,0,0,0)$	$Lh_i = D_6\phi_3 - \frac{1}{6}D_{uuu}G_0\bar\phi_3^3 - D_{uu}G_0\bar\phi_3\bar h_j)$	$D_6 = \left\langle \phi_3^*, \frac{1}{6}D_{uuu}G_0\bar\phi_3^3 + D_{uu}G_0\bar\phi_3\bar h_j)\right\rangle$

Table 13.4 (continued) Coefficients of the cubic order terms in the Taylor expansion of the reduced equations.

14. Homotopy of Boundary Conditions

Boundary conditions influence chemical reactions through diffusion and transportation of the substances, e.g. via convection and fluid flow. Diffusion is described by the Laplace operator Δ, as in the equation

$$\frac{\partial u}{\partial t} = D\Delta u + f(u, \lambda)$$

for unstirred reactions. Diffusion is the underlying mechanism for spatial pattern formations. Properties and spectrum of the Laplacian are decisive for analysis of dynamics and bifurcations of reaction-diffusion equations. As we have seen in previous chapters, linear stability of a solution $u = u_0$ is determined by eigenvalues of the linearized operator

$$D\Delta + D_u f(u_0, \lambda).$$

In this chapter we discuss some properties of boundary conditions and the Laplace operator. We consider a few frequently used types of boundary conditions and summarize the classical Sturm-Liouville theory. We study eigenvalues of the Laplacian and solutions of a linear scalar elliptic differential equation with respect to variable boundary conditions.

14.1 Boundary Conditions

Different boundary conditions may result in distinct bifurcation scenarios in reaction-diffusion equations, see Chapter 15, Ashwin/Böhmer/Mei [21, 22], Ashwin/Mei [25], Dillon/Maini/Othmer [86], Mei/Theil [227]. In fact, posing and realizing appropriate boundary conditions, e.g. for chemical reactions in open and large systems, can be quite delicate (cf. Gray/Scott [273]). Often components of a system of reaction-diffusion equations have different boundary conditions. This makes a big difference in the behavior of systems and scalar equations, e.g. in Hopf bifurcations, spirals and other pattern formations. Even for stationary bifurcation problems multiple bifurcations are expected more likely in systems. Moreover, stability of bifurcating solution branches and critical modes can vary considerably from a scalar equation to a system.

We consider the equation

$$\frac{\partial u}{\partial t} = \Delta u + \lambda u - f(u, \lambda) \quad \text{in} \quad \Omega \tag{14.1}$$

as a representative of scalar equations and

$$\begin{aligned} \frac{\partial u_1}{\partial t} &= \Delta u_1 + f_1(u_1, u_2, \lambda), \\ \frac{\partial u_2}{\partial t} &= d\Delta u_2 + f_2(u_1, u_2, \lambda) \end{aligned} \tag{14.2a}$$

as a model for systems of reaction-diffusion equations.

14.1.1 Homotopy of boundary conditions

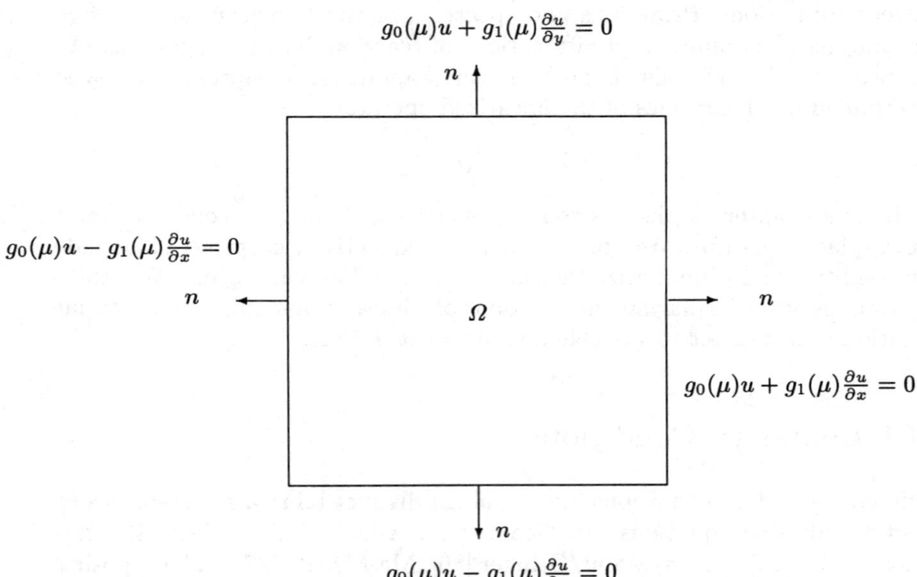

Fig. 14.1. A homotopy between Neumann and Dirichlet boundary conditions.

To investigate influence of boundary conditions in bifurcations of reaction-diffusion problems, we study at first properties of the linear problem

$$-\Delta u + u = f \quad \text{in} \quad \Omega \tag{14.3}$$

on a homotopy of boundary conditions

$$g_0(\mu)u + g_1(\mu)\frac{\partial u}{\partial n} = 0 \quad \text{on } \partial\Omega. \tag{14.4}$$

Here $g_1,\ g_2 :\ [0,1] \to \mathbf{R}$ are smooth functions (C^3-continuous) such that

$$g_0(0) = g_1(1) = 0, \qquad g_0(1)g_1(0)g_0(\mu)g_1(\mu) \neq 0 \quad \text{for all } \mu \in (0,1). \tag{14.5}$$

Furthermore, we assume

$$g_0(\mu)g_1(\mu) > 0 \quad \text{for all } \mu \in (0,1).$$

Homotopy of boundary conditions has been used by Fiedler [102] and Gardner [118] to study global attractors and nonsingular solutions of a class of reaction-diffusion equations. They show that these solution sets are independent of boundary conditions. Nevertheless, if the equilibrium is nonhyperbolic and a bifurcation occurs, the bifurcation scenario, e.g. structure of attractors, may vary with respect to boundary conditions. This has been observed by Dillon/Maini/Othmer [86] in the study of pattern formation in generalized Turing systems in intervals and by Mei/Theil [227] in the analysis of steady state bifurcations as well as by Holder/Schaeffer [161] and Schaeffer/Golubitsky [267] on mode-jumping of von Kármán equations.

14.1.2 Boundary conditions for different components

Turing system consists of reaction-diffusion equations with the identical boundary conditions for all components, e.g.

$$g_0(\mu)u_i + g_1(\mu)\frac{\partial u_i}{\partial n} = 0 \quad \text{on } \partial\Omega, \quad i = 1, 2. \tag{14.6}$$

This kind of boundary conditions simplifies the mathematical analysis of the system. In particular, the analysis of linear stability can be reduced to the discussion of eigenvalue problems of finite dimensional matrices, as we have seen in previous chapters. Nevertheless, different boundary conditions can be imposed for two components in (14.2),

$$g_0(\mu)u_1 + g_1(\mu)\frac{\partial u_1}{\partial n} = 0 \quad \text{on } \partial\Omega, \tag{14.7a}$$

$$h_0(\mu)u_2 + h_1(\mu)\frac{\partial u_2}{\partial n} = 0 \quad \text{on } \partial\Omega, \tag{14.7b}$$

for example, if $h_0 = g_1$ and $h_1 = g_0$, the boundary condition for u_1 goes from Neumann to Dirichlet, while for u_2 from Dirichlet to Neumann when we vary μ from 0 to 1. As noted in Dillon *et al* [86], mixed types of boundary conditions for different components have stabilizing effect on bifurcating solutions (patterns) and reduce the sensitivity of patterns on boundary conditions.

On the other hand, multiplicities of solutions may change as the boundary conditions vary, in particular, if symmetry of the problem is broken.

Linear stability analysis for reaction-diffusion problems with mixed boundary conditions (14.7) is much more complicated than the case (14.6). Recall the linearized operator of (14.2) at a homogeneous steady state (say 0)

$$D_uG(0,\lambda,\mu) = \begin{pmatrix} \Delta + f_{1u_1}(0,\lambda) & f_{1u_2}(0,\lambda) \\ f_{2u_1}(0,\lambda) & d\Delta + f_{2u_2}(0,\lambda) \end{pmatrix}.$$

Let us consider the eigenvalue pair σ and (u_1, u_2) of the problem

$$\begin{aligned} (\Delta + f_{1u_1}(0,\lambda))u_1 + f_{1u_2}(0,\lambda)u_2 &= \sigma u_1, \\ f_{2u_1}(0,\lambda)u_1 + (d\Delta + f_{2u_2}(0,\lambda))u_2 &= \sigma u_2 \end{aligned} \tag{14.8}$$

with the boundary conditions (14.7). Analytical solution of this problem is generally not available and numerical approximation is necessary. We take here spectral method as an example. Denote κ_i^g, ϕ_i^g (resp. κ_i^h, ϕ_i^h) eigenpairs of the Laplacian Δ with the boundary condition (14.7a) (resp. (14.7b)). Since eigenfunctions $\{\phi_i^g,\ i = 1,2,\ldots\}$ and $\{\phi_i^h,\ i = 1,2,\ldots\}$ define a basis of $C^{2,s}(\Omega)$, solutions of (14.8) are of the form

$$\begin{pmatrix} u_1 \\ u_2 \end{pmatrix} = \begin{pmatrix} \sum\limits_{i=1}^{\infty} A_i\phi_i^g \\ \sum\limits_{i=1}^{\infty} B_i\phi_i^h \end{pmatrix}.$$

Substituting this into (14.8) yields

$$\sum_{i=1}^{\infty}\left[(-\kappa_i^g + f_{1u_1}(0,\lambda))A_i\phi_i^g + f_{1u_2}(0,\lambda)B_i\phi_i^h\right] = \sigma\sum_{i=1}^{\infty} A_i\phi_i^g\ ,$$

$$\sum_{i=1}^{\infty}\left[f_{2u_1}(0,\lambda)A_i\phi_i^g + (-d\kappa_i^h + f_{2u_2}(0,\lambda))B_i\phi_i^h\right] = \sigma\sum_{i=1}^{\infty} B_i\phi_i^h\ .$$

Taking inner products with ϕ_j^g, $j = 1,2,\ldots$ respectively, we obtain

$$(-\kappa_j^g + f_{1u_1}(0,\lambda) - \sigma)A_j = -\sum_{i=1}^{\infty} f_{1u_2}(0,\lambda)B_i\alpha_{ij}\ , \tag{14.9a}$$

$$f_{2u_1}(0,\lambda)A_j = \sum_{i=1}^{\infty}(d\kappa_i^h - f_{2u_2}(0,\lambda) + \sigma)B_i\alpha_{ij}\ , \tag{14.9b}$$

where

$$\alpha_{ij} := \frac{\langle\phi_j^g,\ \phi_i^h\rangle}{\langle\phi_j^g,\ \phi_j^g\rangle},\quad i,j = 1,2,\ldots.$$

This is an eigenvalue problem with an infinite matrix. If we choose $h_i = g_i$, $i = 0, 1$ in (14.7b), i.e., the same (so-called scalar) boundary conditions for u_1 and u_2, then $\alpha_{ij} = \delta_{ij}$ and (14.9) can be solved directly. Finite dimensional approximation of (14.9) is chosen by truncating at sufficient large i, see also Mei [223].

14.1.3 Mixed boundary conditions along the sides

To study impact of symmetry-breaking in boundary conditions on bifurcation scenario we may consider a square domain and impose the following conditions on its four sides

$$
\begin{aligned}
g_0(\mu)u(x,0) - g_1(\mu)\frac{\partial u}{\partial y}(x,0) &= 0,\\
g_0(\mu)u(x,\pi) + g_1(\mu)\frac{\partial u}{\partial y}(x,\pi) &= 0,\\
\frac{\partial u}{\partial x}(0,y) &= 0,\\
\frac{\partial u}{\partial x}(\pi,y) &= 0.
\end{aligned}
$$

These boundary conditions break the D_4-symmetry into D_2-symmetry. Variation of bifurcation scenario is discussed in Section 15.6.

14.1.4 Dynamical boundary conditions

In biological and chemical reaction processes, pattern generation and annihilation take place constantly along boundaries of patterns. Dynamics, e.g. spiral waves, inside the physical domain may be induced by dynamical boundary conditions (cf. Dellnitz *et al* [83]), as

$$g_0(\mu,t)u_1 + g_1(\mu,t)\frac{\partial u_1}{\partial n} = 0 \quad \text{on } \partial\Omega, \tag{14.10a}$$

$$h_0(\mu,t)u_2 + h_1(\mu,t)\frac{\partial u_2}{\partial n} = 0 \quad \text{on } \partial\Omega. \tag{14.10b}$$

Here $g_i(\mu,t)$ and $h_i(\mu,t)$ are functions in t. Of particular interests are the periodic functions.

14.2 A Brief Review of Sturm-Liouville Theory

In this section we summarize briefly some well-known properties of the Sturm-Liouville equations and refer to Pinsky [242] for details of the proofs. The general Sturm-Liouville equation is

$$(su')' + (\lambda\rho - q)u = 0 \quad \text{in } \Omega := [a,b] \tag{14.11}$$

where $s(x)$, $\rho(x)$ and $q(x)$ are given continuous functions in Ω, and $u(x)$, λ are the unknowns. λ is called the *eigenvalue parameter*, and ρ is the *weight function*. Equation (14.11) is solved subject to one of the following types of boundary conditions

1. separable (Robin)

$$\begin{aligned} -u(a)\sin\alpha + u'(a)\cos\alpha = 0, \\ u(b)\sin\beta + u'(b)\cos\beta = 0; \end{aligned} \tag{14.12}$$

2. periodic

$$\begin{aligned} u(a) = u(b), \\ u'(a) = u'(b). \end{aligned} \tag{14.13}$$

The one-dimensional eigenvalue problem of the Laplacian

$$u'' + \lambda u = 0$$

is a special type of Sturm-Liouville equations. Another important (Bessel) equation corresponds to

$$s(x) = x, \qquad \rho(x) = x, \qquad q(x) = 0.$$

In particular, we mention that the equation (14.11) is in self-adjoint form. Any equations of the form

$$a(x)u'' + b(x)u'(x) + [\lambda c(x) - d(x)]u(x) = 0$$

can be transformed into the form (14.11) by taking

$$s(x) = e^{\int_0^x \frac{b(t)}{a(t)}dt}, \quad \rho(x) = s(x)\frac{c(x)}{a(x)} \quad \text{and} \quad q(x) = s(x)\frac{d(x)}{a(x)}.$$

We suppose

$$s(x) > 0, \qquad \rho(x) > 0 \quad \text{for } x \in (a,b).$$

If $s(a)s(b)\rho(a)\rho(b) = 0$, then (14.11) is called a *singular Sturm-Liouville problem*. Otherwise, it is said *regular*. A regular Sturm-Liouville equation (14.11) has the following five important properties.

1) Orthogonality of Eigenfunctions. Let (u_1, λ_1) and (u_2, λ_2) be solutions of the Sturm-Liouville equation (14.11), satisfying the boundary condition

$$s(b)[u_1(b)u_2'(b) - u_1'(b)u_2(b)] = s(a)[u_1(a)u_2'(a) - u_1'(a)u_2(a)].$$

If $\lambda_1 \neq \lambda_2$, then ϕ_1 and ϕ_2 are orthogonal with respect to $L^2(\Omega)$-product with the weight function $\rho(x)$.

2) Nondegeneracy of Eigenvalues. Let u_1, u_2 be solutions of the Sturm-Liouville problem (14.11) with the boundary condition (14.12) and the same eigenvalue

$$u_1(x) \not\equiv 0, \quad u_2(x) \not\equiv 0 \quad \text{and} \quad s(a)s(b) \neq 0.$$

Then there exists a constant $C \neq 0$, such that

$$u_1(x) = Cu_2(x).$$

3) Positivity of Eigenvalues. Let (u, λ) be a solution of the Sturm-Liouville equation (14.11) satisfying the conditions

$$s(x)u(x)u'(x)\Big|_a^b = s(b)u(b)u'(b) - s(a)u(a)u'(a) \leq 0.$$

Then

$$\lambda \int_a^b \rho(x)u^2(x)\, dx \geq \int_a^b q(x)u(x)^2\, dx + \int_a^b s(x)(u'(x))^2\, dx. \tag{14.14}$$

If in addition $q(x) \geq 0$, then the terms on the right hand side are nonnegative and all eigenvalues are positive or zero. In particular, this result holds for the following cases

i) *Homogeneous boundary conditions:* $u(a) = u(b) = 0$ or $u'(a) = u'(b) = 0$;
ii) *Mixed boundary conditions:* $u(a) = 0, \quad u'(b) = 0$;
iii) *Singular Sturm-Liouville problem:* $s(a) = 0, \quad s(b) = 0$.

4) Completeness of Eigenfunctions. There exists an infinite sequence of solutions (ϕ_n, λ_n) of the regular Sturm-Liouville problem (14.11) with the separable boundary conditions (14.12). If $f(x)$ is a smooth function satisfying the boundary conditions (14.12), then the following series is uniformly convergent for $x \in [a, b]$

$$f(x) = \sum_{n=1}^{\infty} A_n \phi_n(x),$$

where the Fourier coefficients are given by the formulas

$$A_n \int_a^b \phi_n(x)^2 \rho(x)\, dx = \int_a^b f(x)\phi_n(x)\rho(x)\, dx, \quad n = 1, 2, \ldots.$$

Based on the completeness of the eigenfunctions, one can write, e.g., a solution of the heat equation $u_t = Ku_{xx}$ in the form

$$u(x, t) = U(x) + \sum_{n=1}^{\infty} \phi_n(x) e^{-\lambda_n K t},$$

where $U(x)$ is a steady-state solution. Thereafter, the positivity of eigenvalues guarantees that we approach a steady-state solution when $t \to \infty$.

5) Asymptotic Behavior of Eigenvalues and Eigenfunctions. Let (ϕ_n, λ_n) be the normalized eigenfunctions and eigenvalues of the regular Sturm-Liouville problem

$$\begin{aligned}(su')' + (\lambda\rho - q)u &= 0 \quad \text{in} \quad x \in (a,b),\\ u(a) &= 0,\\ u(b) &= 0,\\ \int_a^b u(x)^2\rho(x)\,dx &= 1.\end{aligned}$$

Then we have the following asymptotic formulas

$$\begin{aligned}\lambda_n &= \left(\frac{n\pi}{\tilde{b}-\tilde{a}}\right) + O(\frac{1}{n}), \qquad n\to\infty,\\ v_n(z) &= \pm\sqrt{\frac{2}{\tilde{b}-\tilde{a}}}\sin\frac{n\pi(z-\tilde{a})}{\tilde{b}-\tilde{a}} + O(\frac{1}{n}), \qquad n\to\infty, \qquad (14.15)\\ v_n'(z) &= \pm n\pi\sqrt{\frac{2}{(\tilde{b}-\tilde{a})^3}}\cos\frac{n\pi(z-\tilde{a})}{\tilde{b}-\tilde{a}} + O(1), \qquad n\to\infty,\end{aligned}$$

where $\tilde{a} = z(a)$, $\tilde{b} = z(b)$ and $z(x), v(z)$ are defined implicitly by the equations

$$\begin{aligned}\phi(x) &= \left(\frac{\rho}{s^3}\right)^{-1/4} v(z),\\ \frac{dz}{dx} &= \left(\frac{\rho}{s}\right)^{1/2},\\ v'' + [\lambda - \tilde{q}(z)]v &= 0, \quad z\in(\tilde{a},\ \tilde{b}).\end{aligned}$$

14.3 Eigenvalues of the Laplacian with Robin Boundary Conditions

To study influence of boundary conditions on the bifurcation scenario of a system of two reaction-diffusion equations

$$\left.\begin{aligned}\frac{\partial u}{\partial t} &= \Delta u + f(u,v,\lambda)\\ \frac{\partial v}{\partial t} &= d\Delta v + g(u,v,\lambda)\end{aligned}\right\} \quad \text{in } \Omega, \qquad (14.16)$$

we connect the Neumann and Dirichlet boundary conditions with the homotopy in (14.4). Linear stability of the trivial solution curve $(0,0,\lambda)$ is described by the linearized operator

$$L := \begin{pmatrix}\Delta + f_u(0,0,\lambda) & f_v(0,0,\lambda)\\ g_u(0,0,\lambda) & d\Delta + g_v(0,0,\lambda)\end{pmatrix}. \qquad (14.17)$$

Assume that $(\sigma,\ \phi)$ is an eigenpair of $-\Delta$ with the homotopy boundary conditions

$$\begin{aligned} -\Delta\phi &= \sigma\phi_1 && \text{in } \Omega, \\ g_0(\mu)\frac{\partial\phi}{\partial n} + g_1(\mu)\phi &= 0 && \text{on } \partial\Omega. \end{aligned}$$

Then the subspace

$$X_\sigma := \left\{ \begin{pmatrix} a \\ b \end{pmatrix} \phi; \quad a, b \in \mathbf{R} \right\}$$

is invariant under the operator L. The restriction of L in X_σ is the 2×2 matrix

$$M_\sigma := \begin{pmatrix} -\sigma + f_u(0,0,\lambda) & f_v(0,0,\lambda) \\ g_u(0,0,\lambda) & -d\sigma + g_v(0,0,\lambda) \end{pmatrix}.$$

Thus the spectrum of L consists of the eigenvalues of the matrices M_σ. It is crucial to determine eigenvalues of the Laplacian with respect to the corresponding boundary conditions. We choose a rectangular domain $\Omega := [0, a_1] \times [0, a_2]$ and consider the eigenvalue problem

$$\begin{aligned} \Delta u + \lambda u &= 0 && \text{in } \Omega, \\ g_0(\mu)u + g_1(\mu)\frac{\partial u}{\partial n} &= 0 && \text{on } \partial\Omega. \end{aligned} \tag{14.18}$$

Recall the well-known separation rule of variables for partial differential equations. We take the ansatz $u(x_1, x_2) = u_1(x_1)u_2(x_2) \not\equiv 0$ and derive from (14.18)

$$\frac{u_1''}{u_1} + \frac{u_2''}{u_2} + \lambda = 0.$$

Hence, there are constants $k_1,\ k_2 \in \mathbf{R}$, such that $\lambda = k_1 + k_2$ and

$$u_1'' + k_1 u_1 = 0, \qquad u_2'' + k_2 u_2 = 0$$

hold with the boundary conditions

$$\begin{aligned} g_0(\mu)u_i(0) - g_1(\mu)u_i'(0) &= 0, \\ g_0(\mu)u_i(a_i) + g_1(\mu)u_i'(a_i) &= 0. \end{aligned} \qquad i = 1, 2,$$

We write these equations into a canonical form

$$\frac{\partial^2 u}{\partial x^2} + ku = 0 \qquad \text{in } \Omega := [0, a] \tag{14.19}$$

with the boundary conditions

$$g_0(\mu)u(0) - g_1(\mu)u'(0) = 0, \qquad g_0(\mu)u(a) + g_1(\mu)u'(a) = 0. \tag{14.20}$$

Evidently, the Sturm-Liouville theory is applicable to this equation. The general solution of (14.19) without boundary conditions is

$$u(x) = \begin{cases} A_1 \cos\sqrt{k}\ x + A_2 \sin\sqrt{k}\ x, & \text{for } k > 0; \\ A_1 x + A_2, & \text{for } k = 0; \\ A_1 e^{\sqrt{-k}\ x} + A_2 e^{-\sqrt{-k}\ x}, & \text{for } k < 0. \end{cases} \tag{14.21}$$

The constants A_1 and A_2 are determined by the boundary conditions (14.20). First of all, if $k = 0$, then

$$\begin{aligned} g_0(\mu)A_2 - g_1(\mu)A_1 &= 0, \\ g_0(\mu)(A_1 a + A_2) + g_1(\mu)A_1 &= 0. \end{aligned}$$

These equations have merely the trivial solution $A_1 = A_2 = 0$ except in the case $g_0(\mu) = 0$, i.e., the homogeneous Neumann boundary conditions, which makes the system singular such that it has a one-dimensional solution space $\{(0, A_2) \mid A_2 \in \mathbf{R}\}$. For $k < 0$ in (14.21), the boundary conditions (14.20) imply

$$\begin{aligned} g_0(\mu)(A_1 + A_2) - g_1(\mu)(A_1\sqrt{-k} - A_2\sqrt{-k}) &= 0, \\ g_0(\mu)(A_1 e^{a\sqrt{-k}} + A_2 e^{-a\sqrt{-k}}) + g_1(\mu)(A_1\sqrt{k}e^{a\sqrt{-k}} - A_2\sqrt{-k}e^{-a\sqrt{-k}}) &= 0. \end{aligned}$$

We rewrite these as a homogeneous system for A_1 and A_2

$$\begin{aligned} (g_0(\mu) - g_1(\mu)\sqrt{-k})A_1 + (g_0(\mu) + g_1(\mu)\sqrt{-k})A_2 &= 0, \\ (g_0(\mu) + g_1(\mu)\sqrt{-k})e^{2a\sqrt{-k}}A_1 + (g_0(\mu) - g_1(\mu)\sqrt{-k})A_2 &= 0. \end{aligned} \tag{14.22}$$

Determinant of the coefficient matrix of this system is

$$\begin{aligned} &(g_0(\mu) - g_1(\mu)\sqrt{-k}\,)^2 - (g_0(\mu) + g_1(\mu)\sqrt{-k}\,)^2 e^{2a\sqrt{-k}} \\ &< (g_0(\mu) - g_1(\mu)\sqrt{-k}\,)^2 - (g_0(\mu) + g_1(\mu)\sqrt{-k}\,)^2 \quad (\text{due to } \; e^{2a\sqrt{-k}} > 1) \\ &= -4g_0(\mu)g_1(\mu)\sqrt{-k}, \end{aligned}$$

which vanishes if and only if $g_0(\mu)g_1(\mu) = 0$. Hence, the system (14.22) has merely the trivial solution $A_1 = A_2 = 0$.

Let us now consider the case $k > 0$ in (14.21). By (14.20) we have

$$\begin{aligned} & g_0(\mu)A_1 - g_1(\mu)\sqrt{k}A_2 = 0, \\ & \Big(g_0(\mu)\cos\sqrt{k}a - g_1(\mu)\sqrt{k}\sin\sqrt{k}a\Big)A_1 \\ &+ \Big(g_0(\mu)\sin\sqrt{k}a + g_1(\mu)\sqrt{k}\cos\sqrt{k}a\Big)A_2 = 0. \end{aligned} \tag{14.23}$$

This homogeneous system for A_1, A_2 has nontrivial solution if and only if determinant of its coefficient matrix vanishes

$$\begin{aligned} & g_0(\mu)\Big(g_0(\mu)\sin\sqrt{k}a + g_1(\mu)\sqrt{k}\cos\sqrt{k}a\Big) \\ &+ g_1(\mu)\sqrt{k}\Big(g_0(\mu)\cos\sqrt{k}a - g_1(\mu)\sqrt{k}\sin\sqrt{k}a\Big) = 0, \end{aligned}$$

which can be simplified as

$$2g_0(\mu)g_1(\mu)\sqrt{k}\;\cos\sqrt{k}a + (g_0(\mu)^2 - g_1(\mu)^2 k)\sin\sqrt{k}a = 0. \tag{14.24}$$

For $\mu \in (0,1)$, the inequality $\sin\sqrt{k}a \neq 0$ holds and the equation (14.24) reduces to

$$\Big(\nu k + \tan(\frac{a}{2}\sqrt{k})\Big)\Big(\nu k - \cot(\frac{a}{2}\sqrt{k})\Big) = 0,$$

where $\nu = g_1(\mu)/g_0(\mu)$. This equation has infinite number of solutions $\{k_i,\ i = 1,\ldots,\infty\}$. For each root k_i, the system (14.23) has a solution

$$\begin{pmatrix} A_1 \\ A_2 \end{pmatrix} = c\begin{pmatrix} g_1(\mu)\sqrt{k_i} \\ g_0(\mu) \end{pmatrix} \quad \text{for all} \quad c \in \mathbf{R}.$$

Correspondingly, the nontrivial solution of (14.19) is given as

$$u_i(x) = \text{const}\cdot\Big(g_1(\mu)\sqrt{k_i}\ \cos\sqrt{k_i}x + g_0(\mu)\sin\sqrt{k_i}x\Big).$$

According to the Sturm-Liouville theory, $\{u_i(x)\ ;\quad i = 1,\ldots,\infty\}$ define a basis for the space $C[0,a]$ of continuous functions which satisfy the boundary conditions (14.20), see Section 14.2, Pinsky [242] and Birkhoff/Rota [40].

For the two-dimensional domain $\Omega := [0,a_1]\times[0,a_2]$, solutions of (14.18) are of the form $(\phi_i(x_1,x_2),\ \lambda_i)$ with

$$\begin{aligned} \lambda_i &= k_i^2 + l_i^2, \\ \phi_i(x_1,x_2) &= [g_0(\mu)\sin k_i x_1 + g_1(\mu)k_i\cos k_i x_1] \\ &\quad\cdot[g_0(\mu)\sin l_i x_2 + g_1(\mu)l_i\cos l_i x_2], \end{aligned} \tag{14.25}$$

where k_i and l_i satisfy the equation

$$2g_0(\mu)g_1(\mu)k\cos ka_1 + (g_0(\mu)^2 - g_1(\mu)^2k^2)\sin ka_1 = 0$$

and

$$2g_0(\mu)g_1(\mu)l\cos la_2 + (g_0(\mu)^2 - g_1(\mu)^2l^2)\sin la_2 = 0,$$

respectively. Moreover, $\{\phi_i\ ;\quad i = 1,\ldots,\infty\}$ define a basis for the space $C(\Omega)$ with the boundary condition (14.18). In other words, solutions of the two-dimensional problem (14.18) with the Robin type boundary conditions are finite linear combinations of $\{(\phi_i(x_1,x_2),\lambda_i)\ ;\ i = 1,\ldots,\infty\}$. Moreover, solution $u(x_1,x_2)$ of the linear problem

$$\begin{aligned} \Delta u + cu &= f \quad \text{in} \quad \Omega := [0,a_1]\times[0,a_2], \\ g_0(\mu)u + g_1(\mu)\tfrac{\partial u}{\partial n} &= 0 \quad \text{on} \quad \partial\Omega \end{aligned}$$

can be written as a uniformly convergent series

$$u(x_1,x_2) = \sum_{i,j=1}^{\infty} A_{i,j}\phi_{i,j}(x_1,x_2).$$

Concluding above, we write eigenpairs of (14.18) as (λ_i,ϕ_i), $i = 1,\ldots,\infty$ with

$$\lambda_i \geq 0 \quad i = 1, 2, \ldots, \infty.$$

The equality holds if and only if $i = 1$ and $g_0(\mu) = 0$, i.e., for the homogeneous Neumann problem. The associated eigenfunctions $\phi_i,\ i = 1, 2, \ldots$ are given as (14.25). Generically the eigenvalue λ_i is simple. Multiple eigenvalues occur if there are $(k_i, l_i) \neq (\tilde{k}_i, \tilde{l}_i)$ such that

$$\lambda_i = k_i^2 + l_i^2 = \tilde{k}_i^2 + \tilde{l}_i^2.$$

Furthermore, if the problem (14.18) has symmetries, multiple eigenvalue can be generic. For example, if $a_1 = a_2$, i.e., the domain Ω is a square with the side length a, then the reflection symmetry

$$S: \ (x_1,\ x_2) \to (x_2,\ x_1) \quad \text{and} \quad Su(x_1, x_2) := u(S(x_1, x_2)) \quad \text{for all} \quad u \in C(\Omega)$$

forces the eigenvalues of (14.18) to be generically double, the second eigenfunction associated to λ_i is

$$\tilde{\phi}_i := \phi_i(x_2, x_1).$$

Figure 14.4 shows curves of eigenvalues of (14.18) in a square domain with respect to the homotopy parameter μ. Note that generically all points on the curves are double eigenvalues. An intersection of two curves gives eigenvalues with the multiplicity four.

14.4 Variational Form

As it is well-known, at corners of polygonal domains Ω, solutions of (14.3) with (14.4) are in $C^2(\Omega)$ only for appropriate f and g. To avoid the technicalities for ensuring classical solutions, we write the problem into variational form and study weak solutions.

Consider the Sobolev space

$$H^1(\Omega) := \left\{ u \in L^2(\Omega) \ \middle| \ \frac{\partial u}{\partial x}, \frac{\partial u}{\partial y} \in L^2(\Omega) \right\}$$

with the norm $\|\cdot\|_{1,\Omega}$ and the semi-norm $|\cdot|_{1,\Omega}$, respectively. For domains with $\partial\Omega \in C^{0,1}$, typically polygons, we have

$$\|w\|_{0,\partial\Omega} \leq C\|w\|_{1,\Omega} \quad \text{for all} \quad H^1(\Omega), \tag{14.26a}$$

$$\left\| \frac{\partial w}{\partial n} \right\|_{0,\partial\Omega} \leq C\|w\|_{2,\Omega} \quad \text{for all} \quad H^2(\Omega), \tag{14.26b}$$

where C is a constant independent of w, see e.g. Hackbusch [146]. Now in the space $H^1(\Omega) \times H^1(\Omega)$ we define a parameter-dependent bilinear form for $\mu \in [0, 1)$

$$a(u,\ v,\ \mu) := \int_\Omega (\nabla u \nabla v + uv)\ dxdy + \frac{g_0(\mu)}{g_1(\mu)} \int_{\partial\Omega} uv\ ds \tag{14.27}$$

for all $u, v \in H^1(\Omega)$. A weak form of (14.3) with (14.4) is

$$\text{Find } u \in H^1(\Omega), \text{ such that } a(u,\ v,\ \mu) = (g, v) \text{ for all } v \in H^1(\Omega). \tag{14.28}$$

In particular, a weak form of the Neumann problem corresponds to $\mu = 0$. Since the bilinear form $a(\cdot, \cdot, \mu)$ is continuous and elliptic over $H^1(\Omega)$ for all $\mu \in [0, 1)$, the problem (14.28) has a unique solution $u(\mu) \in H^1(\Omega)$ for every $g \in H^{-1}(\Omega)$ and $\mu \in [0, 1)$. Moreover, this solution $u(\mu)$ satisfies the boundary condition (14.4). Denote

$$T(\mu):\ g \in H^{-1}(\Omega) \mapsto T(\mu)g = u(\mu) \in H^1(\Omega) \tag{14.29}$$

as the solution operator of (14.28). For any fixed $\mu \in [0, 1)$ the operator $T(\mu)$ is linear and bounded. Furthermore, symmetry of the Laplacian implies that $T(\mu)$ is self-adjoint, see e.g. Golubitsky *et al* [131]. This can be also seen directly from the following equality for all $f,\ g \in H^{-1}(\Omega)$

$$\begin{aligned}
&\langle T(\mu)f,\ g\rangle_{H^1(\Omega)\times H^{-1}(\Omega)} \\
&= \int_\Omega (T(\mu)f)g\ dx \ \Big(\text{Note that } T(\mu)g \text{ satisfies (14.28)}\Big) \\
&= \int_\Omega \Big(\nabla(T(\mu)f)\nabla(T(\mu)g) + (T(\mu)f)(T(\mu)g)\ dx \\
&\quad + \frac{g_0(\mu)}{g_1(\mu)} \int_{\partial\Omega} (T(\mu)f)(T(\mu)g)\ ds \\
&= \int_\Omega (-\Delta + id)(T(\mu)f)\nabla(T(\mu)g)\ dx \\
&\quad + \int_{\partial\Omega} (T(\mu)g)\Big(\nabla(T(\mu)f) + \frac{g_0(\mu)}{g_1(\mu)}(T(\mu)f)\Big)\ ds \\
&= \int_\Omega f(T(\mu)g)\ dx \\
&= \langle T(\mu)g,\ f\rangle_{H^1(\Omega)\times H^{-1}(\Omega)}.
\end{aligned}$$

More importantly is that together with the Gelfand relation

$$H^1(\Omega) \overset{c}{\hookrightarrow} L^2(\Omega) \hookrightarrow H^{-1}(\Omega)$$

the Riese-Schauder theory is applicable to the operator $T(\mu)$ for all $\mu \in [0, 1)$, i.e.,

- For all $\lambda \in \mathbf{C}$, either $T(\mu) - \lambda I$ is invertible or λ is an eigenvalue of $T(\mu)$;
- Eigenvalues of $T(\mu)$ are countable with the property

  $$\lambda_1 \geq \lambda_2 \geq \cdots \to 0$$

  and eigenspace for each $\lambda_i,\ i = 1, 2, \ldots$ is finite dimensional;

- The equation $T(\mu)u - \lambda u = f$ is solvable if and only if $f \perp \mathrm{Ker}(T(\mu) - \lambda I)$.

Same conclusion holds for $\mu = 1$ in the space $H_0^1(\Omega)$.

14.5 Continuity of Solutions along the Homotopy

We consider in this section continuity of the solution $u(\mu)$ of the linear problem (14.28) with respect to the homotopy parameter μ. Since the bilinear form (14.27) is undefined for $\mu = 1$, we restrict the discussion in this section to $\mu \in [0, 1)$. For $\mu = 0$ and 1, i.e., the Neumann and Dirichlet problems respectively, more precise estimates will be given in the next section.

Lemma 14.5.1. *The operator $T(\mu)$ is continuous and differentiable with respect to μ in $[0, 1)$.*

Proof: Let $\epsilon \in \mathbf{R}$ be sufficiently small and $\mu \in [0, 1)$ be arbitrary. Note $u(\mu + \epsilon)$ and $u(\mu)$ are defined implicitly by

$$\begin{aligned} a(u(\mu+\epsilon),\ v,\ \mu+\epsilon) &= (g, v) \quad \text{for all} \quad v \in H^1(\Omega), \\ a(u(\mu),\ v,\ \mu) &= (g, v) \quad \text{for all} \quad v \in H^1(\Omega). \end{aligned}$$

We derive for all $v \in H^1(\Omega)$

$$\begin{aligned} 0 &= a(u(\mu+\epsilon),\ v,\ \mu+\epsilon) - a(u(\mu),\ v,\ \mu) \\ &= \int_\Omega \Big(\nabla(u(\mu+\epsilon) - u(\mu))\nabla v + (u(\mu+\epsilon) - u(\mu))v \Big)\, dxdy \qquad (14.30) \\ &+ \frac{g_0(\mu+\epsilon)}{g_1(\mu+\epsilon)} \int_{\partial\Omega} u(\mu+\epsilon) v\, ds - \frac{g_0(\mu)}{g_1(\mu)} \int_{\partial\Omega} u(\mu) v\, ds. \end{aligned}$$

Rewrite it as

$$\begin{aligned} &\int_\Omega \Big(\nabla(u(\mu+\epsilon) - u(\mu))\nabla v + (u(\mu+\epsilon) - u(\mu))v \Big)\, dxdy \\ &+ \frac{g_0(\mu+\epsilon)}{g_1(\mu+\epsilon)} \int_{\partial\Omega} \Big(u(\mu+\epsilon) - u(\mu) \Big) v\, ds \\ &= \left(\frac{g_0(\mu)}{g_1(\mu)} - \frac{g_0(\mu+\epsilon)}{g_1(\mu+\epsilon)} \right) \int_{\partial\Omega} u(\mu) v\, ds. \end{aligned}$$

Choosing $v = u(\mu + \epsilon) - u(\mu)$ gives immediately

$$\|u(\mu+\epsilon) - u(\mu)\|_{1,\Omega} \le C \left| \frac{g_0(\mu)}{g_1(\mu)} - \frac{g_0(\mu+\epsilon)}{g_1(\mu+\epsilon)} \right| \|u(\mu)\|_{0,\partial\Omega}\,, \qquad (14.31)$$

where we have used Schwarz's inequality and the property (14.26a). This equality shows that $u(\mu) = T(\mu)g$, i.e., the operator $T(\mu)$ is continuous with respect to μ in $[0, 1)$. Furthermore, to show the differentiability of $u(\mu)$ in μ, we rewrite (14.30) as

$$\int_\Omega \left(\frac{\nabla(u(\mu+\epsilon)-u(\mu))}{\epsilon}\nabla v + \frac{u(\mu+\epsilon)-u(\mu)}{\epsilon} v \right) dxdy$$
$$+ \frac{g_0(\mu)}{g_1(\mu)} \int_{\partial\Omega} \frac{u(\mu+\epsilon)-u(\mu)}{\epsilon} v \, ds$$
$$= \frac{1}{\epsilon} \left(\frac{g_0(\mu)}{g_1(\mu)} - \frac{g_0(\mu+\epsilon)}{g_1(\mu+\epsilon)} \right) \int_{\partial\Omega} u(\mu+\epsilon) v \, ds.$$

On the other hand, let $\tilde{u}(\mu)$ be the solution of

$$-\Delta w + w = 0 \quad \text{in } \Omega, \tag{14.32a}$$
$$\frac{\partial w}{\partial n} + \frac{g_0(\mu)}{g_1(\mu)} w = -\frac{d}{d\mu}\left(\frac{g_0(\mu)}{g_1(\mu)} \right) u(\mu) \quad \text{on } \partial\Omega. \tag{14.32b}$$

Then $\tilde{u}(\mu)$ satisfies the variational equation

$$a(\tilde{u}(\mu),\ v,\ \mu) = -\frac{d}{d\mu}\left(\frac{g_0(\mu)}{g_1(\mu)} \right) \int_{\partial\Omega} u(\mu) v \, ds \quad \text{for all } v \in H^1(\Omega).$$

Thereafter,

$$a\left(\frac{u(\mu+\epsilon)-u(\mu)}{\epsilon} - \tilde{u}(\mu),\ v,\ \mu\right)$$
$$= O(\epsilon) \int_{\partial\Omega} u(\mu+\epsilon) v \, ds - \frac{d}{d\mu}\left(\frac{g_0(\mu)}{g_1(\mu)} \right) \int_{\partial\Omega} \Big(u(\mu+\epsilon) - u(\mu)\Big) v \, ds.$$

Choosing

$$v = \frac{u(\mu+\epsilon)-u(\mu)}{\epsilon} - \tilde{u}(\mu),$$

we obtain

$$\left\| \frac{u(\mu+\epsilon)-u(\mu)}{\epsilon} - \tilde{u}(\mu) \right\|_{1,\Omega} \le O(\epsilon) \|u(\mu+\epsilon)\|_{0,\partial\Omega}$$
$$+ \left| \frac{d}{d\mu}\left(\frac{g_0(\mu)}{g_1(\mu)} \right) \right| \|u(\mu+\epsilon) - u(\mu)\|_{0,\partial\Omega}.$$

Let $\epsilon \to 0$. We see that $u(\mu) = T(\mu)g$ is differentiable with respect to μ in $[0, 1)$ and the derivative is $\tilde{u}(\mu)$. ∎

Remark 14.5.1. Define $\tilde{g}(\mu) \in H^{-1}(\Omega)$ implicitly by

$$(\tilde{g}(\mu), v) = -\frac{d}{d\mu}\left(\frac{g_0(\mu)}{g_1(\mu)} \right) \cdot \int_{\partial\Omega} T(\mu) g \cdot v \, ds.$$

We have

$$u'(\mu) = T'(\mu) g = T(\mu)\tilde{g}(\mu) = \tilde{u}(\mu), \tag{14.33}$$

where $\tilde{u}(\mu)$ is the weak solution of (14.32). Higher order derivatives of $u(\mu)$ can be derived in a similar manner.

Since $T(\mu)$ is self-adjoint, so is its derivative $T'(\mu)$.

14.6 Neumann and Dirichlet Problems

We consider in this section behavior of solutions of the linear problem (14.3) with (14.4) as $\mu \to 0$ and $\mu \to 1$, respectively. As indicated above, $u(\mu)$ is continuous at $\mu = 0$. Moreover, the weak form of the Neumann problem corresponds to $\mu = 0$ in (14.28). Let $u(\mu)$ be the solution of (14.28) and $u(0)$ be the solution of Neumann problem for an arbitrarily fixed right hand side g. Similarly to (14.30) we derive from

$$a(u(\mu),\ v,\ \mu) - a(u(0), v, 0) = 0 \quad \text{for all} \quad v \in H^1(\Omega)$$

that

$$a(u(\mu) - u(0),\ v,\ \mu) = \frac{g_0(\mu)}{g_1(\mu)} \int_{\partial\Omega} u(0)v \, ds \quad \text{for all} \quad v \in H^1(\Omega).$$

Choosing $v = u(\mu) - u(0)$ and using Schwarz's inequality as well as (14.26a), we obtain

Theorem 14.6.1. *Let $u(\mu)$ and $u(0)$ be solutions of the linear problem (14.3) with the Robin boundary condition (14.4) and the Neumann boundary condition $\frac{\partial u}{\partial n} = 0$, respectively. Then*

$$\|u(\mu) - u(0)\|_{1,\Omega} \leq \frac{g_0(\mu)}{g_1(\mu)} \|u(0)\|_{1,\Omega}. \tag{14.34}$$

In particular, we see that $u(\mu)$ tends to $u(0)$ as fast as $g_0(\mu)$ goes to 0 when $\mu \to 0$. The solution of linear Neumann problem is well approximated by the solution of Robin problems in the sense that (14.34) is an estimate with respect to the H^1-norm. For Dirichlet problem ($\mu = 1$) in (14.4), we have

Theorem 14.6.2. *Let $u(\mu)$ and $u(1)$ be solutions of the linear problem (14.3) with the Robin boundary condition (14.4) and the Dirichlet boundary condition $u = 0$, respectively. For $\mu \to 1^-$, we have the following estimate*

$$\|u(\mu) - u(1)\|_{1,\Omega} \leq \left(\frac{g_1(\mu)}{g_0(\mu)}\right)^{1/2} \|u(1)\|_{2,\Omega}, \tag{14.35a}$$

$$\|u(\mu) - u(1)\|_{0,\Omega} \leq \frac{g_1(\mu)}{g_0(\mu)} \|u(1)\|_{2,\Omega}. \tag{14.35b}$$

Remark 14.6.1. Theorem 14.6.2 indicates that smoothness of the solution $u(\mu)$ with respect to μ depends on the choice of the homotopy functions $g_0(\mu)$, $g_1(\mu)$, for example, the estimates (14.35) with $g_1(\mu) = (1-\mu)^2$ in (14.35a) are smoother than those with $g_1(\mu) = 1 - \mu$.

Proof: The weak form of the Dirichlet problem

$$\begin{aligned} -\Delta u + u &= g \qquad \text{in } \Omega, \\ u &= 0 \qquad \text{on } \partial\Omega \end{aligned}$$

in the space $H^1(\Omega)$ is

$$a(u, v, 0) - \int_{\partial\Omega} \frac{\partial u}{\partial n} v \, ds = (g, v) \quad \text{for all } v \in H^1(\Omega). \tag{14.36}$$

Note that $u(1)$ satisfies this equation and $u(1) = 0$ on the boundary $\partial\Omega$, and $u(\mu)$ satisfies (14.28). We have

$$a(u(\mu) - u(1),\ v,\ \mu) = \int_{\partial\Omega} \frac{\partial u(1)}{\partial n} v \, ds \quad \text{for all} \ \ v \in H^1(\Omega).$$

Let $v = u(\mu) - u(1)$. We derive

$$\begin{aligned} &\|u(\mu) - u(1)\|_{1,\Omega}^2 + \frac{g_0(\mu)}{g_1(\mu)} \|u(\mu) - u(1)\|_{0,\partial\Omega}^2 \\ &= \int_{\partial\Omega} \frac{\partial u(1)}{\partial n} v \, ds \\ &\le \left\| \frac{\partial u(1)}{\partial n} \right\|_{0,\partial\Omega} \|u(\mu) - u(1)\|_{0,\partial\Omega}. \end{aligned}$$

This implies

$$\|u(\mu) - u(1)\|_{0,\partial\Omega} \le \frac{g_1(\mu)}{g_0(\mu)} \left\| \frac{\partial u(1)}{\partial n} \right\|_{0,\partial\Omega} \le C \frac{g_1(\mu)}{g_0(\mu)} \|u(1)\|_{2,\Omega}, \tag{14.37}$$

which in turn yields

$$\begin{aligned} &\|u(\mu) - u(1)\|_{1,\Omega}^2 \\ &\le \left\| \frac{\partial u(1)}{\partial n} \right\|_{0,\partial\Omega} \|u(\mu) - u(1)\|_{0,\partial\Omega} \\ &\le C \frac{g_1(\mu)}{g_0(\mu)} \|u(1)\|_{2,\Omega}^2. \end{aligned}$$

Then the estimate (14.35a) follows immediately. On the other hand, for any $\eta \in L^2(\Omega)$, the problem

$$\begin{aligned} -\Delta z + z &= \eta \qquad \text{in } \Omega, \\ z &= 0 \qquad \text{on } \partial\Omega \end{aligned}$$

has a unique solution $z \in H^2(\Omega)$ and

$$\|z\|_{2,\Omega} \le C\|\eta\|_{0,\Omega},$$

see e.g. Hackbusch [146] and Oden [238]. We note that $u(1)$ and $u(\mu)$ satisfy (14.36) and (14.28) with $v = z$ respectively. Furthermore, $u(1)$ and z vanish on the boundary $\partial\Omega$. Thereafter, we derive

$$\begin{aligned}
&|\langle u(\mu) - u(1), \ -\Delta z + z\rangle| \\
&= \left| \int_\Omega \Big(\nabla(u(\mu)-u(1))\nabla z + (u(\mu)-u(1))z\Big)\, dxdy - \int_{\partial\Omega} (u(\mu)-u(1))\frac{\partial z}{\partial n}\, ds \right| \\
&= \left| \int_{\partial\Omega} (u(\mu) - u(1))\frac{\partial z}{\partial n}\, ds \right| \\
&\le \|u(\mu) - u(1)\|_{0,\partial\Omega} \left\| \frac{\partial z}{\partial n} \right\|_{0,\partial\Omega} \\
&\le C\frac{g_1(\mu)}{g_0(\mu)}\|u(1)\|^2_{2,\Omega}\|z\|_{2,\Omega}.
\end{aligned}$$

Now the estimate

$$\begin{aligned}
\|u(\mu) - u(1)\|_{0,\Omega} &= \sup_{\eta\in L^2(\Omega)} \frac{|\langle u(\mu) - u(1), \ \eta\rangle_\Omega|}{\|\eta\|_{0,\Omega}} \\
&= \sup_{\eta\in L^2(\Omega)} \frac{|\langle u(\mu) - u(1), \ -\Delta z + z\rangle_\Omega|}{\|\eta\|_{0,\Omega}} \\
&= C\frac{g_1(\mu)}{g_0(\mu)}\|u(1)\|^2_{2,\Omega} \sup_{\eta\in L^2(\Omega)} \frac{\|z\|_{2,\Omega}}{\|\eta\|_{0,\Omega}} \\
&\le C\frac{g_1(\mu)}{g_0(\mu)}\|u(1)\|_{2,\Omega}
\end{aligned}$$

completes our proof. ∎

Remark 14.6.2. Evidently, the above discussion is applicable to linearization of the general semi-linear elliptic equations

$$\sum_{i+j\le 2} D^i\left(a_{ij}(x,y,\lambda)D^j u\right) + f(u,\lambda) = 0 \tag{14.38}$$

on a bounded domain $\Omega \subset \mathbf{R}^2$ with $C^{k,1}$-boundary $\partial\Omega$, $c > 0$ and

$$\sum_{i+j=2} a_{i,j}(x,y,\lambda)\xi^2 \ge c\|\xi\|^2 \quad \text{for all} \ (x,y) \in \Omega, \ \lambda \in \mathbf{R}, \ \xi \in \mathbf{R}^2.$$

It is the coerciveness that ensures similar conclusions as above.

14.7 Properties of Eigenvalues

Analysis of bifurcation requires information on eigenvalues of the linearized problem. To investigate behavior of eigenvalues with respect to variation in boundary conditions, we consider in this section again the eigenvalue problem

$$\begin{aligned} \Delta u + \lambda u &= 0 \qquad \text{in } \Omega := [0, a_1] \times [0, a_2], \\ g_0(\mu) u + g_1(\mu) \tfrac{\partial u}{\partial n} &= 0 \qquad \text{on } \partial\Omega. \end{aligned} \tag{14.39}$$

We are interested in variations of the eigenpairs with respect to the homotopy parameter μ.

14.7.1 One-dimensional problems

We recall that in Section 14.3 separating variables reduces the equation (14.39) to the one-dimensional problem

$$\frac{\partial^2 u}{\partial x^2} + \kappa^2 u = 0 \qquad \text{in } \Omega := [0, a] \tag{14.40}$$

with the boundary conditions

$$g_0(\mu) u(0) - g_1(\mu) u'(0) = 0, \qquad g_0(\mu) u(a) + g_1(\mu) u'(a) = 0. \tag{14.41}$$

Here we have replaced k in (14.19) with κ^2 to simplify the notations. We want to study properties of the wave number κ as a function of μ. The equation for κ is

$$2 g_0 g_1 \kappa \cos(\kappa\pi) + (g_0^2 - g_1^2 \kappa^2) \sin(\kappa\pi) = 0. \tag{14.42}$$

This equation is $\mathbf{Z}_2$-symmetric with respect to the reflection $\kappa \mapsto -\kappa$. Each solution κ has a counterpart $-\kappa$. Since eigenvalues of the Laplacian Δ consist of $-\kappa_1^2 - \kappa_2^2$, negative wave numbers do not make any contribution. We restrict our attention to the positive solutions of (14.42). For Neumann ($\mu = 0$) and Dirichlet boundary conditions ($\mu = 1$) the solutions of (14.42) are integers $\{\kappa \in \mathbf{N}\}$. Eigenfunctions which are independent of the space variable x or y occur only for $\mu = 0$ with zero as the wave number in the corresponding direction.

Lemma 14.7.1. *If the inequality $g_0(1) g_1(0) g_0(\mu) g_1(\mu) \neq 0$ holds for all $\mu \in (0,1)$, then the equation (14.42) has no integer solution κ except 0 for $\mu \in (0,1)$. Furthermore, if $\left(g_1(\mu)/g_0(\mu)\right)' < 0$, then the elements $n \in \mathbf{N} \cup \{0\}$ for $\mu = 0$ and $n+1 \in \mathbf{N} \cup \{0\}$ for $\mu = 1$ are connected monotonously with positive solutions of (14.42).*

Proof: Suppose that for some $\mu \in (0,1)$ there is an integer κ satisfying (14.42). Then

$$g_0(\mu) g_1(\mu) \kappa = 0.$$

The condition on g_i shows immediately $\kappa = 0$. Hence, for any solutions of (14.42) and $\mu \in (0,1)$, we have $\sin \kappa\pi \neq 0$ except $\kappa = 0$. Keeping this in mind, we multiply (14.42) with $\sin(\kappa\pi)$ and factorize it into

$$\Big(g_0 \sin(\kappa\pi) + g_1 \kappa \cos(\kappa\pi) - \kappa g_1\Big)\Big(g_0 \sin(\kappa\pi) + g_1 \kappa \cos(\kappa\pi) + \kappa g_1\Big) = 0.$$

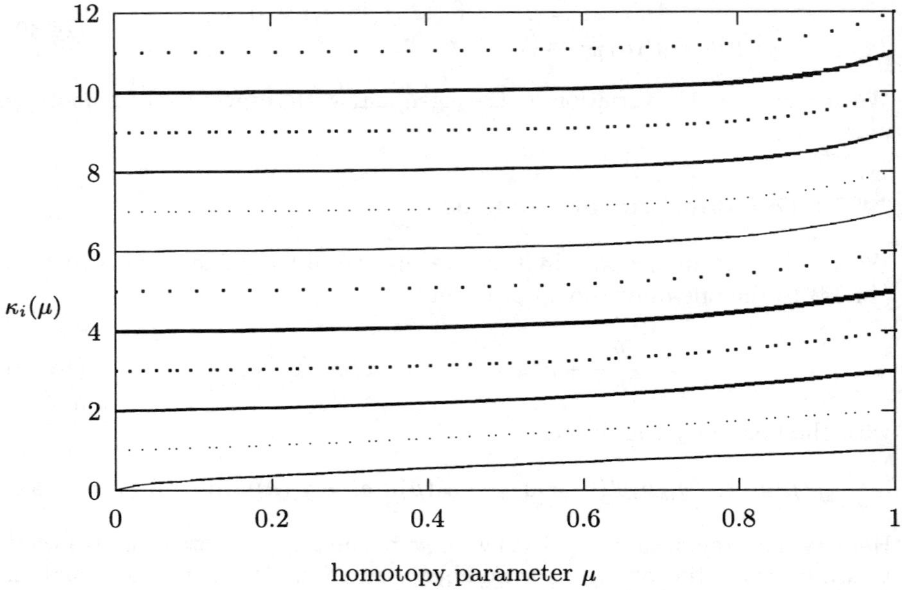

Fig. 14.2. Eigenvalues of $-\frac{d^2}{dx^2}$ in $\Omega := [0,\pi]$ along a homotopy of the boundary conditions $\mu u(0) - (1-\mu)u(0) = 0,\ \ \mu u(\pi) + (1-\mu)u(\pi) = 0$ for μ from 0 to 1.

Equivalently we have either

$$g_0 \sin(\kappa\pi) + g_1\kappa\cos(\kappa\pi) = g_1\kappa \tag{14.43a}$$

or

$$g_0 \sin(\kappa\pi) + g_1\kappa\cos(\kappa\pi) = -g_1\kappa. \tag{14.43b}$$

Note that $\sin(\kappa\pi) \neq 0$ holds for all solutions of (14.42). We rewrite this system as

$$f_1(\kappa,\mu) := \frac{g_1(\mu)}{g_0(\mu)}\kappa - \cot\left(\frac{\pi}{2}\kappa\right) = 0$$

or

$$f_2(\kappa,\mu) := \frac{g_1(\mu)}{g_0(\mu)}\kappa + \tan\left(\frac{\pi}{2}\kappa\right) = 0.$$

Differentiating f_1 and f_2 with respect to κ yields

$$\frac{\partial f_1}{\partial \kappa} = \frac{g_1}{g_0} + \frac{\pi}{2}\left(1+\cot^2\left(\frac{\pi}{2}\kappa\right)\right) > 0,$$
$$\frac{\partial f_2}{\partial \kappa} = \frac{g_1}{g_0} + \frac{\pi}{2}\left(1+\tan^2\left(\frac{\pi}{2}\kappa\right)\right) > 0.$$

Hence, κ can be solved uniquely as a smooth function of μ either from $f_1(\kappa,\mu) = 0$ or from $f_2(\kappa,\mu) = 0$. Furthermore,

$$\frac{d\kappa}{d\mu} = -\left(\frac{\partial f_i}{\partial \kappa}\right)^{-1} \frac{\partial f_i}{\partial \mu} = -\left(\frac{\partial f_i}{\partial \kappa}\right)^{-1} \frac{d}{d\mu}\left(\frac{g_1}{g_0}\right), \quad i = 1 \text{ or } 2.$$

If $(g_1(\mu)/g_0(\mu))' < 0$, which is satisfied by $g_0 = \mu$ and $g_1 = 1 - \mu$, the wave number $\kappa(\mu)$ increases monotonously with respect to μ. In particular, starting from $n \in \mathbf{N} \cup \{0\}$ at $\mu = 0$, it ends at $n + 1$ for $\mu = 1$. ∎

As we have seen before, for Neumann and Dirichlet boundary conditions the wave numbers are integers and their parities are well-defined.

Definition 14.7.1. *Parities of the wave numbers $k_i(\mu)$ for Robin boundary conditions are defined via the statements in (14.43) as*

$$\kappa(\mu) = \begin{cases} \text{EVEN} & \textit{if } g_0(\mu)\sin(\kappa(\mu)\pi) = g_1(\mu)\kappa(\mu)[1 - \cos(\kappa(\mu)\pi)], \\ \text{ODD} & \textit{if } g_0(\mu)\sin(\kappa(\mu)\pi) = -g_1(\mu)\kappa(\mu)[1 + \cos(\kappa(\mu)\pi)] \end{cases} \tag{14.44}$$

for all $\mu \in [0, 1]$.

On the other hand, note that the inequality $\sin(\kappa\pi) \neq 0$ implies $\cos(\kappa\pi/2) \neq 0$. We have the following equivalences for the parity of a wave number.

$$\begin{aligned} \text{even} &\Leftrightarrow g_0 \sin(\kappa\pi) + g_1\kappa\cos(\kappa\pi) = g_1\kappa \Leftrightarrow g_1\kappa\sin(\kappa\pi) - g_0\cos(\kappa\pi) = g_0, \\ \text{odd} &\Leftrightarrow g_0 \sin(\kappa\pi) + g_1\kappa\cos(\kappa\pi) = -g_1\kappa \Leftrightarrow g_1\kappa\sin(\kappa\pi) - g_0\cos(\kappa\pi) = -g_0. \end{aligned}$$

Remark 14.7.1. From Lemma 14.7.1 we know that the integer wave number $k_i \in \mathbf{N} \cup \{0\}$ for $\mu = 0$ is connected to $k_i + 1 \in \mathbf{N}$ for $\mu = 1$. However parities of these integers are different. For the purpose of classification in Robin boundary conditions, either of them can be used. We have defined the parity to be consistent with those of wave numbers of the Neumann problem.

The critical modes and wave numbers move continuously from one to the next higher (lower) mode as the boundary condition goes from Neumann to Dirichlet type. If no mode interaction occurs along the homotopy, the analysis in Mei/Theil [227] shows that bifurcating solution branches vary also continuously except a possible change of bifurcation directions. This explains partially the observation in Dillon/Maini/Othmer [86] about smooth transitions between different spatial patterns, for which secondary bifurcation may be an additional source.

The monotonously increasing connection between eigenvalues of Neumann and Dirichlet problems is ensured by a special homotopy, satisfying the condition

$$\left(\frac{g_1(\mu)}{g_0(\mu)}\right)' < 0.$$

If this condition is violated, the connection will be more complicated. The following two examples illustrate the well-known fact that spectrum of a differential operator depends subtly on boundary conditions.

Example 14.7.1. If we change the boundary conditions (14.41) to be

$$g_0(\mu)u(0) + g_1(\mu)u'(0) = 0, \qquad g_0(\mu)u(a) + g_1(\mu)u'(a) = 0, \tag{14.45}$$

a similar discussion as in Section 14.3 shows that an eigenvalue κ^2 of the operator $-d^2/dx^2$ in (14.40) is given via either $\kappa = g_0(\mu)/g_1(\mu) \to \infty$ $(\mu \to 1)$ or solutions of the equation

$$\sin(\kappa a) = 0,$$

i.e., $\kappa = n\pi/a$, $n = 1, 2, \ldots$, which are independent of the homotopy parameter μ.

Example 14.7.2. (cf. Kato[178, pp.293]) Consider weak form of the operator $-d^2/dx^2$ in $L^2(0, \pi)$ with the boundary condition $u(0) = 0$, $u(\pi) - \mu u'(\pi) = 0$, $\mu \in \mathbf{R}$. Its spectrum consists of eigenvalues $\kappa(\mu)^2$, which are given by solutions of the equation

$$\sin(\kappa\pi) - \mu\kappa\cos(\kappa\pi) = 0.$$

The associated eigenfunction is $\sin(\kappa(\mu)x)$. In addition, there is an eigenvalue $\kappa(\mu)^2$ given by the single solution of

$$1 - e^{2\kappa\pi} + \mu\kappa(1 + e^{2\kappa\pi}) = 0$$

with the associated eigenfunction $\sinh(\kappa(\mu)x)$. The latter is one and the only one negative eigenvalue for $\mu \in (0, \pi)$ and tends to $-\infty$ as $\mu \to 0^+$, while the operator itself is positive definite for $\mu = 0$, see Figure 14.3.

For appropriate functions $g_0(\mu)$, $g_1(\mu)$ the perturbation theory in Kato [178] is applicable to the analysis of continuity of eigenpairs of the Laplacian with respect to the homotopy parameter μ. As an example we study the properties of $\kappa(\mu)$ in the neighborhood of $\mu = 1$, i.e, perturbation of the Dirichlet problem. Suppose $\tilde{g}_1(\mu) := g_1(\mu)/g_0(\mu)$ is analytic. We rewrite (14.40) and (14.41) as

$$\begin{aligned} \frac{\partial^2 u}{\partial x^2} + \kappa^2 u = 0 \qquad &\text{in } \Omega := [0, a], \\ u(0) - \tilde{g}_1(\mu)u'(0) = 0,& \\ u(a) + \tilde{g}_1(\mu)u'(a) = 0.& \end{aligned} \tag{14.46}$$

Let $T(\mu)$ denote the operator d^2/dx^2 with the boundary conditions in (14.46). By Kato [178] it is holomorphic with respect to μ. Furthermore, its eigenvalues are analytic for $\mu \in [0, 1]$ if $g_0(\mu)$, $g_1(\mu)$ satisfy (14.5). They are holomorphic for μ as a complex variable if (14.5) is violated. Generally for a family of self-adjoint homomorphic operators $T(\mu)$, every isolated eigenvalue of $T(\mu_0)$ with finite multiplicity splits into a group of eigenvalues of $T(\mu)$ which are holomorphic at μ_0 in the complex plane. Each of these holomorphic functions

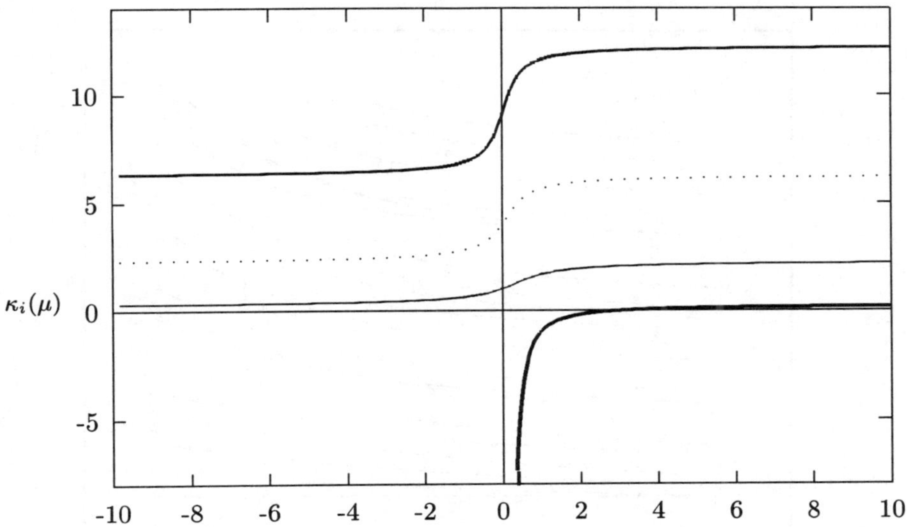

Fig. 14.3. First four eigenvalues of $-u'' + \lambda u = 0$ in $\Omega := [0, \pi]$ with the boundary conditions $u(0) = 0,\ u(\pi) - \mu u'(\pi) = 0$ for $\mu \in \mathbf{R}$.

and its associated eigenfunctions can be continued analytically along the real axis, as long as the eigenvalues remain isolated and have finite multiplicities, even when the graph of eigenvalue crosses the graph of other eigenvalues. Therefore, eigenpairs of the Laplacian with the boundary conditions (14.18) are differentiable with respect to μ.

14.7.2 Two-dimensional problems

Let us come back to the two-dimensional eigenvalue problem (14.39). It follows from (14.25) and Lemma 14.7.1 that eigenvalues of Neumann and Dirichlet problems are connected monotonously by the eigenvalues of (14.39) when μ varies from 0 to 1. More precisely, the connection $\lambda(\mu) = k(\mu)^2 + l(\mu)^2$ extends from $\lambda(0) = -(k^2/a_1^2 + l^2/a_2^2)\pi^2$ to $\lambda(1) = -((k+1)^2/a_1^2 + (l+1)^2/a_2^2)\pi^2$ for $k, l \in \mathbf{N} \cup \{0\}$, see e.g. Fig. 14.4. Correspondingly, bifurcation points of Neumann and Dirichlet problems on the trivial solution curve are connected by continuously differentiable curves $(0, \lambda(\mu), \mu)$ of bifurcation points of nonlinear problem (14.16).

To calculate derivatives of eigenpairs of the Laplacian with respect to the homotopy parameter μ in the boundary conditions (14.18), we include dependence of the eigenpairs on μ into the equation

$$\begin{aligned} \Delta u(\mu) + \lambda(\mu) u(\mu) &= 0 \qquad \text{in } \Omega = [0, a_1] \times [0, a_2] \\ g_0(\mu) u(\mu) + g_1(\mu) \tfrac{\partial u}{\partial n}(\mu) &= 0 \qquad \text{on } \partial\Omega \end{aligned}$$

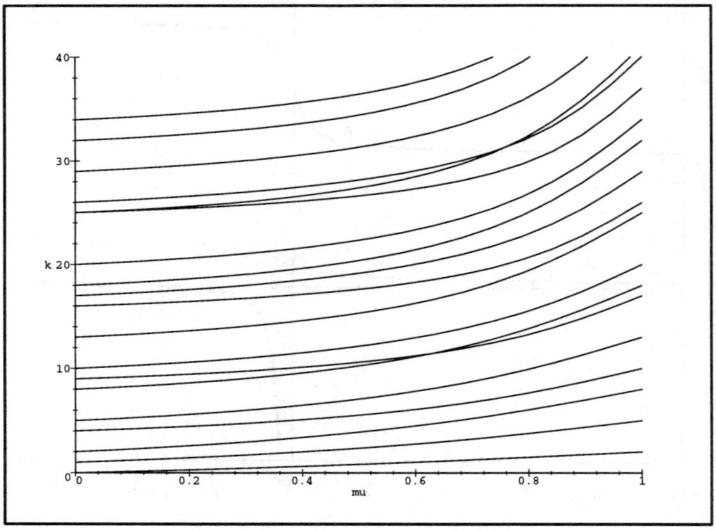

Fig. 14.4. Eigenvalues of $-\Delta$ on the square $\Omega = [0,\pi]\times[0,\pi]$ with the boundary conditions $\mu u + (1-\mu)\frac{\partial u}{\partial n} = 0$ for μ from 0 to 1.

and differentiate it with respect to μ

$$\begin{aligned} \Delta u'(\mu) + \lambda(\mu)u'(\mu) &= -\lambda'(\mu)u(\mu) \qquad \text{in } \Omega = [0,a_1]\times[0,a_2] \\ g_0(\mu)u'(\mu) + g_1(\mu)\tfrac{\partial u'}{\partial n}(\mu) &= -g_0'(\mu)u(\mu) - g_1'(\mu)\tfrac{\partial u}{\partial n}(\mu) \quad \text{on } \partial\Omega. \end{aligned} \tag{14.47}$$

Multiplying the equation in (14.47) with $u(\mu)$ and integrating over Ω yields

$$\begin{aligned} -\lambda'(\mu)\langle u(\mu),\, u(\mu)\rangle &= \int_\Omega u(\mu)\Delta u'(\mu)\,dx + \lambda(\mu)\int_\omega u(\mu)u'(\mu)\,dx \\ &= \int_{\partial\Omega}\left(u(\mu)\frac{\partial u'}{\partial n}(\mu) - u'(\mu)\frac{\partial u}{\partial n}(\mu)\right)\,ds. \end{aligned}$$

With the boundary conditions on $u(\mu)$ and $u'(\mu)$ we obtain for $g_0(\mu)\neq 0$, $g_1(\mu)=0$

$$\lambda'(\mu) = \frac{\displaystyle\int_{\partial\Omega}\frac{\partial u}{\partial n}(\mu)\left(g_0'(\mu)u(\mu) + g_1'(\mu)\frac{\partial u}{\partial n}(\mu)\right)\,ds}{g_0(\mu)\langle u(\mu),\, u(\mu)\rangle} \tag{14.48}$$

and for $g_1(\mu)\neq 0$

$$\lambda'(\mu) = \frac{\displaystyle\int_{\partial\Omega}u(\mu)\left(g_0'(\mu)u(\mu) + g_1'(\mu)\frac{\partial u}{\partial n}(\mu)\right)\,ds}{g_1(\mu)\langle u(\mu),\, u(\mu)\rangle}\,. \tag{14.49}$$

Thereafter, the derivative $u'(\mu)$ can be solved directly from the equation (14.47).

Remark 14.7.2. Alternatively, derivative of $\kappa(\mu)$ can be also calculated as in Lemma 14.7.1 via the defining equations (14.43)

$$\lambda'(\mu) = 2k(\mu)k'(\mu) + 2l(\mu)l'(\mu).$$

15. Bifurcations along a Homotopy of Boundary Conditions

We study in this chapter the impact of boundary conditions on steady state bifurcations of reaction-diffusion problems. To distinguish influence of boundary conditions from that of interactions among the different species (components), we consider a scalar reaction-diffusion equation

$$\Delta u + \lambda u - f(u, \lambda) = 0 \quad \text{in} \;\; \Omega := (0, \pi) \times (0, \pi) \tag{15.1a}$$

with Robin boundary conditions

$$g_0(\mu)u + g_1(\mu)\frac{\partial u}{\partial n} = 0 \quad \text{on} \;\; \partial\Omega. \tag{15.1b}$$

Here $f : \mathbf{R} \to \mathbf{R}$ and $g_1, \; g_2 : [0,1] \to \mathbf{R}$ are smooth functions (C^3-continuous) with

$$f(0, \lambda) = 0, \quad f(u, \lambda) = O(\|u\|^2), \tag{15.2a}$$

$$g_0(0) = g_1(1) = 0, \; g_0(1)g_1(0)g_0(\mu)g_1(\mu) \neq 0 \; \text{ for all } \mu \in (0,1). \tag{15.2b}$$

We are interested in the bifurcation structure of (15.1) and its variation along a homotopy path (15.1b) from Neumann to Dirichlet boundary conditions, for examples, changes of symmetries, numbers and directions of bifurcating solution branches. In this chapter we take generic corank-2 bifurcations of (15.1) as an example and study their variation along homotopy path (15.1b) from Neumann to Dirichlet boundary conditions.

Section 15.1 is a brief introduction of the role of boundary conditions in bifurcation problems. In Section 15.2 we study at first linear stability of the trivial solution $\{(0, \lambda); \; \lambda \in \mathbf{R}\}$ and the induced symmetries in kernel of the linearized operator. Thereafter we derive normal forms of equivariant polynomials for the induced actions of D_4 in Section 15.3. Variation of the bifurcation scenario along the homotopy of boundary conditions is analyzed in Section 15.4 for wave numbers with different parities and demonstrated with a numerical example in Section 15.5. Finally in Section 15.6 we study briefly impact of a forced symmetry-breaking in boundary conditions on the bifurcation scenario of a reaction-diffusion equation.

15.1 Introduction

As a bifurcation process in reaction-diffusion problems, Turing [293] suggests that nonhomogeneous steady states can be stable in appropriate situations and form spatial patterns. Recent experiments have confirmed this prediction (cf. Gray/Scott [273]). Putting the reaction vessel into electrical field, Münster/Watzl [234] have simulated various boundary conditions and observed spatially nonuniform steady state solutions. In particular, mixed boundary conditions seem to be favored by stable nonuniform stationary solutions. Patterns produced in such ways in polymers with large molecules possess specific physical properties, such as conductivity and enforcement in the structure, etc. Generally, influence of boundary conditions on pattern formation, e.g. target patterns of chemical reactions and stripes in the coats of animals, is sophisticated and involves both local and global bifurcations. For example, spiral wave is a typical characteristic of systems of reaction-diffusion equations. However, by dynamical boundary conditions Dellnitz *et al* [83] induce spiral waves in a scalar reaction-diffusion equations.

For a class of reaction-diffusion problems Gardner [118] shows that the major properties of nonsingular solutions, e.g. regularity and uniqueness remain invariant along a homotopy from Neumann to Dirichlet problems. Similar properties for equivalent classes of global attractors of reaction-diffusion equations on the unit interval are established by Fiedler [102]. Hale and Rocha discussed in [149, 150] influence of boundary conditions and diffusion rate on attractors. They decompose the problem in the spirit of center manifold theory and relate dynamics of reaction-diffusion equations near a hyperbolic equilibrium to a system of ordinary differential equations. However, dynamics near nonhyperbolic equilibrium is more complicated. Their dependence on boundary conditions still needs to be studied.

Some Neumann and Dirichlet problems can be embedded into equations with periodic boundary conditions. This reveals underlying symmetries in solutions of the original equations, see e.g. Crawford et al [77], Gomes/Stewart [133] and Mei [217]. In addition, some Dirichlet problems to be shifted to those of Neumann problems via coordinate transformations (cf. Castro [53]).

Boundary conditions have notable impact on the bifurcation scenario of some problems, for example, the steady state problem

$$\Delta u + \lambda f(u) = 0, \qquad f(u) = u + O(\|u\|^2)$$

on a square domain has different bifurcation diagrams for Neumann and Dirichlet problems, even the numbers of bifurcating solution branches are different (cf. Mei [219, 223]). Similar conclusions are derived also for Hopf bifurcations, see e.g. Ashwin/Böhmer/Mei [21, 22]. In [86] Dillon, Maini and Othmer consider a system of reaction-diffusion equations on intervals with mixed boundary conditions, i.e., the components may have different types of boundary conditions. By varying boundary conditions from Neumann to

Dirichlet type in different components, they observed the bifurcation scenarios with various characters. Mode-jumping in the buckling of rectangular plates is another phenomenon which occurs only for specific boundary conditions, see Holder/Schaeffer [161], Schaeffer/Golubitsky [267] for more details.

15.2 Linear Stability Analysis and Symmetries

As we have seen in Chapter 14, spectrum of the Laplacian Δ with Robin boundary conditions (15.1b) consists of eigenvalues

$$\lambda = -\kappa_1(\mu)^2 - \kappa_2(\mu)^2.$$

These eigenvalues are generically of the multiplicity two, and the associated eigenfunctions are $\phi_i := \tilde{\phi}_i/||\tilde{\phi}_i||,\ i = 1, 2$ with

$$\begin{aligned}\tilde{\phi}_1(x,y) &:= \Big(g_0 \sin(\kappa_1 x) + g_1\kappa_1 \cos(\kappa_1 x)\Big)\Big(g_0 \sin(\kappa_2 y) + g_1\kappa_2 \cos(\kappa_2 y)\Big),\\ \tilde{\phi}_2(x,y) &:= \Big(g_0 \sin(\kappa_2 x) + g_1\kappa_2 \cos(\kappa_2 x)\Big)\Big(g_0 \sin(\kappa_1 y) + g_1\kappa_1 \cos(\kappa_1 y)\Big).\end{aligned}$$

The wave numbers κ_1 and κ_2 satisfy the equation

$$2g_0g_1\kappa\cos(\kappa\pi) + (g_0^2 - g_1^2\kappa^2)\sin(\kappa\pi) = 0, \tag{15.3}$$

see also (14.42).

Define

$$X := \left\{u \in H^1(\Omega)\ \Big|\ g_0(\mu)u + g_1(\mu)\frac{\partial u}{\partial n} = 0\right\}. \tag{15.4}$$

We write the problem (15.1) into a weak form with $G:\ X \times \mathbf{R} \mapsto X$

$$G(u,\lambda) := u + (\lambda+1)Tu - Tf(u,\lambda) = 0 \tag{15.5}$$

Here $T:\ L^2(\Omega) \mapsto X$ is the weak inverse of $\Delta - id$. We remind that here $T(\mu)$ is equal to the operator $-T(\mu)$ defined in (14.29). Hence, it has the same properties as discussed in Chapter 14.

A bifurcation point (u_0, λ_0) of (15.5) on the trivial solution $\{(0,\lambda);\ \lambda \in \mathbf{R}\}$ is characterized by the statement $\mathrm{Ker}(D_uG_0) = \mathrm{Ker}(id{+}(\lambda_0{+}1)T) \neq \{0\}$. It follows from properties of the Laplacian that

$$(u_0,\ \lambda_0) = \big(0,\ -\kappa_1(\mu)^2 - \kappa_2(\mu)^2\big).$$

Parity of λ_0 is defined via those of $(\kappa_1(\mu), \kappa_2(\mu))$ as (even, even), (even, odd), (odd, even) or (odd, odd). Denote evaluation of G and its derivatives at (u_0, λ_0) as G_0, D_uG_0, etc. The Fredholm property of $D_uG_0 = I + (\lambda_0 + 1)T$ implies

$$X = \mathrm{Ker}(D_u G_0) \oplus \mathrm{Im}(D_u G_0). \tag{15.6}$$

The bifurcation point (u_0, λ_0) is of *corank-2* if $\mathrm{Ker}(D_u G_0) = \mathrm{span}[\phi_1, \phi_2]$ is two dimensional. In this case any element (u, λ) in $X \times \mathbf{R}$ has a unique decomposition

$$\begin{aligned} u &= \alpha_1 \phi_1 + \alpha_2 \phi_2 + w, \qquad \alpha_i \in \mathbf{R},\ w \in \mathrm{Im}(D_u G_0), \\ \lambda &= \alpha_0 + \beta. \end{aligned} \tag{15.7}$$

The discussion in the sequel is restricted to corank-2 bifurcations.

Consider the D_4-symmetry of the square and its generators S_1, R with

$$S_1(x, y) = (\pi - x, y), \qquad R(x, y) = (\pi - y, x).$$

Alternatively, the reflections S_1, S_2 with

$$S_1(x, y) = (\pi - x, y), \qquad S_2(x, y) = (y, x).$$

can be also used as generators of D_4. Thanks to the boundary conditions (15.1b) the problem (15.1) is *D_4-equivariant*

$$\gamma G(u, \lambda) = G(\gamma u, \lambda) \quad \text{for all} \quad (u, \lambda) \in X \times \mathbf{R},\ \gamma \in D_4.$$

As the homogeneous Dirichlet (Neumann) boundary conditions, actions of D_4 in $\mathrm{Ker}(D_u G_0)$ depend on the parity of $\kappa_1(\mu)$, $\kappa_2(\mu)$, see Definition 14.7.1. More precisely, for all $\phi = \alpha\phi_1 + \alpha\phi_2 \in \mathrm{Ker}(D_u G_0)$, let us consider the induced actions of D_4 on $(\alpha_1, \alpha_2)^T \in \mathbf{R}^2$. By definition we have $S_2\phi_1 = \phi_2$, $S_2\phi_2 = \phi_1$. Hence,

$$S_2 \sim \begin{pmatrix} 0 & 1 \\ 1 & 0 \end{pmatrix}. \tag{15.8}$$

Lemma 15.2.1. *The representations of S_1 in* $\mathrm{Ker}(D_u G_0)$ *are as follows*

if κ_1, κ_2 both are even, then $S_1 \sim I$; (15.9a)

if κ_1 is odd and κ_2 is even, we have $S_1 \sim \begin{pmatrix} -1 & 0 \\ 0 & 1 \end{pmatrix}$; (15.9b)

if κ_1 is even, and κ_2 is odd, we have $S_1 \sim \begin{pmatrix} 1 & 0 \\ 0 & -1 \end{pmatrix}$; (15.9c)

if κ_1, κ_2 both are odd, then $S_1 \sim -I$. (15.9d)

Proof: For all $\phi = \alpha\phi_1 + \alpha\phi_2 \in \mathrm{Ker}(D_u G_0)$ we have

$$S_1\phi = \phi(\pi - x, y) = \alpha_1 S_1 \phi_1 + \alpha_2 S_1 \phi_2,$$

while

$$\begin{aligned}
S_1\phi_1 &= \phi_1(\pi - x, y) \\
&= (g_0 \sin(\kappa_1(\pi - x)) + g_1\kappa_1 \cos(\kappa_1 x)) \cdot (g_0 \sin(\kappa_2 y) + g_1\kappa_2 \cos(\kappa_2 y)) \\
&= \{[g_1\kappa_1 \sin(\kappa_1\pi) - g_0 \cos(\kappa_1\pi)] \sin(\kappa_1 x) \\
&\quad + [g_1\kappa_1 \cos(\kappa_1\pi) + g_0 \sin(\kappa_1\pi)] \cos(\kappa_1 x)\} \\
&\quad \cdot [g_0 \sin(\kappa_2 y) + g_1\kappa_2 \cos(\kappa_2 y)] \\
&= \begin{cases} \phi_1, & \text{if } \kappa_1 \text{ is even;} \\ -\phi_1, & \text{if } \kappa_1 \text{ is odd.} \end{cases}
\end{aligned}$$

On the other hand,

$$\begin{aligned}
S_1\phi_2 &= \phi_2(\pi - x, y) \\
&= \{[g_1\kappa_2 \sin(\kappa_2\pi) - g_0 \cos(\kappa_2\pi)] \sin(\kappa_2 x) \\
&\quad + [g_1\kappa_2 \cos(\kappa_2\pi) + g_0 \sin(\kappa_2\pi)] \cos(\kappa_2 x)\} \\
&\quad \cdot [g_0 \sin(\kappa_1 y) + g_1\kappa_1 \cos(\kappa_1 y)] \\
&= \begin{cases} \phi_2, & \text{if } \kappa_2 \text{ is even;} \\ -\phi_2, & \text{if } \kappa_2 \text{ is odd.} \end{cases}
\end{aligned}$$

Thus the representations of S_1 in $\mathrm{Ker}(D_u G_0)$ is given as (15.9). ∎

Remark 15.2.1. The statements (15.8) and (15.9) show that the action of D_4 in $\mathrm{Ker}(D_u G_0) = \mathrm{span}\{\phi_1, \phi_2\}$ is reducible for $(\kappa_1, \kappa_2) =$ (even, even) or (odd, odd), and irreducible for $(\kappa_1, \kappa_2) =$(odd, even) or (even, odd).

15.3 Normal Forms of Equivariant Polynomials

Thanks to the decompositions of the space X and the unknown (u, λ) in (15.6) and (15.7), respectively, we can apply the Liapunov-Schmidt method to the problem (15.5) and transform it equivalently into a system of two *reduced bifurcation equations* for the unknowns α_1, α_2 and the parameter β

$$\begin{aligned}
&B(\alpha_1, \alpha_2, \beta, \mu) \\
&:= (\lambda_0 + 1) \begin{pmatrix} \langle \phi_1, \; G(\alpha_1\phi_1 + \alpha_2\phi_2 + w(\alpha_1, \alpha_2, \beta), \lambda_0 + \beta)\rangle \\ \langle \phi_2, \; G(\alpha_1\phi_1 + \alpha_2\phi_2 + w(\alpha_1, \alpha_2, \beta), \lambda_0 + \beta)\rangle \end{pmatrix} \\
&= -\beta \begin{pmatrix} \alpha_1 \\ \alpha_2 \end{pmatrix} + \begin{pmatrix} \langle \phi_1, \; f(\alpha_1\phi_1 + \alpha_2\phi_2 + w(\alpha_1, \alpha_2, \beta), \lambda_0 + \beta)\rangle \\ \langle \phi_2, \; f(\alpha_1\phi_1 + \alpha_2\phi_2 + w(\alpha_1, \alpha_2, \beta), \lambda_0 + \beta)\rangle \end{pmatrix} \\
&= 0.
\end{aligned} \tag{15.10}$$

Here $w(\alpha_1, \alpha_2, \beta)$ is the unique solution of

$$QG(\alpha_1\phi_1 + \alpha_2\phi_2 + w, \lambda_0 + \beta) = 0 \tag{15.11}$$

and Q is a projection from X onto $\mathrm{Im}(D_u G_0)$

$$Q(u) := u - \langle \phi_1,\ u\rangle\phi_1 - \langle \phi_2,\ u\rangle\phi_2 \qquad \text{for all} \quad u \in X.$$

Dependence of the equations (15.10) on μ is included implicitly in boundary conditions. Solutions of this finite dimensional system have a one-to-one correspondence with those of the original problem and contain all information of bifurcations of (15.1) at (u_0, λ_0). Furthermore, the reduced equations are also D_4-equivariant under the induced action in $\text{Ker}(D_uG_0)$. The D_4-equivariance induces certain structure in $B(\alpha_1, \alpha_2, \beta, \mu)$ at $(\alpha_1, \alpha_2, \beta) = (0, 0, 0)$.

Recall that $G(0, \lambda) = 0$ holds for all $\lambda \in \mathbf{R}$ and linear terms vanish in the reduced bifurcation equations. The Taylor expansion of $B(\alpha_1, \alpha_2, \beta, \mu)$ is of the form

$$B(\alpha_1, \alpha_2, \beta, \mu) = \begin{pmatrix} \sum\limits_{\substack{i,j,k=0\\ i+j+k\geq 2}} a_{ijk}(\mu)\alpha_1^i\alpha_2^j\beta^k \\ \sum\limits_{\substack{i,j,k=0\\ i+j+k\geq 2}} b_{ijk}(\mu)\alpha_1^i\alpha_2^j\beta^k \end{pmatrix} \tag{15.12}$$

with

$$a_{00k} = b_{00k} = 0, \quad k = 2, 3 \ldots .$$

By including the homotopy parameter μ into the arguments of B we emphasize here that the coefficients a_{ijk}, b_{ijk} vary along the homotopy path. Their expansions with respect to μ will become important for degenerate cases and for mode interactions, see Section 15.6 and Chapter 16. To determine structure of B with actions of D_4, we use the equivariance

$$S_2B(\alpha_1, \alpha_2, \beta, \mu) = B(S_2(\alpha_1, \alpha_2), \beta, \mu)$$

and derive

$$a_{ijk} = b_{jik} \quad \text{for all} \quad i, j, k.$$

The orthogonality $\langle\phi_i, \phi_j\rangle = \delta_{ij}$ of eigenfunctions of the Laplacian implies

$$a_{011} = a_{012} = 0.$$

Consider the expansion (15.12) up to the third order. The following statements are derived directly from the equivariance of $S_1B(\alpha_1, \alpha_2, \beta, \mu) = B(S_1(\alpha_1, \alpha_2), \beta, \mu)$ under the actions in (15.9).

- If $(k_1, k_2) =$ (even, odd) or (odd, even), then

$$a_{010} = a_{200} = a_{020} = a_{011} = a_{210} = a_{030} = a_{201} = a_{021} = a_{012} = 0$$

  and the 3-jet of $B(\alpha_1, \alpha_2, \beta, \mu)$ reduces to the form

$$j_3B(\alpha_1, \alpha_2, \beta, \mu) = -\beta\begin{pmatrix}\alpha_1\\ \alpha_2\end{pmatrix} + \begin{pmatrix} a\alpha_1^3 + b\alpha_1\alpha_2^2 \\ b\alpha_1^2\alpha_2 + a\alpha_2^3\end{pmatrix} = 0, \tag{15.13}$$

where $a, b \in \mathbf{R}$ are constants determined via the Liapunov-Schmidt method. We exclude the case $a = b = 0$ for which the problem is not 3-determined and higher order terms in the Taylor expansion of $B(\alpha_1, \alpha_2, \beta, \mu)$ should be considered.

- If $(k_1, k_2) =$ (odd, odd), then

$$a_{110} = a_{200} = a_{020} = a_{111} = a_{201} = a_{021} = 0$$

and the 3-jet of $B(\alpha_1, \alpha_2, \beta, \mu)$

$$j_3B(\alpha_1, \alpha_2, \beta, \mu) = -\beta \begin{pmatrix} \alpha_1 \\ \alpha_2 \end{pmatrix} + \begin{pmatrix} a\alpha_1^3 + b\alpha_1^2\alpha_2 + c\alpha_1\alpha_2^2 + d\alpha_2^3 \\ d\alpha_1^3 + c\alpha_1^2\alpha_2 + b\alpha_1\alpha_2^2 + a\alpha_2^3 \end{pmatrix} = 0. \tag{15.14}$$

At $\mu = 0$ we know from Crawford *et al* [77], Gomes/Stewart [133] and Mei [217] that the Neumann problem has the hidden symmetries and can be embedded into a periodic problem whose symmetry corresponds to a much larger group. Hence, the 3-jet of B reduces to

$$j_3B(\alpha_1, \alpha_2, \beta, 0) = -\beta \begin{pmatrix} \alpha_1 \\ \alpha_2 \end{pmatrix} + \begin{pmatrix} a\alpha_1^3 + c\alpha_1\alpha_2^2 \\ c\alpha_1^2\alpha_2 + a\alpha_2^3 \end{pmatrix} = 0. \tag{15.15}$$

- If $(k_1, k_2) =$ (even, even), the action of S_1 in $\text{Ker}(D_uG_0)$ is trivial and the 2-jet of $B(\alpha_1, \alpha_2, \beta, \mu)$ is

$$j_2B(\alpha_1, \alpha_2, \beta, \mu) = -\beta \begin{pmatrix} \alpha_1 \\ \alpha_2 \end{pmatrix} + \begin{pmatrix} a\alpha_1^2 + 2b\alpha_1\alpha_2 + b\alpha_2^2 \\ b\alpha_1^2 + 2b\alpha_1\alpha_2 + a\alpha_2^2 \end{pmatrix} = 0. \tag{15.16}$$

Here the coefficients a, b are derived directly from (15.10) as

$$a := \frac{1}{2}\int_\Omega f_{uu}(0, \lambda_0)\phi_i^3 \, dx\, dy,$$

$$b := \frac{1}{2}\int_\Omega f_{uu}(0, \lambda_0)\phi_i^2\phi_j \, dx\, dy, \; i, j = 1, 2, \; i \neq j.$$

Generically the system (15.16) is determined for $\mu \in (0, 1]$. For $\mu = 0$, the Neumann problem gains more symmetry and the 3-jet of $B(\alpha_1, \alpha_2, \beta, \mu)$ is needed. Normal form of the 3-jet is same as (15.15). One can also use μ as an additional bifurcation parameter and consider the following 3-jet in the neighborhood of $\mu = 0$

$$j_3B = -\beta \begin{pmatrix} \alpha_1 \\ \alpha_2 \end{pmatrix} + \begin{pmatrix} (a\alpha_1^2 + 2b\alpha_1\alpha_2 + c\alpha_2^2)\mu + d\alpha_1^3 + e\alpha_1\alpha_2^2 \\ (c\alpha_1^2 + 2b\alpha_1\alpha_2 + a\alpha_2^2)\mu + e\alpha_1^2\alpha_2 + d\alpha_2^3 \end{pmatrix} = 0. \tag{15.17}$$

15.4 Variations of Bifurcations along the Homotopy

The 3-jets of the reduced bifurcation equations in Section 15.3 are generically determined and allow simple calculation of solutions. In this section we study variations of bifurcations along a homotopy between Neumann and Dirichlet problems. These depend on the wave numbers and will be considered separately.

15.4.1 (κ_1, κ_2) = (odd, even) or (even, odd)

Generically the inequality $a \neq \pm b$ holds in the normal form (15.13). Thereafter, the system (15.13) has eight nontrivial solutions

$$\left(\pm(\beta/a)^{1/2},\ 0\right), \qquad \left(0,\ \pm(\beta/a)^{1/2}\right), \tag{15.18a}$$

$$\left(\pm(\beta/(a+b))^{1/2},\ \pm(\beta/(a+b))^{1/2}\right). \tag{15.18b}$$

To ensure real solutions, we adjust the signs of β in accordance with the signs of a and $a+b$ along the homotopy path. In the neighborhood of (0,0) these eight points lead to eight solution branches of (15.10), which in turn via the Liapunov-Schmidt reduction yield eight solution branches of the original problem (15.5). On the other hand, we see that $((\beta/a)^{1/2}, 0)$ and $-((\beta/a)^{1/2}, 0)$ tend to $(0,0)$ for $\beta \to 0$ and lie in the same fixed point space with $\alpha_2 = 0$. Extending the solution curve continuously across the axis $\beta = 0$, we obtain from $((\beta/a)^{1/2}, 0)$ and $-((\beta/a)^{1/2}, 0)$ the same solution curve of (15.5). In other words, the eight solutions (15.18) of (15.13) yield four different nontrivial solution branches of (15.5) bifurcating at (u_0, λ_0). Moreover, these solution branches are of the form

$$\begin{aligned} &u = \alpha_1(\beta)\phi_1 + \alpha_2(\beta)\phi_2 + w(\beta), \quad \alpha_i \in \mathbf{R}, \ w \in \mathrm{Im}(D_u G_0), \\ &\lambda = \alpha_0 + \beta. \end{aligned}$$

In the neighborhood of $\beta = 0$, $\alpha_i(\beta)$ behaves as in (15.18). Symmetries of these solution branches are identical with those of (15.18) under the induced action of D_4 in $\mathrm{Ker}(D_u G_0)$. A change of signs of a or $a+b$ means a change of bifurcation direction of the corresponding solution branch. As conclusions we have

Theorem 15.4.1. *For (κ_1, κ_2) = (odd, even) or (even, odd), if the statement $a \cdot (a \pm b) \neq 0$ holds along the homotopy path $\mu \in [0,1]$, then the problem (15.1) has four nontrivial solution branches bifurcating from the trivial solution curve at a corank-2 bifurcation points. Moreover, the solution branches corresponding to (15.18a) (resp. (15.18b)) bifurcate to the right if $a > 0$ (resp. $a + b > 0$) and to the left if $a < 0$ (resp. $a + b < 0$), i.e., these branches are real for $\beta > 0$ and $\beta < 0$, respectively. All of them are pitchfork solution branches.*

Remark 15.4.1. If $a = 0$ or $a = \pm b$ at $\mu = \mu_0 \in (0,1)$ and no mode-interaction occurs, i.e., $\dim(\mathrm{Ker}(D_u G_0)) = 2$ still holds, the normal form (15.13) becomes

$$j_3 B(\alpha_1, \alpha_2, \beta, \mu) = -\beta \begin{pmatrix} \alpha_1 \\ \alpha_2 \end{pmatrix} + b \begin{pmatrix} \alpha_1 \alpha_2^2 \\ \alpha_1^2 \alpha_2 \end{pmatrix} = 0, \tag{15.19}$$

and

$$j_3B(\alpha_1,\alpha_2,\beta,\mu) = -\beta\begin{pmatrix}\alpha_1\\ \alpha_2\end{pmatrix} + a\begin{pmatrix}\alpha_1^3 \pm \alpha_1\alpha_2^2\\ \pm\alpha_1^2\alpha_2 + \alpha_2^3\end{pmatrix} = 0, \tag{15.20}$$

respectively. Their solutions are

$$(\alpha_1,\alpha_2) = \begin{cases} \{(0,0),\ \pm\sqrt{\beta/b}\,(1,\pm 1)\} & \text{for } a=0;\\ \{(0,0),\ \pm\sqrt{\beta/a}\,(1,\pm 1)\} & \text{for } a=-b;\\ \{(0,0),\ (\alpha_1,\alpha_2);\ \alpha_1^2+\alpha_2^2=\beta\} & \text{for } a=b. \end{cases}$$

For $a = b$, the existence of a continuous orbit of solutions shows that the 3-jet of B is not determined and higher order terms in its Taylor expansion are needed for specifying bifurcation scenario. If $a(\mu_0) = 0$ occurs for some $\mu_0 \in [0,1]$, then as μ approaches μ_0 the solutions (15.18a) reduces to $(0,0)$ with $\beta = 0$, corresponding to the trivial solution curve $u \equiv 0$. By changing sign of β according to sign of the coefficient a at $\mu = \mu_0$ we can ensure the solution (15.18a) to be real. The corresponding solution branches of (15.5) change their bifurcation directions in such a way at $\mu = \mu_0$. Similar conclusion holds for (15.18b) if $a(\mu_0) = -b(\mu_0)$ for some $\mu_0 \in [0,1]$.

If $a + b$ and a do not change signs, structure of the bifurcation diagram and symmetries of the solutions remain invariant along the homotopy path.

15.4.2 (κ_1, κ_2) = (odd, odd)

For $\mu = 0$, the normal form (15.15) for Neumann problems is of the same structure as (15.13). Its solutions are given in (15.18). The Neumann problem has four solution branches at (u_0, λ_0). For $\mu \in (0,1]$ the hidden symmetry is no more present and the normal form is (15.14). Sum and difference of the two components in (15.14) are

$$(\alpha_1+\alpha_2)[-\beta + (a+d)\alpha_1^2 - (a-b-c+d)\alpha_1\alpha_2 + (a+d)\alpha_2^2] = 0,$$
$$(\alpha_1-\alpha_2)[-\beta + (a-d)\alpha_1^2 + (a+b-c-d)\alpha_1\alpha_2 + (a-d)\alpha_2^2] = 0.$$

Consider the cases $\alpha_1 = \pm\alpha_2$. We obtain the following four nontrivial solutions

$$\pm\left((\beta/(a+b+c+d))^{1/2},\quad (\beta/(a+b+c+d))^{1/2}\right), \tag{15.21a}$$
$$\pm\left((\beta/(a-b+c-d))^{1/2},\quad -(\beta/(a-b+c-d))^{1/2}\right). \tag{15.21b}$$

Similarly as in the last subsection, these solutions are real for appropriate β and lead to two solution branches of (15.10) and (15.1), respectively. The bifurcation directions are determined by the signs of $a+b+c+d$ and $a-b+c-d$ respectively. The coefficients b and d tend to zero as μ approaches zero. Solution branches of Robin problem gain translational symmetry at $\mu = 0$ and become solutions of Neumann problem. On the other hand, the system

$$-\beta + (a+d)\alpha_1^2 - (a-b-c+d)\alpha_1\alpha_2 + (a+d)\alpha_2^2 = 0, \quad (15.22a)$$
$$-\beta + (a-d)\alpha_1^2 + (a+b-c-d)\alpha_1\alpha_2 + (a-d)\alpha_2^2 = 0 \quad (15.22b)$$

contains only monomials in (α_1, α_2) with even order. Hence, if (α_1, α_2) is a solution, so are $-(\alpha_1, \alpha_2)$ and $\pm(\alpha_2, \alpha_1)$. Its four solutions yield then another two conjugate solution branches of (15.1) on the same group orbit.

15.4.3 $(\kappa_1, \kappa_2) =$ (even, even)

For the Robin and Dirichlet problems, i.e., $\mu \in (0, 1]$, the reduced bifurcation equations (15.10) are generically 2-determined. We derive from the difference of the two equations in (15.16) that

$$(\alpha_1 - \alpha_2)[-\beta + (a-b)(\alpha_1 - \alpha_2)] = 0.$$

Hence, if $a + 3b \neq 0$, then

$$\frac{\beta}{a+3b}(1,\ 1)$$

is a solution of (15.16). If $a \neq b$ and $(a-b)(a-5b) > 0$, the other two nontrivial solutions are

$$\left(\frac{\beta\left(a-b+\sqrt{(a-b)(a-5b)}\,\right)}{2(a-b)^2}, \frac{\beta\left(a-b-\sqrt{(a-b)(a-5b)}\,\right)}{2(a-b)^2}\right),$$
$$\left(\frac{\beta\left(a-b-\sqrt{(a-b)(a-5b)}\,\right)}{2(a-b)^2}, \frac{\beta\left(a-b+\sqrt{(a-b)(a-5b)}\right)}{2(a-b)^2}\right).$$

These three nontrivial roots of (15.16) lead to three bifurcating solution branches of (15.1) passing through (u_0, λ_0) transcritically. So we do not expect any changes of bifurcation directions along the homotopy path except at $\mu = 0$, where the Neumann problem is 3-determined and the eight solutions of (15.17) (resp. (15.15)) yielding four nontrivial solution branches of (15.1). What remains is to study how these three solutions of the Robin problem merge to the four solution branches of the Neumann problem at $\mu = 0$.

15.5 A Numerical Example

Consider the equation

$$\Delta u + \lambda u - \nu(u^2 + p(x,y)u^3) = 0 \quad \text{in } \Omega := [0,\pi] \times [0,\pi]. \quad (15.23)$$

Namely, we choose $f(u, \lambda) := \nu(u^2 + p(x,y)u^3)$ in (15.1a). Furthermore, let

$$g_0(\mu) = \mu \quad \text{and} \quad g_1(\mu) = 1 - \mu.$$

We study variations of bifurcations along the homotopy $(\kappa_1(\mu), \kappa_2(\mu))$ from $(1,2)$ to $(2,3)$, from $(1,3)$ to $(2,4)$ and from $(2,4)$ to $(3,5)$, respectively.

Let $(\kappa_1(\mu),\ \kappa_2(\mu))$ be a path of wave numbers with the starting points $(\kappa_1(0), \kappa_2(0)) \in \{(1,2), (1,3), (2,4)\}$ and $\lambda_0(\mu) = -\kappa_1(\mu)^2 - \kappa_2(\mu)^2$. Along these three eigenpaths there are no secondary bifurcation and mode-interactions. We perform the Liapunov-Schmidt reduction at $(u_0, \lambda_0(\mu)) = (0, \lambda_0(\mu))$ for an arbitrarily fixed $\mu \in [0,1]$. Since the second component of $B(\alpha_1, \alpha_2, \beta, \mu)$ follows directly from the first one via symmetries, i.e., by exchanging the indices, we calculate only the 3-jet of its first component

$$\begin{aligned} & j_3 b_1(\alpha_1, \alpha_2, \beta, \mu) \\ = & -\beta\alpha_1 + \nu\Big\{ \big[\langle\phi_1,\ \phi_1^2\rangle\alpha_1^2 + 2\langle\phi_1,\ \phi_1\phi_2\rangle\alpha_1\alpha_2 + \langle\phi_1,\ \phi_2^2\rangle\alpha_2^2\big] \\ & + \langle\phi_1,\ p(x,y)\phi_1^3 + 2\phi_1 w_{20}\rangle\alpha^3 \\ & + \langle\phi_1,\ 3p(x,y)\phi_1^2\phi_2 + 2\phi_1 w_{11} + 2\phi_2 w_{20}\rangle\alpha_1^2\alpha_2 \\ & + \langle\phi_1,\ 3p(x,y)\phi_1\phi_2^2 + 2\phi_1 w_{02} + 2\phi_2 w_{11}\rangle\alpha_1\alpha_2^2 \\ & + \langle\phi_1,\ p(x,y)\phi_2^3 + 2\phi_2 w_{02}\rangle\alpha_2^3\Big\} \end{aligned}$$

where $\alpha_1^2 w_{20} + 2\alpha_1\alpha_2 w_{11} + \alpha_2^2 w_{02}$ is the 2-jet of the solution $w(\alpha_1, \alpha_2, \beta)$ of (15.11) and $w_{ij} \in \mathrm{Im}(D_u G_0)$ are solutions of the following equations respectively

$$\begin{aligned} D_u G_0 w_{20} &= \nu Q\phi_1^2, \\ D_u G_0 w_{11} &= 2\nu Q\phi_1\phi_2, \\ D_u G_0 w_{02} &= \nu Q\phi_2^2. \end{aligned}$$

They are solved numerically with finite difference methods. The coefficients $\langle\phi_1,\ \phi_i\phi_j\rangle$ can be calculated analytically, while $\langle\phi_1,\ \phi_i w_{20}\rangle$, $\langle\phi_1,\ \phi_i w_{11}\rangle$ and $\langle\phi_1,\ \phi_i w_{02}\rangle$ are calculated numerically. In all calculations we choose

$$\nu = 1, \qquad p(x,y) = -1/7.$$

15.5.1 Discretization with finite difference methods

We discretize the equation

$$\begin{aligned} \Delta u &= f && \text{in } \Omega := [0,\pi] \times [0,\pi], \\ g_0(\mu)u + g_1(\mu)\tfrac{\partial u}{\partial n} &= 0 && \text{on } \partial\Omega \end{aligned} \tag{15.24}$$

with the five-point star scheme of the finite difference method. We take a uniform grid over Ω

$$\Omega_h := \{(x_i, y_j); \quad x_i = i\cdot h,\ y_j = j\cdot h,\ \ i,\ j = 1, \ldots, N+1\}$$

with the meshsize $h := \pi/N$ for an integer $N > 0$.

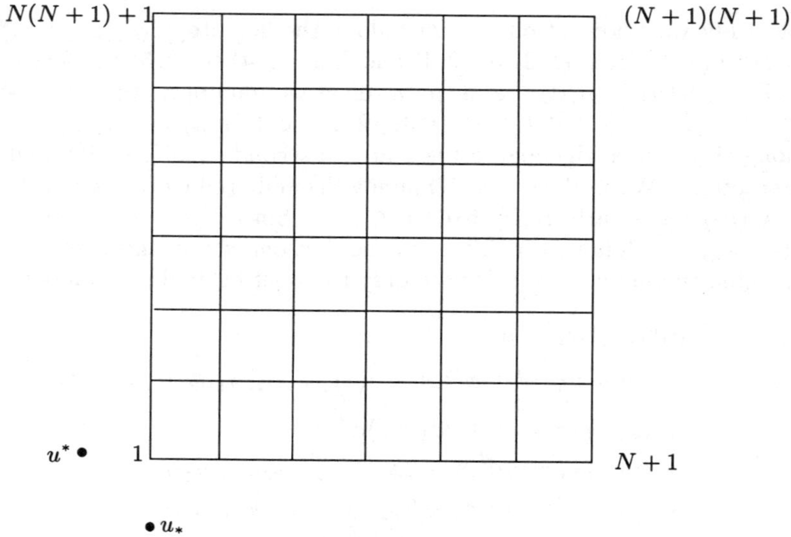

Fig. 15.1. Grid points over the domain Ω.

Let $u := (u_1, \ldots, u_{(N+1)\times(N+1)})$ denote values of the function $u(x, y)$ at the grid points. At the inner grid points, i.e., $i, j \in \{2, \ldots, N\}$, the five-point star discretization of the Laplacian Δ is

$$h^2 \Delta u \approx -4u_{(N+1)i+j} + u_{(N-1)i+j} + u_{(N+1)i+j} + u_{(N+1)i+j+1} + u_{(N+1)i+j-1}.$$

Along the boundary of the square, discretization of the Laplacian requires points outside of the domain, see e.g. u^* and u_* in Figure 15.1. We approximate these values via the boundary conditions in (15.24) as follows.

At the corner for $i = 1$, we discretize the boundary conditions with central difference scheme

$$-g_1 \frac{u_2 - u^*}{2h} + g_0 u_1 = 0,$$
$$-g_1 \frac{u_{N+2} - u_*}{2h} + g_0 u_1 = 0.$$

Here and in the sequel u^* and u_* represent the extrapolations of $u(x, y)$ in x- and y-directions, respectively. It is evident that

$$g_1 u^* = g_1 u_2 - 2h g_0 u_1,$$
$$g_1 u_* = g_1 u_{N+2} - 2h g_0 u_1.$$

At the corner $i = N + 1$ discretizations of the boundary conditions are

$$g_1 \frac{u^* - u_N}{2h} + g_0 u_{N+1} = 0,$$
$$-g_1 \frac{u_{2(N+1)} - u_*}{2h} + g_0 u_{N+1} = 0.$$

We obtain

$$\begin{aligned}g_1 u^* &= g_1 u_N - 2hg_0 u_{N+1},\\ g_1 u_* &= g_1 u_{2N+2} - 2hg_0 u_{N+1}.\end{aligned}$$

Similarly, at corners $i = N(N+1)+1$ and $i = (N+1)(N+1)$ we have

$$\begin{aligned}g_1 u^* &= g_1 u_{N(N+1)+2} - 2hg_0 u_{N(N+1)+1},\\ g_1 u_* &= g_1 u_{(N-1)(N+1)+1} - 2hg_0 u_{N(N+1)+1}\end{aligned}$$

and

$$\begin{aligned}g_1 u^* &= g_1 u_{(N-1)(N+1)-1} - 2hg_0 u_{(N+1)(N+1)},\\ g_1 u_* &= g_1 u_{N(N+1)+1} - 2hg_0 u_{(N+1)(N+1)},\end{aligned}$$

respectively. Along the four sides we derive for $i = 2, \ldots, N$

$$\begin{aligned}&-g_1 \frac{u_{(N+1)+i} - u_*}{2h} + g_0 u_i = 0\\ \Rightarrow\ & g_1 u_* = g_1 u_{(N+1)+i} - 2hg_0 u_i,\\ &g_1 \frac{u^* - u_{i(N+1)-1}}{2h} + g_0 u_{i(N+1)} = 0\\ \Rightarrow\ & g_1 u^* = g_1 u_{i(N+1)-1} - 2hg_0 u_{i(N+1)},\\ &-g_1 \frac{u_{i(N+1)+2} - u^*}{2h} + g_0 u_{i(N+1)+1} = 0\\ \Rightarrow\ & g_1 u^* = g_1 u_{i(N+1)+2} - 2hg_0 u_{i(N+1)+1},\\ &g_1 \frac{u_* - u_{(N-1)(N+1)+i}}{2h} + g_0 u_{N(N+1)+i} = 0\\ \Rightarrow\ & g_1 u_* = g_1 u_{(N-1)(N+1)+i} - 2hg_0 u_{N(N+1)+i}.\end{aligned}$$

Thereafter, we discretize the equation $\Delta u = f$ with the five-point star scheme. To incorporate the homotopy of boundary conditions we multiply the equations at the boundary points with the factor g_1. For example, at the corner $i = 1$, the approximation

$$\begin{aligned}h^2 g_1 \Delta u_1 &\approx -4u_1 + u_2 + u_{N+2} + u^* + u_*\\ &= -4(g_1 + hg_0)u_1 + 2g_1 u_2 + 2g_1 u_{N+2}\end{aligned}$$

yields a discretization of the equation (15.24)

$$-4(g_1 + hg_0)u_1 + 2g_1 u_2 + 2g_1 u_{N+2} = h^2 g_1 f_1.$$

At the other three corners, we have

$$\begin{aligned}&-4(g_1 + hg_0)u_{N+1} + 2g_1 u_N + 2g_1 u_{2N+2}\\ &= h^2 g_1 f_{N+1},\end{aligned}$$

$$\begin{aligned} &-4(g_1 + hg_0)u_{N(N+1)+1} + 2g_1 u_{N(N+1)+2} + 2g_1 u_{(N-1)(N+1)} \\ &= h^2 g_1 f_{N(N+1)+1}, \\ &-4(g_1 + hg_0)u_{(N+1)(N+1)} + 2g_1 u_{(N+1)(N+1)-1} + 2g_1 u_{N(N+1)} \\ &= h^2 g_1 f_{(N+1)(N+1)}. \end{aligned}$$

Along the four sides, we have for $i = 2, \ldots, N$

$$\begin{aligned} &(u_{i+1} + u_{i-1} + 2u_{N+1+i})g_1 - (4g_1 + 2hg_0)u_i \\ &= h^2 g_1 f_i, \\ &(u_{(N+1)(i+1)} + u_{(N+1)(i-1)} + 2u_{(N+1)i-1})g_1 - (4g_1 + 2hg_0)u_{(N+1)i} \\ &= h^2 g_1 f_{(N+1)i}, \\ &(2u_{(N+1)i+2} + u_{(N+1)(i+1)+1} + u_{(N+1)(i-1)+1})g_1 - (4g_1 + 2hg_0)u_{(N+1)i+1} \\ &= h^2 g_1 f_{(N+1)i+1}, \\ &(u_{(N+1)N+i+1} + u_{(N+1)N+i-1} + 2u_{(N+1)(N-1)+i})g_1 - (4g_1 + 2hg_0)u_{(N+1)N+i} \\ &= h^2 g_1 f_{(N+1)N+i}. \end{aligned}$$

At the inner grid points the Laplacian is discretized as

$$u_{(N-1)i+j} + u_{(N+1)i+j} + u_{(N+1)i+j+1} + u_{(N+1)i+j-1} - 4u_{(N+1)i+j} = h^2 f_{(N+1)i+j}$$

for $i,\ j = 2, \ldots, N$. We write these equations into a matrix form

$$\begin{pmatrix} A & 2g_1 I_{N+1} & & & & \\ \tilde{I}_{N+1} & B & \tilde{I}_{N+1} & & & \\ & \ddots & \ddots & \ddots & & \\ & & \ddots & \ddots & \ddots & \\ & & & \tilde{I}_{N+1} & B & \tilde{I}_{N+1} \\ & & & & 2g_1 I_{N+1} & A \end{pmatrix} \begin{pmatrix} u_1 \\ \vdots \\ u_{(N+1)(N+1)} \end{pmatrix}$$

$$= h^2 \mathrm{diag}(2g_1 I_{N+1},\ \tilde{I}_{N+1},\ \ldots,\ \tilde{I}_{N+1},\ 2g_1 I_{N+1}) \begin{pmatrix} f_1 \\ \vdots \\ f_{(N+1)(N+1)} \end{pmatrix},$$

where $\tilde{I} = \mathrm{diag}(2g_1,\ 1, \cdots, 1,\ 2g_1)$ and

$$A := \begin{pmatrix} -4(g_1+hg_0) & 2g_1 & & & & \\ 2g_1 & -(4g_1+2hg_0) & 2g_1 & & & \\ & \ddots & \ddots & \ddots & & \\ & & \ddots & \ddots & \ddots & \\ & & & 2g_1 & -(4g_1+2hg_0) & 2g_1 \\ & & & & 2g_1 & -4(g_1+hg_0) \end{pmatrix}$$

and

$$B := \begin{pmatrix} -(4g_1 + 2hg_0) & 2g_1 & & & & \\ 1 & -4 & 1 & & & \\ & \ddots & \ddots & \ddots & & \\ & & \ddots & \ddots & \ddots & \\ & & & \ddots & \ddots & \ddots \\ & & & 1 & -4 & 1 \\ & & & & 2g_1 & -(4g_1 + 2hg_0) \end{pmatrix}.$$

15.5.2 Homotopy of $(\kappa_1(\mu),\ \kappa_2(\mu))$ from (1,2) to (2,3)

Along the homotopy path from (1,2) to (2,3) we obtain the coefficients a, b in the normal form (15.13) as in Table 15.1. The coefficient a is positive, while $a + b$ changes signs on the homotopy path. Therefore, the solution branches of (15.23) corresponding to (15.18a) bifurcate to the right, while the solution branches corresponding to (15.18b) bifurcate to the left for $\mu \le 0.3$, and change bifurcation direction to the right for certain $\mu \in (0.3, 0.4)$, and then change the direction back to the left again for certain $\mu \in (0.8, 0.9)$. In other words, along the homotopy path of corank-2 bifurcation points starting at (κ_1, κ_2) =(1,2), the problem (15.23) has two nontrivial solution branches bifurcating to the right, the other two bifurcating to the left near Dirichlet and Neumann boundary conditions. In the middle of the homotopy path, all four solution branches bifurcate to the right, see Fig. 15.2.

μ	$\kappa_1(\mu)$	$\kappa_2(\mu)$	a	b	$a+b$
0	1.00	2.00	.095729	-.160914	-.065181
.1	1.07	2.03	.086751	-.128004	-.041249
.2	1.14	2.08	.077249	-.100923	-.023671
.3	1.22	2.13	.067198	-.074699	-.007501
.4	1.30	2.19	.056554	-.052053	.004501
.5	1.40	2.26	.045664	-.034198	.011466
.6	1.50	2.36	.034947	-.022257	.012690
.7	1.61	2.48	.025164	-.014601	.010563
.8	1.74	2.63	.016997	-.011132	.005865
.9	1.87	2.81	.010814	-.013323	-.002509
1.	2.00	3.00	.006069	-.024142	-.018073

Table 15.1. The coefficients a, b in the normal form (15.13) on the homotopy path from (1,2) to (2,3).

15.5.3 Homotopy of $(\kappa_1(\mu),\ \kappa_2(\mu))$ from (1,3) to (2,4)

Along the homotopy path from (1,3) to (2,4) we obtain the coefficients a, b, c, d in the normal form (15.14) as in Table 15.2.

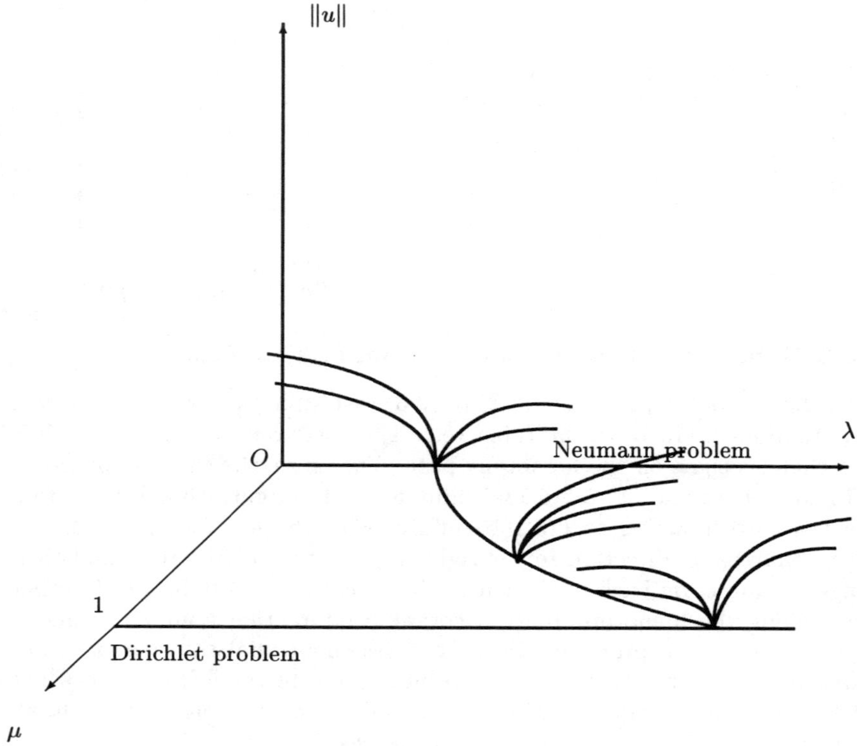

Fig. 15.2. Bifurcation diagram along the curve of corank-2 bifurcation points with (κ_1, κ_2) varying from (1,2) to (2,3).

For the Neumann problem four nontrivial solution branches pass through $(u_0, \lambda_0) = (1, 3)$. Owing to the inequalities $a < 0$ and $a + c > 0$, two branches bifurcate to the left, while the other two to the right. For $\mu \in (0, 1]$, i.e., the Robin boundary conditions, the solutions of (15.14) can be calculated directly from (15.21). The branch corresponding to (15.21b) changes its bifurcation direction immediately after the homotopy parameter moves away from zero, since $a - b + c - d$ changes signs. The system (15.22) has a solution of the form

$$(\alpha_1, \alpha_2) = (\tilde{\alpha}_1 \sqrt{-\beta},\ \tilde{\alpha}_2 \sqrt{-\beta})$$

and the other three solutions are given via $-(\alpha_1, \alpha_2)$ and $\pm(\alpha_2, \alpha_1)$. Table 15.3 states the numerical values of $(\tilde{\alpha}_1, \tilde{\alpha}_2)$ on the homotopy path. These solutions induce two nontrivial branches of (15.23) bifurcating to the left, since they are real only for $\beta \leq 0$. The bifurcation diagram along the homotopy path is illustrated in Fig. 15.3.

μ	$\kappa_1(\mu)$	$\kappa_2(\mu)$	a	b	c	d	$\hat{a}$	$\tilde{a}$
0	1.00	3.00	-.00117	0	.00186	0	.000690	.000690
.1	1.07	3.02	-.00158	.002057	.00259	.000686	.00375	-.00167
.2	1.14	3.05	-.00204	.004192	.00326	.001398	.00681	-.00434
.3	1.22	3.09	-.00248	.006604	.00438	.002201	.01071	-.00691
.4	1.30	3.13	-.00339	.009451	.00564	.003150	.01485	-.01035
.5	1.40	3.19	-.00434	.01254	.00741	.004179	.01979	-.01365
.6	1.50	3.27	-.00588	.01562	.00932	.005203	.02426	-.01738
.7	1.61	3.38	-.00795	.01832	.01097	.006106	.02745	-.02141
.8	1.74	3.54	-.01082	.01995	.01140	.006651	.02718	-.02602
.9	1.87	3.75	-.01425	.02003	.00905	.006676	.02151	-.03191
1.	2.00	4.00	-.01849	.01954	.00366	.006514	.01122	-.04088

Table 15.2. The coefficients a, b, c, d in the normal form (15.14) on the homotopy path connecting the wave numbers (1,3) and (2,4). Here $\hat{a} := a + b + c + d$ and $\tilde{a} := a - b + c - d$.

μ	0	0.1	0.2	0.3	0.4	0.5	0.6	0.7	0.8	0.9	1.0
$\tilde{\alpha}_1$	0.00	4.03	5.22	5.45	5.05	4.47	3.79	3.16	2.60	2.23	2.07
$\tilde{\alpha}_2$	29.3	23.6	18.4	15	12.4	10.8	9.56	8.60	7.82	7.09	6.37

Table 15.3. Solutions $\pm(\tilde{\alpha}_1, \tilde{\alpha}_2)\sqrt{-\beta}$ and $\pm(\tilde{\alpha}_2, \tilde{\alpha}_1)\sqrt{-\beta}$ of the system (15.22) on the homotopy path from (1,3) to (2,4).

15.5.4 Homotopy of $(\kappa_1(\mu),\ \kappa_2(\mu))$ from (2,4) to (3,5)

On the homotopy path from (2,4) to (3,5) all the coefficients a, b in the normal form (15.13) can be calculated directly, see Table 15.4.

Since for all $\mu \in (0, 1]$ the inequalities $a + 3b \neq 0$, $(a - b)(a - 5b) > 0$ hold, the problem (15.23) has three nontrivial solution branches across $(0, \lambda_0(\mu))$ transcritically. All three solutions gain translational symmetry at $\mu = 0$, the Neumann problem. Moreover, the number of nontrivial solution branches increases there from three to four.

μ	$\kappa_1(\mu)$	$\kappa_2(\mu)$	a	b
0.00	2.00	4.00	0	0
0.10	2.03	4.02	0.1815E-03	0.6007E-02
0.20	2.08	4.04	0.8945E-03	0.1420E-01
0.30	2.13	4.07	0.2326E-02	0.2233E-01
0.40	2.19	4.10	0.4369E-02	0.3031E-01
0.50	2.26	4.15	0.7785E-02	0.3903E-01
0.60	2.36	4.22	0.1283E-01	0.4780E-01
0.70	2.48	4.32	0.1885E-01	0.5489E-01
0.80	2.63	4.47	0.2478E-01	0.5849E-01
0.90	2.81	4.69	0.2858E-01	0.5878E-01
1.00	3.00	5.00	0.3058E-01	0.6186E-01

Table 15.4. The coefficients a, b in the normal form (15.13) on the homotopy path from (2,4) to (3,5).

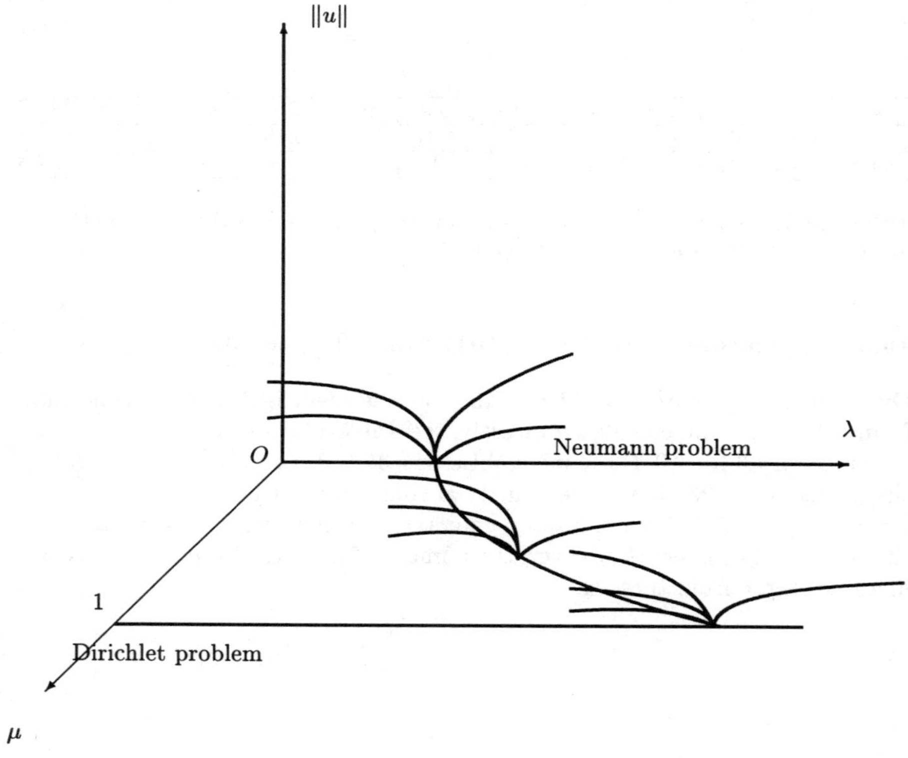

Fig. 15.3. Bifurcation diagram along the homotopy path of corank-2 bifurcation points with (κ_1, κ_2) varying from (1,3) to (2,4).

15.6 Forced Symmetry-Breaking via Boundary Conditions

To study the impact of symmetry-breaking in boundary conditions on bifurcation scenario of reaction-diffusion equations we impose the following boundary conditions to the problem (15.1)

$$\begin{aligned} g_0(\mu)u(x,0) - g_1(\mu)\frac{\partial u}{\partial y}(x,0) &= 0, \\ g_0(\mu)u(x,\pi) + g_1(\mu)\frac{\partial u}{\partial y}(x,\pi) &= 0, \\ \frac{\partial u}{\partial x}(0,y) &= 0, \\ \frac{\partial u}{\partial x}(\pi,y) &= 0. \end{aligned} \tag{15.25}$$

These boundary conditions break the D_4-symmetry into the D_2-symmetry. Moreover, the properties (15.2b) of $g_0(\mu)$, $g_1(\mu)$ make (15.25) a homotopy from the homogeneous Neumann boundary conditions along the four sides of Ω at $\mu = 0$ to the mixed boundary conditions at $\mu = 1$, which are of the Neumann type along the sides $x = 0,\ \pi$, and of the Dirichlet type at $y = 0,\ \pi$.

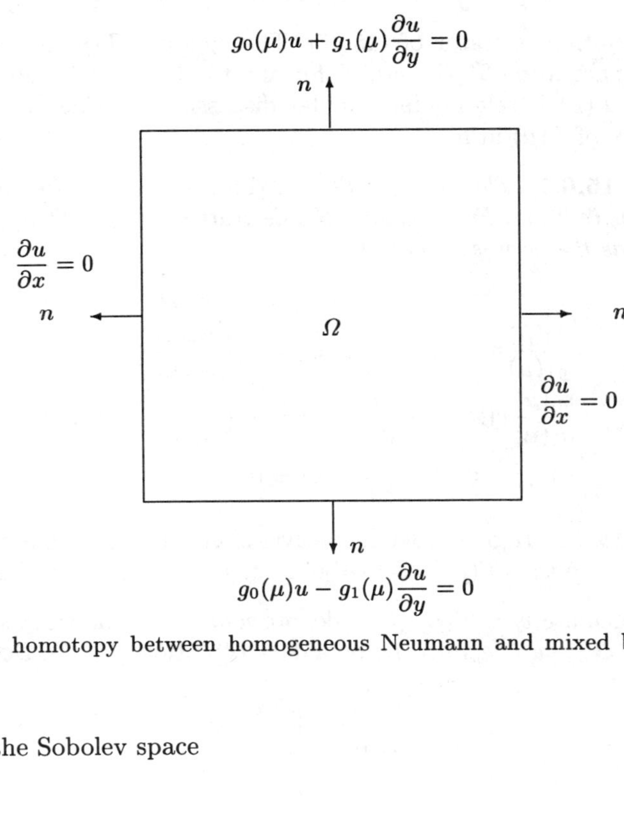

Fig. 15.4. A homotopy between homogeneous Neumann and mixed boundary conditions.

Consider the Sobolev space

$$X := H^1(\Omega) = \left\{ u \in L^2(\Omega);\ \frac{\partial u}{\partial x}, \frac{\partial u}{\partial y} \in L^2(\Omega) \right\}$$

with the norm $\|\cdot\|_{1,\Omega}$. The weak form of the linear problem

$$\Delta u - u = g \quad \text{in } \Omega \tag{15.26}$$

with the boundary conditions (15.25) is

$$\text{Find } u \in H^1(\Omega), \text{ such that } a(u,v,\mu) = (g,v) \text{ for all } v \in H^1(\Omega). \tag{15.27}$$

Here $a(u,v,\mu)$, $\mu \in [0,1)$ is a continuous and coercive bilinear form defined on $X \times X$ by

$$a(u,v,\mu) := -\int_\Omega (\nabla u \nabla v + uv)\,dxdy - \frac{g_0(\mu)}{g_1(\mu)} \int_0^\pi [u(x,0)v(x,0) + u(x,\pi)v(x,\pi)]\,dx.$$

For $\mu = 1$ we choose the bilinear form

$$\tilde{a}(u,v,\mu) := -\int_\Omega (\nabla u \nabla v + uv)\,dxdy.$$

Denote

$$T(\mu):\ g \in H^{-1}(\Omega) \longmapsto T(\mu)g = u(\mu) \in H^1(\Omega) \tag{15.28}$$

as the solution operator of (15.27). The operator $T(\mu)$ is linear, bounded and compact from $L^2(\Omega)$ into X for any fixed $\mu \in [0,1)$. Furthermore, the operator $T(\mu)$ is self-adjoint. Similar discussions as Chapter 14 ensure the continuity of $T(\mu)$ in μ.

Lemma 15.6.1. *The operator $T(\mu)$ is continuous and differentiable with respect to μ in $[0,1)$. Furthermore, the derivative $u'(\mu) = T'(\mu)g$ for all $g \in Y$ is given as the weak solution of*

$$\begin{aligned}
&\Delta v - v = 0 \quad \text{in } \Omega, \\
&\frac{g_0(\mu)}{g_1(\mu)} v(x,0) - \frac{\partial v}{\partial y}(x,0) = -\left(\frac{g_0(\mu)}{g_1(\mu)}\right)' u(x,0), \\
&\frac{g_0(\mu)}{g_1(\mu)} v(x,\pi) + \frac{\partial v}{\partial y}(x,\pi) = -\left(\frac{g_0(\mu)}{g_1(\mu)}\right)' u(x,\pi), \\
&\frac{\partial v}{\partial x}(0,y) = 0, \qquad \frac{\partial v}{\partial x}(\pi,y) = 0.
\end{aligned} \tag{15.29}$$

Remark 15.6.1. Higher order derivatives of $u(\mu)$ can be calculated in a similar manner. Note that $T(\mu)$ is self-adjoint, so is its derivative $T'(\mu)$.

To calculate $v := T(\mu)'g$, we denote $u = T(\mu)g$ the weak solution of the equation $\Delta u - u = g$ with the boundary condition (15.25). Define

$$\hat{v} := \left(\frac{g_0(\mu)}{g_1(\mu)}\right)' \left(-\frac{y^2}{\pi} + y\right) u(x,y). \tag{15.30}$$

It is easy to verify that $\hat{v}$ satisfies the boundary conditions in (15.29). Let $v = w + \hat{v}$ and substitute it into (15.29). We obtain the equation

$$\Delta w - w = -(\Delta\hat{v} - \hat{v})$$

with the boundary conditions (15.25). Furthermore,

$$-(\Delta\hat{v} - \hat{v}) = -\left(\frac{g_0(\mu)}{g_1(\mu)}\right)'\left[(-\frac{y^2}{\pi} + y)g - \frac{2}{\pi}u + 2(-\frac{2y}{\pi} + 1)\frac{\partial u}{\partial y}\right].$$

Hence, the weak solution $v = T'(\mu)g$ of the equation (15.29) is

$$\begin{aligned} T'(\mu)g &= -T(\mu)(\Delta\hat{v} - \hat{v}) + \hat{v} \qquad (15.31)\\ &= \left(\frac{g_0(\mu)}{g_1(\mu)}\right)'\left\{T(\mu)\left[\frac{2}{\pi}T(\mu)g + 2(\frac{2y}{\pi} - 1)\frac{\partial}{\partial y}(T(\mu)g) + \left(\frac{y^2}{\pi} - y\right)g\right]\right.\\ &\quad \left. + \left(-\frac{y^2}{\pi} + y\right)T(\mu)g\right\}. \end{aligned}$$

In the sequel we consider the weak form

$$G(u, \lambda, \mu) := u + (\lambda + 1)T(\mu)u - T(\mu)f(u, \lambda) = 0 \qquad (15.32)$$

of the problem (15.1). The mapping $G:\ X \times R \to X$ is as smooth as f in (u, λ). It is also continuously differentiable in μ.

Owing to the boundary condition we consider group

$$D_2 = \{S_1,\ R^2;\ S_1R^2,\ I\}. \qquad (15.33)$$

Let

$$\Gamma := \begin{cases} D_4 & \text{if } f(u, \lambda) \text{ is not odd in } u \text{ and } \mu = 1; \\ D_2 & \text{if } f(u, \lambda) \text{ is not odd in } u \text{ and } \mu \neq 1. \end{cases}$$

The *Γ-equivariance* of the mapping G,

$$G(\gamma u, \lambda, \mu) = \gamma G(u, \lambda, \mu) \qquad \text{for all} \quad \gamma \in \Gamma,\ u \in X,\ \lambda \in \mathbf{R}.$$

can be verified directly via the generators of Γ.

15.6.1 Bifurcation points

Since $D_\lambda G(0, \lambda, \mu) = D_\mu G(0, \lambda, \mu) \equiv 0$ and $D_u G(0, \lambda, \mu) = I + (\lambda+1)T(\mu)$ for all $\lambda,\ \mu \in \mathbf{R}$, a bifurcation of (15.32) occurs at a point $(0, \lambda, \mu)$ on the trivial solution manifold $\{(0, \lambda, \mu);\ \lambda \in \mathbf{R},\ \mu \in [0, 1]\}$ if the linearized problem

$$D_u G(0, \lambda, \mu)u = u + (\lambda + 1)T(\mu)u = 0 \qquad (15.34)$$

has nontrivial solutions. By the definition (15.28) of $T(\mu)$ we see this equation is the weak form of the eigenvalue problem

$$\Delta u + \lambda u = 0 \qquad \text{in } \Omega = (0, \pi) \times (0, \pi)$$

with the boundary condition (15.25). Solving this problem with the rule of separating variables we obtain eigenvalues of the Laplacian $-\Delta$

$$\lambda = n^2 + k(\mu)^2 \tag{15.35}$$

with the corresponding eigenfunction

$$\phi(\mu) := \frac{\tilde{\phi}}{||\tilde{\phi}||} \quad \text{with} \quad \tilde{\phi} := (\cos nx)[g_0(\mu)\sin(k(\mu)y) + g_1(\mu)k(\mu)\cos(k(\mu)y)]. \tag{15.36}$$

Here $k(\mu)$ is a solution of (15.3). We conclude that bifurcation points of (15.32) on the trivial solution manifold are of the form

$$\left\{(0,\ \lambda(\mu),\ \mu) \quad | \quad \lambda(\mu) = n^2 + k(\mu)^2,\ n \in \mathbf{N},\ \text{and } k(\mu) \ \text{ satisfying (15.3)}\right\},$$

see e.g., Figure 15.5. The kernel $\text{Ker}(D_uG(0, \lambda(\mu), \mu))$ is generically one-dimensional and

$$\text{Ker}(D_uG(0, \lambda(\mu), \mu)) = \text{span}[\phi(\mu)].$$

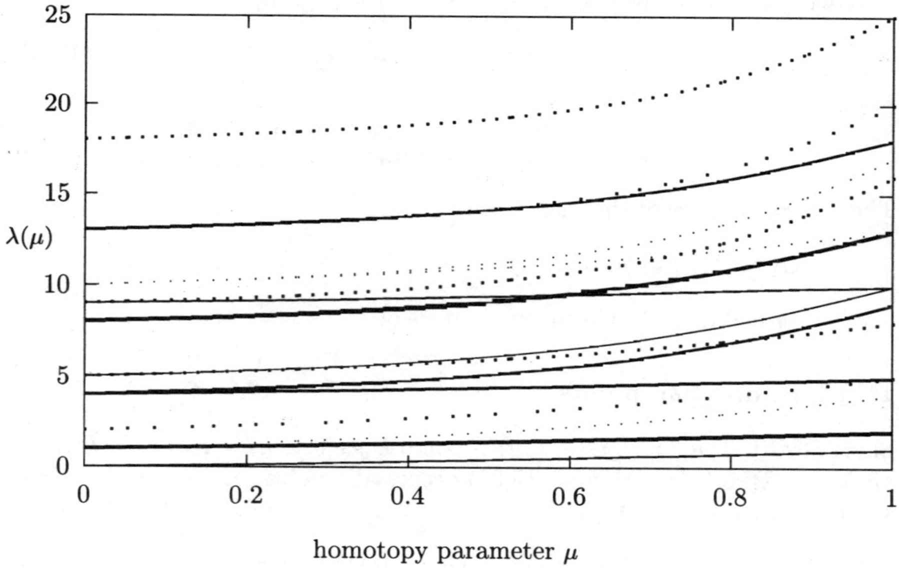

Fig. 15.5. Bifurcation points of (15.32) in the parameter space (λ, d). Here we have chosen $g_0(\mu) = \mu,\ g_1(\mu) = 1 - \mu$.

Note that at $\mu = 0, 1$ solutions of the equation (15.3) are integers. Furthermore, from the statement (15.35) and Figure 15.5 we see that two curves of bifurcation points can intersect, in particular, when μ approaches zero and one, respectively. An intersection point corresponds to a multiple bifurcation of (15.32).

At $\mu = 0$ the boundary conditions (15.25) reduce to the homogeneous Neumann type. Correspondingly, an eigenvalues of $-\Delta$ is of the form

$$\lambda_0 = n^2 + k^2, \quad n,\ k \in \mathbf{N} \cup \{0\}.$$

If $n \neq k$, such a eigenvalue is generically double and the associated eigenspace is

$$\mathrm{Ker}(D_u G(0, \lambda_0, 0)) = \mathrm{span}[\phi_1, \phi_2],$$

where

$$\phi_1 = \frac{2}{\pi}\cos(nx)\cos(ky), \qquad \phi_2 = \frac{2}{\pi}\cos(kx)\cos(ny).$$

As $\mu \to 0$, two different curves of simple bifurcation points approach the same double bifurcation point. With λ as the bifurcation parameter, there is exactly one solution branch bifurcating from the trivial solution curve at a simple bifurcation point, while at a double bifurcation point for $\mu = 0$ the Neumann problem has up to four different nontrivial solution branches (cf. Mei [223]). Moreover, the D_4-symmetry of the Neumann problem is broken into D_2-symmetry as μ moves away from 0. This leads to the question how the bifurcation scenario reacts to such a symmetry-breaking. We want to investigate which solution branches of the Neumann problem and what symmetries of the solutions persist as the parameter μ varies. To this end we write all elements $(u, \lambda, \mu) \in X \times \mathbf{R} \times \mathbf{R}$ as

$$\begin{aligned} u &= \sum_{i=1}^{\ell} \alpha_i \phi_i + w = \alpha \cdot \phi + w, \\ \lambda &= \lambda_0 + \beta, \\ \mu &= \mu_0 + \nu, \end{aligned}$$

where $\ell = 1$ or 2, $\alpha = (\alpha_1, \ldots, \alpha_\ell)$, $\phi = (\phi_1, \ldots, \phi_\ell)$, $\alpha_i,\ \beta,\ \nu \in \mathbf{R}$ and $w \in \mathrm{Im}(D_u G(0, \lambda_0, \mu_0))$. Applying the Liapunov-Schmidt method to (15.32) at $(0, \lambda_0, \mu_0)$ reduces (15.32) equivalently to a system of ℓ algebraic equations

$$B(\alpha, \beta, \nu) := \Big(\langle \phi_i,\ G(\alpha \cdot \phi + w(\alpha, \beta, \nu),\ \lambda_0 + \beta,\ \mu_0 + \nu) \rangle \Big)_{i=1}^{\ell} = 0. \quad (15.37)$$

Here $w(\alpha, \beta, \nu) \in \mathrm{Im}(D_u G(0, \lambda_0, \mu_0))$ with $w(0,0,0) = 0$ and $D_\alpha w(0,0,0) = 0$ is the unique solution of

$$QG(\alpha \cdot \phi + w,\ \lambda_0 + \beta,\ \mu_0 + \nu) = 0,$$

and $Q := I - \sum_{i=1}^{\ell} \langle \phi_i, \cdot \rangle \phi_i$ is a projection from X onto $\mathrm{Im}(D_u G(0, \lambda_0, \mu_0))$. We take the Taylor expansion of the components of $(\lambda_0 + 1) B(\alpha, \beta, \nu) = 0$ at $(\alpha, \beta, \nu) = (0, 0, 0)$ and consider the truncated form

$$0 = -\beta \alpha_i + (\lambda_0 + 1)^2 \langle \phi_i, \, T'(\mu_0)(\alpha \cdot \phi) \rangle \nu \tag{15.38}$$
$$+ \Big\langle \phi_i, \frac{1}{2} D_{uu} f_0 (\alpha \cdot \phi)^2 + D_{uu} f_0 (\alpha \cdot \phi) \Big(\frac{1}{2} \sum_{|\alpha|=2} D^{\alpha} w_0 \alpha^{\alpha} \Big) + \frac{1}{6} D_{uuu} f_0 (\alpha \cdot \phi)^3 \Big\rangle$$

for $i = 1, \ldots, \ell$. Here and in the sequel $D_{uu} f_0$ and $D_{uuu} f_0$ denote the derivatives of f at $(u, \lambda) = (0, \lambda_0)$; $\alpha \in \mathbf{N}^{\ell}$ is a multi-index and

$$\frac{1}{2} \sum_{|\alpha|=2} D^{\alpha} w_0 \alpha^{\alpha}$$

represents the second order terms in the Taylor expansion of w at $(\alpha, \beta, \nu) = (0, 0, 0)$.

With the knowledge of the bifurcation point $(0, \lambda_0, \mu_0)$ and the null space $\mathrm{Ker}(D_u G_0)$, the terms $\langle \phi_i, \, \frac{1}{2} D_{uu} f_0 (\alpha \cdot \phi)^2 \rangle$ and $\langle \phi_i, \frac{1}{6} D_{uuu} f_0 (\alpha \cdot \phi)^3 \rangle$ in (15.38) can be calculated directly. The other terms involve the derivatives $T'(\mu_0)$ and $D_{\alpha_i \alpha_j} w_0$. Since the function $w(\alpha, \beta, \nu)$ is defined implicitly by the equation (15.11), the term $D_{\alpha_i \alpha_j} w_0$ is described as the unique solution of the linear problem

$$D_u G_0 v = Q T(\mu_0) D_{uu} f_0 \phi_i \phi_j, \quad v \in \mathrm{Im}(D_u G_0). \tag{15.39}$$

Although the term $T'(\mu_0)(\alpha \cdot \phi)$ is also directly available via (15.31), we calculate

$$\begin{aligned}
&\langle \phi_i, \, T'(\mu_0)(\alpha \cdot \phi) \rangle \\
&= \Big(\frac{g_0}{g_1} \Big)' (\mu_0) \Big\langle \phi_i, \, \Big(-\frac{y^2}{\pi} + y \Big) T(\mu_0)(\alpha \cdot \phi) \\
&\quad + T(\mu_0) \Big[\frac{2}{\pi} T(\mu_0)(\alpha \cdot \phi) + 2 \Big(\frac{2y}{\pi} - 1 \Big) \frac{\partial}{\partial y} \Big(T(\mu_0)(\alpha \cdot \phi) \Big) + \Big(\frac{y^2}{\pi} - y \Big) (\alpha \cdot \phi) \Big] \Big\rangle \\
&= \Big(\frac{g_0}{g_1} \Big)' (\mu_0) \frac{1}{(\lambda_0 + 1)^2} \Big\langle \phi_i, \frac{2}{\pi} (\alpha \cdot \phi) - 2 (\lambda_0 + 1) \Big(\frac{2y}{\pi} - 1 \Big) \frac{\partial}{\partial y} \Big(T(\mu_0)(\alpha \cdot \phi) \Big) \Big\rangle \\
&= \Big(\frac{g_0}{g_1} \Big)' (\mu_0) \frac{2}{(\lambda_0 + 1)^2} \Big[\frac{\alpha_i}{\pi} + \Big\langle \phi_i, \, \Big(\frac{2y}{\pi} - 1 \Big) \frac{\partial}{\partial y} (\alpha \cdot \phi) \Big\rangle \Big].
\end{aligned} \tag{15.40}$$

15.6.2 Bifurcation scenarios

1) Simple bifurcations for $\mu \in (0, 1)$.

Let $\lambda(\mu) = n^2 + k(\mu)^2$ be a homotopy of simple eigenvalues of the Laplacian and $\lambda(0) = n^2 + m^2$, $\lambda(1) = n^2 + (m + 1)^2$. Then $(0, \lambda(\mu), \mu)$ is a curve of simple bifurcation points of (15.32) and the kernel

$$\mathrm{Ker}(D_u G(0, \lambda(\mu), \mu)) = \mathrm{span}[\phi]$$

is 1-dimensional and the basis function ϕ is given in (15.36). For an arbitrary $\mu_0 \in (0,1)$ the bifurcation equation (15.38) at $(0, \lambda(\mu), \mu)$ reduces to

$$\begin{aligned} 0 = & -\beta\alpha + (\lambda_0 + 1)^2 \langle \phi,\ T'(\mu_0)\phi \rangle \nu\alpha + \left\langle \phi,\ \frac{1}{2} D_{uu} f_0 \phi^2 \right\rangle \alpha^2 \\ & + \left\langle \phi,\ \frac{1}{2} D_{uu} f_0 (D_{\alpha\alpha} w_0)\phi + \frac{1}{6}(D_{uuu} f_0)\phi^3 \right\rangle \alpha^3. \end{aligned} \tag{15.41}$$

Here $\alpha \in \mathbf{R}$ is a scalar and

$$\langle \phi,\ T'(\mu_0)\phi \rangle = \left(\frac{g_0}{g_1}\right)'(\mu_0) \frac{2}{(\lambda_0+1)^2} \left[\frac{1}{\pi} + \left\langle \phi,\ \left(\frac{2y}{\pi} - 1\right)\frac{\partial}{\partial y}\phi \right\rangle \right].$$

Theorem 15.6.1. *The problem (15.32) undergoes a pitchfork bifurcation at all points on the curve $(0, \lambda(\mu), \mu)$, $\mu \in (0,1)$ of simple bifurcation points. Moreover, the 3-jet of the bifurcation equation (15.41) is*

$$j_3[(\lambda_0 + 1) B(\alpha, \beta, \nu)] = (-\beta + a\nu)\alpha + c\alpha^3 = 0 \tag{15.42}$$

with

$$a = 2\left(\frac{g_0}{g_1}\right)'(\mu_0) \left[\frac{1}{\pi} + \left\langle \phi,\ \left(\frac{2y}{\pi} - 1\right)\frac{\partial}{\partial y}\phi \right\rangle \right], \tag{15.43a}$$

$$c = \left\langle \phi,\ \frac{1}{2} D_{uu} f_0 (D_{\alpha\alpha} w_0)\phi + \frac{1}{6} D_{uuu} f_0 \phi^3 \right\rangle. \tag{15.43b}$$

Proof: Since $\langle \phi,\ \phi^2 \rangle = 0$, the second order term vanishes. The conclusion follows directly from equations (15.41) and (15.42) consecutively. ∎

The nontrivial solution of (15.42) is given as

$$\alpha = \left(\frac{\beta - a\nu}{c}\right)^{1/2}.$$

2) Double bifurcations of the Neumann problem.

For the Neumann problem ($\mu = 0$) a generic double bifurcation point $(0, \lambda_0, 0)$ has the property $\lambda_0 = n^2 + k^2(0)$ with the wave numbers n, $k := k(0)$ $(n \neq k)$ as integers. Furthermore,

$$\mathrm{Ker}(D_u G_0) = \mathrm{span}[\phi_1,\ \phi_2]$$

with

$$\phi_1 := \begin{cases} \dfrac{2}{\pi}\cos(nx)\cos(ky), & \text{for } n \cdot k \neq 0; \\ \dfrac{\sqrt{2}}{\pi}\cos(nx), & \text{for } n \neq 0,\ k = 0; \\ \dfrac{\sqrt{2}}{\pi}\cos(ky), & \text{for } n = 0,\ k \neq 0, \end{cases}$$

and

$$\phi_2 := \phi_1(y, x).$$

As the homotopy parameter μ in the boundary conditions varies, this double bifurcation point is split into two simple bifurcation points $(0, \lambda_i(\mu), \mu)$, $i = 1, 2$ with $\lambda_1 = n^2 + k^2(\mu)$ and $\lambda_2 = k^2 + n^2(\mu)$ for $\mu \neq 0$. We study how the bifurcation scenario at $(0, \lambda_0, 0)$ changes as μ moves away from zero.

Note that $\langle \phi_i,\ \phi_j\phi_l \rangle = 0$ for all $i, j, l = 1, 2$. The equation (15.39) can be solved analytically (cf. Mei [223]). Together with the statements

$$\langle \phi_i^4,\ 1 \rangle = \begin{cases} \dfrac{9}{4\pi^2} & \text{for } n \cdot k \neq 0, \\[2ex] \dfrac{3}{2\pi^2} & \text{for } n \cdot k = 0,\ n^2 + k^2 \neq 0; \end{cases}$$

and $\langle \phi_i^2,\ \phi_j^2 \rangle = \dfrac{1}{\pi^2}$ for $i \neq j$, we simplify the equations (15.38) into

$$\begin{aligned} -\beta\alpha_1 + (\lambda_0 + 1)^2 \langle \phi_1,\ T'(\mu_0)(\alpha_1\phi_1 + \alpha_2\phi_2) \rangle \nu + c_1\alpha_1^3 + c_2\alpha_1\alpha_2^2 = 0, \\ -\beta\alpha_2 + (\lambda_0 + 1)^2 \langle \phi_2,\ T'(\mu_0)(\alpha_1\phi_1 + \alpha_2\phi_2) \rangle \nu + c_2\alpha_1^2\alpha_2 + c_1\alpha_2^3 = 0, \end{aligned} \tag{15.44}$$

where $c_1,\ c_2$ are constants. If $n \cdot k \neq 0$, then

$$\begin{aligned} c_1 &= \frac{1}{6\pi^2} \left[\frac{9}{4} D_{uuu} f_0 - \frac{1}{4} (D_{uu} f_0)^2 \frac{45(k^2 - n^2)^2 + 4k^2n^2}{(k^2 - 3n^2)(n^2 - 3k^2)(n^2 + k^2)} \right], \\ c_2 &= \frac{1}{6\pi^2} \left[3D_{uuu} f_0 - 6(D_{uu} f_0)^2 \frac{1}{n^2 + k^2} \left(\frac{(k^2 - n^2)^2 - 4k^2n^2}{[(k^2 + n^2)^2 - 16k^2n^2]} - \frac{1}{2} \right) \right]. \end{aligned}$$

If $n = 0,\ k \neq 0$, then

$$\begin{aligned} c_1 &= \frac{1}{6\pi^2} \left[\frac{3}{2} D_{uuu} f_0 + \frac{5}{2k^2} (D_{uu} f_0)^2 \right], \\ c_2 &= \frac{1}{2\pi^2} D_{uuu} f_0. \end{aligned}$$

If $\nu = 0$, the equations (15.44) coincide with those in Mei [223], and yield four nontrivial solutions of (15.32) with the symmetries of isotropy subgroups of $\phi_1,\ \phi_2$ and $\phi_1 \pm \phi_2$, respectively. For $\nu \neq 0$, the forced symmetry-breaking in boundary conditions introduce in (15.44) the terms

$$\begin{aligned} &(\lambda_0 + 1)^2 \langle \phi_i,\ T'(\mu_0)(\alpha_1\phi_1 + \alpha_2\phi_2) \rangle \nu \\ &= 2\Big(\frac{g_0}{g_1}\Big)'(0) \left[\frac{\alpha_i}{\pi} + \Big\langle \phi_i,\ (\frac{2y}{\pi} - 1)\frac{\partial}{\partial y}(\alpha_1\phi_1 + \alpha_2\phi_2) \Big\rangle \right] \\ &=: d_i\alpha_i, \quad i = 1, 2. \end{aligned}$$

Here,

$$d_1 = d_2 = \frac{4}{\pi}\Big(\frac{g_0}{g_1}\Big)'(0), \quad \text{if } n \cdot k \neq 0,$$

$$d_1 = \frac{4}{\pi}\Big(\frac{g_0}{g_1}\Big)'(0), \qquad d_2 = 0, \quad \text{if } n = 0,\ k \neq 0.$$

Thereafter, the system (15.44) becomes

$$\begin{aligned} &\Big[-\beta + d_1\nu + c_1\alpha_1^2 + c_2\alpha_2^2\Big]\alpha_1 = 0, \\ &\Big[-\beta + d_2\nu + c_2\alpha_1^2 + c_1\alpha_2^2\Big]\alpha_2 = 0, \end{aligned} \tag{15.45}$$

Solutions of this system are

$$a)\ \Big(\pm\Big(\frac{\beta - d_1\nu}{c_1}\Big)^{1/2}, 0\Big), \qquad \Big(0,\ \pm\Big(\frac{\beta - d_2\nu}{c_1}\Big)^{1/2}\Big) \tag{15.46}$$

$$b)\ \Big(\pm\Big(\frac{(c_1-c_2)\beta-(c_1d_1-c_2d_2)\nu}{c_1^2 - c_2^2}\Big)^{1/2}, \pm\Big(\frac{(c_1-c_2)\beta+(c_2d_1-c_1d_2)\nu}{c_1^2 - c_2^2}\Big)^{1/2}\Big).$$

Remark 15.6.2. The coefficients a and c in (15.43) have properties

$$\lim_{\mu\to 0} a = d_i \quad \text{for } \phi = \phi_i,\ i = 1, 2$$

$$\lim_{\mu\to 0} c = c_1.$$

The solutions in (15.46a) are pure mode solution branches with the isotropy groups of ϕ_1, ϕ_2, respectively. They correspond to those bifurcating at the simple bifurcation points on the curves $(0, \lambda_i(\mu), \mu)$, $i = 1, 2$. The solutions in (15.46b) are mixed mode branches involving both ϕ_1 and ϕ_2 modes. They are generated as a result of the mode interaction. Moreover, if $n \cdot k = 0$, the terms $d_1\nu$, $d_2\nu$ break the D_4-symmetry of the Neumann problem and the mixed mode solution branches have merely the trivial symmetry. The pure mode and mixed mode solution branches may intersect at

$$\beta = \frac{(c_1d_1 - c_2d_2)\nu}{(c_1 - c_2)} \quad \text{or} \quad \beta = \frac{(c_2d_1 - c_1d_2)\nu}{(c_1 - c_2)},$$

and induce a secondary bifurcation.

If $n \cdot k \neq 0$, then $d_1 = d_2$ and the D_4-symmetry is preserved in (15.45) and the mixed mode solutions (15.46b) becomes

$$\pm\Big(\Big(\frac{\beta - d_1\nu}{c_1 + c_2}\Big)^{1/2},\ \pm\Big(\frac{\beta - d_1\nu}{c_1 + c_2}\Big)^{1/2}\Big).$$

Symmetries of these solutions are the isotropy groups of $\phi_1 + \phi_2$ and $\phi_1 - \phi_2$, respectively.

Example 15.6.1. Choose

$$f(u, \lambda) = \lambda(u^2 + u^3). \tag{15.47}$$

We consider bifurcation scenarios at the corank-2 bifurcation points $u_0 = 0,\ \mu_0 = 0$ and $\lambda_0 = 5,\ 1$, respectively. Moreover, we examine variations of the bifurcation scenarios as the homotopy parameter μ moves away from zero, i.e., as the homogeneous Neumann boundary conditions with D_4-symmetry are perturbed. To simplify the discussion, we take $g_0(\mu) = \mu,\ g_1(\mu) = 1 - \mu$. Note that at $\mu = 0$

$$\operatorname{Ker}(D_u G_0) = \operatorname{span}[\phi_1,\ \phi_2].$$

It can be verified that $c_1 \neq 0,\ c_2 \neq 0$ and $c_1^2 - c_2^2 \neq 0$ for any $n,\ k \in \mathbf{N} \cup \{0\}$.

(1) Wave numbers $n = 1,\ k = 2$**:** For $\lambda(0) = 5$ and $n = 1,\ k = 2$ we have

$$\phi_1 = \frac{2}{\pi} \cos x \cos 2y, \qquad \phi_2 = \frac{2}{\pi} \cos 2x \cos y.$$

Furthermore, $d = \dfrac{4}{\pi},\quad c_1 = \dfrac{5695}{132\pi^2},\quad c_2 = \dfrac{110220}{132\pi^2}$. Solutions $(\alpha_1,\ \alpha_2)$ in (15.46) becomes

$$\left(\pm\left(\frac{132\pi^2(\beta - 4\nu/\pi}{5695}\right)^{1/2},\ 0\right), \qquad \left(0,\ \pm\left(\frac{132\pi^2(\beta - 4\nu/\pi)}{5695}\right)^{1/2}\right)$$
$$\pm\left(\left(\frac{132\pi^2(\beta - 4\nu/\pi)}{115915}\right)^{1/2},\ \pm\left(\frac{132\pi^2(\beta - 4\nu/\pi)}{115915}\right)^{1/2}\right).$$

(2) Wave numbers $n = 0,\ k = 1$**:** We have $\lambda(0) = 1$ for $n = 0,\ k = 1$, moreover,

$$\phi_1 = \frac{\sqrt{2}}{\pi} \cos y, \qquad \phi_2 = \frac{\sqrt{2}}{\pi} \cos x.$$

Simple calculations show $d_1 = \dfrac{4}{\pi},\ d_2 = 0\ c_1 = \dfrac{19}{6\pi^2}$, and $c_2 = \dfrac{3}{\pi^2}$. The solution branches described by $(\alpha_1,\ \alpha_2)$ in (15.46) are

$$\left(\pm\left(\frac{6\pi^2\,(\beta - 4\nu/\pi)}{19}\right)^{1/2},\ 0\right), \qquad \left(0,\ \pm\left(\frac{6\pi^2\beta}{19}\right)^{1/2}\right),$$
$$\pm\left(\left(\frac{6(\pi^2\beta - 76\pi\nu)}{37}\right)^{1/2},\ \pm\left(\frac{6(\pi^2\beta + 72\pi\nu)}{37}\right)^{1/2}\right).$$

The pure ϕ_2-mode solution branch meets a mixed mode solution at $\beta = 76\nu/\pi$ and induces a secondary bifurcation.

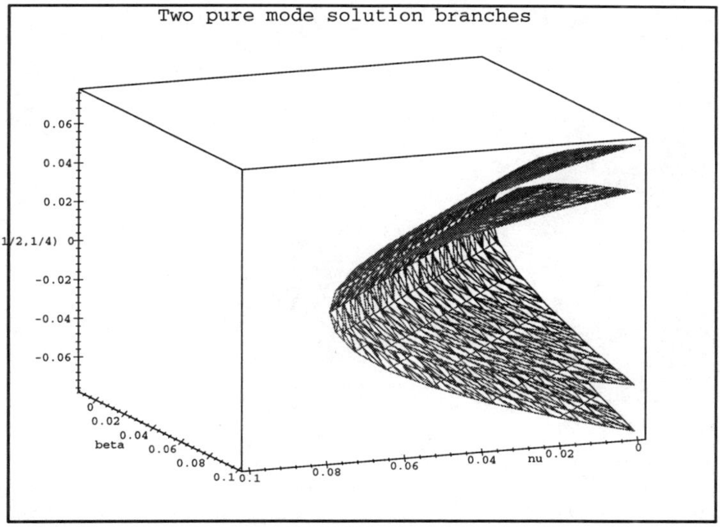

Fig. 15.6. Two solution branches of pure ϕ_1 and ϕ_2 modes.

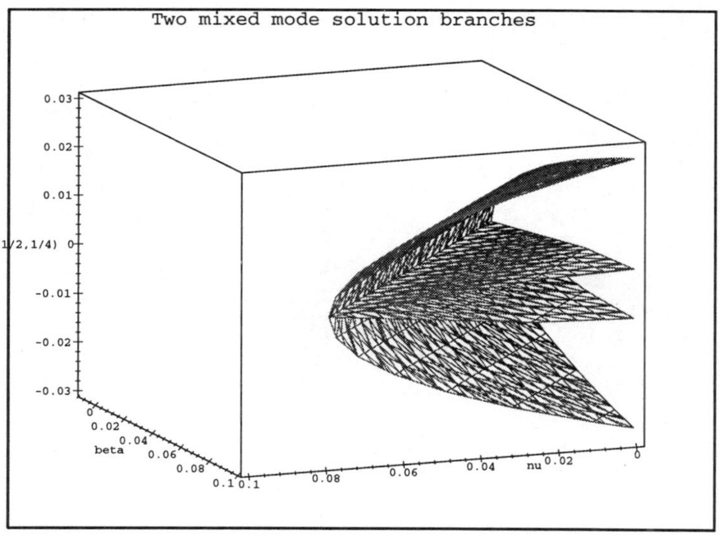

Fig. 15.7. Two solution branches of mixed ϕ_1 and ϕ_2 modes.

16. A Mode Interaction on a Homotopy of Boundary Conditions

In this chapter we study a steady/steady state mode interaction of the equation

$$\frac{\partial u}{\partial t} = \Delta u + \lambda u - f(u,\lambda) \quad \text{in} \quad \Omega := [0,\pi] \times [0,\pi] \tag{16.1a}$$

on a homotopy from Neumann to Dirichlet boundary conditions

$$g_0(\mu)u + g_1(\mu)\frac{\partial u}{\partial n} = 0 \quad \text{on} \quad \partial\Omega. \tag{16.1b}$$

Here $g_1,\ g_2:\ [0,1] \to \mathbf{R}$ are smooth functions with the properties

$$g_0(0) = g_1(1) = 0, \quad g_0(1)g_1(0)g_0(\mu)g_1(\mu) \neq 0 \quad \text{for all} \quad \mu \in (0,1), \tag{16.1c}$$

see also the statements (14.4) and (14.5), respectively. The following discussion can be seen as a case study on mode interactions in Chapter 12. After a brief introduction of a special mode interaction in Section 16.1 we address D_4-symmetry and normal forms of the reduced bifurcation equations in Section 16.2. Based on the normal forms we study generic bifurcations in Section 16.3 and scales of bifurcating solution branches in Section 16.4, respectively. Furthermore, we exploit the possibilities of secondary bifurcations in Section 16.5. In Section 16.6 we discuss calculations of the truncated bifurcation equations. Section 16.7 is devoted to the reduced stability of bifurcating solution branches. Finally in Section 16.8 we illustrate the whole discussion with a numerical example.

16.1 Introduction

As we have seen in previous chapters, a mode interaction takes place at intersection of two curves of bifurcation points. For local bifurcations we have steady/steady state, Hopf/steady state and Hopf/Hopf mode interactions. Global bifurcations may also interact with local ones, e.g. saddle-node homoclinic orbits and Hopf-homoclinic (Hopf-Silnikov) orbits etc. By choosing the scalar equation (16.1a) we restrict the discussion to steady/steady state mode-interactions. This separates the influence of boundary conditions on bifurcation scenarios from those of interactions of different substances in a reaction process.

We assume f to be smooth and

$$f(0, \lambda) = 0 \quad \text{for all } \lambda \in \mathbf{R} \quad \text{and} \quad f(x, \lambda) = O(|x|^2) \quad \text{for } x \in \mathbf{R},\ |x| \ll 1.$$

Furthermore, let f and its derivatives have a polynomial growth at infinite, i.e., there exist constants $c_1, c_2, r \geq 0$, such that

$$\left| \frac{\partial^i}{\partial x^i} f(x, \lambda) \right| \leq c_1 + c_2 |x|^r \qquad \text{for} \quad |x| \to \infty, \quad i = 1, 2, 3.$$

As we have seen in Chapter 15, bifurcation points of (16.1a) on the trivial solution $\{(0, \lambda, \mu);\ \ \lambda,\ \mu \in \mathbf{R}\}$ are of the form $(0, \lambda(\mu), \mu)$, where

$$\lambda(\mu) = -(\kappa_1(\mu)^2 + \kappa_2(\mu)^2)$$

is an eigenvalue of the Laplacian Δ. The corresponding eigenspace is generically two dimensional and spanned by $\phi_1,\ \phi_2$ with $\phi_1 := \phi(\kappa_1, \kappa_2),\ \ \phi_2 := \phi(\kappa_2, \kappa_1)$ and

$$\phi(\kappa_1, \kappa_2) := C(g_0 \sin(\kappa_1 x) + g_1 \kappa_1 \cos(\kappa_1 x))(g_0 \sin(\kappa_2 y) + g_1 \kappa_2 \cos(\kappa_2 y)). \tag{16.2}$$

Here C is a constant such that ϕ has the unit norm. The wave numbers κ_1 and κ_2 satisfy

$$2 g_0 g_1 \kappa \cos(\kappa\pi) + (g_0^2 - g_1^2 \kappa^2) \sin(\kappa\pi) = 0,$$

see also (14.42). The special form $(0, \lambda(\mu), \mu)$ of bifurcation points yields mode interactions along the homotopy of boundary conditions at the intersection points of two curves of eigenvalues of the Laplacian. Alternatively, such a mode interaction can be also interpreted as induced by diffusion and boundary conditions.

Variation of bifurcations is discussed in Chapter 15 along a curve $(\lambda(\mu), \mu)$ on which no intersection with other curves of bifurcation points occurs, i.e., no mode-interactions. Nevertheless, some curves of bifurcation points do intersect for $\mu \in [0, 1]$, see Figure 14.4 in Chapter 14. At an intersection point nonlinear interactions of different modes may yield new solutions.

In this chapter we choose the functions $g_0(\mu)$ and $g_1(\mu)$ as

$$g_0(\mu) = \mu, \qquad g_1(\mu) = 1 - \mu$$

and consider two curves $(0, \lambda_i(\mu), \mu)$, $i = 1, 2$, of bifurcation points, corresponding to the homotopy $\lambda_1(\mu)$ from $\lambda_1(0) = 8$ to $\lambda_1(1) = 18$, and the homotopy $\lambda_2(\mu)$ from $\lambda_2(0) = 9$ to $\lambda_2(1) = 17$. Since the homotopies start with $\lambda_1(0) = 8 < 9 = \lambda_2(0)$ and end with $\lambda_1(1) = 18 > 17 = \lambda_2(1)$, there is a intersection of these two curves for some $\mu \in (0, 1)$. More precisely, by path following we find at $\mu_0 := 0.6284333$ that

$$\lambda_1(\mu_0) = 0.7380507^2 + 3.3014006^2 = 2 * 2.39206655^2 = \lambda_2(\mu_0) \tag{16.3}$$

Correspondingly, eigenspace of the Laplacian is three dimensional and spanned by

$$\begin{aligned} \phi_1 &:= \phi(0.7380507, 3.3014006), \\ \phi_2 &:= \phi(3.3014006, 0.7380507), \\ \phi_3 &:= \phi(2.39206655, 2.39206655), \end{aligned} \tag{16.4}$$

see (16.2) for the definition of ϕ.

Let $T(\mu)$ be the (weak) inverse of $\Delta - id$ with respect to the boundary conditions (16.1b), see also (14.29). We consider the problem

$$G(u, \lambda, \mu) := u - T(\mu)[(\lambda + 1)u + f(u, \lambda)] = 0. \tag{16.5}$$

The operator G maps $H^1(\Omega) \times \mathbf{R} \times \mathbf{R}$ into $H^1(\Omega)$ and is obviously smooth in u, λ and μ for $\mu \in (0, 1)$. The equation (16.5) is a weak form of the stationary equation

$$\Delta u + \lambda u - f(u, \lambda) = 0 \quad \text{in} \quad \Omega := [0, \pi] \times [0, \pi]$$

with the boundary conditions (16.1b). Define

$$\lambda_0 := \lambda_1(\mu_0) = \lambda_2(\mu_0).$$

The kernel of the operator $D_u G_0 = I - (\lambda_0 + 1)T(\mu_0)$ is three dimensional and

$$\mathrm{Ker}(D_u G_0) := \mathrm{span}[\phi_1,\ \phi_2,\ \phi_3]. \tag{16.6}$$

On the other hand, since $T(\mu) :\ L^2(\Omega) \to H^1(\Omega)$ is self-adjoint and compact, we have

$$H^1(\Omega) = \mathrm{Im}(D_u G_0) \oplus \mathrm{Ker}(D_u G_0). \tag{16.7}$$

By $D_\lambda G(0, \lambda, \mu) = 0$ and $D_\mu G(0, \lambda, \mu) = 0$ for $\lambda, \mu \in \mathbf{R}$ we see that $(0, \lambda_0, \mu_0)$ is a corank-3 bifurcation point of (16.5).

16.2 Symmetries and Normal Forms

To study interactions of the modes ϕ_i, $i = 1, 2, 3$ we reduce (16.5) to a three dimensional algebraic system via the Liapunov-Schmidt method. Before doing this we consider symmetry of this problem and normal form of the reduced bifurcation equations.

We choose again the generators

$$S_1(x, y) = (\pi - x, y), \qquad S_2(x, y) = (y, x) \quad \text{for all } (x, y) \in \Omega$$

for D_4-symmetry of the square Ω and define the action

$$\gamma u(x,y) = u\big(\gamma^{-1}(x,y)\big) \quad \text{for all} \quad (x,y) \in \Omega,\ \gamma \in D_4$$

for functions u in the space

$$X := H^1(\Omega).$$

As we have seen before, the mapping $G(u,\lambda,\mu)$ is D_4-equivariant

$$\gamma G(u,\lambda,\mu) = G(\gamma u,\lambda,\mu) \quad \text{for all} \quad (u,\lambda,\mu) \in X \times \mathbf{R} \times \mathbf{R},\ \gamma \in D_4.$$

Furthermore, the spaces $\mathrm{Im}(D_uG_0)$ and $\mathrm{Ker}(D_uG_0)$ are D_4-invariant. If we identify $v \in \mathrm{Ker}(D_uG_0)$ with its coordinates $\alpha := (\alpha_1,\alpha_2,\alpha_3) \in \mathbf{R}^3$ via $v = \sum_{i=1}^{3} \alpha_i \phi_i$, the induced action of D_4 in $\mathbf{R}^3$ is generated by

$$S_1 = \begin{pmatrix} 1 & 0 & 0 \\ 0 & -1 & 0 \\ 0 & 0 & 1 \end{pmatrix}, \qquad S_2 = \begin{pmatrix} 0 & 1 & 0 \\ 1 & 0 & 0 \\ 0 & 0 & 1 \end{pmatrix}, \tag{16.8}$$

see also Table 12.3.2 for wave numbers with the parity (odd, even, odd). Define a projection $Q:\ X \to \mathrm{Im}(D_uG_0)$ as

$$Qu = u - \sum_{i=1}^{3} \langle \phi_i,\ u\rangle \phi_i \quad \text{for all } u \in X$$

and decompose elements $(u,\lambda,\mu) \in X \times \mathbf{R} \times \mathbf{R}$ as

$$(u,\lambda,\mu) = (v + w,\ \lambda_0 + \beta,\ \mu_0 + \nu), \quad w \in \mathrm{Im}(D_uG_0),\ v \in \mathrm{Ker}(D_uG_0).$$

We apply the Liapunov-Schmidt reduction to (16.5) at $(0,\lambda_0,\mu_0)$ and obtain a system of algebraic equations

$$\begin{aligned} & B(\alpha,\beta,\nu) \\ := & \left(\Big\langle \phi_j,\ G\big(\sum_{i=1}^{3} \alpha_i\phi_i + w(\sum_{i=1}^{3} \alpha_i\phi_i,\beta,\nu),\lambda_0+\beta,\nu_0+\nu\big) \Big\rangle,\ j = 1,2,3 \right) \\ = & \ 0. \end{aligned} \tag{16.9}$$

This system is also D_4-equivariant with respect to the induced D_4-action (16.8)

$$\gamma B(\alpha,\beta,\nu) = B(\gamma(\alpha),\beta,\nu) \quad \text{for all } (\alpha,\beta,\nu) \in \mathbf{R}^3 \times \mathbf{R}^2. \tag{16.10}$$

With this property we determine the structure of the reduced bifurcation equations a prior to the Liapunov-Schmidt reduction. We write $B(\alpha,\beta,\nu)$ formally as

$$B(\alpha,\beta,\nu) = \begin{pmatrix} B_1(\alpha_1,\alpha_2,\alpha_3,\beta,\nu) \\ B_2(\alpha_1,\alpha_2,\alpha_3,\beta,\nu) \\ B_3(\alpha_1,\alpha_2,\alpha_3,\beta,\nu) \end{pmatrix}.$$

The equality $S_1B(\alpha,\beta,\nu) = B(S_1\alpha,\beta,\nu)$ implies that B_1, B_3 are even functions in α_2, while B_2 is odd in α_2. Hence

$$\begin{aligned} B_i(\alpha_1,\alpha_2,\alpha_3,\beta,\nu) &= \tilde{B}_i(\alpha_1,\alpha_2^2,\alpha_3,\beta,\nu), \quad i=1,3,\\ B_2(\alpha_1,\alpha_2,\alpha_3,\beta,\nu) &= \tilde{B}_2(\alpha_1,\alpha_2^2,\alpha_3,\beta,\nu)\alpha_2. \end{aligned}$$

On the other hand, it follows from $S_2B(\alpha,\beta,\nu) = B(S_2\alpha,\beta,\nu)$ that

$$\begin{aligned} B_1(\alpha_1,\alpha_2,\alpha_3,\beta,\nu) &= B_2(\alpha_2,\alpha_1,\alpha_3,\beta,\nu),\\ B_3(\alpha_1,\alpha_2,\alpha_3,\beta,\nu) &= B_3(\alpha_2,\alpha_1,\alpha_3,\beta,\nu). \end{aligned}$$

We rewrite $B(\alpha,\beta,\nu)$ as

$$B(\alpha,\beta,\nu) = \begin{pmatrix} B_1(\alpha_1^2,\alpha_2^2,\alpha_3,\beta,\nu)\alpha_1 \\ B_1(\alpha_2^2,\alpha_1^2,\alpha_3,\beta,\nu)\alpha_2 \\ \frac{1}{2}[B_3(\alpha_1^2,\alpha_2^2,\alpha_3,\beta,\nu) + B_3(\alpha_2^2,\alpha_1^2,\alpha_3,\beta,\nu)] \end{pmatrix}. \tag{16.11}$$

By singularity theory, see Section 6.8 and Golubitsky/Stewart/Schaeffer [131], a D_4-equivariant problem is generically 3-determined. We consider the 3-jet

$$j_3B(\alpha,\beta,\nu) = \begin{pmatrix} a_1\alpha_1 + a_2\alpha_1\alpha_3 + a_3\alpha_1^3 + a_4\alpha_1\alpha_2^2 + a_5\alpha_1\alpha_3^2 \\ a_1\alpha_2 + a_2\alpha_2\alpha_3 + a_3\alpha_2^3 + a_4\alpha_1^2\alpha_2 + a_5\alpha_2\alpha_3^2 \\ b_1\alpha_3 + b_2(\alpha_1^2+\alpha_2^2) + b_3\alpha_3^2 + b_4\alpha_3^3 + b_5(\alpha_1^2+\alpha_2^2)\alpha_3 \end{pmatrix}. \tag{16.12}$$

Here the coefficients $a_i,\ b_i, i = 1,\ldots,5$ are functions of the parameters $\beta,\ \nu$. These functions are defined implicitly via the function $w(v,\beta,\nu)$ in the Liapunov-Schmidt reduction. In combination with scales of the components α_i and detection of secondary bifurcations we take Taylor expansions of $a_i,\ b_i,\ i = 1,\ldots,5$ in $\beta,\ \nu$ at (0,0), e.g.

$$a_i = a_i(0,0) + \frac{\partial a_i}{\partial \beta}(0,0)\beta + \frac{\partial a_i}{\partial \nu}(0,0)\nu + O(\|(\beta,\nu)\|^2). \tag{16.13}$$

The statement $D_\alpha B(0,0,0) = 0$ implies

$$a_1(0,0) = 0, \qquad b_1(0,0) = 0. \tag{16.14}$$

16.3 Generic Bifurcation Behavior

The kernel $\mathrm{Ker}(D_uG_0)$ is reducible under the induced D_4-actions (16.8). If we vary μ and consider the subspaces

$$V_1 := \mathrm{span}[\phi_1,\phi_2], \qquad V_2 := \mathrm{span}[\phi_3]$$

on the curves $(0,\lambda_i(\mu),\mu)$ of bifurcation points $i = 1,2$, these subspaces are D_4-invariant and irreducible. Since the D_4-action in V_2 is trivial, the

2-jet of the reduced bifurcation equation describes qualitatively the bifurcation scenario at points on $(0, \lambda_2(\mu), \mu)$. We truncate the third component of $j_3 B(\alpha, \beta, \nu)$ at the quadratic order and consider

$$\left(a_1 + a_2\alpha_3 + a_3\alpha_1^2 + a_4\alpha_2^2 + a_5\alpha_3^2\right)\alpha_1 = 0, \tag{16.15a}$$

$$\left(a_1 + a_2\alpha_3 + a_3\alpha_2^2 + a_4\alpha_1^2 + a_5\alpha_3^2\right)\alpha_2 = 0, \tag{16.15b}$$

$$b_1\alpha_3 + b_2(\alpha_1^2 + \alpha_2^2) + b_3\alpha_3^2 = 0. \tag{16.15c}$$

It follows from Bezout's theorem for polynomial equations that this system has $3^2 \cdot 2 = 18$ isolated solutions. Thus the original problem (16.5) may have up to 17 nontrivial solution branches at $(0, \lambda_0, \mu_0)$. On the other hand, it is obviously that if $(\alpha_1, \alpha_2, \alpha_3)$ is a solution of (16.15), so is $(-\alpha_1, -\alpha_2, \alpha_3)$.

Along the curves $(0, \lambda_i(\mu), \mu)$, $i = 1, 2$, nontrivial solutions of (16.5) involve merely either the modes ϕ_1, ϕ_2, or ϕ_3. At $(0, \lambda_0, \mu_0)$ all three modes may contribute to the solutions. To study such interactions, we assume the following nondegeneracy condition

$$a_3 \cdot b_3 \cdot (a_3 \pm a_4) \cdot (b_3 a_3 - b_2 a_5) \cdot [b_3(a_3 + a_4) - 2b_2 a_5] \neq 0. \tag{16.16}$$

This condition ensures that the system (16.15) have nonsingular solutions.

16.3.1 Solutions with the modes ϕ_1, ϕ_2

To see whether solutions at points on $(0, \lambda_1(\mu), \mu)$ with the modes ϕ_1, ϕ_2 persist at $\mu = \mu_0$, we set $\alpha_3 = 0$ and obtain from the equation (16.15c)

$$b_2(\alpha_1^2 + \alpha_2^2) = 0.$$

Case 1) If $b_2 \neq 0$, then $\alpha_1 = \alpha_2 = 0$ is the only solution of (16.15c). The problem (16.5) has no ϕ_1-mode or ϕ_2-mode or mixed ϕ_1 and ϕ_2 at $(0, \lambda_0, \mu_0)$. In other words, the solution branches coming along $(0, \lambda_1(\mu), \mu)$ are annihilated at $\mu = \mu_0$, the mode interaction point, see also Subsection 16.3.3.
Case 2) If $b_2 = 0$, e.g. in the case $D_{uu}f(0, \lambda) = 0$, then the system (16.15) reduces to

$$\left(a_1 + a_3\alpha_1^2 + a_4\alpha_2^2\right)\alpha_1 = 0, \tag{16.17a}$$

$$\left(a_1 + a_3\alpha_2^2 + a_4\alpha_1^2\right)\alpha_2 = 0. \tag{16.17b}$$

This coincides with the 3-jet of the reduced bifurcation equations at the points $(0, \lambda_1(\mu), \mu)$ with $\mu \neq \mu_0$. The system (16.17) has eight nontrivial solutions

$$(\pm\sqrt{-a_1/a_3},\ 0), \qquad (0,\ \pm\sqrt{-a_1/a_3}\ ); \tag{16.18a}$$

$$\sqrt{-a_1/(a_3 + a_4)}\ (\pm 1,\ \pm 1). \tag{16.18b}$$

The solutions in (16.18a) lead to pure ϕ_1- and ϕ_2-mode solutions of (16.5), while (16.18b) yield mixed mode solutions involving both modes ϕ_1 and ϕ_2.

16.3.2 Pure ϕ_3-mode solutions

The solution branch bifurcating at $(0, \lambda_2(\mu), \mu)$, $\mu \neq \mu_0$ is a pure ϕ_3-mode solution. For $\mu = \mu_0$, suppose $\alpha_3 \neq 0$ and $\alpha_1 = \alpha_2 = 0$. The system (16.15) is reduced to

$$b_1 + b_3\alpha_3 = 0.$$

We obtain

$$\alpha_3 = -b_1/b_3 = -b_3^{-1}(0,0)\Big(\frac{\partial b_1}{\partial \beta}(0,0)\beta + \frac{\partial b_1}{\partial \nu}(0,0)\nu\Big) + \cdots. \tag{16.19}$$

Thus (16.5) has a pure ϕ_3-mode solution branch at $(0, \lambda_0, \mu_0)$. This branch has the full D_4-symmetry of the mode ϕ_3. In other words, the D_4-branch persists at the mode interaction.

16.3.3 Interactions of three modes

We study all possible interactions of the modes ϕ_1, ϕ_2 with ϕ_3.
i) Let $\alpha_1 = 0$, $\alpha_2 \cdot \alpha_3 \neq 0$. We derive from (16.15)

$$a_1 + a_2\alpha_3 + a_3\alpha_2^2 + a_5\alpha_3^2 = 0, \tag{16.20a}$$
$$b_1\alpha_3 + b_2\alpha_2^2 + b_3\alpha_3^2 = 0. \tag{16.20b}$$

It has four solutions

$$\alpha_2 = \pm\left(\frac{-b_3a_1 + (b_1a_5 - b_3a_2)\alpha_3^{\pm}}{b_3a_3 - b_2a_5}\right)^{1/2}, \tag{16.21a}$$

and

$$\alpha_3^{\pm} = \frac{b_2a_2 - a_3b_1 \pm \Big((b_2a_2 - a_3b_1)^2 + 4a_1b_2(b_3a_3 - b_2a_5)\Big)^{1/2}}{2(b_3a_3 - b_2a_5)}. \tag{16.21b}$$

These give four solution branches involving the ϕ_2- and ϕ_3-modes. Since the mode ϕ_3 has full D_4-symmetry, these branches have the same symmetry as the pure ϕ_2-modes. Recall (16.14) and that a_1 and b_1 are of the order $O(\|(\beta,\nu)\|)$. If $a_2b_2 \neq 0$, then α_3^+ corresponds to a secondary bifurcation on the pure ϕ_3-mode branch, see Sections 16.4 and 16.5 below.
ii) Similarly, if $\alpha_2 = 0$, $\alpha_1 \cdot \alpha_3 \neq 0$, the system (16.15) becomes

$$a_1 + a_2\alpha_3 + a_3\alpha_1^2 + a_5\alpha_3^2 = 0, \tag{16.22a}$$
$$b_1\alpha_3 + b_2\alpha_1^2 + b_3\alpha_3^2 = 0. \tag{16.22b}$$

It has four solutions

$$\alpha_1 = \pm\left(\frac{-b_3a_1 + (b_1a_5 - b_3a_2)\alpha_3^{\pm}}{b_3a_3 - b_2a_5}\right)^{1/2} \tag{16.23}$$

and $\alpha_3^\pm$ is same as in (16.21b). They are on the group orbit of the solutions in (16.21). The corresponding solution branches of (16.5) have symmetry of the ϕ_1-mode.
iii) Finally, if $\alpha_1 \cdot \alpha_2 \cdot \alpha_3 \neq 0$, the system (16.15) reduces to

$$a_1 + a_2\alpha_3 + a_3\alpha_1^2 + a_4\alpha_2^2 + a_5\alpha_3^2 = 0, \tag{16.24a}$$

$$a_1 + a_2\alpha_3 + a_3\alpha_2^2 + a_4\alpha_1^2 + a_5\alpha_3^2 = 0, \tag{16.24b}$$

$$b_1\alpha_3 + b_2(\alpha_1^2 + \alpha_2^2) + b_3\alpha_3^2 = 0. \tag{16.24c}$$

Generically $a_3 - a_4 \neq 0$, hence we derive $\alpha_1 = \pm\alpha_2$ from the first two equations. Thereafter, similarly as the cases ii), iii), we obtain

$$\alpha_2 = \pm\left(\frac{-b_3a_1 + (b_1a_5 - b_3a_2)\alpha_3^\pm}{b_3(a_3 + a_4) - 2b_2a_5}\right)^{\frac{1}{2}}, \tag{16.25a}$$

and

$$\alpha_3^\pm = \frac{2b_2a_2 - (a_3+a_4)b_1 \pm \left((2b_2a_2 - (a_3+a_4)b_1)^2 + 8a_1b_2(b_3(a_3+a_4) - 2b_2a_5)\right)^{\frac{1}{2}}}{2[b_3(a_3 + a_4) - 2b_2a_5]}. \tag{16.25b}$$

Here we obtain eight different solutions with $\alpha_1 \cdot \alpha_2 \cdot \alpha_3 \neq 0$. The relation $\alpha_1 = \pm\alpha_2$ indicates that symmetries of the corresponding solution branches are given via those of $\phi_1 \pm \phi_2$ respectively.

16.4 Scales of Solution Branches

Solutions of (16.15) which vanish at $(\beta, \nu) = (0, 0, 0)$ induce solution branches of the original problem (16.5) bifurcating at $(0, \lambda_0, \mu_0)$. Substituting Taylor expansions of the coefficients a_i, b_i at $(\alpha, \beta, \nu) = (0, 0, 0)$ into (16.15) yields scales of components of the solutions. For the determination of qualitative behavior of solution branches it is sufficient to consider leading terms of the coefficients a_i, b_i in the system (16.15). Namely, the coefficients in (16.15) can be replaced as follows

$$\begin{aligned} a_1 &= a_{1\beta}^0\beta + a_{1\nu}^0\nu, & b_1 &= b_{1\beta}^0\beta + b_{1\nu}^0\nu, \\ a_2 &= a_2^0 + a_{2\beta}^0\beta + a_{2\nu}^0\nu, & b_2 &= b_2^0 + b_{2\beta}^0\beta + b_{2\nu}^0\nu, \\ a_i &= a_i^0, \quad i = 3, 4, 5, & b_3 &= b_3^0. \end{aligned} \tag{16.26}$$

Here and in the sequel the super-index ^{0} stands for evaluations of functions at $(\beta, \nu) = (0, 0)$ and the sub-indices $_\beta$, $_\nu$ represent partial derivatives with respect to β and ν, respectively. For examples,

$$a_i^0 := a_i(0,0), \qquad a_{1\beta}^0 := \frac{\partial a_1}{\partial \beta}(0,0), \qquad a_{1\nu}^0 := \frac{\partial a_1}{\partial \nu}(0,0$$

The pure ϕ_1-mode solution in (16.18a) can be rewritten as

$$\alpha_1 = \pm\Big(- \frac{a_{1\beta}^0\beta + a_{1\nu}^0\nu}{a_3^0}\Big)^{1/2}, \qquad \alpha_2 = 0.$$

Depending on signs of the coefficients $a_{1\beta}^0$, $a_{1\nu}^0$ and a_3^0, we adjust the variables β, ν such that

$$\frac{a_{1\beta}^0\beta + a_{1\nu}^0\nu}{a_3^0} \le 0. \tag{16.27}$$

This ensures α_1 to be real and controls stepsizes in the continuation of solution manifolds in two dimensional parameter space.

Similarly, the other solution branches in (16.18) and (16.19) are as

$$\alpha_i = \text{const} \cdot (a_{1\beta}^0\beta + a_{1\beta}^0\nu)^{1/2} + h.o.t., \quad i = 1, 2 \tag{16.28a}$$

and

$$\alpha_3 = -\frac{1}{b_3^0}(b_{1\beta}^0\beta + b_{1\nu}^0\nu) + h.o.t. \tag{16.28b}$$

However, the component α_3 in (16.21b) and (16.25b) is of the scale

$$\alpha_3^+ \sim \begin{cases} \dfrac{a_2^0 b_2^0}{a_3^0 b_3^0 - a_5^0 b_2^0}, & \text{if } a_2^0 b_2^0 \ne 0; \\ (c\alpha_1 + db_1)^{1/2} & \text{if } a_2^0 = 0,\ b_2^0(a_3^0 b_3^0 - a_5^0 b_2^0) \ne 0; \\ c\alpha_1 + db_1 & \text{otherwise,} \end{cases} \tag{16.29a}$$

$$\alpha_3^- \sim c\alpha_1 + db_1. \tag{16.29b}$$

where $c,\ d \in \mathbf{R}$ are constants depending on a_i^0 and b_j^0, $i = 1, \ldots, 5,\ j = 1, 2, 3$. Since only those solutions which vanish at $(\beta, \nu) = (0, 0)$ lead to bifurcating solution branches of the original problem, the solution (16.29a) with $a_2^0 b_2^0 \ne 0$ becomes irrelevant. To ensure real values of the component $\alpha_3^\pm$ in (16.21b) we require that

$$\begin{aligned} &(a_2b_2 - a_3b_1)^2 + 4a_1b_2(a_3b_3 - a_5b_2) \\ &= (a_2^0b_2^0)^2 + [2a_2^0b_2^0(a_2^0b_{2\beta}^0 + a_{2\beta}^0b_2^0 - a_3^0b_{1\beta}^0) + 4a_{1\beta}^0b_2^0(a_3^0b_3^0 - a_5^0b_2^0)]\beta \\ &\quad +[2a_2^0b_2^0(a_2^0b_{2\nu}^0 + a_{2\nu}^0b_2^0 - a_3^0b_{1\nu}^0) + 4a_{1\nu}^0b_2^0(a_3^0b_3^0 - a_5^0b_2^0)]\nu + O(\|(\beta, \nu)\|^2) \\ &\ge 0. \end{aligned}$$

Similar conditions are also necessary for α_2 to be real in (16.21a) and (16.23) as well as (16.25) respectively.

Remark 16.4.1. To trace primary branches with a parameterization in β we can treat the coefficients a_i and b_j as functions of β and fix temporally the parameter ν. Conversely, one may also continue the solution in ν for fixed β and examine how a solution varies with respect to the boundary conditions. The moving frame scheme in Chapter 2 can be used for continuation of solution manifold both in β and ν.

16.5 Secondary Bifurcations

Let us examine the mixed modes (16.21)-(16.25). Under the nondegeneracy conditions (16.16) the component $\alpha_3^{\pm}$ in (16.21b) and (16.25b) does not vanish except at $a_1 = 0$ for which $\alpha_1 = \alpha_2 = 0$ held. In other words, thanks to its D_4-symmetry the ϕ_3-mode penetrates in the solution branches coming along $(0, \lambda_1(\mu), \mu)$. If $b_2 \neq 0$, all nontrivial branches bifurcating at $(0, \lambda_0, \mu_0)$ involve the ϕ_3-mode.

The mixed mode branches in (16.21), (16.23) become pure ϕ_3-mode solution in (16.19) if the following equations are satisfied

$$\frac{a_1 b_3}{b_1 a_5 - b_3 a_2} = -\frac{b_1}{b_3},$$

$$\frac{b_2 a_2 - a_3 b_1 \pm \Big((b_2 a_2 - a_3 b_1)^2 + 4 a_1 b_2 (b_3 a_3 - b_2 a_5)\Big)^{1/2}}{2(b_3 a_3 - b_2 a_5)} = -\frac{b_1}{b_3}.$$

With the coefficients in (16.26) this system can be solved for the unknowns β and ν. Let β_0, ν_0 be such a solution. The corresponding point on the pure ϕ_3-mode solution is a secondary bifurcation point. This can be also seen from the Jacobian of the left hand side of (16.15) with respect to α_1, α_2, α_3 along the ϕ_3-mode solution

$$\begin{pmatrix} a_1 + a_2\alpha_3 + a_5\alpha_3^2 & 0 & 0 \\ 0 & a_1 + a_2\alpha_3 + a_5\alpha_3^2 & 0 \\ 0 & 0 & b_1 + 2b_3\alpha_3 \end{pmatrix}.$$

Since all eigenvalues of this Jacobian are real, only steady state secondary bifurcations are possible. Note that $b_1 + 2b_3\alpha_3 = -b_1 \neq 0$ holds on the pure ϕ_3-mode branch. The condition for a secondary bifurcation is

$$a_1 + a_2\alpha_3 + a_5\alpha_3^2 = 0.$$

This equation coincides to (16.20) and (16.22) for $\alpha_1 = \alpha_2 = 0$.

Similarly, if β_0 and ν_0 are chosen such that the coefficients in (16.26) satisfy

$$\frac{a_1 b_3}{b_1 a_5 - b_3 a_2} = -\frac{b_1}{b_3},$$

$$\frac{2b_2 a_2 - (a_3 + a_4) b_1 \pm \Big((2b_2 a_2 - (a_3 + a_4) b_1)^2 + 8 a_1 b_2 (b_3(a_3 + a_4) - 2b_2 a_5)\Big)^{1/2}}{2[b_3(a_3 + a_4) - 2b_2 a_5]}$$

$$= -\frac{b_1}{b_3},$$

the mixed mode solutions in (16.25) intersect the ϕ_3-mode solution and secondary bifurcation occurs. A generic bifurcation diagram is illustrated in Fig. 16.1.

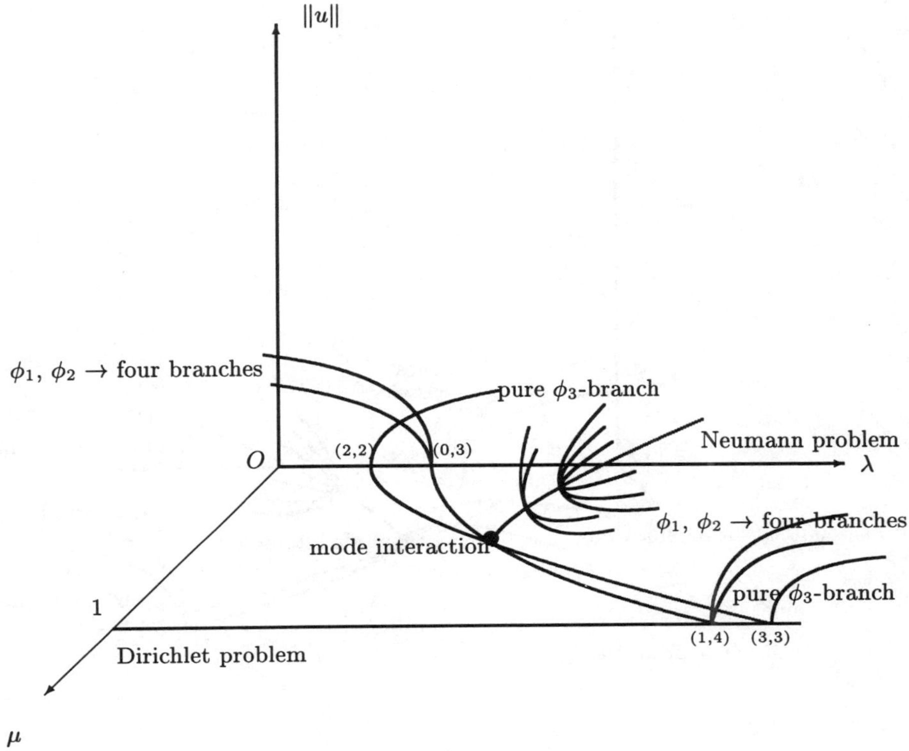

Fig. 16.1. Bifurcation diagram along the curves of bifurcation points $(0, \lambda_i(\mu), \mu)$, $i = 1, 2$, which are homotopies from $(0, 8, 0)$ to $(0, 18, 1)$, and from $(0, 9, 0)$ to $(0, 17, 1)$, respectively. Note that there is an intersection point under the condition $b_2 \neq 0$.

For $b_2 = 0$, there is no secondary bifurcation on the branches corresponding to α_3^+, while on the branches corresponding to α_3^- the conditions for secondary bifurcation reduce to

$$\frac{a_1 b_3}{b_1 a_5 - b_3 a_2} = -\frac{b_1}{b_3}.$$

As indicated Subsection 16.3.1, the nontrivial solution branches coming along $(0, \lambda_1(\mu), \mu)$ are preserved at $(0, \lambda_0, \mu_0)$, but they interact with the ϕ_3-mode via secondary bifurcations on the ϕ_3-branch, see Fig. 16.2 for a generic bifurcation diagram.

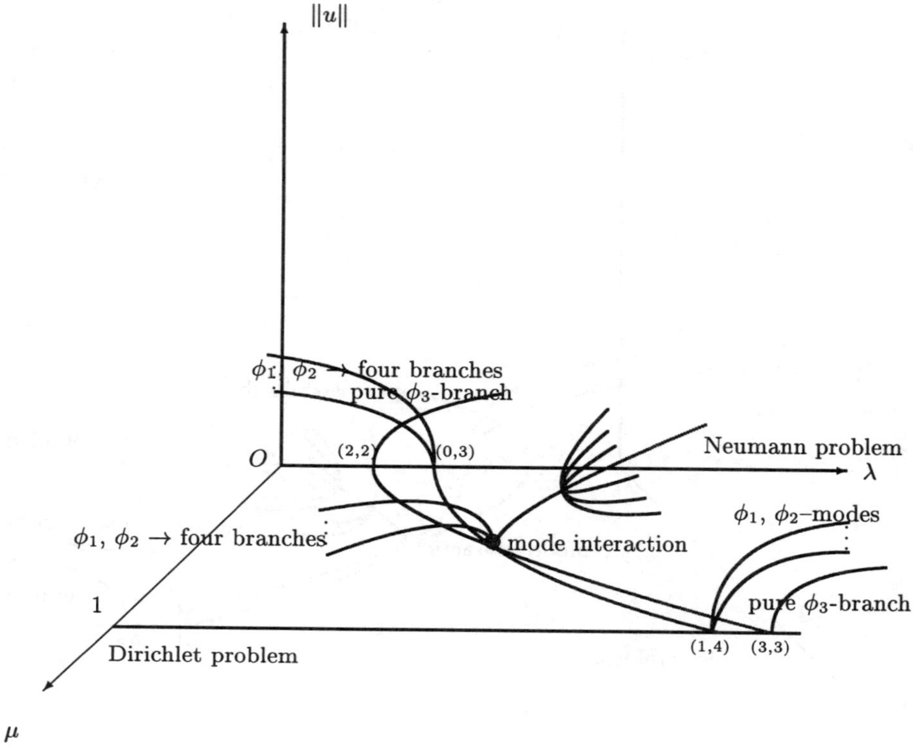

Fig. 16.2. Same situation as in Fig. 16.2 except $b_2 = 0$. The solutions coming along $(0, \lambda_i(\mu), \mu)$, $i = 1, 2$ persist at mode interaction and new solutions occurs as secondary branches.

16.5.1 Secondary Hopf bifurcations

Hopf bifurcations occur generically in systems of reaction diffusion equations. For scalar equations (16.1), if the nonlinearity involves the gradient ∇u such as transportation terms, or nonlinear boundary conditions are imposed, then the linearized operator may have complex eigenvalues, leading to Hopf bifurcation as parameters cross critical values.

The truncated bifurcation equations (16.15) are derived merely by symmetry properties and can origin from systems of reaction diffusion equations. We consider here briefly possibilities of secondary Hopf bifurcations. A Hopf bifurcation may occur on branches of mixed modes with either $\alpha_2 \cdot \alpha_3 \neq 0$ or $\alpha_1 \cdot \alpha_3 \neq 0$. As an example, let us consider the branch in (16.21) with $\alpha_1 = 0$, $\alpha_2 \cdot \alpha_3 \neq 0$. The Jacobian of the left hand side of (16.15) along this branch is

$$\begin{pmatrix} a_1 + a_2\alpha_3 + a_4\alpha_2^2 + a_5\alpha_3^2 & 0 & 0 \\ 0 & a_1 + a_2\alpha_3 + 3a_3\alpha_2^2 + a_5\alpha_3^2 & 2a_5\alpha_2\alpha_3 \\ 0 & 2b_2\alpha_2 & b_1 + 2b_3\alpha_3 \end{pmatrix}$$

$$= \begin{pmatrix} (a_4 - a_3)\alpha_2^2 & 0 & 0 \\ 0 & 2a_3\alpha_2^2 & 2a_5\alpha_2\alpha_3 \\ 0 & 2b_2\alpha_2 & b_1 + 2b_3\alpha_3 \end{pmatrix}.$$

Hopf bifurcation occurs at the points where this Jacobian has pure imaginary eigenvalues. This corresponds to a vanishing trace of the 2×2 submatrix, while its determinant is positive, i.e.,

$$b_1 + 2a_3\alpha_2^2 + 2b_3\alpha_3 = 0,$$
$$2[a_3b_1 + 2(a_3b_3 - a_5b_2)\alpha_3]\alpha_2^2 > 0.$$

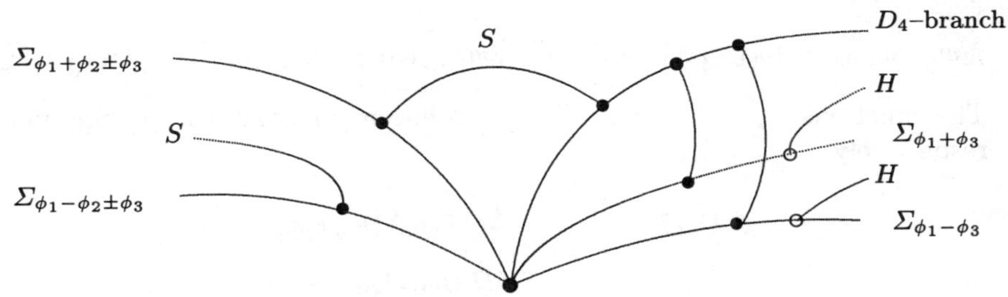

Fig. 16.3. Generic bifurcation diagram at the mode interaction

16.6 Calculation of Truncated Bifurcation Equations

To calculate the coefficients a_i, b_j of (16.15), we substitute the operator G (16.5) into the reduced bifurcation equations (16.9). Denote

$$\alpha \cdot \phi := \sum_{i=1}^{3} \alpha_i\phi_i, \qquad w := w(\alpha, \beta, \nu)$$

and

$$j_3B(\alpha, \beta, \nu) = \begin{pmatrix} B_1(\alpha, \beta, \nu) \\ B_2(\alpha, \beta, \nu) \\ B_3(\alpha, \beta, \nu) \end{pmatrix}.$$

We have

$$
\begin{aligned}
&B_i(\alpha,\beta,\nu)\\
&= \alpha_j - j_3\Big\langle \phi_i,\ T(\mu_0+\nu)\Big[(\lambda_0+\beta+1)(\alpha\cdot\phi+w) + f(\alpha\cdot\phi+w,\lambda_0+\beta)\Big]\Big\rangle\\
&= -\frac{1}{\lambda_0+1}\Big[\beta\alpha_j + \Big\langle \phi_i,\ j_3 f(\alpha\cdot\phi+w,\lambda_0+\beta)\Big\rangle\Big]\\
&\quad -\Big\langle \phi_i,\ T'(\mu_0)\Big[(\lambda_0+\beta+1)(\alpha\cdot\phi+j_2w)+j_2 f(\alpha\cdot\phi+w,\lambda_0+\beta)\Big]\Big\rangle\cdot\nu+O(\nu^2)\\
&= -\frac{1}{\lambda_0+1}\Big[\beta\alpha_j + \Big\langle \phi_i,\ \frac{1}{2}D^2 f_0(\alpha\cdot\phi,\beta)^2\\
&\quad +D_uDf_0(\alpha\cdot\phi,\beta)(j_2w) + \frac{1}{6}D_{uuu}f_0(\alpha\cdot\phi)^3) + \frac{1}{2}D_{uu\lambda}f_0(\alpha\cdot\phi)^2\beta\Big\rangle\Big]\\
&\quad -\Big\langle \phi_i,\ T'(\mu_0)[(\lambda_0+\beta+1)(\alpha\cdot\phi+j_2w) + \frac{1}{2}D^2 f_0(\alpha\cdot\phi,\beta)^2]\Big\rangle\cdot\nu+O(\nu^2)\\
&= 0.
\end{aligned}
$$

Consider the 2-jet of $w(\alpha,\beta,\nu)$

$$j_2w = w_{2,0,0}\alpha_1^2+w_{0,2,0}\alpha_2^2+w_{0,0,2}\alpha_3^2+2w_{1,1,0}\alpha_1\alpha_2+2w_{1,0,1}\alpha_1\alpha_3+2w_{0,1,1}\alpha_2\alpha_3.$$

The functions $w_{i,j,k} \in \mathrm{Im}(D_uG_0)$ are solutions of the following equations respectively

$$
\begin{aligned}
D_uG_0w_{2,0,0} &= -\frac{1}{2}QT(\mu_0)D_{uu}f_0\phi_1^2,\\
D_uG_0w_{1,1,0} &= -QT(\mu_0)D_{uu}f_0\phi_1\phi_2,\\
D_uG_0w_{0,2,0} &= -\frac{1}{2}QT(\mu_0)D_{uu}f_0\phi_2^2,\\
D_uG_0w_{1,0,1} &= -QT(\mu_0)D_{uu}f_0\phi_1\phi_3,\\
D_uG_0w_{0,0,2} &= -\frac{1}{2}QT(\mu_0)D_{uu}f_0\phi_3^2.
\end{aligned}
$$

Since the operator D_uG_0 is singular and has rank-3 deficiency, we solve the generic equation

$$D_uG_0w = Qg, \qquad w\in \mathrm{Im}(D_uG_0)$$

via the enlarged system

$$
\begin{aligned}
D_uG_0w + \sum_{i=1}^{3} c_i\phi_i &= g,\\
\langle \phi_i,\ w\rangle = 0,\quad & i=1,2,3.
\end{aligned}
$$

The component w in the solutions of both problems are identical. However, the enlarged system is nonsingular and can be solved approximately with the standard numerical methods. In particular, if D_uG_0 is coercive, so is the linear operator of the enlarged system. An example of discretization of the enlarged system is

$$\begin{pmatrix} D_uG_0^h & \phi_1^h & \phi_2^h & \phi_3 \\ {\phi_1}^h & 0 & 0 & 0 \\ {\phi_2}^h & 0 & 0 & 0 \\ {\phi_3}^h & 0 & 0 & 0 \end{pmatrix} \begin{pmatrix} w \\ c_1 \\ c_2 \\ c_3 \end{pmatrix} = \begin{pmatrix} g \\ 0 \\ 0 \\ 0 \end{pmatrix}. \tag{16.30}$$

Theorem 16.6.1. *(Bai/Mei [31]) The solution w of (16.30) can be calculated as*

$$w = \tilde{w} - \sum_{i=1}^{3} ({\phi_i^h}^*, \tilde{w})\phi_i^h,$$

where $\tilde{w}$ is the unique solution of the nonsingular system

$$\left(D_uG_0^h + C_1 e_j e_j^T + C_2 e_k e_k^T + C_3 e_l e_l^T\right)\tilde{w} = g,$$

and (j,k,l) is chosen in $ID \cap ID^*$ with

$$ID := \{(i_1,i_2,i_3) \in \mathbf{N}^3;\ \text{vectors } (\phi^h_{j,i_1}, \phi^h_{ji_2}, \phi^h_{ji_3}), j = 1,2,3 \text{ are independent}\}$$

$$ID^* := \{(i_1,i_2,i_3) \in \mathbf{N}^3;\ \text{vectors } ({\phi^h_{ji_1}}^*, {\phi^h_{j,i_2}}^*, {\phi^h_{ji_3}}^*), j = 1,2,3 \text{ are indep.}\}.$$

Here $C_i \in \mathbf{R},\ i = 1,2,3$ are arbitrary nonvanishing constants.

The main advantage of this method is that it avoids changes in computer codes for solving nonsingular problems, which allows an easy exploitation of efficient solvers for sparse and band matrices in many available software packages.

If approximations of w_{ijk} and ϕ_i are known, we take into account the fact $D^l f(0,\lambda) = 0,\ l = 0,1,\ldots$ and obtain easily the coefficients $a_i,\ b_j$ of (16.15) as follows

$$\frac{\partial a_1}{\partial \beta}(0,0) = -\frac{1}{\lambda_0+1}, \qquad \frac{\partial a_1}{\partial \nu}(0,0) = -(\lambda_0+1)\langle \phi_1,\ T'(\mu_0)\phi_1\rangle,$$

$$\frac{\partial b_1}{\partial \beta}(0,0) = -\frac{1}{\lambda_0+1}, \qquad \frac{\partial b_1}{\partial \nu}(0,0) = -(\lambda_0+1)\langle \phi_3,\ T'(\mu_0)\phi_3\rangle,$$

$$a_2(0,0) = -\frac{1}{\lambda_0+1}\Big\langle \phi_1,\ D_{uu}f_0\phi_1\phi_3\Big\rangle,$$

$$\frac{\partial a_2}{\partial \beta}(0,0) = -\frac{1}{\lambda_0+1}\Big\langle \phi_1,\ D_{u\lambda}f_0 w_{1,0,1} + D_{uu\lambda}f_0\phi_1\phi_3\Big\rangle,$$

$$\frac{\partial a_2}{\partial \nu}(0,0) = -\Big\langle \phi_1,\ T'(\mu_0)[(\lambda_0+1)w_{1,0,1} + D_{uu}f_0\phi_1\phi_3]\Big\rangle,$$

$$a_3(0,0) = -\frac{1}{\lambda_0+1}\Big\langle \phi_1,\ \frac{1}{6}D_{uuu}f_0\phi_1^3 + \frac{1}{2}D_{uu}f_0\phi_1 w_{2,0,0}\Big\rangle,$$

$$a_4(0,0) = -\frac{1}{\lambda_0+1}\Big\langle \phi_1,\ \frac{1}{2}D_{uuu}f_0\phi_1\phi_2^2 + \frac{1}{2}D_{uu}f_0(\phi_1 w_{0,2,0} + \phi_2 w_{1,1,0})\Big\rangle,$$

$$a_5(0,0) = -\frac{1}{\lambda_0+1}\Big\langle \phi_1,\ \frac{1}{2}D_{uuu}f_0\phi_1\phi_3^2 + \frac{1}{2}D_{uu}f_0(\phi_1 w_{0,0,2} + \phi_3 w_{1,0,1})\Big\rangle,$$

$$
\begin{aligned}
b_2(0,0) &= -\frac{1}{\lambda_0+1}\Big\langle \phi_3,\ \frac{1}{2}D_{uu}f_0\phi_1^2\Big\rangle,\\
\frac{\partial b_2}{\partial \beta}(0,0) &= -\frac{1}{\lambda_0+1}\Big\langle \phi_3,\ D_{u\lambda}f_0 w_{0,0,2} + \frac{1}{2}D_{uu\lambda}f_0\phi_3^2\Big\rangle,\\
\frac{\partial b_2}{\partial \nu}(0,0) &= -\Big\langle \phi_3,\ T'(\mu_0)[(\lambda_0+1)w_{2,0,0} + \frac{1}{2}D_{uu}f_0\phi_1^2]\Big\rangle,\\
b_3(0,0) &= -\frac{1}{\lambda_0+1}\Big\langle \phi_3,\ \frac{1}{2}D_{uu}f_0\phi_3^2\Big\rangle.
\end{aligned}
$$

16.6.1 Derivatives with respect to homotopy parameter

The operator $T'(\mu)$ is self-adjoint. Hence

$$\langle \phi_j,\ T'(\mu_0)v\rangle = \langle T'(\mu_0)\phi_j,\ v\rangle \quad \text{for all} \quad v \in X.$$

Let

$$\tilde{\phi}_j := T'(\mu_0)\phi_j, \qquad h_0 := \frac{d}{d\mu}\left(\frac{g_0(\mu)}{g_1(\mu)}\right)\Big|_{\mu=\mu_0}.$$

Similarly as (14.33), we see that $\tilde{\phi}_j$ is the weak solution of

$$
\begin{aligned}
\Delta u - u = 0 &\qquad \text{in } \Omega,\\
\frac{\partial u}{\partial n} + \frac{g_0(\mu)}{g_1(\mu)}u = -\frac{d}{d\mu}\left(\frac{g_0(\mu)}{g_1(\mu)}\right)u(\mu) &\quad \text{on } \partial\Omega.
\end{aligned}
$$

at μ_0. Since $T(\mu_0)\phi_j = \phi_j/(\lambda_0+1)$, we have $a(\phi_k, v, \mu_0) = (\lambda_0+1)\langle \phi_k, v\rangle$ for all $v \in X$ and

$$a(\tilde{\phi}_j,\ v,\ \mu_0) = -\frac{h_0}{(\lambda_0+1)}\int_{\partial\Omega}\phi_j v\ ds \qquad \text{for all} \quad v \in H^1(\Omega).$$

Here $-a(\cdot,\cdot,\mu)$ is the bilinear form defined in (14.27). Thereafter,

$$
\begin{aligned}
\langle \phi_k,\ T'(\mu_0)\phi_j\rangle &= \langle \tilde{\phi}_j,\ \phi_k\rangle\\
&= \frac{1}{\lambda_0+1}a(\tilde{\phi}_j,\ \phi_k,\ \mu_0)\\
&= -\frac{h_0}{(\lambda_0+1)}\int_{\partial\Omega}\phi_j\phi_k\ ds.
\end{aligned}
\tag{16.31}
$$

Similarly, let $v = j_2 w$ and recall that

$$D_u G_0(j_2 w) = -\frac{1}{2}QD_{uu}f_0\Big(\sum_{i,j=1}^{3}\alpha_i\alpha_j\phi_i\phi_j\Big).$$

We obtain

$$
\begin{aligned}
&-\Big\langle \phi_k,\ \frac{1}{2}T'(\mu_0)D_{uu}f_0\Big(\sum_{i,j=1}^{3}\alpha_i\alpha_j\phi_i\phi_j\Big)\Big\rangle \\
&= -\frac{1}{2}\Big\langle \tilde{\phi}_k,\ D_{uu}f_0\Big(\sum_{i,j=1}^{3}\alpha_i\alpha_j\phi_i\phi_j\Big)\Big\rangle \\
&= a(\tilde{\phi}_k, j_2w, \mu) \qquad\qquad (16.32)\\
&= -\frac{h_0}{(\lambda_0+1)}\int_{\partial\Omega}\phi_k(j_2w)\ ds.
\end{aligned}
$$

By the statement (14.33) we have

$$
T'(\mu_0)(j_2w) = T(\mu_0)\tilde{w},
$$

where $\tilde{w}$ is defined via

$$
\begin{aligned}
\langle \tilde{w},\ v\rangle &= -h_0\int_{\partial\Omega} T(\mu_0)(j_2w)\cdot v\ ds \\
&= -\frac{h_0}{(\lambda_0+1)}\int_{\partial\Omega}\Big[j_2w + \frac{1}{2}D_{uu}f_0\Big(\sum_{i,j=1}^{3}\alpha_i\alpha_j\phi_i\phi_j\Big)\Big]v\ ds.
\end{aligned}
$$

Hence, for $k = 1, 2, 3$, we have

$$
\begin{aligned}
\langle \phi_k,\ T'(\mu_0)(j_2w)\rangle &= \langle \phi_k, T(\mu_0)\tilde{w}\rangle \\
&= \frac{1}{\lambda_0+1}\langle \phi_k, \tilde{w}\rangle \qquad\qquad (16.33)\\
&= -\frac{h_0}{(\lambda_0+1)^2}\int_{\partial\Omega}\Big[j_2w + \frac{1}{2}D_{uu}f_0\Big(\sum_{i,j=1}^{3}\alpha_i\alpha_j\phi_i\phi_j\Big)\Big]\phi_k\ ds.
\end{aligned}
$$

Substituting (16.31) and (16.32) into (16.33) we derive

$$
\begin{aligned}
\frac{\partial a_1}{\partial \nu}(0,0) &= \frac{h_0}{(\lambda_0+1)}\int_{\partial\Omega}\phi_1^2\ ds, \\
\frac{\partial b_1}{\partial \nu}(0,0) &= \frac{h_0}{(\lambda_0+1)}\int_{\partial\Omega}\phi_3^2\ ds, \\
\frac{\partial a_2}{\partial \nu}(0,0) &= \frac{h_0}{(\lambda_0+1)}\int_{\partial\Omega}D_{uu}f_0\phi_1^2\phi_3\ ds, \\
\frac{\partial b_2}{\partial \nu}(0,0) &= \frac{h_0}{(\lambda_0+1)}\int_{\partial\Omega}\frac{1}{2}D_{uu}f_0\phi_1^2\phi_3\ ds = \frac{1}{2}\frac{\partial a_2}{\partial \nu}(0,0).
\end{aligned}
$$

These quantities are also well defined for $\mu = 1$, since on the boundary $\partial\Omega$ the eigenfunction ϕ_i contains the factor $g_1(\mu)$, see the definition of ϕ in (16.2) and the defining equations (14.43) for the wave number κ. For example, at $x = \pi$ we have

$$
\begin{aligned}
\phi_1 &= C(g_0(\mu)\sin(\kappa_1\pi) + g_1(\mu)\kappa_1\cos(\kappa_1\pi))(g_0(\mu)\sin(\kappa_2 y) + g_1(\mu)\kappa_2\cos(\kappa_2 y)) \\
&= \pm C g_1(\mu)\Big(g_0(\mu)\sin(\kappa_2 y) + g_1(\mu)\kappa_2\cos(\kappa_2 y)\Big) \qquad \text{(due to (14.43))}.
\end{aligned}
$$

16.7 Reduced Stability of Bifurcating Solution Branches

We study in this section stabilities of the stationary bifurcating solution branches via the concepts of reduced stability in Kielhöfer/Lauterbach [187].

Since the two curves of bifurcation points $(0, \lambda_i(\mu), \mu)$ are already beyond the first eigenvalue of the Laplacian, all bifurcating branches along $(0, \lambda_i(\mu), \mu)$, $i = 1, 2$ are unstable in the sense of linear stability for the scalar reaction diffusion equation (16.1a). Nevertheless, on one side, the reduced stability becomes relevant for the cases that the first few unstable modes do not contribute to the solutions, e.g. if the problem is restricted to the subdomains or forced to have certain symmetries and nodal line properties, etc. In fact, scales of solutions vary constantly as a process of self-organizations, there is a continuous change of stabilities in reactions. On the other side, systems of reaction diffusion equations with multiple parameters often invlove several modes in forming stable spatial patterns, see e.g. [129, pp.312] and [293]. Every mode can become critical for appropriate parameters. For these cases the concept of reduced stability is very useful. For example, let us consider again the Brusselator equations

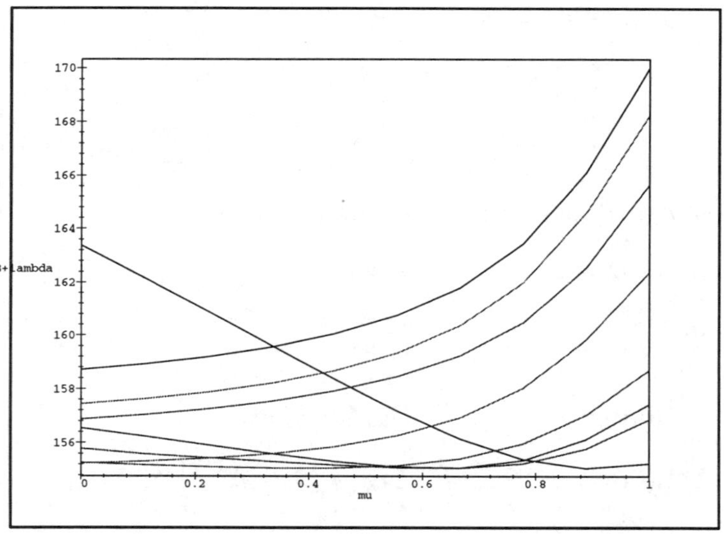

Fig. 16.4. Graph of critical B with respect to μ from 0 to 1. Different curves correspond to different eigenvalues $\kappa(\mu)$ of the Laplacian $-\Delta$.

$$\frac{\partial u_1}{\partial t} = \Delta u_1 + A - (B + \lambda + 1)u_1 + u_1^2 u_2$$
$$\text{in} \quad \Omega = [0,\pi] \times [0,\pi], \qquad (16.34)$$
$$\frac{\partial u_2}{\partial t} = d\Delta u_2 + (B + \lambda)u_1 - u_1^2 u_2$$

with the boundary conditions

$$g_0(\mu)\begin{pmatrix} u_1 \\ u_2 \end{pmatrix} + g_1(\mu)\begin{pmatrix} \frac{\partial u_1}{\partial n} \\ \frac{\partial u_2}{\partial n} \end{pmatrix} = g_0(\mu)\begin{pmatrix} A \\ (B+\lambda)/A \end{pmatrix} \quad \text{on} \quad \partial\Omega. \qquad (16.35)$$

We fix A, B and d and vary λ and μ as bifurcation parameters. Linear stability of the homogeneous state

$$(u_1, u_2) = \left(A, \frac{B+\lambda}{A}\right) \qquad (16.36)$$

with respect to the parameters λ, μ can be discussed similarly as in Section 10.3, where λ and the diffusion rate d are treated as parameter. In Figure 16.4 we depict the first few curves of steady state and Hopf bifurcation points in the parameter space $(B,\ \mu)$. The envelope of these curves describes a change of stability of the homogeneous solution. Intersection points of the curves on the envelope lead to generic mode interactions.

16.7.1 Stability of solution branches at $(0, \lambda_1(\mu), \mu)$

As we have seen in Section 16.3.1, for $\mu \neq \mu_0$ bifurcating solution branches on $(0, \lambda_1(\mu), \mu)$ are described by the system (16.17). Denote

$$B_3(\alpha_1, \alpha_2, \beta, \nu) := \begin{pmatrix} \left(a_1 + a_3\alpha_1^2 + a_4\alpha_2^2\right)\alpha_1 \\ \left(a_1 + a_3\alpha_2^2 + a_4\alpha_1^2\right)\alpha_2 \end{pmatrix} = 0.$$

On its solutions in (16.18) we calculate derivatives of B_3 with respect to α_1, α_2 and examine its eigenvalues.

a) For the pure mode solutions $(\alpha_1, \alpha_2) = (\pm\sqrt{-a_1/a_3},\ 0)$, and $(0,\ \pm\sqrt{-a_1/a_3})$ we have

$$\frac{\partial B_3}{\partial(\alpha_1, \alpha_2)}(\alpha_1, \alpha_2, \beta, \nu) = \begin{pmatrix} -2a_1 & 0 \\ 0 & a_1 - \dfrac{a_1 a_4}{a_3} \end{pmatrix}.$$

If $a_1 > 0$ (implying $a_3 < 0$ for real solutions) and $a_1/a_3(a_3 - a_4) < 0$, then the pure mode branches are stable. Otherwise, they are unstable.

b) For the mixed mode solutions $(\alpha_1, \alpha_2) = \sqrt{-a_1/(a_3 + a_4)}(\pm 1,\ \pm 1)$, we have

$$\frac{\partial B_3}{\partial(\alpha_1, \alpha_2)} = \begin{pmatrix} a_3 & \pm a_4 \\ \pm a_4 & a_3 \end{pmatrix}.$$

Its eigenvalues are

$$-\frac{2a_1}{a_3 + a_4}(a_3 \pm a_4), \qquad -\frac{2a_1}{a_3 + a_4}(a_3 \mp a_4),$$

which determine the stability of the mixed mode solutions.

16.7.2 Stability of solution branches at $(0, \lambda_2(\mu), \mu)$

The reduced bifurcation equation along $(0, \lambda_2(\mu), \mu)$ is two and three determined for for $\mu \neq 0$ and $\mu = 0$, respectively. Correspondingly, we truncate the Taylor expansion of the reduced bifurcation equation at the quadratic and cubic orders, respectively

$$B_2(\alpha_3, \beta, \nu) := b_1\alpha_3 + b_3\alpha_3^2 \qquad \text{for } \mu \neq 0$$

and

$$B_3(\alpha_3, \beta, \nu) := b_1\alpha_3 + b_4\alpha_3^3 \qquad \text{for } \mu = 0.$$

At the nontrivial solution branch $\alpha_3 = -b_1/b_3$ for $\mu \neq 0$, and $\alpha_3 = \sqrt{-b_1/b_4}$ for $\mu = 0$, we have

$$\frac{\partial B_2}{\partial \alpha_3} = -b_1 \qquad \text{for } \mu \neq 0$$

and

$$\frac{\partial B_3}{\partial \alpha_3} = -2b_1 \qquad \text{for } \mu = 0.$$

Hence, if $b_1 > 0$, the corresponding solution is stable.

16.7.3 Stability of solution branches at mode interaction

Consider the mode interaction at the point $\mu = \mu_0$ and the truncated Taylor expansion (16.15) of the reduced bifurcation equations. We define

$$B_3(\alpha_1, \alpha_2, \alpha_3, \beta, \nu) := \begin{pmatrix} (a_1 + a_2\alpha_3 + a_3\alpha_1^2 + a_4\alpha_2^2 + a_5\alpha_3^2)\,\alpha_1 \\ (a_1 + a_2\alpha_3 + a_3\alpha_2^2 + a_4\alpha_1^2 + a_5\alpha_3^2)\,\alpha_2 \\ b_1\alpha_3 + b_2(\alpha_1^2 + \alpha_2^2) + b_3\alpha_3^2 \end{pmatrix}.$$

To study the reduced stabilities of the bifurcating solution branches in (16.18), (16.19), (16.21), (16.23) and (16.25), respectively, we examine the derivatives of B_3. For example, for (16.19) we have $(\alpha_1, \alpha_2, \alpha_3) = (0, 0, -b_1/b_3)$ and

$$\frac{\partial B_3}{\partial(\alpha_1, \alpha_2, \alpha_3)} = \begin{pmatrix} a_1 & 0 & 0 \\ 0 & a_1 & 0 \\ 0 & 0 & -b_1 \end{pmatrix}.$$

Obviously this branch is stable if $a_1 < 0$ and $b_1 > 0$.

If $b_2 = 0$, then bifurcating solution branches along $(0, \lambda_1(\mu), \mu)$ persist at $\mu = \mu_0$, see (16.18). Stabilities of these solutions are determined by eigenvalues of the matrices

$$\frac{\partial B_3}{\partial(\alpha_1, \alpha_2, \alpha_3)} = \begin{pmatrix} -\frac{2a_1}{a_3+a_4} & 0 & 0 \\ 0 & -\frac{2a_1}{a_3+a_4} & 0 \\ 0 & 0 & 1 \end{pmatrix} \quad \text{and} \quad \begin{pmatrix} a_3 & \pm a_4 & 0 \\ \pm a_4 & a_3 & 0 \\ 0 & 0 & b_1 \end{pmatrix},$$

respectively.

On the solutions in (16.21) we have

$$\frac{\partial B_3}{\partial(\alpha_1, \alpha_2, \alpha_3)} = \begin{pmatrix} a_1 & 0 & 0 \\ 0 & a_1 + a_2\alpha_3 + 3a_3\alpha_2 + a_5\alpha_3 & a_2\alpha_2 + 2a_5\alpha_2\alpha_3 \\ 0 & 2b_2\alpha_2 & b_1 + 2b_3\alpha_3 \end{pmatrix}.$$

Thus stability of the solution is determined by the sign of a_1 and those of the eigenvalues of the matrix

$$\begin{pmatrix} a_1 + a_2\alpha_3 + 3a_3\alpha_2 + a_5\alpha_3 & a_2\alpha_2 + 2a_5\alpha_2\alpha_3 \\ 2b_2\alpha_2 & b_1 + 2b_3\alpha_3 \end{pmatrix}.$$

Stability of other mixed mode solutions in (16.23) and (16.25) can be analyzed similarly.

16.8 A Numerical Example

We consider the equation

$$\Delta u + \lambda u - \xi(u^2 + p(x,y)u^3) = 0 \quad \text{in } \Omega = [0,\pi] \times [0,\pi], \tag{16.37}$$

with ξ as an additional parameter and $p(x,y)$ a D_4-equivariant function. In the numerical computations we set

$$\xi = 1, \quad p(x,y) = -\frac{4}{3}. \tag{16.38}$$

16.8.1 Solution branches along $(0, \lambda_1(\mu), \mu)$

On the homotopy $(0, \lambda_1(\mu), \mu)$ with $\mu \neq \mu_0$, by discussions in Chapter 15 bifurcation scenario of (16.37) is described by the following truncation of the reduced bifurcation equations

$$\left(\frac{\partial a_1}{\partial \beta}\beta + a_3\alpha_1^2 + a_4\alpha_2^2\right)\alpha_1 = 0,$$
$$\left(\frac{\partial a_1}{\partial \beta}\beta + a_3\alpha_2^2 + a_4\alpha_1^2\right)\alpha_2 = 0.$$

Table 16.1 shows values of the coefficients a_1, a_3, a_4 along the homotopy. Mixed modes solution branches

$$\left(\pm\left(\frac{\partial a_1}{\partial \beta}\frac{\beta}{a_3 + a4}\right)^{1/2}, \ \pm\left(\frac{\partial a_1}{\partial \beta}\frac{\beta}{a_3 + a4}\right)^{1/2}\right)$$

bifurcate to the right along the whole homotopy, while the pure ϕ_1-mode branch

$$\left(\pm\left(\frac{\partial a_1}{\partial \beta}\frac{\beta}{a_3}\right)^{1/2},\ 0\right)$$

and the pure ϕ_2-mode branch

$$\left(0,\ \pm\left(\frac{\partial a_1}{\partial \beta}\frac{\beta}{a_3}\right)^{1/2}\right)$$

change directions along the homotopy.

μ	$\kappa_1(\mu)$	$\kappa_2(\mu)$	$\frac{\partial a_1}{\partial \beta}$	a_3	a_4	$a_3 + a_4$
0	.000	3.00	-0.10000	0.003189	-0.068674	-0.065483
.1	.258	3.02	-0.097966	0.002756	-0.067203	-0.064447
.2	.375	3.05	-0.095646	0.002427	-0.065562	-0.063135
.3	.470	3.08	-0.092973	0.002079	-0.064131	-0.062052
.4	.556	3.13	-0.089858	0.001726	-0.062598	-0.060872
.5	.638	3.19	-0.086177	0.001499	-0.060336	-0.058837
.6	.716	3.27	-0.081771	0.003676	-0.054445	-0.050769
.7	.792	3.38	-0.076452	-0.004030	-0.056769	-0.060799
.8	.864	3.54	-0.070072	-0.004109	-0.050205	-0.054314
.9	.934	3.75	-0.062791	-0.005335	-0.045280	-0.050615
1.	1.00	4.00	-0.055556	-0.006773	-0.042546	-0.049318

Table 16.1. The coefficients a_3, a_4 on the homotopy path from (0,9,0) to (0,17,1).

16.8.2 Solution branches along $(0, \lambda_2(\mu), \mu)$

Along the homotopy $(0, \lambda_2(\mu), \mu)$ from $(0, 8, 0)$ to $(0, 18, 1)$ the kernel $\mathrm{Ker}(D_uG(0, \lambda_2(\mu), \mu))$ is one dimensional and the bifurcation is simple for all $\mu \neq \mu_0$. For $\mu \neq \mu_0$ and $\mu \neq 0$, the 2-jet of the reduced bifurcation equation is

$$\frac{\partial b_1}{\partial \beta}\beta\alpha_3 + b_3\alpha_3^2 = 0$$

Here $\frac{\partial b_1}{\partial \beta} = -1/(\lambda_2(\mu) + 1)$ and

$$b_3 = -\frac{1}{\lambda_2(\mu) + 1}\langle\phi_3,\ \frac{1}{2}D_{uu}f_0\phi_3^2\rangle.$$

Table 16.2 summarizes the values of these coefficients along the homotopy. There exists exactly one solution branch passing through the trivial solution curve $u \equiv 0$. The bifurcation is transcritical for $\mu \neq 0$. If we fix μ and use λ to parameterize the solution branch, it is of the form

$$u(t) = t\alpha_3(t)\phi_3 + t^2 w(t), \qquad w \perp \phi_3$$
$$\lambda(t) = \lambda_2(\mu) + t, \qquad \alpha_3(0) = -\frac{1}{b_3}\frac{\partial b_1}{\partial \beta}.$$

μ	λ_2	$\frac{\partial b_1}{\partial \beta}$	b_3
0	8.0	-.1111	0.00
0.1	8.2802	-.1077	.0003788
0.2	8.6219	-.1039	.001581
0.3	9.0449	-.09955	.003647
0.4	9.5770	-.09454	.006465
0.5	10.258	-.08883	.009693
0.6	11.143	-0.08235	.01269
0.7	12.306	-.07515	.01466
0.8	13.831	-.06742	.01507
0.9	15.764	-.05965	.01430
1.0	18.00	-.05263	.01341

Table 16.2. The coefficient b_3 on the homotopy path $(u(\mu), \lambda(\mu), \mu)$ from (0,8,0) to (0,18,1).

For $\mu = 0$ we consider the 3-jet of the reduced bifurcation equation

$$\frac{\partial b_1}{\partial \beta}\beta\alpha_3 + b_4\alpha_3^3 = 0,$$

where $b_4 = -\xi/(\lambda_2(0) + 1)\langle\phi_3,\ p(x, y)\phi_3^3 - 5\phi_3 w_2\rangle = -0.01062$ with

$$w_2 = \frac{5}{\pi^2}\left[\frac{1}{4} - \frac{1}{12}\cos(4x) - \frac{1}{12}\cos(4y) - \frac{1}{28}\cos(4x)\cos(4y)\right]$$

as the solution of the equation

$$D_u G(0, \lambda_3(0), 0)w_2 = 5T(0)\phi_3^2.$$

The bifurcation is pitchfork and the bifurcating solution branch is of the form

$$u(t) = t\alpha_3(t)\phi_3 + t^2 w(t), \quad w \perp \phi_3$$

$$\lambda(t) = \lambda_2(\mu) - t^2, \quad \alpha_3(0) = \sqrt{\frac{1}{b_4}\frac{\partial b_1}{\partial \beta}} = 3.234569.$$

16.8.3 Mode interaction

At $\mu = \mu_0$, we have

$$\lambda_1(\mu_0) = 0.7380507^2 + 3.3014006^2 = 2 * 2.392066550^2 = \lambda_2(\mu_0),$$

i.e., the curve $(0, \lambda_1(\mu_0), \mu_0)$ and $(0, \lambda_2(\mu_0), \mu_0)$ intersect and yield a mode interaction. Since bifurcation scenario of (16.37) at $(0, \lambda_1(\mu_0), \mu_0)$ is described by equations (16.15), we calculate coefficients in (16.15) as follows

$$
\begin{aligned}
a_1(0,0) &= 0, \quad \frac{\partial a_1}{\partial \beta}(0,0) = -\frac{1}{\lambda_0+1} = -0.0804, \quad \frac{\partial a_1}{\partial \nu}(0,0) = 0.6345; \\
b_1(0,0) &= 0, \quad \frac{\partial b_1}{\partial \beta}(0,0) = -\frac{1}{\lambda_0+1} = -0.0804, \quad \frac{\partial b_1}{\partial \nu}(0,0) = 0.8780; \\
a_2(0,0) &= \frac{10\xi}{\lambda_0+1}(\phi_1,\ \phi_1\phi_3) = -0.01007; \\
a_3(0,0) &= -\frac{\xi}{\lambda_0+1}(\phi_1,\ p(x,y)\phi_1^3 - 5\phi_1 w_{2,0,0}) = -0.000078; \\
a_4(0,0) &= -\frac{\xi}{\lambda_0+1}(\phi_1,\ 3p(x,y)\phi_1\phi_2^2 - 5\phi_1 w_{0,2,0} - 5\phi_2 w_{1,1,0}) = -0.05549; \\
a_5(0,0) &= -\frac{\xi}{\lambda_0+1}(\phi_1,\ 3p(x,y)\phi_1\phi_3^2 - 5\phi_1 w_{0,0,2} + \phi_3 w_{1,0,1}) = 0.01848; \\
b_2(0,0) &= -\frac{2\xi}{\lambda_0+1}(\phi_3,\ \phi_1^2) = 0.005039; \\
\frac{\partial b_2}{\partial \beta}(0,0) &= 0, \quad \frac{\partial b_2}{\partial \nu}(0,0) = 0.06163; \\
b_3(0,0) &= -\frac{5\xi}{\lambda_0+1}(\phi_3,\ \phi_3^2) = 0.01345.
\end{aligned}
$$

Multiplying the equations in (16.15) by 100, we obtain

$$
\begin{aligned}
(-8.04\beta + 63.45\nu - 1.007\alpha_3 - 0.0078\alpha_1^2 - 5.549\alpha_2^2 + 1.848\alpha_3^2)\alpha_1 &= 0, \\
(-8.04\beta + 63.45\nu - 1.007\alpha_3 - 5.549\alpha_1^2 - 0.0078\alpha_2^2 + 1.848\alpha_3^2)\alpha_2 &= 0, \\
(-8.04\beta + 87.80\nu)\alpha_3 + (0.5039 + 6.163\nu)(\alpha_1^2 + \alpha_2^2) + 1.345\alpha_3^2 &= 0.
\end{aligned}
$$

By the discussions in Section 16.3 these equations have the following nontrivial solutions

1. Pure ϕ_3-mode:

$$(\alpha_1, \alpha_2, \alpha_3) = (0,\ 0,\ -5.9778\beta + 65.279\nu).$$

2. Mixed ϕ_1- and ϕ_3- modes:
   - Primary branches with $\alpha_2 = 0$ and

$$
\begin{aligned}
\alpha_1 &= \pm\Big(-.27 * 10^{-7}\nu - 2.96 * 10^2\beta^2 + 5.08 * 10^3\nu\beta - 2.16 * 10^4\nu^2\Big)^{1/2}, \\
\alpha_3 &= 63.00893748\nu - 7.984111222\beta + O(\|(\beta,\nu)\|^2);
\end{aligned}
$$

   - Secondary branches with $\alpha_2 = 0$ and

$$
\begin{aligned}
\alpha_1 &= \Big(-.7750007155 + 98.71855269\nu - 14.56062400\beta\Big)^{1/2}, \\
\alpha_3 &= .5388428056 - 63.66275672\nu + 8.050705806\beta.
\end{aligned}
$$

3. Mixed ϕ_2- and ϕ_3- modes:

- Primary branches with $\alpha_1 = 0$ and

$$\alpha_2 = \pm\Big(- .27 * 10^{-7}\nu - 2.96 * 10^2\beta^2 + 5.08 * 10^3\nu\beta - 2.16 * 10^4\nu^2\Big)^{1/2},$$
$$\alpha_3 = 63.00893748\nu - 7.984111222\beta + O(\|(\beta,\nu)\|^2);$$

- Secondary branches with $\alpha_1 = 0$ and

$$\alpha_2 = \Big(- .7750007155 + 98.71855269\nu - 14.56062400\beta\Big)^{1/2},$$
$$\alpha_3 = .5388428056 - 63.66275672\nu + 8.050705806\beta.$$

4. Mixed ϕ_1-, ϕ_2- and ϕ_3-modes (here $\alpha_1 = \pm\alpha_2$):
   - Primary branches:

$$\alpha_2 = \pm(-.59 * 10^{-9}\beta - 1.49 * 10^2\beta^2 + 2.54 * 10^3\nu\beta - 1.08 * 10^4\nu^2)^{1/2} + O(\|(\beta,\nu)\|),$$
$$\alpha_3 = 63.00893744\,\nu - 7.984111219\,\beta + O(\|(\beta,\nu)\|^2);$$

   - Secondary branches

$$\alpha_2 = \pm\,(-.0157690 + 23.857263\,\nu - 2.8377137\,\beta)^{1/2} + O(\|(\beta,\nu)\|);$$
$$\alpha_3 = .10869975 - 114.20162\,\nu + 12.769371\,\beta + O(\|(\beta,\nu)\|^2).$$

16.8.4 Switching and continuation of solution branches

We are interested in how a solution branch bifurcates at the mode interaction point and how it extends to a solution of the Dirichlet problem. By varying the homotopy parameter μ from 0.6284333 to 1 we examine variation of solutions and bifurcation scenario. We proceed path following and branch switching of bifurcating solutions as follows.

1. Based on the above analysis on existence of bifurcating solution manifolds, we determine ranges of parameters μ, β.
2. We choose an appropriate path in the parameter space, e.g. $\beta = 0$ or $\nu = 0$ or β as a function of ν and parametrize the solution curve on the manifold along this path with an enlarged system as in Chapter 4. By doing so we obtain one-parameter problem and continue the bifurcating solution branch with the enlarged problem.
3. Alternatively, we can use the moving frame scheme in Chapter 4 to continue the bifurcating in both λ and μ directions. We adjust stepsizes in μ-direction, i.e., homotopy of boundary conditions, towards Dirichlet problem. In λ-direction, i.e., the bifurcation parameter, we use arclength for the parametrization of solution branches. The stepsize for λ is adjusted implicitly according to the error estimates and evaluation of the defects.

4. When the homotopy in μ arrives the Dirichlet (resp. Neumann) boundary conditions, we check how the solution is incorporated with the bifurcation scenario of the Dirichlet boundary conditions.

Figures 16.5–16.9 show some solution branches on the homotopy from $\mu = 0.6284333$ to $\mu = 1$. Figure 16.6 is a secondary branch at the mode interaction and merges to a primary branch of Dirichlet problem when μ approaches 1, while the other four branches are primary branches at mode interactions.

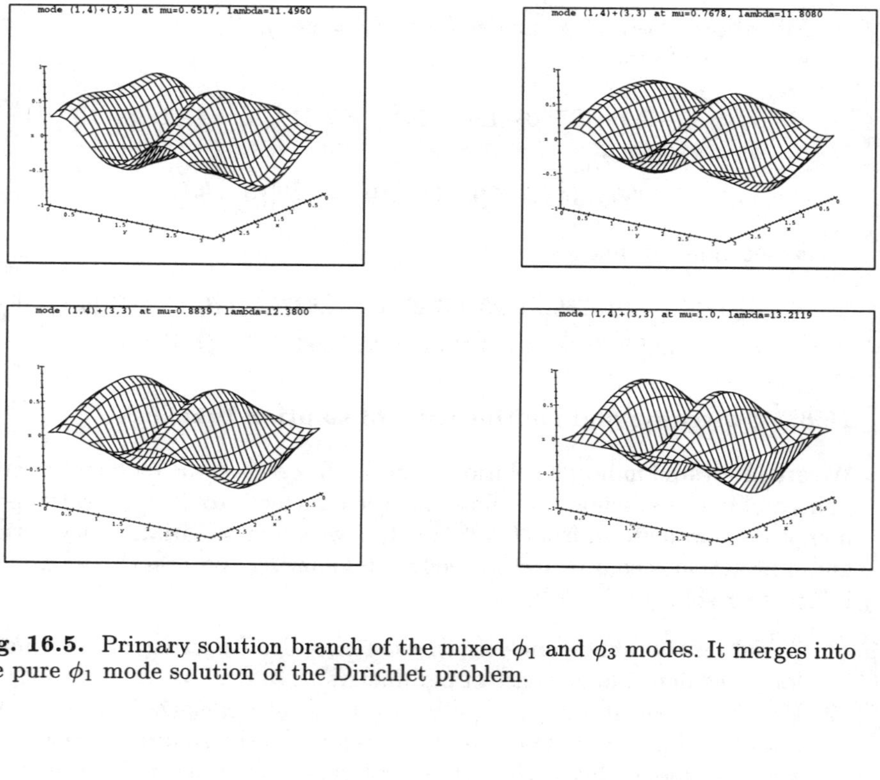

Fig. 16.5. Primary solution branch of the mixed ϕ_1 and ϕ_3 modes. It merges into the pure ϕ_1 mode solution of the Dirichlet problem.

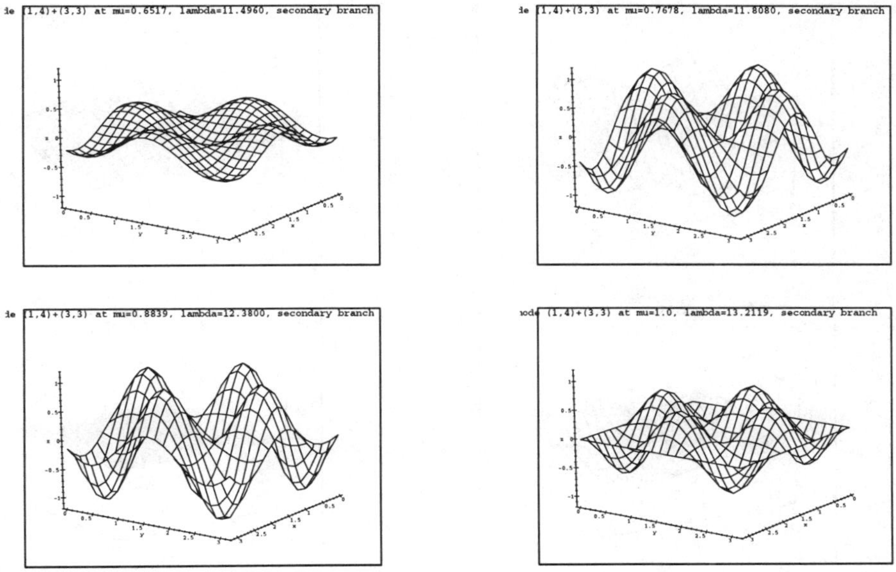

Fig. 16.6. Secondary solution branch of the mixed ϕ_1 and ϕ_3 modes. It connects with the mixed ϕ_1 and ϕ_2 mode solution of the Dirichlet problem.

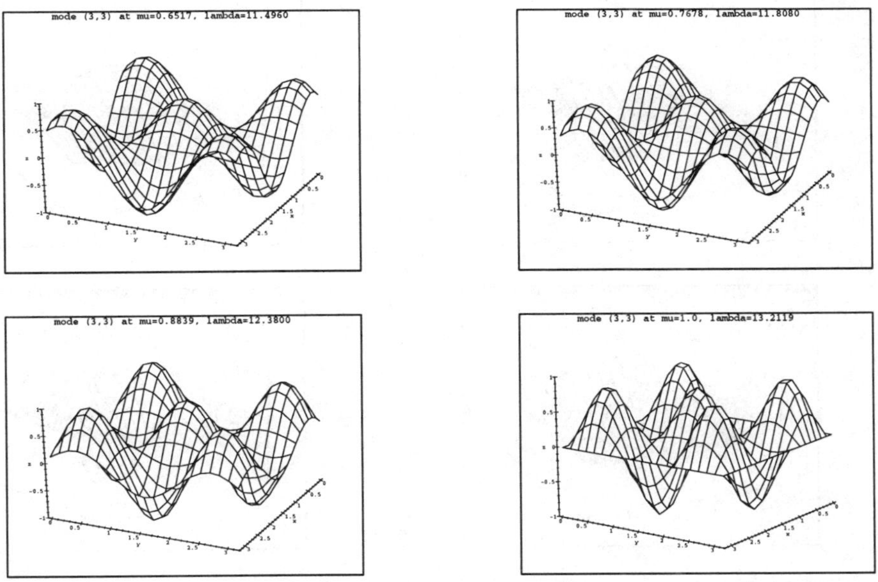

Fig. 16.7. Primary solution branch of the pure ϕ_3 modes. This branch persists along the homotopy of boundary conditions.

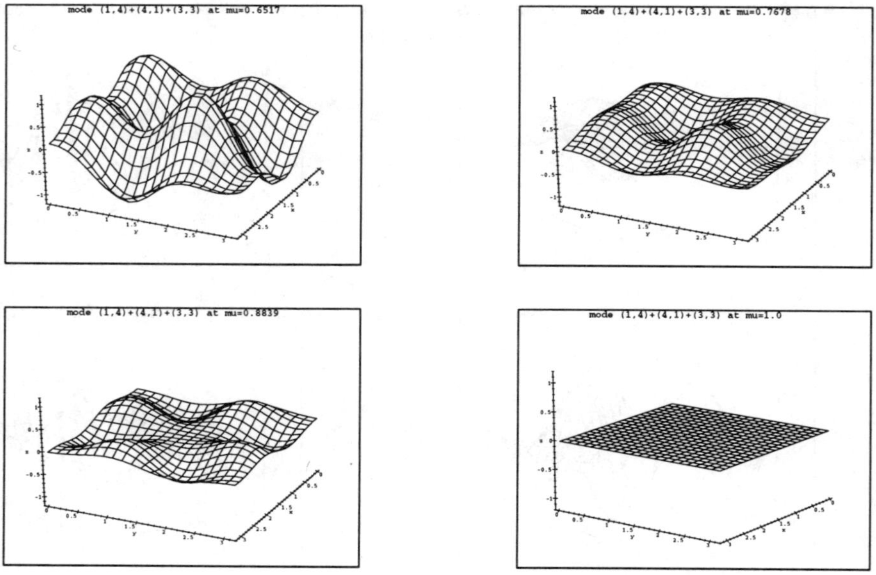

Fig. 16.8. Primary solution branch of the mixed ϕ_1, ϕ_2 and ϕ_3 modes. I

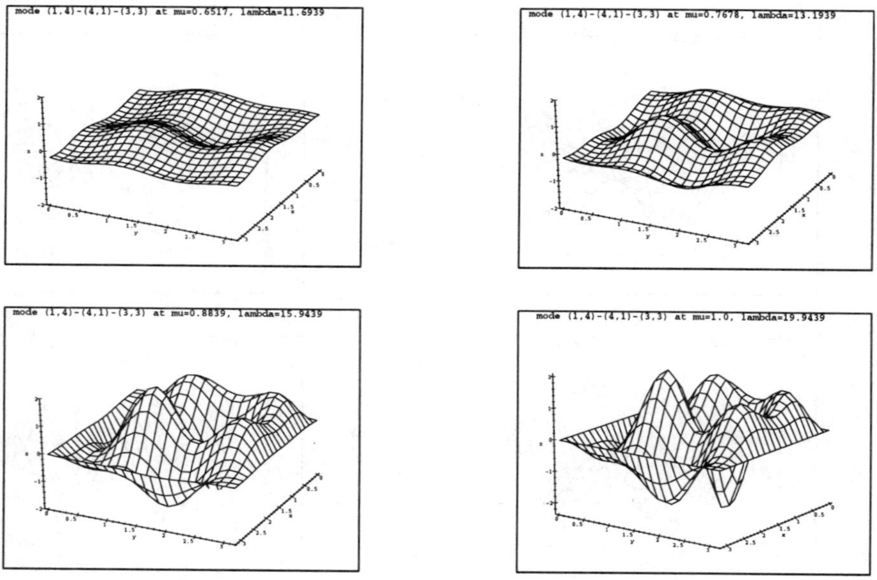

Fig. 16.9. Primary solution branch of the mixed ϕ_1, $-\phi_2$ and $-\phi_3$ modes. It connects with the mixed ϕ_1 and $-\phi_2$ mode solution of the Dirichlet problem.

List of Figures

List of Tables

Bibliography

1. J. Alexander and G. Auchmuty: Bifurcation analysis of reaction-diffusion equations, VI: Multiply periodic travelling waves. *SIAM J. Math. Anal.*, 19:100–109, 1988.
2. E.L. Allgower, P. Ashwin, K. Böhmer and Z. Mei: Liapunov-Schmidt reduction for a bifurcation problem with periodic boundary conditions on a square domain. In: E.L. Allgower, K. Georg and R. Miranda (Eds.), *Exploiting Symmetry in Applied and Numerical Analysis*, Lectures in Applied Mathematics 29, pp. 11–22. AMS, Providence, 1993.
3. E.L. Allgower and Ph. H. Schmidt: An algorithm for piecewise-linear approximation of an implicitly defined manifold. *SIAM J. Numer. Anal.*, 22:322–346, 1985.
4. E.L. Allgower and K. Böhmer: Resolving singular nonlinear equations. *Rocky Mountain J. Math.*, 18:225–268, 1988.
5. E.L. Allgower, K. Böhmer, K. Georg and R. Miranda: Exploiting symmetry in boundary element methods. *SIAM J. Numer. Anal.*, 29:534–552, 1992.
6. E.L. Allgower, K. Böhmer and M. Golubitsky: *Bifurcations and Symmetry: Cross Influences between Mathematics and Applications*, ISNM 104, Birkhäuser Verlag, Basel, Boston, 1992.
7. E.L. Allgower, K. Böhmer, A. Hoy and V. Janovsky: On a class of direct methods for solving nonlinear equations. Submitted to *ZAMM*, 1998.
8. E.L. Allgower, K. Böhmer and Z. Mei: A generalized equibranching lemma with application to $D_4 \times Z_2$ symmetric elliptic problems, Technical Report **9**, Department of Mathematics, University of Marburg 1990.
9. E.L. Allgower, K. Böhmer and Z. Mei: A complete bifurcation scenario for the 2d-nonlinear Laplacian with Neumann boundary conditions on the unit square. In: R. Seydel, F.W. Schneider, T. Küpper and H. Troger (Eds.), *Bifurcation and Chaos: Analysis, Algorithms, Applications*, ISNM 87, pp. 1–18. Birkhäuser Verlag, Basel, 1991.
10. E.L. Allgower, K. Böhmer and Z. Mei: On a problem decomposition for semilinear nearly symmetric elliptic problems. In: W. Hackbusch (Ed.), *Parallel Algorithms for Partial Differential Equations*, pp. 1–17. Vieweg Verlag, Braunschweig, 1991.
11. E.L. Allgower, K. Böhmer and Z. Mei: An extended equivariant branching theory. *Math. Meth. in Appl. Sci.*, 15:433–451, 1992.
12. E.L. Allgower, K. Böhmer and Z. Mei: Branch switching at a corank-4 bifurcation point of semi-linear elliptic problems with symmetry. *IMA J. Numer. Anal.*, 14:161–182, 1994.
13. E.L. Allgower and C.S. Chien: Continuation and local perturbation for multiple bifurcations. *SIAM J. Sci. Stat. Comput.*, 7:1265–1281, 1986.
14. E.L. Allgower and K. Georg: *Numerical Continuation Methods: An Introduction.* Springer-Verlag, Berlin, Heidelberg, New York, 1990.

15. E.L. Allgower, K. Georg and R. Miranda: *Exploiting Symmetry in Applied and Numerical Analysis*, Lectures in Applied Mathematics 29. AMS, Providence, 1993.
16. H. Amann: Nonhomogeneous linear and quasilinear elliptic and parabolic boundary value problems. In: H.J. Schmeisser, H. Triebel (Eds.) *Function Spaces, Differential Operators and Nonlinear Analysis*, Teubner-Texte Math. 133, pp. 9-126. B. G. Teubner Verlag, 1993.
17. W.E. Arnoldi: The principle of minimized iterations in the solution of the matrix eigenvalue problem. *Quart. Appl. Math.*, 9:17–29, 1951.
18. U.M. Ascher and R.J. Spiteri: Collocation software for boundary value differential-algebraic equations. *SIAM J. Sci. Comput.*, 15:938–952, 1994.
19. P. Ashwin: Steady state mode interaction in the Kuramoto-Sivashinky equation on a rectangle. Dissertation, Math. Institute, University of Warwick, 1991.
20. P. Ashwin: High corank mode interactions on a rectangle. In: E.L. Allgower, K. Böhmer and M. Golubitsky (Eds.), *Bifurcation and Symmetry*, ISNM 104, pp. 23–33. Birkhäuser, Basel, 1992.
21. P. Ashwin, K. Böhmer and Z. Mei: A numerical Liapunov-Schmidt method for finitely determined problems. In: E.L. Allgower, K. Georg and R. Miranda (Eds.), *Exploiting Symmetry in Applied and Numerical Analysis*, Lectures in Applied Mathematics 29, pp. 49–69. AMS, Providence, 1993.
22. P. Ashwin, K. Böhmer and Z. Mei: A numerical Liapunov-Schmidt method with applications to Hopf bifurcation on a square. *Math. Comp.*, 64:649–670 and S19–S22, 1995.
23. P. Ashwin, K. Böhmer and Z. Mei: Forced symmetry breaking of homoclinic cycles in a PDE with $O(2)$-symmetry. *J. Comp. & Appl. Math.*, 70:297–310, 1996.
24. P. Ashwin and Z. Mei: Liapunov-Schmidt reduction at Hopf bifurcation of the Brusselator equations on a square. Technical report, University of Warwick, 1992.
25. P. Ashwin and Z. Mei: A Hopf bifurcation with Robin boundary conditions. *J. Dynamics & Differential Equations*, 6:487–503, 1994.
26. P. Ashwin and Z. Mei: A numerical bifurcation function for homoclinic orbits. *SIAM J. Numer. Anal.*, 35:2055-2069, 1998.
27. P. Ashwin and Z. Mei: Approximations of a bifurcation function for homoclinic orbits of large systems. Technical report, University of Marburg, 1997.
28. P.J. Aston: Local and global aspects of the $(1, n)$ mode interaction for capillary-gravity waves. *Physica D*, 52:415–428, 1991.
29. Y. Bai: A perturbed collocation method for boundary-value probelms in differential-algebraic equations. *Appl. Math. & Comp.*, 45:269–291, 1991.
30. Y. Bai: A modified Lobatto collocation method for linear boundary value probelms in differential-algebraic equations. *Computing*, 49:139–150, 1992.
31. Y. Bai and Z. Mei: Efficient calculation of nullspace of a singular matrix. Submitted, 1999.
32. R. Barrett *et al*: *Templates for th Solution of Linear Systems: Building Blocks for Iterative Methods.* SIAM, Philadelphia, 1993.
33. D.L. Barrow: The slow wave solution to the FitzHugh-Nagumo equations. *Appl. Anal.*, 30:137–163, 1988.
34. B.P. Belousov: An oscillating reaction and its mechanism. *Sborn. referat. radiat. med. Medgiz*, Moscow (Collection of Abstracts on Radiation Medicine), 1959.
35. M.S. Berger: *Nonlinearity and Functional Analysis.* Academic Press, New York, 1977.

36. W.J. Beyn: Defining equations for singular solutions and numerical applications. In: T. Küpper, H. D. Mittelmann and H. Weber (Eds.), *Numerical Methods for Bifurcation Problems*, ISNM 70, pp. 42–56. Birkhäuser Verlag, Basel, 1984.
37. W.J. Beyn: Global bifurcations and their numerical computation. In: D. Roose, B. De Dier and A Spence (Eds.), *Continuation and Bifurcations: Numerical Techniques and Application*, NATO Adv. Sci. Inst. Math. Phys. Sci. C313, pp. 169–181. Kluwer Acad. Publ., Dordrecht, 1990.
38. W.J. Beyn: The numerical computation of connecting orbits in dynamical systems. *IMA J. Numer. Anal.*, 10:379–405, 1990.
39. W.J. Beyn: Numerical analysis of homoclinic orbits emanating from a Takens-Bogdanov Point. *IMA J. Numer. Anal.*, 14:381–410, 1994.
40. G. Birkhoff and G.C. Rota: *Ordinary Differential Equations.* John Wiley & Sons, New York, 1978.
41. K. Böhmer: On a numerical Lyapunov-Schmidt method for operator equations. *Computing*, 51:185–207, 1993.
42. K. Böhmer and Z. Mei: On a numerical Lyapunov-Schmidt method. In: E.L. Allgower and K.Georg (Eds.), *Computational Solutions of Nonlinear Systems of Equations.* Lectures in Applied Mathematics 26, pp. 79–88. AMS, Providence, 1990.
43. K. Böhmer and Z. Mei: Regularization and computation of a bifurcation problem with corank-2. *Computing*, 41:307–316, 1989.
44. K. Böhmer and Z. Mei: Branch switching and path following of a simplified buckling problem at a corank-4 bifurcation point. *ZAMM*, 71:T748–T751, 1991.
45. K. Böhmer and Z. Mei: Mode interactions of an elliptic system on the square. In: E.L. Allgower, K. Böhmer and M. Golubitsky (Eds.), *Bifurcation and Symmetry*, ISNM 104, pp. 49–58. Birkhäuser Verlag, Basel, 1992.
46. A. Bossavit: Symmetry, group, and boundary value problems–A progressive introduction to noncommutative harmonic analysis of partial differential equations in domains with geometrical symmetry. *Comput. Meths in Appl. Mech. Engrg*, 56:167–215, 1986.
47. G. Bouligand: A une etape decisive de l'algebre: L'oeuvre scientifique et l'oeuvre didactique d'Etienne Bezout. *Rev. Gen. Sci. Pures. Appl. N. S.*, 55:121–123, 1948.
48. F. Brezzi, I. Rappaz and P.A. Raviart: Finite dimensional approximation of nonlinear problems, Part I: Branches of nonsingular solutions. *Numer. Math.*, 36:1–25, 1980.
49. F. Brezzi, I. Rappaz and P.A. Raviart: Finite dimensional approximation of nonlinear problems, Part III: Simple bifurcation points. *Numer. Math.*, 38:1–30, 1981.
50. P. Brunovsky and B. Fiedler: Connecting orbits in scalar reaction diffusion equations II. The complete solution. *J. Differential Equations*, 81:106–135, 1989.
51. V. Canale: The computation of paths of homoclinic orbits. Ph.D. Thesis, Department of Computer Science, University of Toronto, Canada, 1994.
52. J. Carr: *Applications of Centre Manifold Theory.* App. Math. Sci. 35, Springer-Verlag, New York, 1981.
53. S. Castro: *Symmetry and bifurcation of periodic solutions in Neumann boundary value problems.* MSC Thesis, Mathematics Institute, University of Warwick, 1990.
54. N. Chafee and E. Infante: A bifurcation problem for a nonlinear parabolic equation. *J. Applicable Anal.*, 4:17–37, 1974

55. A.R. Champneys and Y.A. Kuznetsov: Numerical detection and continuation of codimension-two homoclinic bifurcations. *Int. J. Bifur. Chaos Appl. Sci. Engrg.*, 4:795–822, 1944.
56. A.R. Champneys, Yu. A. Kuznetsov and B. Sandstede: A numerical toolbox for homoclinic bifurcation analysis. Report 1.95, Dept. of Engineering Maths., University of Bristol, 1995.
57. T.F. Chan: Techniques for large sparse systems arising from continuation methods. In: T. Küpper, H.D. Mittelmann and H. Weber (Eds.), *Numerical Methods for Bifurcation Problems*, ISNM 70, pp. 116–128. Birkhäuser Verlag, Basel, 1984.
58. T.F. Chan: Deflation techniques and block elimination algorithms for solving bordered singular systems. *SIAM J. Sci. Stat. Comput*, 5:121–134, 1984.
59. C.S. Chien and M.S. Chen: Multiple bifurcation in the von Kármán equations. *SIAM J. Sci. Comput.*, 18:1737-1766, 1997.
60. C.S. Chien, W.W. Lin and Z. Mei: Conjugate gradient type methods for semilinear elliptic problems with symmetry. *Comput. Math. App.*, 37:3-22, 1999.
61. C.S. Chien, Z.L. Weng and C.L. Shen: Lanczos type methods for continuation problems. *J. Numer. Lin. Alg. with Appl.*, 4:23-41, 1997.
62. P. Chossat: Forced symmetry breaking of an $O(2)$-symmetric homoclinic cycle. *Nonlinearity*, 6:723–731, 1993.
63. P. Chossat and M. Golubitsky: Hopf bifurcation in the presence of symmetry, center manifold and Liapunov-Schmidt reduction. In: W.F. Langford, F.V. Atkinson and A.B. Mingarelli (Eds.), *Oscillation, Bifurcation and Chaos*, CMS Conf. Proc. 8, pp. 343–352. AMS, Providence, 1987.
64. S.N. Chow and K. Lu: Invariant manifolds for flows in Banach spaces. *J. Differential Equations*, 74:285–317, 1988.
65. K.W. Chu, W. Govaerts and A. Spence: Matrices with rank deficiency two in eigenvalue problems and dynamical systems. *SIAM J. Numer. Anal.*, 31:524–539, 1994
66. G. Cicogna: A nonlinear version of the equivariant bifurcation lemma. *J. Phys. A: Math. Gen.*, 23:600–602, 1981.
67. G. Cicogna: Symmetry breakdown from bifurcation. *Lett. Nuovo Cimento*, 31:600–602, 1981.
68. D.S. Cohen, F.C. Hoppensteadt and R.M. Miura: Slowly-modulated oscillations in nonlinear diffusion process. *SIAM J. Appl. Math.*, 33:217–229, 1977.
69. E.D. Conway and J.A. Smoller: Diffusion and the classical ecological interactions: asymptotics. In: W.E. Fitzgibbon and H.F. Walker (Eds.), *Nonlinear Diffusion*, Research Notes in Math. 14, pp. 53–69. Pitman, London, 1977.
70. P. Coullet and E. Spiegel: Amplitude equations for systems with competing instabilities. *SIAM J. Appl. Math.*, 43:774–819, 1983.
71. M.D. Crandall and P.H. Rabinowitz: Bifurcations from simple eigenvalues. *J. Func. Anal.*, 8:321–340, 1971.
72. M.D. Crandall and P.H. Rabinowitz: The Hopf bifurcation theorem in infinite dimension. *Arch. Rat. Mech. Anal.*, 67:53–72, 1977.
73. J.D. Crawford: Normal forms for driven surface waves. *Physica D*, 52:429–457, 1991.
74. J.D. Crawford: $D_4 \dot{+} \mathbf{T}^2$ mode interactions with hidden rotation symmetry. *Nonlinearity*, 7:697–740, 1993.
75. J.D. Crawford: D_4-symmetric maps with hidden Euclidean symmetry. In: J. Chadam, M. Golubitsky, W. Langford and B. Wetton (Eds.), *Pattern Formation: Symmetry Methods and Applications*, pp. 93–124. AMS, Providence, 1996.
76. J.D. Crawford, J.P. Gollub and D. Lane: Hidden symmetries of parametrically forced waves. *Nonlinearity*, 6:119–164, 1993.

77. J.D. Crawford, M. Golubitsky, E. Knobloch M.G.M. Gomes and I.N. Stewart: Boundary conditions as symmetry constraints. In: M. Roberts and I. Stewart (Eds.) *Singularity Theory and its Applications*, pp. 63–79. Lecture Notes in Math. 1463, Springer-Verlag, New York, 1991.
78. R.-X. Dai and W.C. Rheinboldt: On the computation of manifolds of fold points for parameter-dependent problems. *SIAM J. Numer. Anal.*, 27:437–446, 1990.
79. G. Dangelmayr: Degenerate bifurcations near a double eigenvalue in the Brusselator. *J. Austral. Math. Soc. Ser. B.*, 28:486–535, 1987.
80. F.R. de Hoog and R. Weiss: An approximation theory for boundary value problems on infinite intervals. *Computing*, 24:227–239, 1980.
81. D.W. Decker and H.B. Keller: Multiple limit point bifurcation. *J. Math. Anal. App.*, 75:417–430, 1980.
82. D.W. Decker and H.B. Keller: Path following near bifurcation. *Comm. on Pure. and Appl. Math.*, 34:149–175, 1981.
83. M. Dellnitz, M. Golubitsky, A. Hohmann and I. Stewart: Spirals in scalar reaction-diffusion equations. *Internat. J. Bifur. Chaos Appl. Sci. Engrg.*, 5:1487–1501, 1995.
84. M. Dellnitz and B. Werner: Computational methods for bifurcation problems with symmetry–with special attention to steady state and Hopf bifurcation points. *J. Comp. Appl. Math.*, 26:97-123, 1989.
85. P. Deuflhard, B. Fiedler and P. Kunkel: Efficient numerical path-following beyond critical points. *SIAM J. Numer. Anal.*, 24:912–927, 1987.
86. R. Dillon, P.K. Maini and H.G. Othmer: Pattern formation in generalized Turing systems, I. steady-state patterns in systems with mixed boundary conditions. *J. Math. Biol.*, 32:345–393, 1994.
87. E.J. Doedel: AUTO, a program for the automatic bifurcation analysis of autonomous systems. *Cong. Numer.*, 30:265–384, 1981.
88. E.J. Doedel, A.R. Champneys, T.F. Fairgrieve, Y.A. Kuznetsov, B. Sandstede and X.J. Wang: *AUTO97: Continuation and Bifurcation Software for Ordinary Differential Equations.* Available by FTP from ftp.cs.concordia.ca in directory pub/doedel/auto, 1997.
89. E.J. Doedel and M.J. Friedman: Numerical computation of heteroclinic orbits. *J. Comput. Appl. Math.*, 26:155–170, 1989.
90. E.J. Doedel, M.J. Friedman and B.I. Kunin: Successive continuation for locating connecting orbits. *Numer. Algorithms*, 14:103–124, 1997.
91. E.J. Doedel, M.J. Friedman and A.C. Monteiro: On locating connecting orbits. *Appl. Math. Comput.* , 65:231–239, 1994.
92. E.J. Doedel, H.B. Keller and J.P. Kernévez: Numerical analysis and control of bifurcation problems, part I: bifurcation in finite dimensions. *Int. J. Bifur. and Chaos*, 1:493-520, 1991.
93. E.J. Doedel, H.B. Keller and J.P. Kernévez: Numerical analysis and control of bifurcation problems, part I: bifurcation in finite dimensions. *Int. J. Bifur. and Chaos*, 1:745-772, 1991.
94. E.J. Doedel and J.P. Kernévez: AUTO. Software for continuation and bifurcation problems in ordinary differential equations. Applied Mathematics Technical Report, CALTECH, Pasadena, 1986.
95. C.C. Douglas and J. Mandel: An abstract theory for the domain reduction method. *Computing*, 43:75–96, 1992.
96. K. Duncan and J.C. Eilbeck: Numerical studies of symmetry-breaking bifurcations in reaction-diffusion systems. In: L.M. Ricciardi (Ed.), *Biomathematics and Related Computational Problems*, Naples, 1987, pp. 439–448. Kluwer Acad. Publ, Dordrecht, 1988.

97. J.C. Eilbeck and J.E. Furter: Understanding steady-state bifurcation diagrams for a model reaction-diffusion system. In: D. Roose, B. De Dier and A. Spence (Eds.), *Numerical Techniques and Applications*, NATO ASI Series C313, pp. 25–42. Kluwer Academic Publishers, 1990.
98. J. Elezgaray and A. Arneodo: Modeling reaction-diffusion pattern formation in the Couette flow reactor. *J. Chem. Phys.*, 95:323–350, 1991.
99. C. Elphick, E. Tirapegni, M.E. Brachet, P. Coullet and G. Iooss: A simple characterisation for normal forms of singular vector fields. *Physica D*, 29:95–127, 1987.
100. B. Fiedler: Global Hopf bifurcation in porous catalysts. In: *Equadiff 82*, Lecture Notes in Math. 1017, pp. 177–184. Berlin, New York, 1983.
101. B. Fiedler: *Global bifurcations of periodic solutions with symmetry*, Lecture Notes in Math. 1309, Springer-Verlag, Berlin, 1988.
102. B. Fiedler: Do global attractors depend on boundary conditions? *Doc. Math. J. DMV.*, 1:215–228, 1996.
103. B. Fiedler and J. Mallet-Paret: A Poincare-Bendixson theorem for scalar reaction diffusion equations. *Arch. Rational Mech. Anal.*, 107:325–345, 1989.
104. M. Field and M. Golubitsky: *Symmetry in Chaos*. University Press, Oxford, 1992.
105. P.C. Fife: Asymptotic states for equations of reaction and diffusion. *Bull. Amer. Math. Soci.*, 84:693–726, 1978.
106. G. Fischer: *Ebene algebraische Kurven*. Vieweg Verlag, Brauschweig, 1994.
107. C. Foias, B. Nicolaenko, G.R. Sell and R. Temam: Inertial manifolds for the Kuramoto-Sivashinsky equations and an estimate of their lowest dimension. *J. Math. pures et appl.*, 67:197–226, 1988.
108. R.W. Freund, G.H. Golub and N.M. Nachtigal: Iterative solution of linear systems. *Acta Numerica*, pp. 57–100. 1992.
109. R.W. Freund and N.M. Nachtigal: QMR : A quasi minimal residual method for non-hermitian linear systems. *Numer. Math.*, 60:315–339, 1991.
110. M.J. Friedman: Numerical analysis and accurate computation of heteroclinic orbits in the case of center manifolds. *J. Dynamics Differential Equations.*, 5:59–87, 1993.
111. M.J. Friedman and E.J. Doedel: Numerical computation and continuation of invariant manifolds connecting fixed points. *SIAM J. Numer. Anal.*, 28:789–808, 1991.
112. M.J. Friedman and E.J. Doedel: Computational methods for global analysis of homoclinic and heteroclinic orbits: a case study. *J. Dynamics Differential Equations.*, 5:37–57, 1993.
113. H. Fujii, M. Mimura and Y. Nishiura: A picture of the global bifurcation diagram in ecological interacting and diffusing systems. *Physica D*, 5:1–42, 1982.
114. H. Fujii, Y. Nishiura and Y. Hosono: On the structure of multiple existence of stable stationary solutions in systems of reaction-diffusion equations. In: Y. Nishida, M. Mimura and H. Fujii (Eds.), *Pattern and Waves–Qualitative Analysis of Nonlinear Differential Equations*, Studies in Maths. and its Appl., 18, pp. 157–219, 1986.
115. J.E. Furter and J. López-Gómez: On the existence and uniqueness of coexistence states for the Lotka-Volterra competition model with diffusion and spatially dependent coefficients. *Nonlinear Analysis: Theory, Analysis and Applications*, 25:363–398, 1995.
116. U. Garbotz: Numerical Analysis of Dynamics of Kuramoto-Sivashinsky Equation. Thesis, University of Marburg, 1997.
117. R.A. Gardner: Existence of multidimensional travelling wave solutions of an initial-boundary value problem. *J. Differential Equations*, 61:335–379, 1986.

118. R.A. Gardner: Global continuation of branches of nondegenerate solutions. *J. Differential Equations*, 61:321–334, 1986.
119. T. J.Garratt, G. Moore and A. Spence: Two methods for the numerical detection of Hopf bifurcations. In: R. Seydel, F.W. Schneider, T. Küpper and H. Troger (Eds.), *Bifurcations and Chaos: Analysis, Algorithms, Applications*, ISNM 97, pp. 129–148. Birkhäuser Verlag, Basel, 1991.
120. K. Gatermann and F. Guyard: The symmetry package in Maple. Konrad-Zuse-Zentrum, Preprint SC96, 1996.
121. K. Gatermann and A. Hohmann: Symbolic exploitation of symmetry in numerical pathfollowing. *Impact of Computing in Sci. & Engig.*, 3:330–365, 1991.
122. K. Gatermann and B. Werner: Group theoretical mode interaction with different symmetries. *Internat. J. Bifur. Chaos Appl. Sci. Engrg.*, 4:177–191, 1994
123. K. Gatermann and B. Werner: Secondary Hopf bifurcation caused by steady-state/steady state mode interaction. In: J. Chadam, M. Golubitsky, W. Langford and B. Wetton (Eds.), *Pattern Formation: Symmetry Methods and Applications*, pp. 209–224. AMS, Providence, 1996.
124. C.W. Gear: Differential-algebraic equation index transformations. *SIAM J. Sci. Stat. Comput.*, 9:39–47, 1988.
125. K. Georg and R. Miranda: Exploiting symmetry in solving linear equations. In: E.L. Allgower, K. Böhmer and M. Golubitsky (Eds.), *Bifurcation and Symmetry*, ISNM 104, pp. 157–168. Birkhäuser Verlag, Basel, 1992.
126. P. Glendinning: Global bifurcation in flows. In: T. Bedford and J. Swift (Eds.), *New Directions in Dynamical Systems*, London Mathematical Society Lecture Notes Series 127, pp.120-149. Cambridge University Press,1988.
127. L. Goldwasser, P.K. Maini and J.D. Murray: Splitting of cell clusters and bifurcation of bryozoan branches. *J. Theoret. Biol.*, 137:271–279, 1989.
128. G.H. Golub and C.F. Van Loan: *Matrix Computations*, second Edition. The Johns Hopkins University Press, 1989.
129. M. Golubitsky and D. Schaeffer: *Singularities and Groups in Bifurcation Theory I.* App. Math. Sci. 51. Springer-Verlag, New York, 1986.
130. M. Golubitsky and I.N. Stewart: Hopf bifurcation in the presence of symmetry. *Arch. Rat. Mech. Anal.*, 87:107–165, 1985.
131. M. Golubitsky, I.N. Stewart and D. Schaeffer: *Singularities and Groups in Bifurcation Theory II.* App. Math. Sci. 69. Springer-Verlag, New York, 1988.
132. M.G.M. Gomes and I.N. Stewart: Hopf bifurcations on generalised rectangles with Neumann boundary conditions. In: P. Chossat (Ed.), *Dynamics, Bifurcations and Symmetry*, pp. 139–158. Kluwer Academic Publisher, the Netherlands, 1994.
133. M.G.M. Gomes and I.N. Stewart: Steady PDEs on generalized rectangles: a change of genericity in mode interactions. *Nonlinearity*, 7:253–272, 1994.
134. W. Govaerts: Bordered augmented linear systems in numerical continuation and bifurcation. *Numer. Math.*, 58:353–368, 1990.
135. W. Govaerts: Stable solvers and block-elimination for bordered systems. *SIAM J. Matrix. Anal. Appl.*,12:469-483, 1991.
136. W. Govaerts: *Numerical Methods for Bifurcations of Dynamical Equilibria.* SIAM, Philadelphia, 2000.
137. W. Govaerts, J. Guckenheimer and A. Khibnik: Defining functions for multiple Hopf bifurcations. *SIAM J. Numer. Anal.*, 34:1269-1288, 1997.
138. W. Govaerts and A. Spence: Detection of Hopf points by counting sectors in the complex plane. *Numer. Math.* 75:43-58, 1996.
139. E. Griepentrog and R. März: *Differential-Algebraic Equations and Their Numerical Treatment.* Teubner-Texte zur Mathematik 88. Teubner Verlag, Leipzig, 1986.

140. A. Griewank: On solving nonlinear equations with simple singularities or nearly singular solutions. *SIAM Review*, 27:537–563, 1985.
141. A. Griewank: Automatic evaluation of first and higher-derivative vectors. In: R. Seydel, F.W. Schneider, T. Küpper and H. Troger (Eds.), *Bifurcation and Chaos: Analysis, Algorithms, Applications*, ISNM 97, pp. 135–148. Birkhäuser Verlag, Basel, 1991.
142. A. Griewank and G.W. Reddien: The calculation of Hopf bifurcation points by a direct method. *IMA J. Numer. Anal.*, 3:295–303, 1983.
143. A. Griewank and G.W. Reddien: Characterization and computation of generalized turning points. *SIAM J. Numer. Anal.*, 21:176–185, 1984.
144. J. Guckenheimer and P. Holmes: *Nonlinear Oscillations, Dynamical Systems and Bifurcations of Vector Fields.* App. Math. Sci. 42. Springer-Verlag, New York, 1983.
145. J. Guckenheimer, M. Myers and B. Sturmfels: Computing Hopf bifurcations: three examples from neurophysiology. *SIAM J. Sci. Comput.*, 17:1275–1301, 1996.
146. W. Hackbusch: *Theorie und Numerik elliptischer Differentialgleichungen.* Teubner Verlag, Stuttgart, 1986.
147. P. Hadeler and H.I. Freedman: Predator-prey populations with parasitic infectiton. *J. Math. Bio.*, 27:609–631, 1989.
148. J. Hale: Introduction to dynamic bifurcation. In: L. Salvadori (Ed.), *Bifurcation Theory and Applications*, Lecture Notes in Math., 1075, pp. 285–300. Springer-Verlag, Berlin, 1983.
149. J. Hale and C. Rocha: Interaction of diffusion and boundary conditions. *Nonlinear Analysis: Theory, Method & Applications*, 11:633–649, 1987.
150. J. Hale and C. Rocha: Varying boundary conditions with large diffusivity. *J. Math. pure et appl.*, 66:139–158, 1987
151. B. Hassard: Computation of invariant manifolds. In: P. Holmes (Ed.), *New Approaches to Nonlinear Problems*, pp. 27–42. SIAM, Philadelphia, 1980.
152. B.D. Hassard, N.D. Kazarinoff and Y-H. Wan: *Theory and Applications of Hopf Bifurcation.* Cambridge University Press, Cambridge, 1981.
153. T.J. Healey: Global bifurcation and continuation in the presence of symmetry with an application to solid mechanics. *SIAM J. Math. Anal.*, 19:824–840, 1988.
154. T.J. Healey: A group-theoretic approach to computational bifurcation problems with symmetry. *Comput. Meths. in Appl. Mech. Engrg.*, 67:257–295, 1988.
155. T.J. Healey and H. Kielhöfer: Symmetry and nodal properties in global bifurcation analysis of quasi-linear elliptic equations. *Arch. Rational Mech. Anal.*, 113:299–311, 1991.
156. R.F. Heinenman and A.B. Poore: Multiplicity, stability and oscillatory dynamics of the tubular reaction. *Chem. Engrg. Sci.*, 36:1411–1419, 1981.
157. D. Henry: *Semilinear Parabolic Equations.* Lecture Notes in Math. 840, Springer-Verlag, Berlin, 1978.
158. J. Hersch: Ueber Eigenwerte symmetrischer Membranen. In: *Numerical Treatment of Eigenvalue Problems*, ISNM 69, pp. 105–113. Birkhäuser Verlag, Basel, 1983.
159. F. Hirzebruch and W. Scharlau: *Einführung in die Funktionale Analysis.* Hochschultaschenbücher Band 296, B. I.-Wissenschaftsverlag, 1974.
160. G. Hochschild: *The Structure of Lie Groups.* Holden-Day, San Francisco, 1965.
161. E.J. Holder and D.G. Schaeffer: Boundary condition and mode jumping in the von Kármán equations. *SIAM J. Math. Anal.*, 15:446–458, 1984.

162. B. Hong: A manifold solver with bifurcation and symmetry. In: E.L. Allgower, K. Georg and R. Miranda (Eds.), *Exploiting Symmetry in Applied and Numerical Analysis*, Lectures in Applied Math. 29, pp. 259–272, AMS. Providence, 1993.

163. A. Hoy: An efficiently implementable Gauss-Newton-like method for solving singular nonlinear equations. *Computing*, 41:107–122, 1989.

164. A. Hoy: A remark concerning Mei Zhen's paper on singular nonlinear equations. *Computing*, 44:357–364, 1990.

165. J. Huitfeldt and A. Ruhe: A new algorithm for numerical path following applied to an example from hydrodynamical flow. *SIAM J. Sci. Stat. Comput.*, 11:1181–1192, 1990.

166. G. Iooss and M. Adelmeyer: *Topics in Bifurcation Theory and Applications.* World Scientific Publishing Co., Singapore, New Jersey, London, 1992.

167. G. Iooss and A. Mielke: Time-periodic Ginzburg-Landau equations for one dimensional patterns with large wave length. *Z. angew. Math. Phys.*, 43:125–138, 1992.

168. E. Isaacson and H.B. Keller: *Analysis of Numerical Methods.* John Wiley & Sons Inc., New York, London, Sydney, 1969.

169. A.D. Jepson: Numerical Hopf Bifurcation. Thesis, California Institute of Technology, Pasadena, California, 1981.

170. A.D. Jepson and A.W. Decker: Convergence cones near bifurcation. *SIAM J. Numer. Anal.*, 23:959–975, 1986.

171. A.D. Jepson and H.B. Keller: Steady state and periodic solution paths: their bifurcations and computations. In: T. Küpper, H. D. Mittelmann and H. Weber (Eds.), *Numerical Methods for Bifurcation Problems*, ISNM 70, pp. 219–246. Birkhäuser Verlag, Basel, 1984.

172. A.D. Jepson and A. Spence: Folds in solutions of two parameter systems and their calculations, part I. *SIAM J. Numer. Anal.* 22:347–368, 1985.

173. A.D. Jepson and A. Spence: The numerical solution of nonlinear equations having several parameters. I: Scalar equations. *SIAM J. Numer. Anal.* 22:736–759, 1985.

174. A.D. Jepson and A. Spence: Numerical methods for bifurcation problems. In: A. Iserles and M.J.D. Powell (Eds.) *The State of the Art in Numerical Analysis.* pp. 273–298, Inst. Math. Appl. Conf. Ser. 9, Clarendon Press, Oxford 1987.

175. A.D. Jepson and A. Spence: On a reduction process for nonlinear equations. *SIAM J. Math. Anal.* 20:39–56, 1989.

176. A.D. Jepson, A. Spence and K.A. Cliffe: The numerical solution of nonlinear equations having several parameters. III: Equations with Z_2-symmetry. *SIAM J. Numer. Anal.* 28:809–832, 1991.

177. C. Jones and T. Küpper: On the infinitely many solutions of a semilinear elliptic equation. *SIAM J. Math. Anal.*, 17:803–835, 1986.

178. T. Kato: *Perturbation Theory for Linear Operators.* Grundlehren Mathematischen Wissenschaften, Band 132. Springer-Verlag, Heidelberg, New York, 1976.

179. H.B. Keller: Numerical solution of bifurcation and nonlinear eigenvalue problems. In: P.H. Rabinowitz (Ed.), *Application of Bifurcation Theory*, pp. 359–384. Academic Press, New York, 1977.

180. H.B. Keller: The bordering algorithm and path following near singular points of higher nullity. *SIAM J. Sci. Stat. Comp.* 4:573–582, 1983.

181. H.B. Keller: *Numerical Methods in Bifurcation Problems.* Tata Institute of Fundamental Research. Springer-Verlag, Bombay, 1987.

182. A. Kelley: The stable, center-stable, center-unstable and unstable manifolds. *J. Differential Equations*, 3:546–570, 1988.

183. J.P. Kernevez: *Enzyme Mathematics.* Studies in Mathematics and its Applications, 10. North-Holland Publishing Co., Amsterdam, New York, 1980.
184. I.G. Kevrekedis, B. Nicolaenko and J.C. Scovel: Back in the saddle again: a computer assisted study of the Kuramoto-Sivashinsky equation. *SIAM J. Appl. Math.*, 50:760–790, 1990.
185. A. Khibnik: LINLBF: A program for continuation and bifurcation analysis of equilibria up to codimension three. In: D. Roose, B. De Dier and A. Spence (Eds.), *Continuation and Bifurcations: Numerical Techniques and Applications.* NATO ASI Series, C313, pp. 283–296. Kluwer Academic Publishers, 1990.
186. A.I. Khibnik, Y.A. Kuznetsov, V.V. Levitin and E.V. Nikolaev: Continuation techniqes and interactive software for bifurcation analysis of ODE and iterated maps. *Physica D*, 62:360–371, 1993.
187. H. Kielhöfer and R. Lauterbach: On the principle of reduced stability. *J. of Func. Anal.*, 53:99–111, 1983.
188. S. Kim and J. Guckenheimer: The two-dimensional torus: numerical explorations and mathematical conjectures. In: D.K. Campbell (Ed.), *Chaos*, pp. 145–152. New York, 1990. Amer. Inst. Phys.
189. K. Kirchgässner: Exotische Lösungen des Benardschen Problems. *Math. Meth. in Appl. Sci.*, 1:453–467, 1979.
190. K. Kirchgässner: Waves in weakly-coupled parabolic systems. In: C. Vinti (Ed.), *Nonlinear Analysis and Optimization*, Lecture Notes in Math. 1107, pp. 154–180. Springer-Verlag, Berlin, Heidelberg, New York, 1984.
191. M. Kleczka, W. Kleczka and E. Kreuzer: Bifurcation analysis: a combined numerical and analytical approach. In: D. Roose, B. De Dier and A. Spence (Eds.), *Continuation and Bifurcations: Numerical Techniques and Applications*, NATO ASI Series C313, pp. 123–137. Kluwer Academic Publishers, 1990.
192. M. Kubicék and I. Schreiber: Continuation of heteroclinic and homoclinic orbits—application to pulse and front wave. *ZAMM*, 78:S981-982, 1998.
193. P. Kunkel: Efficient computation of singular points. *IMA J. Numer. Anal.*, 9:421–433, 1989.
194. T. Küpper, H.D. Mittelmann and H. Weber: *Numerical Methods for Bifurcation Problems*, ISNM 70. Birkhäuser Verlag, Basel, 1984.
195. T. Küpper, R. Seydel and H. Troger: *Analysis: Algorithm, Applications: Analysis, Algorithm, Applications*, ISNM 79. Birkhäuser Verlag, Basel, 1987.
196. J.R. Kuttler and V.G. Sigillito: Eigenvalues of the Laplacian in two dimensions. *SIAM Rev.*, 26:163–193, 1984.
197. Y. A. Kuznetsov: *Elements of Applied Bifurcation Theory*, Applied Mathematical Sciences, 112. Springer-Verlag, Berlin, 1995.
198. O. Lanford: Bifurcation of periodic solutions into invariant tori. In: *Nonlinear Problems in the Physical Sciences*, Lecture Notes in Math. 322, Springer-Verlag, New York, 1973.
199. P. Laure and Y. Demay: Symbolic computation and equation on the center manifold: application to the Couette-Taylor problem. *Comput. & Fluids*, 16:229–238, 1988.
200. R. Lauterbach: An example of symmetry breaking with submaximal isotropy. In: M. Golubitsky and J. Guckenheimer (Eds.), *Multiparameter Bifurcation Theory*, Contemporary Mathematics, 56, AMS, Providence, 1986.
201. R. Lefever and I. Prigogine: Symmetry-breaking instabilitites in dissipative systems II. *J. Chem. Phys.*, 48:1695–1700, 1968.
202. M. Lentini and H.B. Keller: Boundary value problems on semi-infinite intervals and their numerical solution. *SIAM J. Numer. Anal.*, 17:577–604, 1980.
203. K.T. Li and Z. Mei: A splitting iterative method for computation of TB-points. *Chinese J. Contemporary Mathematics*, 15:37–38, 1994.

204. L.G. Li and A. Ghoreishi: On positive solutions of general nonlinear elliptic symbiotic interacting systems. *Appl. Anal.*, 40:281–295, 1991.

205. X.B. Lin: Using Melnikov's method to solve Silnikov's problems. *Proc. Roy. Soc. Edingburgh*, 116A:295–325, 1990.

206. J. López-Gómez, J.C. Eilbeck, M. Molina and J. Duncan: Structure of solution manifolds in a strongly coupled elliptic system. *IMA J. Numer. Anal.*, 12:405–428, 1992.

207. K. Lust, D. Roose, A. Spence and A.R. Champneys: An adaptive Newton-Picard algorithm with subspace iteration for computing periodic solutions. *SIAM J. Sci. Comput.* 19:1188–1209, 1998.

208. B.A. Malomed and K.E. Gedalin: Stability of cellular wave structures in oscillatory pattern-forming systems near onset. *Physica D*, 40:33–41, 1989.

209. J.B. McLeod and D.H. Sattinger: Loss of stability and bifurcation at a double eigenvalue. *J. Func. Anal.*, 14:62–84, 1973.

210. Z. Mei: Computing local solution branches with a direct method. *Appl. Anal.*, 30:263–277, 1988.

211. Z. Mei: *Numerical Approximations of Corank-2 Bifurcation Problems.* Dissertation, Dept. of Math., University of Marburg, 1989.

212. Z. Mei: Splitting iteration method for simple singular points and simple bifurcation points. *Computing*, 41:87–96, 1989.

213. Z. Mei: A numerical approximation for the simple bifurcation problems. *Numer. Func. Anal. Optimiz.*, 10:383–400, 1989.

214. Z. Mei: A Newton-like method for simple bifurcation problems with application to large sparse systems. In: D. Roose, B. De Dier and A. Spence (Eds.), *Continuation and Bifurcations: Numerical Techniques and Applications*, NATO ASI Series C313, pp. 247–260. Kluwer Academic Publishers, 1990.

215. Z. Mei: A special extended system and a Newton-like method for simple singular nonlinear equations. *Computing*, 45:157–167, 1990.

216. Z. Mei: Bifurcations of a simplified buckling problem and the effect of discretizations. *Manuscripta Mathematica*, 71:225–252, 1991.

217. Z. Mei: Path following around corank-2 bifurcation points of a semi-linear elliptic problem with symmetry. *Computing*, 46:492–508, 1991.

218. Z. Mei: Solution branches at corank-2 bifurcation points with symmetry. In: R. Seydel, F.W. Schneider, T. Küpper and H. Troger (Eds.), *Bifurcation and Chaos: Analysis, Algorithms, Applications*, ISNM 97, pp. 251–255. Birkhäuser Verlag, Basel, 1991.

219. Z. Mei: Solution branches of a semilinear elliptic problem at corank-2 bifurcation points. *Manuscripta Mathematica*, 75:395–409, 1992.

220. Z. Mei: Utilization of scaling laws and symmetries in the path following of a semilinear elliptic problem. In: E.L. Allgower, K. Böhmer and M. Golubitsky (Eds.), *Bifurcation and Symmetry*, ISNM 104, pp. 263–173. Birkhäuser Verlag, Basel, 1992.

221. Z. Mei: *Bifurcation Problems and Numerical Analysis.* Lecture Note. Center for Applied Mathematics, Xi'an Jiaotong University, 1993.

222. Z. Mei: Exploiting symmetries in numerical analysis of bifurcations of nonlinear elliptic PDEs. In: J.Z. Cui, Z.C. Shi and D.L. Wang (Eds), *Proc. of Conference on Scientific and Engineering Computing for Young Chinese Scientists*, pp. 52-57. National Defence Industry Press, Beijing, 1994.

223. Z. Mei: Solution branches of a semilinear elliptic problem at corank-2 bifurcation points with Neumann boundary conditions. *Proceedings the Royal Society of Edinburgh* , 123A:479–495, 1993.

224. Z. Mei: Bifurcation of semilinear elliptic PDEs with symmetries and their numerical approximations. In: D. Bainov and A. Dishliev (Eds.), *Proc. Fifth. Inter. Coll. on Diff. Eqns.*, pp. 170–178. SCT Publishing, Singapore, 1995.
225. Z. Mei: *Numerische Methoden für nichtlineare Probleme mit Parametern.* Lecture Note, Dept. of Math., University of Marburg, 1996.
226. Z. Mei and A. Schwarzer: Scaling solution branches of one-parameter bifurcation problems. *J. Math. Anal. & Appl.*, 204:102–123, 1996.
227. Z. Mei and F. Theil: Variation of bifurcations along a homotopy from Neumann to Dirichlet problems. *Nonlinear Analysis: Theory, Analysis and Applications*, 27:1381–1395, 1996.
228. A. Mielke: On nonlinear problems of mixed type: a qualitative theory using infinite dimensional center manifolds. *J. Dynamics and Differential Equations*, 4:419–443, 1992.
229. M. Mimura: Pattern-dynamics in excitable reaction-diffusion systems. In: L.M. Ricciardi (Ed.), *Biomathematics and related computational problems*, pp. 669–676. Kluwer Acad. Publ, Dordrecht, 1988.
230. G. Moore, T.J. Garratt and A. Spence: The numerical detection of Hopf bifurcation points. In: D. Roose, B. De Dier and A. Spence (Eds.), *Continuation and Bifurcations: Numerical Techniques and Applications*, NATO ASI Series C313, pp. 227–246. Kluwer Academic Publishers, 1990.
231. G. Moore and A. Spence: The calculation of turning points of nonlinear equations. *SIAM J. Numer. Anal.*, 17:567–576, 1980.
232. G. Moore: The numerical treatment of non-trivial bifurcation points. *Numer. Funct. Anal. Optimiz.*, 2:441–472, 1980.
233. G. Moore: Computation and parametrization of periodic and connecting orbits. *IMA J. of Numer. Anal.*, 15:245–263, 1995.
234. A. Münster and M. Watzl: Stationary spatial patterns in a chemical reaction-diffusion system and effects of an electrical field. Technical report, Inst. of Physical Chemistry, University of Würzburg, 1994.
235. J.D. Murray: *Mathematical Biology*, Biomathematics 13, Springer-Verlag, Berlin, 1989.
236. G. Nicolis and P. Gaspard: Bifurcations, chaos and self-organization in reaction diffusion systems. In: D. Roose, B. De Dier and A. Spence (Eds.), *Continuation and Bifurcations: Numerical Techniques and Applications*, NATO ASI Series C313, pp. 43–70. Kluwer Academic Publishers, 1990.
237. G. Nicolis and I. Prigogine: *Self-Organization in Nonequilibrium Systems.* John Wiley & Sons, New York, 1977.
238. J.T. Oden and J.N. Reddy: *An Introduction to the Mathematical Theory of Finite Elements.* John Wiley & Sons, New York, 1976.
239. J.M. Ortega and W.C. Rheinboldt: *Iterative Solution of Nonlinear Equations in Several Variables.* Academic Press, New York, 1970.
240. C.C. Paige and M.A. Saunders: Solution of sparse indefinite systems of linear equations. *SIAM J. Numer. Anal.*, 12:617–629, 1975.
241. B.N. Parlett: A new look at the Lanczos algorithm for solving symmetric systems of linear equations. *Lin. Alg. Appl.*, 29:323–346, 1980.
242. M.S. Pinsky: *Partial Differential Equations and Boundary-Value Problems with Applications.* Second edition. McGraw-Hill Inc., New York, 1991.
243. V. Pliss: Principle reduction in the theory of stability of motion (russian). *Izv. Akad. Nauk. SSSR Mat. Ser.*, 28:1297–1324, 1964.
244. G. Pönisch: An indirect approach to computing origins of Hopf bifurcations and its applications to problems with symmetry. In: E.L. Allgower, K. Böhmer and M. Golubitsky (Eds.), *Bifurcation and Symmetry*, ISNM 104, pp. 285–294. Birkhäuser Verlag, Basel, 1992.

245. C.B. Price, P. Wambacq and A. Oosterlinck: Computing with reaction-diffusion systems: applications in image processing. In: D. Roose, B. De Dier and A. Spence (Eds.), *Continuation and Bifurcations: Numerical Techniques and Applications*, NATO ASI Series C313, pp. 379–388. Kluwer Academic Publishers, 1990.
246. R.H. Rand and D. Armbruster: *Perturbation Methods, Bifurcation Theory and Computer Algebra.* Applied Math. Sciences, 65, Springer-Verlag, New York, 1987.
247. W.C. Rheinboldt: On a moving-frame algorithm and the triangulation of equilibrium manifolds. In: T. Küpper, R. Seydel and H. Troger (Eds.), *Bifurcation: Analysis, Algorithms, Applications*, ISNM 79, pp. 256–267. Birkhäuser Verlag, Basel, 1987.
248. W.C. Rheinboldt: On the computation of multi-dimensional solution manifolds of parametrized equations. *Numer. Math.*, 53:165–182, 1988.
249. W.C. Rheinboldt: *Numerical Analysis of Parameterized Nonlinear Equations.* John Wiley & Sons, New York, 1986.
250. W.C. Rheinboldt and J.V. Burkardt: A locally parameterized continuation process. *ACM Trans. Software*, 9:215–235, 1983.
251. J. Rinzel: Bursting oscillations in an excitable membrane model. In: B.D. Sleeman and R.J. Jarvis (Eds.) *Proc. 8th Dundee Conf. on the Theory of Ordinary and Partial Differential Equations.* Lecture Notes in Math. 1181, pp.304-316, Springer-Verlag, 1985.
252. R. Rodi: *Turbulence Models and Their Application in Hydraulics.* Third edition. A. Balkema Publisher, Rotterdam, 1988.
253. A.J. Rodríguez-Luis, E. Freire and E. Ponce: A method of homoclinic and heteroclinic continuation in two and three dimensions. In: D. Roose, B. De Dier and A. Spence (Eds.), *Continuation and Bifurcations: Numerical Techniques and Applications.* NATO ASI Series, C313, pp. 197-210. Kluwer Academic Publishers, 1990.
254. A.J. Rodriguez-Luis, E. Freire and E. Ponce: A case study for homoclinic chaos in an autonomous electronic circuit: A trip from Takens-Bogdanov to Hopf-Shilnikov. *Physica D*, 62:230–253, 1993.
255. D. Roose: Numerical computation of origin for Hopf bifurcation in a two-parameter problem. In: T. Küpper, R. Seydel and H. Troger (Eds.), *Bifurcation: Analysis, Algorithms, Applications*, ISNM 79, pp. 268–276. Birkhäuser Verlag, Basel, 1987.
256. D. Roose and B. De Dier: Numerical determination of an emanating branch of Hopf bifurcation points in a two-parameter problem. *SIAM J. Sci. Stat. Comput.*, 10:671–685, 1989.
257. D. Roose and V. Hlavacek: A direct method for the computation of Hopf bifurcation points. *SIAM J. Appl. Math.*, 45:879–893, 1985.
258. W.H. Ruan and C.V. Pao: Asymptotic behavior and positive solutions of a chemical reaction diffusion system. *J. Math. Anal. & Appl.*, 169:157–178, 1992.
259. Y. Saad: Variations on Arnoldi's method for computing eigenelements of large unsymmetric matrices. *Lin. Alg. Appl.*, 34:269–295, 1980.
260. Y. Saad: Projection method for solving large sparse eigenvalue problems. In: B. Kagstroem and A. Ruhe (Eds.), *Matrix Pencils*, Lecture Notes in Math. 973, pp. 121–144. Springer-Verlag, Berlin, 1982.
261. Y. Saad: Chebyshev acceleration techniques for solving nonsymmetric eigenvalue problems. *Math. Comp.*, 42:567–588, 1984.
262. Y. Saad: Least squares polynomials in the complex plane and their use for solving nonsymmetric linear systems. *SIAM J. Numer. Anal.*, 24:155–169, 1987.

263. Y. Saad: Numerical solution of large nonsymmetric eigenvalue problems. *Comput. Phys. Comm.*, 53:71–90, 1989.
264. B.Sandstede: Convergence estimates for the numerical approximation of homoclinic solutions. *IMA J. Numer. Anal.* 17:437–462, 1997.
265. B.Sanstede: Stability of multiple-pulse solutions. *Trans. Amer. Math. Soc.* 350:429–472, 1998.
266. N. Sassmannshausen: Stabilitätaussagen für erweiterte Systeme und ihre Anwendung in der Verzweigungstheorie Dissertation, University of Marburg, 1998.
267. D. Schaeffer and M. Golubitsky: Boundary conditions and mode jumping in the buckling of a rectangular plate. *Commun. Math. Phys.*, 69:209–236, 1979.
268. D. Schaeffer and M. Golubitsky: Bifurcation analysis near a double eigenvalue of a model chemical reaction. *Arch. Rational Mech. Anal.*, 75:315–347, 1981.
269. S. Schecter: Rate of convergence of numerical approximations to homoclinic bifurcation points. *IMA J. Numeri. Anal.*, 15:23–60, 1995.
270. G.M. Shroff and H.B. Keller: Stabilization of unstable procedures: the recursive projection method. *SIAM J. Numer. Math.* 30:1099-1120, 1993
271. A. Schwarzer: Verzweigungsverhalten der nichtlinearen Laplace-Gleichung auf dem Hexagon. MSC Thesis, Dept. of Math., University of Marburg, 1993.
272. A. Schwarzer: Skalierungstechniken für k-bestimmte Verzweigungsprobleme und Iterationsmethoden für große, dünnbesetzte Eigenwertprobleme. Dissertation, Dept. of Math., University of Marburg, 1997.
273. S.K. Scott and P. Gray: Chemical reaction in isothermal system. In: *Nonlinear Phenomena and Chaos: Oscillations and Instabilities.* Malvern Phys. Ser., pp. 70-95. Hilger, Bristol, 1986.
274. R. Sebastian: Anwendung von Krylov-Verfahren auf Verzweigungs- und Fortsetzungsprobleme. Dissertation, Dept. of Math., University of Marburg, 1995.
275. L. Segel: *Mathematical Models in Molecular and Cellular Biology.* Cambrige, 1980.
276. R. Seydel: *From Equilibrium to Chaos—Practical Bifurcation and Stability Analysis.* Elsevier Publishing, New York, Amsterdam, London, 1988.
277. R. Seydel: On a class of bifurcation test functions. Preprint, University of Ulm, Germany, 1996.
278. G.B. Shaw: Degeneracy in the particle–in–a–box problem. *J. Phys. A: Math., Nucl. Gen.*, 7:1537–1546, 1974.
279. G.I. Sivashinsky: Nonlinear analysis of hydrodynamic instability in flames–I: Derivation of the basic equations. *Acta Astronautica*, 4:1177–1206, 1977.
280. G.I. Sivashinsky: Instabilities, pattern formation and turbulence in flames. *Ann. Rev. Fluid Mechanics*, 15:179–199, 1983.
281. J. Smoller: *Shock Waves and Reaction-Diffusion Equations.* Springer-Verlag, New York, 1983.
282. P. Sonneveld: CGS, a fast Lanczos-type solver for nonsymmetric linear systems. *SIAM J. Sci. Stat. Comp.*, 10:36–52, 1989.
283. A. Spence and A. Jepson: The numerical computations of turning points of nonlinear equations. In: C. Baker and G. Miller (Eds.), *Treatment of Integral Equations by Numerical Methods*, pp. 169–183. Academic Press, New York, 1982.
284. A. Steindl: Hopf/Steady-state mode interaction for a fluid conveying elastic tube with D_4-symmetric support. In: E.L. Allgower, K. Böhmer and M. Golubitsky (Eds.), *Bifurcation and Symmetry*, ISNM 104, pp. 301–315. Birkhäuser Verlag, Basel, 1992.
285. A. Steindl: Hopf/Steady-state mode interaction for a fluid conveying elastic tube with D_3-symmetric support. In: E.L. Allgower, K. Georg and R. Miranda

(Eds.), *Exploiting Symmetry in Applied and Numerical Analysis*, Lectures in Applied Mathematics 29, pp. 373–387. AMS, Providence, 1993.
286. E. Stiefel and A. Fässler: *Gruppetheoretische Methoden und ihre Anwendung.* Teubner Verlag, Stuttgart, 1979.
287. P. Stork and B. Werner: Symmetry adapted block diagonalization in equivariant steady state bifurcation problems and its numerical applications. *Advances in Mathematics*, 20:455-487, 1991.
288. J.W. Swift: Hopf bifurcation with the symmetry of the square. *Nonlinearity*, 1:333–377, 1988.
289. R. Szilard: *Theory and Analysis of Plates– Classical and Numerical Methods.* Civil Engrg. & Engrg. Mech. Series. Prentice Hall, INC, New Jersey, 1974.
290. R. Temam: *Infinite-Dimensional Dynamical Systems in Mechanics and Physics.* Applied Mathematical Sciences, 68. Springer-Verlag, Berlin, New York, 1988.
291. F. Theil: Stationäre Verzweigungen für äquivariante ellptische Differentialgleichungen mit Robin Randbedingungen, MSC Thesis, Fachbereich Mathematik der Universität Marburg 1994.
292. L.N. Trefethen and D. Bau: *Numerical Linear Algebra.* SIAM, Philadelphia, USA, 1997.
293. A.M. Turing: The chemical basis of morphogenesis. *Phil. Trans. Roy. Soc. Lond.*, B237:37–72, 1952.
294. S. van Gils and M. Silber: On the uniqueness of invariant tori in $D_4 \times \mathbf{S}^1$ symmetric systems. *Nonlinearity*, 8:615–628, 1995.
295. A. Vanderbauwhede: *Local Bifurcation and Symmetry*, Res. Notes in Math., 75, Pitman, Boston, 1982.
296. A. Vanderbauwhede: Centre manifolds, normal forms and elementary bifurcations. In: U. Kirchgraber and H. O. Walther (Eds.), *Dynamics Reported* 2, pp. 89–169. J. Wiley & Sons, Chichester, 1989.
297. A. Vanderbauwhede and B. Fiedler: Homoclinic period blow-up in reversible and conservative systems. *Z. angew. Math. Phys*, 43:192–318, 1992.
298. A. Vanderbauwhede and G. Iooss: Center manifold theory in infinite dimensions. In: C. K. Johnes, U. Kirchgraber and H. O. Walther (Eds.), *Dynamics Reported: Expositions in Dynamical Systems* 1, pp. 125–163. Springer-Verlag, Berlin, 1992.
299. Y. H. Wan: On the uniqueness of invairant manifolds. *J. Differential Equations*, 24:268–273, 1977.
300. H. Weber and W. Werner: On the accurate determination of non-isolated solutions of nonlinear equations. *Computing*, 26:315–326, 1981.
301. B. Werner: Regular systems for bifurcation points with underlying symmetries. In: T. Küpper, H.D. Mittelmann and H. Weber (Eds.), *Numerical Methods for Bifurcation Problems*, ISNM 70, pp. 562–574. Birkhäuser Verlag, Basel, 1984.
302. B. Werner: Computational methods for bifurcation problems with symmetries and applications to steady states of n-box reaction-diffusion models. In: D.F. Griffiths and G.A. Watson (Eds.), *Numerical Analysis 1987*, pp. 279–293. John Wiley & Sons, New York, 1988.
303. B. Werner: Eigenvalue problems with the symmetry of a group and bifurcations. In: D. Roose, B. De Dier and A. Spence (Eds.), *Continuation and Bifurcation: Numerical Techniques and Applications*, pp. 71–88. Kluwer Academic Publishers, 1990.
304. B. Werner: Test functions for bifurcation points and Hopf points in problems with symmetry. In: E.L. Allgower, K. Böhmer and M. Golubitsky (Eds.), *Bifurcations and Symmetry*, ISNM 104, pp. 317–327. Birkhäuser Verlag, Basel, 1992.

305. B. Werner: Computation of Hopf bifurcation with bordered matrices. *SIAM J. Numer. Anal.*, 33:435–455, 1996.

306. B. Werner and V. Janovsky: Computation of Hopf branches bifurcating from Taken-Bogdanov points for problems with symmetries. In: R. Seydel, F.W. Schneider, T. Küpper and H. Troger (Eds.), *Bifurcations and Chaos*, ISNM 97, pp. 377–388. Birkhäuser Verlag, Basel, 1991.

307. B. Werner and A. Spence: The computation of symmetry-breaking bifurcation points. *SIAM J. Numer. Anal.*, 21:388–399, 1984.

308. A.M. Zhabotinskii: A study of self-oscillatory chemical reaction III: Space behavior. In: B. Chance et al (Eds.), *Biochemical Oscillators*, Academic Press, New York, 1973.

Index